Merrill

Pre-Algebra

A Transition to Algebra

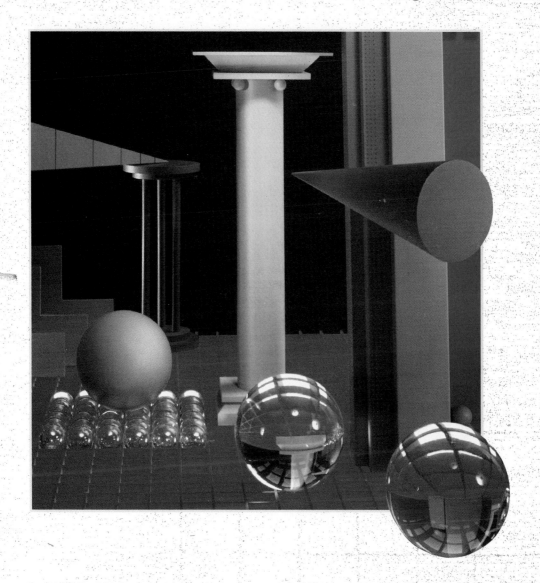

McGraw-Hill

New York, New York   Columbus, Ohio   Mission Hills, California   Peoria, Illinois

Send all inquiries to:
Glencoe/McGraw-Hill
936 Eastwind Drive
Westerville, OH 43081

ISBN: 0-02-824361-7 (Student Edition)

7 8 9 10 11 12 13 14 15   VH/LP 02 01 00 99 98 97 96 95

Authors

Jack Price taught mathematics and science for 13 years in the Detroit Public Schools, where he served for three years as mathematics department head. After receiving his Ed.D. in Mathematics Education from Wayne State University, he became coordinator of mathematics/science for the San Diego County (California) Department of Education. Dr. Price is a past director of the National Council of Teachers of Mathematics, a former member of the National Advisory Committee on Mathematical Education, and a founding member of the Mathematical Sciences Education Board of the National Research Council. He is presently assistant director of the Center for Science and Mathematics Education at California Polytechnic University at Pomona, California, where he teaches mathematics and methods courses for preservice teachers and consults with school districts on curriculum change.

James N. Rath is former chairperson of the mathematics department at Darien High School, Darien, Connecticut. He has taught mathematics at every level of the high school curriculum and has over 30 years of classroom teaching experience. Mr. Rath received his B.A. from Catholic University of America and his M.Ed. and M.A. in mathematics from Boston College. Mr. Rath is co-author of the *Merrill Algebra* program, and is an active member of several local, state, and national professional organizations.

William Leschensky teaches mathematics at Glenbard South High School in Glen Ellyn, Illinois. He has served as mathematics department chairperson in four schools and is a former textbook editor and editorial director. Mr. Leschensky received his B.A. in mathematics from Cornell College and his M.A. in mathematics from the University of Northern Iowa. He has participated in special institutes in mathematics and physics as well as several computer workshops.

Contributing Authors

Olene H. Brame, Ph.D.
Dallas Independent School District
Dallas, Texas

David D. Molina, Ph.D.
Department of Education
Trinity University
San Antonio, Texas

Consultant

Donald W. Collins
Department of Mathematics and Informational Sciences
Sam Houston State University
Huntsville, Texas

Reviewers

Robinette Bowden
Central Davis Junior High School
Layton, Utah

Nevin Engle
Cumberland Valley High School
Mechanicsburg, Pennsylvania

Richard D. Hammann
Carlisle Area School District
Carlisle, Pennsylvania

Gary D. Hawsey
Fulton County Schools
Atlanta, Georgia

Lucy Koors
Emmerich Manual High School
Indianapolis, Indiana

Betty Kreutzer
Pleasanton Middle School
Pleasanton, California

Susan F. Meagher
High School for Law Enforcement and Criminal Justice
Houston, Texas

Barbara Smith
Unionville-Chadds Ford School District
Unionville, Pennsylvania

Peter Smith
Omaha Public Schools
Omaha, Nebraska

Karen Steele
Northeast Junior High School
Charlotte, North Carolina

Larry Stott
Bryant Intermediate School
Salt Lake City, Utah

Nick Vinciguerra
Gettysburg Junior High School
Gettysburg, Pennsylvania

Deborah Volpe
Westerville South High School
Westerville, Ohio

Jan Wilson
Arsenal Technical High School
Indianapolis, Indiana

Hazel Wright
University of North Carolina
Charlotte, North Carolina

Table of Contents

Hints for Success
Chapters 1 and 2
Calculator   15, 80
Estimation   30, 43
Mental Math   25, 74

Solving One-Step Equations

Factors and Fractions

Hints for Success
Chapters 3 and 4
Calculator   131, 137, 146
Estimation   93, 102, 119

Explorations

Did you know there is a method to estimate how many fish are in a certain lake? It is called the capture-recapture method and you may use this method in the **Exploration** on page 330. There are 28 such **Explorations** in your textbook. You may investigate how to solve equations, make predictions, and discover many mathematical patterns. **Explorations** let you *do* mathematics and use your knowledge of arithmetic to find success in algebra.

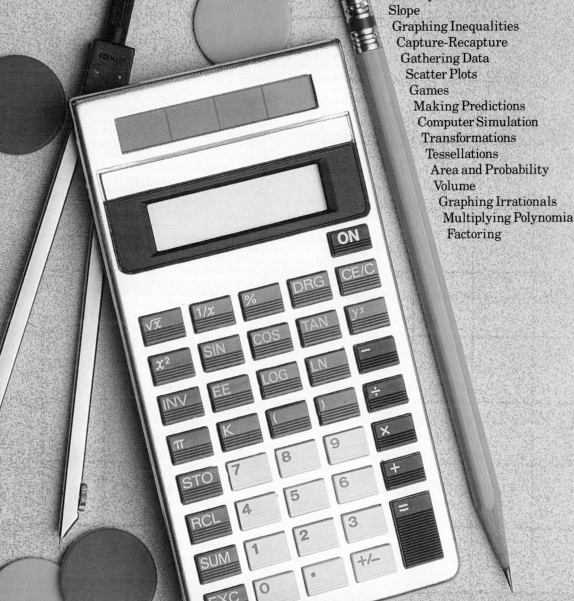

Rationals: Adding and Subtracting Patterns

Rationals: Multiplying and Dividing Patterns

Algebra in Action

What does the Mona Lisa have in common with the Great Pyramid of Egypt? Read the **Algebra in Action** feature on page 326 to find out. In each chapter you will have an opportunity to *use* mathematics in real-world situations.

ix

Solving Equations and Inequalities

Hints for Success
Chapters 7 and 8
Calculator   264
Estimation   256
Mental Math   275

Graphing Equations and Inequalities

History, Biography, Careers

Why is this part of mathematics called *algebra*? Read the **Biography** feature on page 21 to find out. These features contain information about mathematics in the *past* and about career opportunities for the *future*. You may be surprised by what you read.

Proportion and Percent

Hints for Success
Chapters 9 and 10

Calculator	342
Estimation	346, 349, 354
Mental Math	328, 339, 340, 375

Statistics and Graphs

Connections

"When am I ever going to use this stuff?" It may be sooner than you think. You'll find mathematics in most of the subjects you study in school. In the **Music Connection** on page 127 you'll see that mathematics and music are related. The **Geography Connection** on page 293 relates latitude and longitude lines to mathematics. These and other **Connections** will help you *see* mathematics in your life.

Great Expectations

ckens (1812–1870) may be the best-remembered novelist d, and many people believe that *Great Expectations* is l his works. The works include the classi l (1843) and the well-loved cholas Nickleby

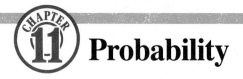

Probability

Hints for Success
Chapters 11 and 12
Calculator   407
Estimation   412, 456

Applying Algebra to Geometry

Measuring Area and Volume

Applying Algebra to Right Triangles

Hints for Success
Chapters 13 and 14
Calculator   511, 529
Estimation   486, 492, 502, 512, 538

Polynomials

Extended Projects

As you gear up to study pre-algebra, you probably have two big questions on your mind: How is pre-algebra different from other mathematics courses I've already taken? How will it prepare me to learn algebra and geometry?

A good way to explain pre-algebra is to describe it as a bridge between arithmetic and algebra. As you cross this bridge, you will move from working problems that are mostly numerical to solving those that require more advanced reasoning skills that require working with variables, or unknowns. You will be introduced to a variety of practical and exciting reasoning skills. These reasoning skills are practical because you will use them in other classes and outside of school. They are exciting because you will find them useful throughout your life.

Here are just two of your "tickets to success."

- **You will learn to solve particular kinds of problems called equations.**
Once you master basic algebra problem-solving methods, you can use them in your daily life. Using formulas in science, deciding how to spend your money, and keeping sports statistics are among the many practical uses for algebra.

- **You will learn to use percent.** Sometimes an estimate will do in figuring percentages — for example, determining roughly what percent of your day is spent in school. In other situations, such as determining the interest your money can earn in a savings account, an exact answer will be required.

Featuring 4 Simple Steps to Problem Solving

The Steps in Action

You can use these four simple steps to solve real-life problems as well as mathematical problems. For example, suppose you and some classmates are on the entertainment committee in charge of organizing a school dance. None of you has ever served on this type of committee. Together you must decide whether to provide a deejay or a live band at the dance, and then hire one. Here is how you could apply the four steps to this hypothetical real-life situation.

The Steps in Brief

Step 1 Explore

- Identify the problem to be solved.
- Ask yourself these questions:
  "What do I *already know* that can help me solve this problem?"
  "What do I *need to find out* to solve this problem?"

Step 2 Plan

- Figure out how all the facts are related.
- Choose the most practical method for solving the problem.
- Make an estimate of the answer.

Step 3 Solve

- Use the method you have chosen to actually solve the problem.

Step 4 Examine

- Examine each step you have taken to solve the problem.
- Ask yourself these questions:
  "Based on the facts, does this answer seem reasonable?"
  "If not, where is my reasoning faulty?"
  "Is there a faster or better way to solve the problem?"

Step 1 Explore

Identify the problem to be solved.
"Should we hire a deejay or a live band for the school dance?"

What do we already know that can help us solve this problem?
"A lot of kids said the deejay used at the last dance was boring."

"Live bands cost more than deejays."

"We don't have much money to spend on entertainment."

Step 2 Plan

Figure out how all the facts are related.
"Let's brainstorm a list of live bands and find out how much each costs."

"Let's do the same to come up with a list of deejays."

"No matter how cheap the boring deejay is, let's agree not to make that mistake again."

Choose the most practical method for solving the problem.
"It would be nice to take a survey of other students' opinions, but this would not be practical because we must make a decision soon."

"Instead, let's narrow our lists of live bands and deejays to ones we can afford."

Make an estimate of the answer.
"Let's take a preliminary, informal vote to see which of the affordable bands and deejays we like."

Step 3 Solve

Use the method you have chosen to actually solve the problem.
"After discussing the reasons for our choices, let's vote again and agree to let the majority rule. We will hire the band or deejay that gets the most votes."

Step 4 Examine

Examine your solution. Is it reasonable? Is there a better way to solve the problem?
"Our committee has several representatives from each grade. So, our vote gave us a good idea of what most students like."

"Since we didn't have much time, a committee decision was better than letting all the students vote on entertainment for the dance."

"The band we voted for will be within our budget since they were on our list."

It's Your Turn
Follow the same problem-solving process used by the entertainment committee to explain how a group of students on the refreshments committee would decide what to serve at the dance.

In Your Estimation

Estimation is a skill you use nearly every day. Suppose you plan to be home all evening and invite a friend to come over after she eats dinner. She correctly interprets this to mean that she does not need to arrive at a precise time. You can probably expect her to arrive anytime between 5:30 and 7:30 P.M

You and a friend are going to a movie. You need to find out exactly when it begins. Then you must be in your seats by 7:15 P.M., for example, so you don't miss anything on the screen.

In the first situation, you can safely estimate — or calculate approximately — when your friend will arrive. In the second, it would be foolish to make an estimate rather than to find out exactly when the movie begins.

In pre-algebra, you will develop your estimating skills to solve some problems and to check others. After you explore the problem to be solved (step 1, remember?), always estimate your answer (step 2) before solving the problem (step 3).

When Is It Safe to Estimate?

Use your reasoning skills, along with some good old-fashioned common sense, to determine whether the following situations require an exact answer or an estimate.

1. Your mother agrees to give you an advance on your allowance to buy stereo speakers, but says you must pay her back that amount plus 8 percent interest.

2. Michael wants to figure out how much his store discount coupon is worth on the sweater he plans to buy.

3. Angela needs to figure out how many sequins she needs to buy to sew on her costume for the school play.

4. Tanya's father asks her teacher to give him a general idea of where Tanya's grades rank her in a class of 187 students.

5. When they receive their $6 dinner check, Jessica and Andrea decide to reward their friendly waiter by giving him a 20% tip.

6. John asks his mom how many hours it will take them to drive from North Carolina to Florida on their family vacation.

Putting Your Estimating Skills to Work

Estimating skills are especially handy when you find yourself without a pencil and paper or calculator and you must literally "use your head." To make estimating easier, you can round numbers.

For example, let's say Karen works 12 hours a week at a part-time job and earns $4.08 an hour. No taxes are withheld from her small paychecks. She is saving for a mountain bike that costs $279. She wants to estimate her weekly pay to see how long she needs to save to buy the bike. Because it ends in 0, it is easier to multiply by 10 than by 12. So, to make her estimate, Karen will use 10 for the number of hours she works.

1. Which number would be reasonably accurate and easier to use for Karen's hourly pay?

A. $4.00   B. $4.10   C. $4.15   D. $4.25

2. Multiply the number of estimated hours Karen works each week by her estimated hourly pay. Choose the most accurate estimate of how much Karen will earn in one week.

A. $40   B.$48   C. $49   D. $50

3. About how long will Karen need to work to save enough money to buy the $279 bike?

A. 4 weeks   B. 7 weeks   C. 10 weeks   D. 12 weeks

Choosing the Method of Computation

One of the most important steps in problem solving is deciding what method you will use to find the answer. This process is also called "choosing the method of computation." Sometimes, as with Karen saving to buy a bike, estimating is the best method for finding the answer.

Rely on your estimating skills whenever an *approximate* answer is appropriate to the situation. When an exact answer is required, you will need to use mental math, pencil and paper, or a calculator to carry out the computation. Use this handy chart to help you decide which method to use to solve the problems on page 7.

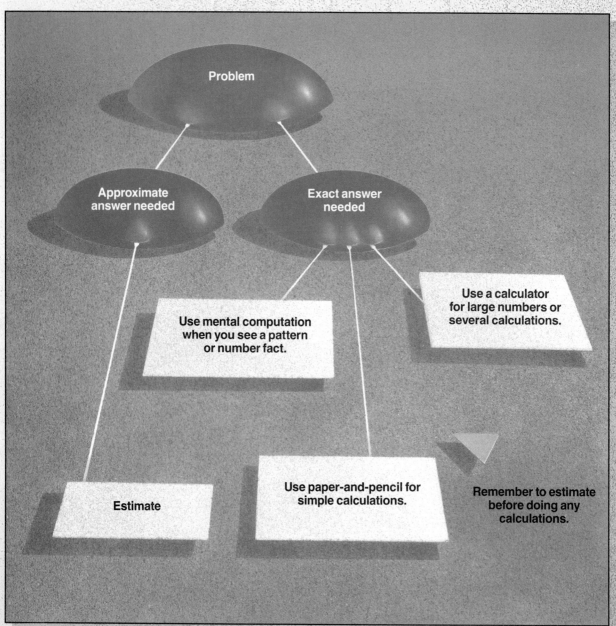

Problem

Approximate answer needed

Exact answer needed

Use mental computation when you see a pattern or number fact.

Use a calculator for large numbers or several calculations.

Estimate

Use paper-and-pencil for simple calculations.

Remember to estimate before doing any calculations.

Choosing the Method of Computation

Choose the best method of computation to solve each problem.

1. Is $7 enough money to buy a loaf of bread for $0.98, 1 pound of ground beef for $2.29, and luncheon meat for $3.29?

2. Each of 25 persons donated $10 to the school scholarship fund. How many dollars were donated in all?

3. Lucia worked 32.5 hours and earned $149.50. How much was she paid for each hour?

4. Rich buys 4 cassettes for $6.95 each. What is the total cost?

5. How many seconds are in one week?

6. Luis has $100.75 in his checking account. He writes checks for $21.78, $43, and $7.08. What is the new balance?

7. Jane's lunch cost $3.64. She gives the cashier a $10-bill. How much change should Jane receive?

8. Helen needs to save $300 for the class trip. She has already saved $265. She plans to earn $20 mowing grass and $25 babysitting. Will she have enough money for the trip?

9. Ted wants to buy a drum set for $359 and drum cases for $49. He has saved $259. How much more does he need to save?

10. It costs $283 to plant an acre of trees in a national forest. What is the cost of planting 640 acres?

Let's Be Reasonable

In an ideal world, there would be time to re-think and re-work every problem you encounter. In real life, this usually isn't possible. Instead, you will need to use your reasoning skills to determine if your answer seems reasonable. Sometimes your estimating skills will help you determine if an answer is reasonable. Sometimes your common sense will be your guide. Often, you will rely on a combination of these two skills to quickly judge whether an answer is reasonable or unreasonable.

Practice your reasoning skills by determining whether the answers to the following situations or problems are reasonable or unreasonable. The first one is done for you.

1. Marty measures his room. He found that one wall was 43 meters long. *Oops! Measure again, Marty! A meter is a little longer than a yard, or 3 feet. Multiply 3 by 43. Marty's wall can't possibly be 43 meters long because that figure is about 129 feet.*

2. Elizabeth asked her mother for $5 for school supplies and lunch. She plans to buy 5 markers for art class that cost 55 cents each. Will she have enough left to buy the standard school lunch?

3. Jonathan is in charge of ordering 40 folders, one for each member of the science club. The catalog showed the total price for the folders to be $12. He calculates that each club member needed to give him 3 cents to pay for his or her folder.

4. Alison wants to save enough money to buy a compact disc player that costs $129.95 plus 5 percent sales tax. She plans to save $12 a week from her allowance and her regular babysitting job. She tells her friend Jessica that in three months she'll have enough to buy the CD player.

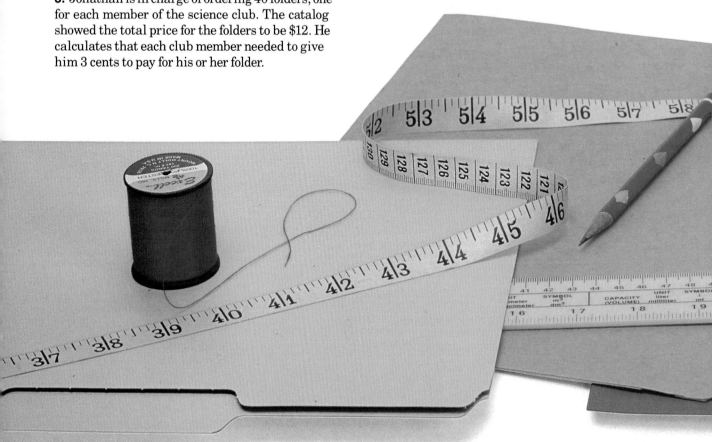

5. To make fruit juice from frozen concentrate, Matthew reads the directions that say to add three cans of cold water for each can of concentrate. To serve a crowd, Matthew decides he must measure out 12 cans of cold water to add to 6 cans of concentrate.

6. Anne's father loaned her $30. She is repaying $5 a week with no interest. She tells her friends that since all her spending money is going toward repaying the loan, it will be at least two months before she can afford to go to a movie.

7. David's mother gives him a $10 bill to buy the following items at the grocery: a gallon of milk for $2.09, 4 containers of yogurt that cost 79¢ each and a package of frozen beans that cost $1.39. David asks his mother if he can use the change to buy a movie ticket.

8. Leah and Kacy decide to split the 15% tip on their $7.25 lunch tab. Kacy says 40¢ each is enough to cover the tip.

9. John spent a total of $6.25 on school lunches last week. He was sick Tuesday and missed school. He says his average cost for lunch last week was about a dollar a day.

10. Sarah buys a pattern for a short dress that requires 3 yards of fabric. She decides to make it into a long dress by adding another 24 inches to the length. Sarah buys an extra ¼ yard of fabric so she will have enough to make a longer dress.

Getting Good at Graphs

Some people are nervous about reading graphs. They shouldn't be! Graphs provide a quick and easy way to look at data. Data is simply information that has been gathered. A graph displays data in an easy-to-understand format. It enables you to easily make comparisons.

Four common types of graphs are bar graphs, pictographs, line graphs, and circle graphs. Use the following graphs to answer the questions. The first question that relates to each kind of graph is answered for you.

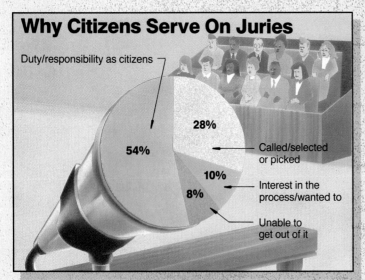

Why Citizens Serve On Juries

Duty/responsibility as citizens

28%

54%

Called/selected or picked

10%

8%

Interest in the process/wanted to

Unable to get out of it

Circle Graphs

Circle graphs are often used to show how something is divided into parts. The sum of the parts of the circle adds up to 100%. This circle graph shows the reasons citizens serve on juries.

1. Which reason did most people give as a reason for serving on a jury? *Clearly, they feel it is their duty as citizens.*
2. What percentage of the people were interested in the process?
3. More than one-half of the people felt it was their duty to serve on a jury. How is this shown on the graph?

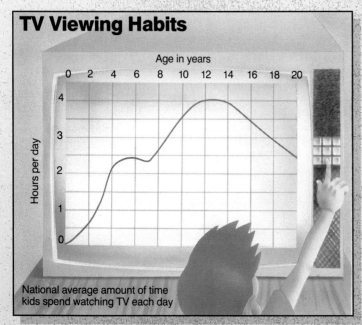

TV Viewing Habits

Age in years

0 2 4 6 8 10 12 14 16 18 20

4

3

2

1

0

Hours per day

National average amount of time kids spend watching TV each day

Line Graphs

Line graphs provide another method for displaying data. They are often used to show trends or how something changes over time.

The line graph in this example shows the national average amount of time kids spend watching television each day.

1. Which age group watched the most television each day? *Clearly, kids between age 12 and 14 watch the most television.*
2. What reasons could account for the decrease in television watching after age 14?
3. Which two age groups watch about 3 hours of television each day?

Bar Graph

The bar graph in this example shows the change in sales for various music products. Each bar stands for a gain or a loss.

1. Which music product has had the greatest percent increase?
   You can easily see that the greatest change is with music videos.

2. Which product has had the greatest percent decrease?

3. Which product has had the least amount of change?

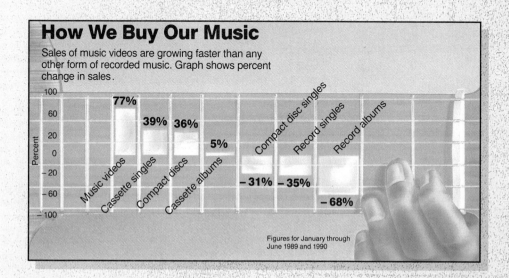

How We Buy Our Music

Sales of music videos are growing faster than any other form of recorded music. Graph shows percent change in sales.

Figures for January through June 1989 and 1990

Pictographs

Often you will see pictographs like this one in newspapers and magazines. It is actually a type of bar graph. The bars on a graph may be vertical, like the one showing the change in music products, or horizontal, like this pictograph. Both are correct.

The pictograph in this example shows the methods people use to fall asleep. The methods are listed on the left. Each "Z" stands for 2%. Half of a "Z" stands for 1%.

1. Which method is used by the most people? *Clearly, it's watching television.*

2. What percent of the people did nothing to fall asleep?

3. Which method is used by 13% of the people?

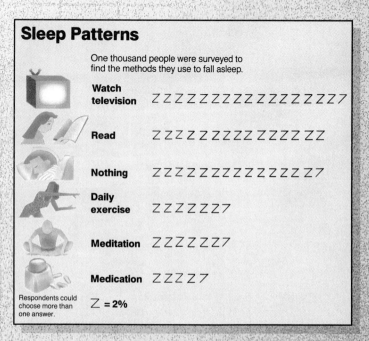

Sleep Patterns

One thousand people were surveyed to find the methods they use to fall asleep.

Watch television Z Z Z Z Z Z Z Z Z Z Z Z Z Z Z Z Z

Read Z Z Z Z Z Z Z Z Z Z Z Z Z Z

Nothing Z Z Z Z Z Z Z Z Z Z Z Z Z

Daily exercise Z Z Z Z Z Z

Meditation Z Z Z Z Z Z

Medication Z Z Z Z Z

Respondents could choose more than one answer.

Z = 2%

11

CHAPTER 1

CHAPTER OBJECTIVES

In this chapter you will learn to:

- evaluate expressions using order of operations
- use properties of whole numbers to simplify expressions
- solve equations and inequalities mentally
- translate verbal phrases into algebraic expressions
- solve problems using the guess-and-check strategy

The Language of Algebra

Picture yourself standing in an endless line of people. The line stretches from Washington, D.C., to San Francisco, CA. Your challenge: determine how long it will take to pass a basketball from one end to the other.

What information do you need to solve this problem? Here are some hints:

How far is it from Washington, D.C., to San Francisco?

How many people are in line?

How many seconds, on average, does it take for one person to pass a basketball to the next person in line?

The answers to these questions may *vary.* In algebra, you use *variables* to stand for unknown quantities. Variables are the muscles of algebra! In this chapter and throughout this book you'll work with variables. Let's get started!

Estimate how long it will take to pass a basketball from Washington, D.C., to San Francisco.

⇒ Geography ⇐ Connection

Class Project

Suppose the population of your state was evenly scattered throughout the state. About how many people would be standing in each square mile?

Discuss the meaning of the phrase *population density* with a classmate. Compute the population density for two other states. Display your data in a bar graph.

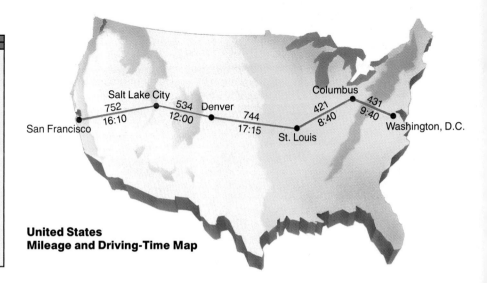

United States Mileage and Driving-Time Map

Naming Numbers

Materials: calculator

Every number can be renamed as the sum, difference, product, or quotient of two or more other numbers. For example, 7 can be renamed as the sum 4 + 3; 15 can be renamed as the quotient 30 ÷ 2. In this Exploration, you will try to rename numbers using only 2s, 5s, and 9s.

Explore: Suppose that these are the only keys that work on your calculator.

[2] [5] [9] [+] [−] [×] [÷] [=]

Can you rename every number from 1 to 25 in two different ways using only these keys?

Here are some examples:

▶ One way to rename 15 is 9 [+] 2 [+] 2 [+] 2 [=].

▶ One way to rename 19 is 2 [×] 5 [+] 9 [=].

Your Turn: Try to rename every number from 1 to 25 in two different ways. Keep track of your results in a chart.

Analysis

1. Were you able to rename every number in two different ways?

2. Explain how you can rename numbers greater than 25 using only 2s, 5s, and 9s. Give examples using 47, 63, and 56.

Wrap-Up

3. Leonhard Euler, the great Swiss mathematician, died in 1783 at the age of 76. Using only the numbers 1, 7, 8, and 3, rename 76.

1-1   Order of Operations

Objective:
Evaluate expressions using the order of operations.

Key Terms:
order of operations
parentheses
brackets

If you play on any sports team, you know that there are certain rules that govern the play. These rules help avoid confusion and disagreements.

Rules are also important in algebra. Why? One reason is that numerical expressions often contain more than one operation. An expression like $9 + 2 \times 2$ might be computed in two ways.

$9 + 2 \times 2 = 9 + 4$ Multiply,
$9 + 2 \times 2 = 13$ then add.

$9 + 2 \times 2 = 11 \times 2$ Add, then
$9 + 2 \times 2 = 22$ multiply.

Just as in sports, we need a rule to avoid confusion and tell us which value is correct. This rule is called the **order of operations.**

First: Do all multiplications and divisions from left to right.
Second: Do all additions and subtractions from left to right.

The order of operations rule guarantees that each numerical expression has a *unique* value. Using this rule, the correct value of $9 + 2 \times 2$ is 13.

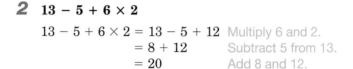

Examples

Find the value of each expression.

1 $5 \times 6 + 4$

$5 \times 6 + 4 = 30 + 4$ Multiply 5 and 6.
$\qquad\qquad = 34$ Add 30 and 4.

2 $13 - 5 + 6 \times 2$

$13 - 5 + 6 \times 2 = 13 - 5 + 12$ Multiply 6 and 2.
$\qquad\qquad\qquad = 8 + 12$ Subtract 5 from 13.
$\qquad\qquad\qquad = 20$ Add 8 and 12.

3 $9 \div 3 + 4 \times 7 - 20 \div 5$

$9 \div 3 + 4 \times 7 - 20 \div 5 = 3 + 4 \times 7 - 20 \div 5$ Divide 9 by 3.
$\qquad\qquad\qquad\qquad = 3 + 28 - 20 \div 5$ Multiply 4 and 7.
$\qquad\qquad\qquad\qquad = 3 + 28 - 4$ Divide 20 by 5.
$\qquad\qquad\qquad\qquad = 31 - 4$ Add 3 and 28.
$\qquad\qquad\qquad\qquad = 27$ Subtract 4 from 31.

Calculator Hint

Calculators that follow the order of operations rule are called scientific calculators. To find the value of $15 + 3 \times 7$, enter

$15 \ \boxed{+} \ 3 \ \boxed{\times} \ 7.$

If your calculator displays 36, your calculator is a scientific calculator.

The order of operations can be changed by using grouping symbols, such as **parentheses,** (), and **brackets,** []. The value of the expression $(2 + 6) \times 3$ is found by performing the operation within the parentheses first.

$$(2 + 6) \times 3 = 8 \times 3 \quad \text{Compute } 2 + 6 \text{ first.}$$
$$= 24 \quad \text{Now multiply 8 and 3.}$$

The order for performing the operations in an expression is summarized as follows.

Order of Operations	1. Do all operations within grouping symbols first; start with the innermost grouping symbols. 2. Next, do all multiplications and divisions from left to right. 3. Then, do all additions and subtractions from left to right.

In algebra there are many ways to indicate multiplication and division. A raised dot or parentheses can be used to indicate multiplication.

Can you find these notations in the examples below?

$8 \cdot 7$ **means** 8×7

$3\,(4), (3)\,4,$ or $(3)\,(4)$ **means** 3×4

A fraction bar can be used to indicate division.

$\dfrac{26 - 2}{4 + 8}$ **means** $(26 - 2) \div (4 + 8)$

Examples

Find the value of each expression.

4 $2(3 + 5) - 3 \cdot 5 = 2(8) - 3 \cdot 5 \quad 3 \cdot 5 \text{ means } 3 \times 5.$
$$= 16 - 15$$
$$= 1$$

5 $4[(9 + 12) - 6(2)] = 4[(21) - 6(2)] \quad \text{Do the operations within}$
$$= 4[21 - 12] \quad \text{the innermost grouping symbols.}$$
$$= 4[9]$$
$$= 36$$

6 $\dfrac{17 + 7}{19 - 13} = (17 + 7) \div (19 - 13) \quad \text{Rewrite the expression}$
$$= (24) \div (6) \quad \text{using parentheses.}$$
$$= 4$$

Checking for Understanding

Communicating Algebra

1. If an expression contained only two operations, division and subtraction, which operation would you do first?

2. Explain how $5 \cdot 3 + 5$ and $5(3 + 5)$ are different.

3. In the expression $2 \cdot 3 + 5 \cdot 3$, which multiplication would you do first?

Guided Practice

Name the operation that should be done first. Then find the value.

4. $5 + 3 \cdot 7$

5. $54 - 42 \div 7$

6. $36 \div 9 \times 3$

7. $24 \div 6 - 9 \div 3$

8. $4 \cdot (5 + 7)$

9. $3 \cdot (4 + 5) - 7$

10. $71 - (17 + 4)$

11. $(25 + 7) - (12 + 3)$

12. $(8 - 5)(4 + 2)$

13. $\dfrac{46 - 4}{6}$

14. $\dfrac{20 - 8}{9 - 5}$

15. $\dfrac{16 + 8}{15 - 7}$

Exercises

Independent Practice

Find the value of each expression.

16. $12 \div 3 + 12 \div 4$

17. $21 \div 7 + 4 \cdot 11$

18. $6 \cdot 3 \div 9 - 1$

19. $12 \div 6 - 2$

20. $15 \div 5 \times 3$

21. $24 \div 6 \cdot 4$

22. $(25 \cdot 3) + (15 \cdot 3)$

23. $(40 \cdot 2) - (6 \cdot 11)$

24. $40 \cdot (6 - 2)$

25. $72 \div 9 \cdot 4 \div 2$

26. $96 \div (12 \cdot 4) \div 2$

27. $144 \div 16 \cdot 9 \div 3$

28. $\dfrac{86 - 11}{9 + 6}$

29. $\dfrac{72 + 12}{35 + 7}$

30. $\dfrac{37 + 38}{30 - 5}$

31. $2[5(4 + 6) - 3]$

32. $3[6(12 - 3)] - 17$

33. $4[12(22 - 19) - 3 \cdot 6]$

34. $4[3(21 - 17) + 3]$

35. $3[(18 - 3) + 4(5 + 7)]$

36. $7[(12 + 5) - 3(19 - 14)]$

***True* or *false*.**

37. $12 + 20 \div 4 - 5 = 3$

38. $36 - 6 \cdot 5 = 6$

39. $3(34 - 19) = 45$

40. $6 \div 2 + 5 \times 4 = 32$

41. $30 \div 6 \cdot 5 = 1$

42. $14 - 5 - 2 \times 2 = 6$

Calculator

Copy each sentence below. Experiment with your calculator to find where to insert parentheses to make each sentence true. You may need to use the parentheses keys.

43. $18 + 4 \times 3 = 66$

44. $18 \div 3 + 6 = 2$

45. $24 \div 2 - 4 + 8 = 0$

46. $15 \div 21 - 18 - 4 = 1$

Application

47. **Consumer Awareness**   Write an expression that would help you find the total cost of fifty $6 movie tickets and eight $12 play tickets. Then find the total cost.

Critical Thinking Wrap-Up

48. Use six toothpicks to represent the number 17.

49. **Make Up a Problem**   Write an expression involving addition and multiplication in which you should add first.

Algebraic Expressions

Materials: cups, counters

In this Exploration, you will model algebraic expressions.

▶ Consider the phrase *the sum of 5 and some number*. This phrase contains a value that you know, 5, and a value that you do not know, some number. You use counters to represent 5 and an empty cup to represent the unknown value.

▶ This cup can hold any number of counters. Suppose you place 3 counters in the cup. Instead of an unknown value, you know that the cup has a value of 3. When you empty the cup and count all the counters, the expression has a value of 8.

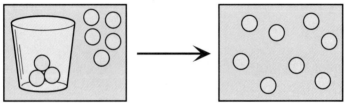

▶ Consider the phrase *three times some number*. Because you do not know the value of the number, let an empty cup hold this value. In this case, you need three empty cups.

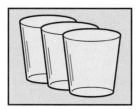

Your Turn: Model the phrase *5 more than 3 times a number*. Place 4 counters in each cup. How many counters are there in all?

Model each phrase.
1. 6 more than a number
2. four times a number
3. 4 more than 2 times a number
4. 3 times 5 more than a number
5. Model the phrase *2 more than 2 times a number*. Fill each cup with 3 counters. How many counters are there in all?

Analysis
6. Write a sentence that describes the meaning of the cup.

1-2   Variables and Expressions

Objective:
Use variables.

Key Terms:
variable
algebraic expression

Did you know that you can estimate the temperature in degrees Fahrenheit by counting the number of times a cricket chirps in one minute? Just count the number of chirps, divide by 4 and then add 37.

Like any language, algebra is a language of symbols. You already know symbols for division and addition. So, one way to write the temperature relationship is: *chirps* $\div$ 4 + 37.

Using arithmetic, you might write ■ $\div$ 4 + 37, where ■ is the number of times the cricket chirped. The ■ acts as a placeholder.

In algebra, a placeholder is called a **variable** because the value can change or vary. Variables are usually letters. Probably the most popular letter is x, but in this case you choose c to stand for chirps or cricket. Now the temperature relationship is $c \div 4 + 37$.

This is called an **algebraic expression** because it is a combination of variables, numbers, and at least one operation.

An expression such as $c \div 4 + 37$ can be evaluated by replacing variables with numbers and then finding the value of the numerical expression. If you counted 80 chirps, you could estimate the temperature by evaluating the expression $80 \div 4 + 37$.

What is an estimate of the temperature if you count 110 chirps?

$$80 \div 4 + 37 = 20 + 37 \quad \text{Divide first, then add.}$$
$$= 57 \qquad \text{The temperature is about } 57° \text{ F.}$$

When you replaced c with 80, you were using an important property of numbers.

Substitution Property of Equality	For all numbers a and b, if $a = b$, then a may be replaced with b.

Example

1 **Evaluate $c + b - 23$ if $c = 25$ and $b = 16$.**

$c + b - 23 = 25 + 16 - 23$ Replace c with 25 and b with 16.
$\qquad\qquad = (25 + 16) - 23$
$\qquad\qquad = 41 - 23 \qquad 25 + 16 = 41$
$\qquad\qquad = 18 \qquad\quad 41 - 23 = 18$

Just as with numerical expressions, mathematicians agree on special notation for multiplication and division with variables.

$2a$ means $2 \times a$

mn means $m \times n$

$6bc$ means $6 \times b \times c$

$\frac{k}{5}$ means $k \div 5$

Examples

2 **Evaluate $3b + 4c$ if $b = 9$ and $c = 8$.**

$3b + 4c = 3(9) + 4(8)$ Replace b with 9 and c with 8.

$= 27 + 32$ Use the order of operations.

$= 59$

3 **Evaluate $\frac{xz}{3}$ if $x = 5$ and $z = 6$.**

$\frac{xz}{3} = (5)(6) \div 3$ $\frac{xz}{3} = xz \div 3$

$= 30 \div 3$ Multiply 5 and 6.

$= 10$ Divide 30 by 3.

4 **Evaluate $2x + (3y - z) + 7$ if $x = 5$, $y = 2$, and $z = 4$.**

$2x + (3y - z) + 7 = 2 \cdot 5 + (3 \cdot 2 - 4) + 7$

$= 2 \cdot 5 + (6 - 4) + 7$ Multiply inside parentheses first.

$= 10 + 2 + 7$

$= 19$

Checking for Understanding

Communicating Algebra

1. Write two different expressions that mean the same as $3y$.

2. Write an expression that means the same as $\frac{a}{b}$.

3. Explain in your own words how variables are used.

Guided Practice

Evaluate each expression if $a = 8$, $k = 7$, $m = 5$, $p = 6$, and $s = 11$.

4. $16 + k$

5. $s - 8$

6. $a - 3$

7. $k + 18$

8. $17 - m$

9. $24 - a$

10. $p + s$

11. $a + k$

12. $s - p + 37$

13. $a + s - 9$

14. $48 + m - k$

15. $p + 104 - s - 26$

Exercises

Independent Practice

Evaluate each expression if $a = 4$, $b = 2$, and $c = 3$.

16. $a + b \cdot c$
17. $6a - b \cdot c$
18. $ab - bc$
19. $4a + b \cdot b$
20. $4a - (b + c)$
21. $6a + 6b$
22. $\frac{a}{b} + c$
23. $\frac{6a}{c}$
24. $\frac{ac}{b}$
25. $7a - (2c + b)$
26. $9a - (4b + 2c)$
27. $2c + 3a + 6b$
28. $\frac{3(4a - 3c)}{a - 3}$
29. $\frac{6(a + b)}{3c}$
30. $\frac{15ac}{3c + 6}$

Computer

In the BASIC computer language, LET statements are used to assign values to the variables. Find the value assigned to A by each LET statement if X = 5 and Y = 15. Multiplication is shown by the symbol $\star$ and division by /.

31. LET A = 2 $\star$ X + 9
32. LET A = 25 + X − 10
33. LET A = 40/X $\star$ 8
34. LET A = (19 + X)/12
35. LET A = 12/(31 − 2 $\star$ Y)
36. LET A = 3 $\star$ X $\star$ Y

Mixed Review

37. If an expression contains only an addition and a division, which would you do first? (Lesson 1-1)

38. Name the operation you would do first in the expression $(6 - 3)(5 + 1)$. (Lesson 1-1)

39. Evaluate the expression $2[3(11 - 7) + 1]$. (Lesson 1-1)

40. *True* or *false*: $9 \div 3 + 5 - 4 = 4$. (Lesson 1-1)

41. Use different symbols to write the expression 8×7 in two other ways. (Lesson 1-1)

Application

42. **Health** The expression $110 + \frac{A}{2}$ is used to estimate a person's normal blood pressure. In this expression, A stands for the person's age in years. Use this expression to estimate the blood pressure of a person who is 18 years old.

Critical Thinking

43. Juan and Maria have calculators. Juan starts at zero and adds 2. Maria starts at 100 and subtracts 3. If they push their keys at the same time, will their displays ever show the same number at the same time? If so, what number is it?

Wrap-Up

 44. Write a sentence that explains why the substitution property is important.

Biography

Al-Khowarizmi

One of the most influential giants of mathematics was an Arabian mathematician, Al-Khowarizmi. In A.D. 820, at the age of forty, Al-Khowarizmi wrote *Hisab al-jabr w'al-muqabala,* which freely translates into the "Science of Transposition and Cancellation." Note that our word *algebra* looks and sounds very much like *al-jabr.* Al-Khowarizmi did not originate algebra, but he brought organization to the mathematics we now call algebra.

1-3 Properties

Objective:
Identify and use
properties of addition
and multiplication.

Key Terms:
commutative
associative
identity

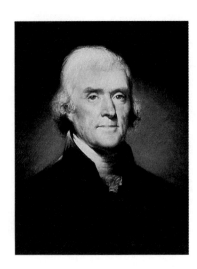

When Thomas Jefferson wrote the Declaration of Independence, he stated certain rights that could be applied to all people. Among these are life, liberty, and the pursuit of happiness.

In mathematics, there are certain properties that are true for addition and multiplication of whole numbers. You may remember the commutative, associative, and identity properties from arithmetic. These properties are summarized in the chart below.

Commutative Properties of Addition and Multiplication	
The order in which numbers are added does not change the sum. $$4 + 7 = 7 + 4$$ For any numbers a and b, $$a + b = b + a.$$	The order in which numbers are multiplied does not change the product. $$6 \cdot 4 = 4 \cdot 6$$ For any numbers a and b, $$a \cdot b = b \cdot a.$$
Associative Properties of Addition and Multiplication	
The way in which addends are grouped does not change the sum. $$(4 + 3) + 5 = 4 + (3 + 5)$$ For any numbers a, b, and c, $$(a + b) + c = a + (b + c).$$	The way in which factors are grouped does not change the product. $$(4 \cdot 2) \cdot 3 = 4 \cdot (2 \cdot 3)$$ For any numbers a, b, and c, $$(a \cdot b) \cdot c = a \cdot (b \cdot c).$$
Identity Properties of Addition and Multiplication	
The sum of an addend and zero is the addend. $$5 + 0 = 5$$ For any number a, $a + 0 = a$.	The product of a factor and one is the factor. $$7 \cdot 1 = 7$$ For any number a, $a \cdot 1 = a$.
Multiplicative Property of Zero	
The product of a factor and zero is zero. $$2 \cdot 0 = 0$$ For any number a, $a \cdot 0 = 0$.	

You can use these properties to compute sums and products mentally. In the following examples, look for sums and products that are multiples of 10.

Examples

Compute mentally.

1   14 + 5 + 16 + 25 = ?

14 + 16 + 5 + 25 = ?

30   +   30   = 60

2   25 · 9 · 4 = ?

25 · 4 · 9 = ?

100 · 9 = 900

CONNECTION TO ALGEBRA

You can also use these properties to rewrite and simplify algebraic expressions.

Examples

Rewrite each expression using the commutative property.

Is subtraction commutative?

3 $x + 3$

$x + 3 = 3 + x$ Change the order.

4 $x \cdot y$

$x \cdot y = y \cdot x$

Rewrite each expression using the associative property. Then simplify.

Is subtraction associative?

5 $(y + 2) + 3$

$(y + 2) + 3 = y + (2 + 3)$ Change the
$= y + 5$ grouping

6 $4 (5m)$

$4 (5m) = (4 \cdot 5)m$
$= 20m$

7   **Evaluate abc if $a = 6$, $b = 7$, and $c = 5$.**

$abc = 6 \cdot 7 \cdot 5$ Replace a with 6, b with 7, and c with 5.

$= 7 \cdot 6 \cdot 5$ Use the commutative property to change the order of 7 and 6.

$= 7 \cdot (6 \cdot 5)$ Group 6 and 5 to make the multiplication easier to do mentally.

$= 7 \cdot 30$

$= 210$

Checking for Understanding

Communicating Algebra

1. Explain the difference between the commutative and associative properties.

2. Using whole numbers, write a mathematical sentence that illustrates the associative property of multiplication.

3. Explain how the properties can help you add and multiply mentally.

Name the property shown by each statement.

4. $(2 + 5) + 3 = 2 + (5 + 3)$ **5.** $7 + 4 = 4 + 7$

6. $9 + 0 = 9$ **7.** $3 \cdot 5 \cdot 0 = 0$

8. $7 \cdot 32 = 32 \cdot 7$ **9.** $(6 \cdot 15) \cdot 3 = 6 \cdot (15 \cdot 3)$

10. $(9 + 7) + 4 = 4 + (9 + 7)$ **11.** $8 \cdot 1 = 8$

12. $(6a)b = 6(ab)$ **13.** $17m = m \cdot 17$

14. $(12 \cdot a) \cdot 5 = (a \cdot 12) \cdot 5$ **15.** $0 = a \times 0$

Exercises

Name the property shown by each statement.

16. $6 + a = a + 6$ **17.** $4(bc) = (4b)c$ **18.** $z \cdot 1 = z$

19. $0 \cdot 91 = 0$ **20.** $7ab = 7ba$ **21.** $0 + xy = xy + 0$

22. $21 + 0 = 21$ **23.** $9a + b = b + 9a$ **24.** $(4 + 7)\,0 = 0$

Compute mentally.

25. $18 + 13 + 2 + 7$ **26.** $4 + 13 + 26 + 5$ **27.** $5 \cdot 13 \cdot 2$

28. $10 \cdot 11 \cdot 6$ **29.** $23 + 37 + 82 + 58$ **30.** $98 \cdot 24 \cdot 0$

Rewrite each expression using a commutative property.

31. $9 + 12$ **32.** $3x + 5$ **33.** $7 + 2z$

Rewrite each expression using an associative property. Then simplify.

34. $(a + 3) + 4$ **35.** $12 + (15 + x)$ **36.** $5\,(3y)$

37. $(n \cdot 6)\,5$ **38.** $(x + 25) + 18$ **39.** $3\,(2z)$

40. $11 + (13 + m)$ **41.** $5\,(4x)$ **42.** $(m \cdot 4)\,7$

43. Evaluate $a + b + c$ if $a = 25$, $b = 49$, and $c = 75$.

44. Name the operation you would do first in the expression $\dfrac{18 - 3}{5}$.   (Lesson 1-1)

45. Evaluate the expression $5 \cdot 6 \div 2 + 1$.   (Lesson 1-1)

46. Write an expression that means the same as $\dfrac{ab}{2}$.   (Lesson 1-2)

47. Evaluate the expression $b + 6$ if $b = 11$.   (Lesson 1-2)

48. Evaluate the expression $2s - 3p$ if $p = 5$ and $s = 12$.   (Lesson 1-2)

49. **Chemistry** When chemists dilute acid, they always pour the acid into the water. Pouring water into acid could produce spattering and burns. These two actions are not commutative. Give another example from your life of actions that are not commutative.

50. Given the expression $(5a)(3b)(4c)(0)$, explain how you can find its value.

51. Find the value of ▮ in the sentence $(50 + 18) + 20 = (▮ + 50) + 20$.

1-4   The Distributive Property

Objective:
Simplify algebraic
expressions using the
distributive property.

Key Terms:
distributive
terms
like terms
simplest form

In the previous lesson, you reviewed some of the properties of operations
on whole numbers. There is another important property that ties
addition and multiplication together. It is the **distributive property.**
To see how this property works, study a situation that may occur in your
own life.

Suppose you baby-sit for the Lopez family to earn spending money. You
charge a fee of $3 per hour. If you baby-sit 5 hours on Friday night and 7
hours on Saturday night, how much will you earn?

There are two ways to figure how much money you will earn.

| fee per hour | total hours | money earned on Friday | money earned on Saturday |

$$3(5+7) = 3(12)$$
$$= 36$$

$$3 \cdot 5 \quad + \quad 3 \cdot 7 = 15 + 21$$
$$= 36$$

Is this a true statement:
$4(8 - 2) = 4 \cdot 8 - 4 \cdot 2$?
What might you
conclude?

Each method shown above has the
same result, 36.

$$3(5 + 7) = 3 \cdot 5 + 3 \cdot 7$$

Your fee is distributed over both days.
The 3 is distributed as a multiplier
over both 5 and 7. This is an example
of the distributive property of
multiplication over addition.

> **Distributive Property**
>
> For any numbers a, b, and c,
> $a(b + c) = ab + ac$ and $(b + c)a = ba + ca$.

Examples

1 **Rewrite $8 \cdot 6 + 8 \cdot 7$ using the distributive property.**

$8 \cdot 6 + 8 \cdot 7$ 8 is a factor of both $8 \cdot 6$ and $8 \cdot 7$.
$8 \cdot 6 + 8 \cdot 7 = 8(6 + 7)$

2 **Compute $15 \cdot 12$ mentally. Use the distributive property.**

$15 \cdot 12 = 15(10 + 2)$ Use $10 + 2$ for 12.
$\qquad = 15 \cdot 10 + 15 \cdot 2$ THINK: $150 + 30$
$\qquad = \quad 150 \quad + \quad 30$ or 180

Mental Math Hint

The Distributive
Property allows you to
break apart one of the
factors into a sum.
You can then add the
two products mentally.

CONNECTION TO ALGEBRA

Let's return to the baby-sitting situation. If you're like most baby-sitters, you vary your hourly rate depending on the number of children or the amount of responsibility you have. You can represent your hourly rate in dollars for a particular family with a variable like x.

If you baby-sit 5 hours on Friday night, you will earn $5x$ dollars. If you baby-sit 7 hours on Saturday night at the same rate, you will earn $7x$ dollars.

Your total earnings can be represented by the expression $5x + 7x$.

The expression $5x + 7x$ has two **terms** with the same variable. These terms are called **like terms.** Some other pairs of like terms are $5a$ and $13a$, $2rs$ and $6rs$, and y and $7y$. You can use the distributive property to simplify expressions that have like terms. Example 3 shows how to simplify $5x + 7x$.

Examples

Simplify each expression.

3 $5x + 7x$ $\qquad$ x is a factor of both $5x$ and $7x$.

$5x + 7x = (5 + 7)x$ $\qquad$ Distributive property

$\qquad\qquad = 12x$ $\qquad$ Substitution property

4 $m + 8m$

$m + 8m = 1m + 8m$ $\qquad$ $m = 1 \cdot m$; Multiplicative identity

$\qquad\qquad = (1 + 8)m$ $\qquad$ Distributive property

$\qquad\qquad = 9m$ $\qquad$ Substitution property

One more time, let's return to the baby-sitting situation. Suppose that in addition to your hourly earnings, you receive a tip of $10. Your total earnings can now be represented by the expression $5x + 7x + 10$ or $12x + 10$. The expression $12x + 10$ is in **simplest form** because it has no like terms and no parentheses.

Examples

Simplify each expression.

5 $14a + 7 + 21a$

$$14a + 7 + 21a = 14a + 21a + 7 \quad \text{Commutative property of addition}$$
$$= (14 + 21)a + 7 \quad \text{Distributive property}$$
$$= 35a + 7 \quad \text{Simplest form}$$

6 $r + 3(s + 7r)$

$$r + 3(s + 7r) = r + 3s + 3 \cdot 7r \quad \text{Distributive property}$$
$$= r + 3s + 21r \quad \text{Substitution property of equality}$$
$$= r + 21r + 3s \quad \text{Commutative property of addition}$$
$$= 1r + 21r + 3s \quad \text{Multiplicative identity}$$
$$= (1 + 21)r + 3s \quad \text{Distributive property}$$
$$= 22r + 3s \quad \text{Simplest form}$$

Checking for Understanding

Communicating Algebra

1. Name the like terms in $7s + 9y + y$.
2. Explain why $7a + 8z - 9x$ is in simplest form.
3. Explain why $4 + 2(x - 3)$ is not in simplest form.

Guided Practice

Restate each expression using the distributive property. Do not compute.

4. $7(3 + 6)$
5. $8(5 + 9)$
6. $6(2 + 4)$

7. $5(11 + 12)$
8. $(c + d)3$
9. $(x + y)5$

10. $4 \cdot 6 + 4 \cdot 12$
11. $8 \cdot 3 + 8 \cdot 6$
12. $10r + 10s$

13. $9x + 9y$
14. $2a + 4b$
15. $3x + 6y$

Exercises

Independent Practice

Simplify each expression.

16. $4p + p$
17. $x + 7x$
18. $r + r$

19. $k + 12k + 23$
20. $6d + d + 15$
21. $17x + 21x + 45$

22. $6m + 4m + 3$
23. $13a + 9a + 8$
24. $14bx + 31bx + 9bx$

25. $y + 9 + 14 + 2y$
26. $24a + a + 16$
27. $4f + 7q + 11f + 8g$

28. $9(r + 7) + 12r$
29. $18y + 5(7 + 3y)$
30. $22c + 4(2 + 4c)$

31. $14(b + 3) + 8b$
32. $3(8 + a) + 7(6 + 4a)$
33. $6(x + y) + 4(2x + 3y)$

Write an expression to find the total for each situation.

34. You buy eight $12-tickets on Monday. On Tuesday, you buy 5 more.

35. Joe bought 3 soft drinks for y cents each. Sally bought 5 soft drinks for y cents each.

36. You charge x dollars per hour to baby-sit. You baby-sit 8 hours on Saturday and 3 hours on Sunday. Then you receive a $5 tip.

37. Sara bought 2 cassettes for $7 each, then she decided to buy 1 more. The tax was $1.25.

Mental Math

Compute each product mentally. Use the distributive property.

38. 3×27 **39.** 4×46 **40.** 4×93 **41.** 6×26

42. 7×42 **43.** 5×86 **44.** 12×14 **45.** 15×16

Mixed Review

46. Evaluate the expression $16 + 10 \div 2 + 3$.   (Lesson 1-1)

47. Write two different expressions that mean the same as $2ab$.   (Lesson 1-2)

48. Evaluate the expression $\dfrac{9(a+b)}{3c}$ if $a = 1$, $b = 3$, and $c = 2$.   (Lesson 1-2)

49. Rewrite $(x + 5) + 1$ using the associative property.   (Lesson 1-3)

50. What property says that any number multiplied by zero will equal zero? (Lesson 1-3)

51. Simplify the expression $4(3z)$.   (Lesson 1-3)

Connection

52. Geometry   In the triangle shown at the right, one angle measures 90°, the second angle measures x, and the third measures $2x$. Write an expression for the sum of the measures of the angles. Simplify the expression.

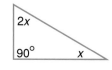

Critical Thinking

53. Use the distributive property to write two expressions for the figure at the right.

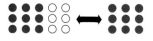

Wrap-Up

54. Make Up a Problem   Write a problem that is similar to Exercise 37.

Team Problem Solving

Marcus, the Math Magician, likes to amaze his audiences with this mind-reading trick.

> Think of a number.
> Add 7 to the number.
> Multiply the result by 2.
> Subtract 4.
> Divide by 2.
> Subtract your original number.

The audience is amazed when Marcus reveals that the result is 5.

Explain why this trick works. Then write one of your own.

Distributive Property

Materials: tiles

In this Exploration, you will look at the distributive property from a geometric point of view.

▶ Let $\square$ be a 1×1 square. That is, the length and width are each 1 unit. What is the area of this square?

▶ Let $\boxed{}$ be a $1 \times x$ rectangle. The width is 1 unit and the length is x units. What is the area of this rectangle?

▶ Then $\boxed{}$ is a rectangle that has a width of 1 unit and a length of $x + 1$ units. What is the area of this rectangle?

Your Turn: Using your tiles, make rectangles with areas of $x + 2$, $2x$, $2x + 1$, and $2x + 2$ square units.

Explore: You can use a geometric idea to check the distributive property. Is it true that $2(x + 1) = 2x + 1$?

$2(x + 1)$ means

$2x + 1$ means

Therefore, $2(x + 1)$ is not the same as $2x + 1$. From the drawing above, $2(x + 1) = 2x + 2$.

Tell whether each statement is *true* or *false*. Justify your answer with tiles or a drawing.

1. $2x + 3 = 6x$
2. $2x = x + x$
3. $3x + 3 = 3(x + 1)$
4. $3x + 3 = 3(x + 3)$
5. $3x + 2x = 10x$
6. $3x + 2x = x(3 + 2)$

Analysis

7. A classmate decides that $3(x + 2) = 3x + 2$. How would you show your classmate that $3(x + 2) = 3x + 6$? Write your solution in paragraph form, complete with drawings.

1-5   Variables and Equations

Objective:
Identify and solve
open sentences.

Key Terms:
equation
open sentence
solve
solution

Suppose you were asked to complete this sentence on your last
history test.

_____?_____ was President of the United States during the Civil War.

This sentence is neither *true* nor *false* until you substitute a name for ?.
If you choose *Abraham Lincoln,* the sentence is true.
If you choose *George Washington,* the sentence is false.

In mathematics, a sentence that contains an equals sign, $=$, is called an
equation. Some equations contain only numbers.

$25 - 15 = 10$ This equation is true.

$8 + 15 = 22$ This equation is false.

An equation that contains at least one variable is called an **open
sentence.** For example, $x + 5 = 20$ is an open sentence. It is neither
true nor false until x is replaced with some number. When you replace
the variable so that the equation is true, you have **solved the
equation.** Any number that makes the equation true is called a
solution. For example, the solution of $x + 5 = 20$ is 15.

*Name a number that is
not a solution of
$x + 5 = 20$.*

Example

1 **Which of the numbers 18, 26, or 28 is the solution of
$112 + x = 140$?**

$112 + 18 = 140$ Try $x = 18$. $112 + 26 = 140$ Try $x = 26$.
$\quad\quad 130 = 140$ $\quad\quad 138 = 140$

This sentence is false.   This sentence is false.

$112 + 28 = 140$ Try $x = 28$.
$\quad\quad 140 = 140$

This sentence is true. The solution is 28.

In this course, you will learn many ways to solve equations. Some
equations can be solved mentally by using basic facts or
arithmetic skills.

Estimation Hint

Always estimate
before you solve an
equation. This allows
you to check that
your solution
is reasonable.
THINK:
$110 + x = 140$

Examples

Solve each equation mentally.

2 $12 = y - 4$
$12 = 16 - 4$
$y = 16$

The solution is 16.

3 $2x = 24$
$2 \cdot 12 = 24$
$x = 12$

The solution is 12.

Checking for Understanding

Communicating Algebra

1. Write an equation that is always true.
2. Write an equation that is always false.
3. Write an open sentence and change it to one that is true.
4. Write an open sentence and change it to one that is false.

Guided Practice

Name the number that is the solution of the given equation.

5. $5 - x = 2$; 3, 5, 7
6. $y + 27 = 58$; 29, 30, 31
7. $3x = 87$; 19, 29, 39
8. $110 = 145 - m$; 35, 40, 45
9. $7 = \frac{14}{a}$; 2, 21, 98
10. $11 = \frac{b}{3}$; 30, 31, 32
11. $2x + 1 = 7$; 3, 4, 5
12. $7 = 5b + 2$; 0, 1, 2

Exercises

Independent Practice

Solve each equation mentally.

13. $r + 11 = 17$
14. $y + 2 = 33$
15. $19 - q = 5$
16. $17 = k - 3$
17. $5x = 25$
18. $\frac{18}{a} = 9$
19. $\frac{y}{3} = 8$
20. $42 - y = 0$
21. $63 = 7q$

Mixed Review

22. What property is shown by the statement $(9 + 4) + 2 = 2 + (9 + 4)$? (Lesson 1-3)

Simplify each expression. (Lesson 1-4)

23. $3m + 7m + 1$
24. $5y + 3(7 + 2y)$

Connection

25. **Geometry** If you want to find the area of a rectangular garden, you use the formula $A = \ell w$, where ℓ is the length and w is the width. Solve the equation if $\ell = 15$ feet and $w = 12$ feet.

Critical Thinking

26. Write an open sentence that has no whole number solution.

Wrap-Up

27. Write an open sentence that has 5 as its solution.

Algebra in Action–Cryptology

Codes and Ciphers

Can you read this message?

JBBQ JB XQ QEB IFYOXOV

The words appear to be nonsense; but if you have the key, you can decipher the message.

Cryptology, the science of writing and interpreting secret codes or ciphers, can be traced back to ancient Egypt. One of the easiest codes is the **substitution cipher,** in which each letter of the regular, or plaintext, alphabet is replaced by another letter. One such cipher was used by the Roman Emperor, Julius Caesar. In the Caesar cipher, the plaintext alphabet is shifted three letters to the right. The new alphabet is called the ciphertext.

Ciphertext	X	Y	Z	A	B	C	D	E	F	G	H	I	J	K	L	M	N	O	P	Q	R	S	T	U	V	W
Plaintext	A	B	C	D	E	F	G	H	I	J	K	L	M	N	O	P	Q	R	S	T	U	V	W	X	Y	Z

Using this key, DOG in Plaintext becomes ALD in ciphertext.

When you think of secret codes, you probably think of how they are used by secret agents or in military operations. Today, cryptology is a rapidly growing profession due to the increase in the need for computer security. It is important for businesses to protect the information they have stored in their computers. What better way is there than to design codes that only certain people in their businesses understand? That's what cryptology is all about.

1. The message above was written using the Caesar cipher. Decipher the message into plaintext.

2. If $c = p + 3$ means that the plaintext was shifted 3 places to the right, what does $c = p - 3$ mean?

3. The following message was written using a substitution cipher. Can you decipher it? Hint: The plaintext alphabet was shifted either five, seven, nine, or ten letters to the right.

TEZXUKT BL XTLR.

4. **Make Up a Problem**   Design your own substitution cipher. Ask a friend to decipher a message written in your ciphertext.

1-6   Strategy: Guess and Check

Objective:
Solve problems using the
guess-and-check strategy.

Key Term:
guess and check

The sports section of Anita's newspaper is torn and part of the article about the Franklin High School football game is missing. Anita can read that the team scored 34 points and made a total of 7 touchdowns and field goals. The team made just 1 extra point after the touchdowns. Anita wants to know how many touchdowns were made in the game. She can use a strategy called **guess and check.**

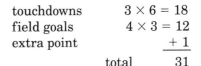 **Explore**

Anita knows the number of points scored, the total number of touchdowns and field goals, and the number of extra points. She also knows that each touchdown is worth 6 points and each field goal is worth 3 points. She wants to know the number of touchdowns.

If Anita's guess results in a total score of 37, how should she change her guess?

Plan

Anita can guess the number of touchdowns. She can then calculate the total number of points. If her answer is 34, her guess is correct. If her answer is not 34, she must change her guess and try again.

Solve

She guesses 3 touchdowns and 4 field goals.

touchdowns	$3 \times 6 = 18$
field goals	$4 \times 3 = 12$
extra point	$+\ 1$
total	31

Since the total is too small, Anita knows she must try a greater number of touchdowns. She tries 4 touchdowns and 3 field goals.

touchdowns	$4 \times 6 = 24$
field goals	$3 \times 3 =\ 9$
extra point	$+\ 1$
total	34

There were 4 touchdowns.

*How can you solve
$y + 230 = 850$ using
guess-and-check?*

Examine

Try several other combinations to see if any have a total score of 34. For example, 5 touchdowns, 2 field goals and 1 extra point has a total score of 37, which is too great. Since no other combinations have a total of 34, Anita is sure that the team scored 4 touchdowns.

Checking for Understanding

Communicating Algebra

1. Explain why it is important to use estimation with the guess-and-check strategy.

Guided Practice

Solve using the guess-and-check strategy.

2. The sum of a number and 126 is 171. Find the number.

3. The product of two consecutive numbers is 930. Find the numbers.

Exercises

Independent Practice

Solve using mental math skills, estimation, paper and pencil, or a calculator.

4. The product of a number and 15 is 105. Find the number.

5. The sum of a number and 89 is 122. Find the number.

6. The product of a number and itself is 196. Find the number.

7. Amos has a total of 10 nickels and quarters. If the value of the coins is $1.70, how many quarters does he have?

8. The Jackson Middle School choir buys 30 tickets to a musical play. The total cost of the tickets is $98.00. If student tickets cost $3.00 and adult tickets cost $5.00, how many adults are going with the choir?

9. The Middletown Junior High School basketball team made 21 baskets to score 36 points. If two of the baskets were 3-point shots, how many baskets were worth 2 points and how many were worth 1 point?

Wrap-Up

10. **Make Up a Problem**   Write a problem that can be solved by using the guess-and-check strategy.

Mid-Chapter Quiz

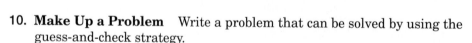

Find the value of each expression.   (Lesson 1-1)

1. $4(6) + 9$

2. $20 - 6 \div 3 + 4 \cdot 8$

3. $\dfrac{15 + 35}{21 + 4}$

Evaluate each expression if $a = 36$ and $b = 12$.   (Lesson 1-2)

4. $a \div b$

5. $3(a + b)$

Simplify each expression.   (Lessons 1-3, 1-4)

6. $(x + 5) + 7$

7. $5(3x)$

8. $9x + 3x$

Solve each equation mentally.   (Lesson 1-5)

9. $24 = y - 5$

10. $3a = 30$

1-7 Solving Equations Using Inverse Operations

Objective:
Solve equations using inverse operations.

Key Term:
Inverse operations

Here's a puzzle for you.

> I'm thinking of a number—
> If you multiply it by 15 and then subtract 17, you get 73.
> What's the number?

You could use the guess-and-check strategy to find the number. But that strategy might take too much time. Another way to solve this problem is to start with 73 and undo the operations.

$$73 + 17 = 90 \quad \text{Add 17.}$$

$$90 \div 15 = 6 \quad \text{Divide by 15.}$$

The number is 6. Check this solution.

What is the inverse of depositing $20 in a bank account?

You were able to undo the operations because addition and subtraction are **inverse operations.** This can be shown by these related sentences.

$$4 + 13 = 17 \qquad \begin{array}{l} 17 - 13 = 4 \\ 17 - 4 \ = 13 \end{array}$$

Which sentence could help you find the value of t?

$$15 + t = 48 \qquad \begin{array}{l} 48 - 15 = t \\ 48 - t = 15 \end{array}$$

These sentences suggest the following definition.

Definition of Subtraction	For all numbers a, b, and c, $a - b = c$ and $a - c = b$ if $b + c = a$.

In a similar way, multiplication and division are inverse operations.

$$6 \cdot 2 = 12 \qquad \begin{array}{l} 12 \div 2 = 6 \text{ or } \frac{12}{2} = 6 \\ 12 \div 6 = 2 \text{ or } \frac{12}{6} = 2 \end{array}$$

$$15r = 60 \qquad \begin{array}{l} 60 \div r = 15 \text{ or } \frac{60}{r} = 15 \\ 60 \div 15 = r \text{ or } \frac{60}{15} = r \end{array}$$

These sentences suggest the following definition.

Definition of Division	For all numbers a, b, and c, with $b \neq 0$ and $c \neq 0$, $\frac{a}{b} = c$ and $\frac{a}{c} = b$ if $bc = a$.

CONNECTION TO ALGEBRA

You have already solved equations by using arithmetic skills and the guess-and-check strategy. You can also use inverse operations to solve equations.

Examples

Solve each equation by using inverse operations.

1 $m - 5 = 21$

$m = 21 + 5$ Write the related addition sentence.

$m = 26$

Estimation Hint

Always estimate before you solve the equation.
THINK: $a + 14 = 27$

The solution is about 13.

2 $a + 13.7 = 27.3$

$a = 27.3 - 13.7$ Write the related subtraction sentence.

$$27.3 \;\boxminus\; 13.7 \;\boxminus\; 13.6$$

$a = 13.6$

3 $\frac{r}{5} = 31$

$r = 31 \cdot 5$ Write the related multiplication sentence.

$r = 155$

4 $1.3b = 39.0$

$b = 39.0 \div 1.3$ Write the related division sentence.

$$39.0 \;\boxdiv\; 1.3 \;\boxminus\; 30$$

$b = 30$

36   *The Language of Algebra*

Checking for Understanding

Communicating Algebra

1. What is the inverse of gaining five yards in football?

2. Write an addition sentence. Then write a related subtraction sentence.

3. Write an open sentence using multiplication. Then write a related division sentence.

Guided Practice

For each sentence, write a related sentence using the inverse operation.

4. $4 + 7 = 11$

5. $21 + 15 = 36$

6. $23 = 17 + m$

7. $14 - 3 = 11$

8. $21 - 17 = 4$

9. $c - 18 = 6$

10. $8 \cdot 9 = 72$

11. $b = (6)(7)$

12. $4y = 16$

13. $28 \div 7 = 4$

14. $12 = \frac{x}{11}$

15. $\frac{a}{19} = 7$

Exercises

Independent Practice

Solve each equation by using the inverse operation. Use a calculator where necessary.

16. $5 + r = 12$

17. $13 + z = 25$

18. $a + 11 = 17$

19. $x - 6 = 18$

20. $16 = z - 25$

21. $17 = t - 3$

22. $34 = a \cdot 2$

23. $16r = 48$

24. $153 = 9x$

25. $\frac{r}{9} = 15$

26. $7 = \frac{g}{12}$

27. $144 = x \cdot 12$

28. $15b = 210$

29. $45z = 315$

30. $\frac{m}{15} = 8$

Calculator

31. $n + 4.73 = 5.56$

32. $4.37 = y - 9.32$

33. $\frac{b}{5.4} = 3.7$

34. $2.5r = 37.5$

35. $19.75 = s + 15.98$

36. $7.2 = 0.36y$

Mixed Review

37. Evaluate the expression $10a - (7b + c)$ if $a = 2$, $b = 1$, and $c = 12$. (Lesson 1-2)

38. Restate $5x + 5y$ using the distributive property. (Lesson 1-4)

Simplify each expression. (Lessons 1-3, 1-4)

39. $5(6y)$

40. $23c + 6(1 + 5c)$

41. Solve mentally: $7x = 63$. (Lesson 1-5)

42. Write an open sentence that has zero for a solution. (Lesson 1-5)

Challenge

Solve for x by using the inverse operation.

43. $x + a = 3$

44. $x + 3 = m$

45. $k = x - 21$

46. $\frac{x}{a} = b$

47. $12 = x + r$

48. $ax = q$

49. **Statistics**   The average of three test scores is found by adding the scores and dividing the sum by 3. Suppose you score 86 and 88 on two tests. What must you score on the third test to have an average of 90?

50. **Geometry**   The sum of the measures of two angles is 90°. One angle is 48°. Solve the equation $y + 48 = 90$ to find the measure of the second angle.

Applications

51. **Sports**   A skateboarder travels 40 meters in 16 seconds. Solve the equation $40 = r \cdot 16$ to find the speed in meters per second.

52. **Sports**   The attendance at the Houston AstroDome was 35,892 on Friday. This was 3285 less than the attendance on Saturday. Solve the equation $35,892 = a - 3285$ to find Saturday's attendance.

Critical Thinking

53. Consider the division sentence $\frac{4}{0} = a$ and its related multiplication sentence $4 = 0 \cdot a$. Use these statements to explain why division by zero is not possible.

Wrap-Up

54. **Make Up a Problem**   Write a puzzle like the one presented at the beginning of the lesson. Ask a friend to solve it.

Reading Algebra

Study the expressions below. How do the parentheses change the meaning of the expression?

Words	Symbols
three times x plus y	$3x + y$
three times the sum of x and y	$3(x + y)$

In the second expression, parentheses are used to show that the *sum, $x + y$,* is multiplied by three. In algebraic expressions, terms enclosed by parentheses are treated as one quantity. The expression $3(x + y)$ is sometimes read *three times the quantity x plus y.*

In verbal problems, look for key words that indicate that parentheses are to be used. Sometimes the words *sum, difference, quantity,* and *total* signal the use of parentheses. These key words should be helpful as you study the next lesson.

1-8 Writing Expressions

Objective:
Translate verbal phrases into algebraic expressions.

Chemists use a system of symbols to write formulas for compounds. For example, H_2O is the chemical formula for water. It means that each water molecule is made up of two atoms of hydrogen (H) and one atom of oxygen (O). This shorthand method saves time and space and shows the pattern that exists between the hydrogen and oxygen molecules.

In the same way, mathematicians can use variables to show patterns in expressions. Suppose that in a certain rectangle, the length is 3 inches more than the width. The chart below shows several possibilities for the width and length.

Width	Length
2	2 + 3 or 5
4	4 + 3 or 7
5	5 + 3 or 8
10	10 + 3 or 13
w	w + 3

The algebraic expressions w and $w + 3$ summarize the pattern that exists between the length and the width.

There are many words and phrases that suggest addition and subtraction. The following phrases all suggest the algebraic expressions $n + 4$ or $4 + n$.

$n + 4$ or $4 + n$
a number plus 4
4 more than a number
a number increased by 4
the sum of a number and 4

The following phrases all suggest subtraction. The phrases on the left suggest the algebraic expression $n - 3$. The phrases on the right suggest $3 - n$. Notice the difference in the phrases.

$n - 3$	$3 - n$
a number decreased by 3	3 decreased by a number
a number minus 3	3 minus a number
subtract 3 from a number	subtract a number from 3
3 less than a number	3 less a number

Chemistry Connection

The formula for common table sugar is $C_{12}H_{22}O_{11}$, where C represents carbon atoms.

The formula for common table salt is NaCl, where Na represents sodium and Cl represents chlorine.

Examples

Translate each phrase into an algebraic expression.

1 six more baskets than team B

Let b represent the number of baskets made by team B.
The words *more than* suggest addition.
The algebraic expression is $b + 6$ or $6 + b$.

2 one less foul than Adita has

Let m represent the number of fouls Adita has.
The words *less than* suggest subtraction.
The algebraic expression is $m - 1$.

Likewise, there are many words and phrases that suggest multiplication and division. The following words and phrases all suggest the algebraic expressions $2x$ or $x \cdot 2$.

2x or x • 2
2 times a number
the product of 2 and a number
2 multiplied by a number
twice a number

The following phrases suggest division. The phrases on the left suggest the expression $n \div 3$. The phrases on the right suggest $3 \div n$.

n ÷ 3		3 ÷ n
a number divided by 3		3 divided by a number
the quotient of *n* and 3		the quotient of 3 and *n*

Examples

Translate each phrase into an algebraic expression.

3 the product of four and *y*

The words *product of* suggest multiplication.
The expression is $4y$ or $y \cdot 4$.

4 the number of batteries divided by two

Let b represent the number of batteries.
The words *divided by* suggest division.
The expression is $b \div 2$ or $\frac{b}{2}$.

Examples

5 **twice the sum of *x* and nine**

In this case, parentheses must be used for the word *sum*.
The expression is $2(x + 9)$.

6 **twice a number increased by eight**

Let *n* represent the number.
No parentheses are needed.
The expression is $2n + 8$.

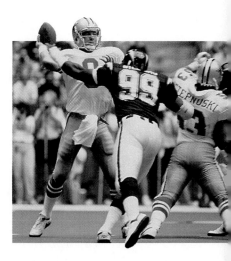

Checking for Understanding

**Communicating
Algebra**

1. Write two verbal phrases for the algebraic
expression $x + 4$.

2. Write two verbal phrases for the algebraic
expression $5y$.

3. If *m* is the number of games the Cowboys won,
what does the expression $m - 5$ represent?

4. Explain the difference between the expressions
$x - 5$ and $5 - x$.

**Guided
Practice**

Translate each phrase into an algebraic expression.

5. four more than *w*.

6. the sum of *y* and 3

7. nine less than *t*

8. *p* decreased by 7

9. eleven increased by *x*

10. five less than *r*

11. the sum of 6 and *z*

12. the difference of *y* and 9

13. twenty minus *r*

14. twelve plus *b*

15. *x* more than eighty

16. twice a certain number

17. a number divided by seven

18. fifteen less than *m*

19. product of *z* and six

20. quotient of forty-six and *y*

Exercises

**Independent
Practice**

Translate each phrase into an algebraic expression.

21. ten more points than Jill scored

22. 3 years younger than Quint

23. Amy's salary plus a $300 bonus

24. twice the sum of a number and 7

25. twice as many fouls as the
Celtics had

26. three times as many dollars as
Tom has

27. three times a number decreased
by 8

28. Marty's age divided by four

Write a verbal phrase for each algebraic expression.

29. $x + 3$ 30. $15 - y$ 31. $9r$

32. $x \div 5$ 33. $2(x + 3)$ 34. $2x + 3$

35. Evaluate the expression $5a - (4b + 2c)$ if $a = 6$, $b = 7$, and $c = 1$. (Lesson 1-2)

36. Simplify the expression $3(4 + a) + 9(2 + 5a)$. (Lesson 1-4)

37. Replace the variable to make the sentence $r + 3 = 7$ true. (Lesson 1-5)

38. Use inverse operations to solve $7x = 182$. (Lesson 1-7)

39. True or false: $14 + 3 = 17$ and $17 - 3 = 14$ are related sentences. (Lesson 1-7)

40. **Geometry**   In a certain rectangle, the length is 20 inches more than the width. Copy and complete the table below.

Width (in.)	5	11	■	16	w	■	x + 1
Length (in.)	■	■	35	■	■	z	■

Statistics   Use the information in the graph at the right to answer each question.

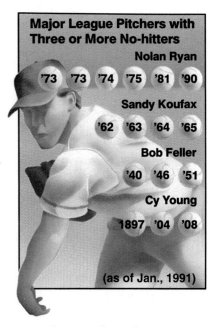

Major League Pitchers with Three or More No-hitters

Nolan Ryan
'73 '73 '74 '75 '81 '90

Sandy Koufax
'62 '63 '64 '65

Bob Feller
'40 '46 '51

Cy Young
1897 '04 '08

(as of Jan., 1991)

41. If r represents the number of no-hitters that Nolan Ryan pitched, which pitcher can be represented by the expression $r - 2$?

42. If y represents the number of no-hitters Cy Young pitched, which other pitcher can be represented by the expression y?

43. If k represents the number of no-hitters Sandy Koufax pitched, write an expression that represents the number of no-hitters Nolan Ryan pitched.

44. **Space Exploration**   Suppose you are an astronaut who travels to the moon to collect rock samples. When you get there, you find that you weighed 6 times as much on Earth as you weigh on the moon. Copy and complete the table below.

Weight on the moon (lb)	10		50	n	2n	
Weight on Earth (lb)		180				y

45. If T represents Tom's age and J represents Jim's age, explain what the sentence $T = J + 5$ means.

46. If a represents your present age in years, how old will you be in six years?

1-9   Use an Equation

Objective:
Solve problems by using
an equation.

Key Term:
defining the variable

In a recent season, Michael Jordan was the leading scorer in the
National Basketball Association with 2633 points. He scored 307 points
more than the second leading scorer, Karl Malone. How many points did
Karl Malone score?

Selecting and trying different strategies are an important part of
problem solving in mathematics. You have already used the guess-and-
check strategy in this chapter. Another method is to use an equation.

In the following example, the problem above is solved by using
an equation.

Example

1   Find the number of points scored by Karl Malone.

| **Explore** | Skim the problem for the general idea. You need to find the number of points that Malone scored. Let p represent the number of points that Malone scored. This is called **defining the variable.** |

| **Plan** | Translate the words into an equation using the variable. *307 more than* suggests addition. Write the equation. |

Estimation Hint

Think:
$2600 - 300 = 2300$
The answer should
be close to 2300
points.

| **Solve** |

points scored by Malone	plus	307	equals	points scored by Jordon
p	$+$	307	$=$	2633

$$p + 307 = 2633$$
$$p = 2633 - 307 \qquad \text{Write the related}$$
$$p = 2326 \qquad \text{subtraction sentence.}$$

Karl Malone scored 2326 points.

| **Examine** | Check the answer against your estimate. Then check the answer in the words of the problem. 2633 points is 307 points more than 2326 points. |

Example

How can you estimate the answer?

2 **Sixteen times some number is equal to 96. Find the number.**

Explore
You need to find a number. Let n represent the number. When you multiply the number by 16, you get 96.

Plan
Translate the words into an equation using the variable.

Sixteen	times	number	equals	ninety-six
16	·	n	=	96

Solve
$$16n = 96$$
$$n = 96 \div 16 \quad \text{Write the related}$$
$$n = 6 \quad \text{division sentence.}$$

The number is 6.

Examine
Check your answer against your estimate. Then check the answer in the words of the problem. 16 times 6 is equal to 96.

Checking for Understanding ⎯⎯⎯⎯⎯⎯⎯⎯⎯⎯⎯⎯⎯⎯⎯

Communicating Algebra

1. Explain what it means to *define a variable*.

2. Why should you check your answer in the words of the problem instead of in the equation?

Guided Practice

Translate each sentence into an equation.

3. Twice a certain number is 6.

4. A number divided by 5 is 3.

5. Four more than a number is 12.

6. Six less than a number is zero.

7. When a number is decreased by 5, the result is 7.

8. Twenty-seven more than a number is thirty-one.

9. Forty-two less than a number is sixty.

10. A number divided by three is 13.

Exercises

Define a variable, write an equation, then solve.

11. Joe paid $15 for a shirt on sale. It was reduced by $9. What was the regular price?

12. Together two items cost $16. One item costs $7. What is the cost of the other item?

13. One season the Pittsburgh Pirates lost 72 games. They won 18 games more than they lost. How many games did they win?

14. The 4-person relay team ran a race in 48 seconds. What was the average time for each person?

15. Clara Hardin earned $175 last week. She worked 35 hours. How much did she earn per hour?

16. In a recent year 4635 runners started the Fairfield Marathon. Only 2338 finished the race. How many runners did not finish?

17. One season Roger Maris and Mickey Mantle hit a total of 117 home runs. Maris hit 61 that year. How many did Mantle hit?

Mixed Review

18. Use inverse operations to solve $14 = z - 31$.   (Lesson 1-7)

19. Write an algebraic expression for the phrase *twice a certain number*. (Lesson 1-8)

20. Write a verbal phrase for the algebraic expression $3(x + 2)$.   (Lesson 1-8)

Decision Making

Solve this problem using mental math skills, estimation, or paper and pencil. Explain your method.

21. Felicia wants to buy 4 chrome wheels for her car. She can spend no more than $74 per wheel. A local newspaper advertises 4 wheels on sale for $320. They usually cost $359. Can Felicia afford to buy the wheels?

Critical Thinking

22. One-half of Celia's age increased by 12 is her sister's age. If her sister is 18 years old, how old is Celia?

Wrap-Up

23. **Make Up a Problem**   Write a problem based on this information: Let x = the number of students present in Mr. Wyatt's class. Use the equation $x + 5 = 33$.

1-10   Inequalities

Objective:
Use inequalities.

Key Term:
inequality

Name a number that is a solution of x < 4. Then name a number that is not a solution.

Under normal conditions, water is in the form of a liquid at temperatures between 32° F and 212° F. When the temperature is less than 32° F, water is in the form of ice. At temperatures above 212° F, water is in the form of steam.

If we let the variable t represent the temperature, we can write mathematical sentences to describe the temperatures for ice or steam.

Say: *t is less than* 32
Write: $t < 32$ Water is in the form of ice.

Say: *t is greater than* 212
Write: $t > 212$ Water is in the form of steam.

Any mathematical sentence that contains < or > is called an **inequality.** Inequalities, just like equations, can be true, false, or open.

$15 > 12$ This sentence is true.

$8 < 3$ This sentence is false.

$x < 4$ This is an open sentence. It is neither true nor false until x is replaced with a number.

The symbols ≤ and ≥ can also be used in inequalities. They are combinations of the equal sign and the inequality symbol.

Say: *x is less than or equal to* 5
Write: $x \le 5$

Say: *y is greater than or equal to* 7
Write: $y \ge 7$

Examples

For the given value, state whether each inequality is *true* or *false*.

1 $y - 9 > 6, y = 12$

$12 - 9 > 6$

$3 > 6$

This sentence is false.

2 $12 \ge \frac{2k}{9} + 8, k = 18$

$12 \ge \frac{2 \cdot 18}{9} + 8$

$12 \ge 4 + 8$

$12 \ge 12$

This sentence is true.

FYI

Americans use more than 400 billion gallons of water daily.

Checking for Understanding

1. Which symbol is read *is greater than or equal to*, $>$ or $\geq$?
2. Name a value for x in which $2x > 1$ is false.
3. Write a sentence that explains the difference between an equation and an inequality.

Guided Practice

State whether each inequality is *true, false* or *open*.

4. $7 > 2$
5. $5 < 3$
6. $x + 2 > 7$
7. $3 \geq 2$
8. $0 \leq 6$
9. $3x - 7 \geq 5$
10. $x - 2 \leq 9$
11. $5 > 6(2)$

State whether each inequality is *true* or *false* for the given value.

12. $r + 2 \geq 7, r = 6$
13. $14 \geq m + 6, m = 4$
14. $3t \geq 7, t = 0$
15. $2 < 2x - 5, x = 3$

Exercises

Independent Practice

State whether each inequality is *true* or *false* for the given value.

16. $5k + 10 < 20, k = 2$
17. $15 < 3m + 7, m = 4$
18. $0 < 3r - 5, r = 2$
19. $2r - 7 \leq 7, r = 7$
20. $3x + 2x - 9 \geq 17, x = 6$
21. $3x - 12 > 0, x = 4$

Evaluate each expression if $a = 3$, $b = 2$, and $c = 7$. Then replace each ▦ with <, >, or = to make a true sentence.

22. ab ▦ c
23. $a + 2b$ ▦ $c + 7$
24. $3a - 2b$ ▦ $c + b$
25. $2c$ ▦ $3a + 2b + 1$
26. $5a - 2c$ ▦ $3b$
27. $4a + b - c$ ▦ 0

Mixed Review

28. Use inverse operations to solve $17 = n + 8$. (Lesson 1-7)
29. Write an algebraic expression for the phrase *the quotient of 18 and y*. (Lesson 1-8)
30. Write an equation for the sentence: *Six less than a number is 15.* (Lesson 1-9)
31. Write an equation and solve: Four swimmers swam a relay race in 56 seconds. What was the average time for each swimmer? (Lesson 1-9)

Applications

32. **Sports** The Mets are losing to the Astros by a score of 5 to 3. Write an inequality that would tell how many runs would give the Mets the lead.

33. **Recreation** During the last few years, in-line skating has become a popular pastime. Today there are about 400,000 people who participate in this sport. This is more than twice as many people as there were several years ago. Which inequality, $s > 200,000$ or $s < 200,000$, describes how many skaters there were several years ago?

Critical Thinking

34. If $x > 3$ and $x < 15$, explain what the open sentence $3 < x < 15$ means.

Wrap-Up

35. Find at least two values for x that make $2x + 3 \geq 7$ true. Find two values for x that make it false.

Review

Language and Concepts

Choose the correct term to complete each sentence.

1. When you replace a variable in an equation so that a true sentence results, you have (evaluated, solved) the equation.

2. Mathematical sentences like $x + 3 = 10$ and $2y = 14$ are examples of (equations, inequalities).

3. If an expression contains only the operations of multiplication and addition, you would (add, multiply) first.

4. If an expression contains an addition operation in parentheses and a multiplication operation not in parentheses, you would (add, multiply) first.

5. The inequality $(x \geq 2, x \leq 2)$ means x *is less than or equal to 2.*

Skills

Find the value of each expression.   (Lesson 1-1)

6. $47 - 7 \times 5$
7. $72 \div 12 \times 3$
8. $18 + 12 \div 3 - 6$
9. $4 \cdot 3 + 8 \div 2 - 5$
10. $(36 \div 6) \div (6 - 4)$
11. $7[(29 + 11) - 3(16 - 9)]$

Evaluate each expression if $a = 12$, $b = 11$, $k = 8$, and $s = 14$.   (Lesson 1-2)

12. $29 + a - s$
13. $2a - 9$
14. $a - k$
15. $b + (a - k)$
16. $3a + 2b - k$
17. $\frac{a}{3} + b$

Name the property shown by each statement.   (Lessons 1-3, 1-4)

18. $16 + 4 = 4 + 16$
19. $317 + 0 = 317$
20. $(7 + y) + 2 = 7 + (y + 2)$
21. $r \cdot 1 = r$
22. $8(6 \cdot 9) = (8 \cdot 6)9$
23. $2(2 + 5) = 2 \cdot 2 + 2 \cdot 5$

Simplify each expression.   (Lessons 1-3, 1-4)

24. $2(3y)$
25. $3 + (4 + x)$
26. $a + 9a$
27. $2a + 3a + 10$
28. $8 + 36y - 14y$
29. $4(a + 3) + 10a$

Solve each equation mentally.   (Lesson 1-5)

30. $y + 3 = 15$
31. $2x = 10$
32. $16 = 25 - t$
33. $\frac{m}{4} = 9$
34. $a + 15 = 25$
35. $4x = 20$

Solve each equation by using the inverse operation. (Lesson 1-7)

36. $b + 48 = 55$ **37.** $y - 57 = 72$ **38.** $y + 12 = 33$

39. $\frac{x}{2} = 18$ **40.** $32m = 384$ **41.** $672 = 21t$

Translate each phrase into an algebraic expression. (Lesson 1-8)

42. *five more than x*   **43.** *the product of b and 5*

44. *ten less than y*   **45.** *a number divided by 6*

State whether each inequality is true or false for the given value. (Lesson 1-10)

46. $3x > 15, x = 6$ **47.** $m + 9 < 15, m = 3$

48. $3x - 12 \geq 0, x = 4$ **49.** $2y + y > 21, y = 7$

Applications and Problem Solving

50. The sum of a number and itself is 92. Use the guess-and-check strategy to find the number. (Lesson 1-6)

ortfolio Suggestion

A portfolio is representative samples of your work, collected over a period of time. Begin your portfolio by selecting an item that shows something new you learned in this chapter.

Curriculum Connection

- **Science** Find an example of an equation in your science book. Explain how you use the substitution property in science.

- **Social Studies** In the chapter opener, you were asked to estimate how long it would take to pass a basketball across the country. Find the distance in miles across your state. Estimate how long it would take to pass the basketball across your state.

Read More About It

Berry, Marilyn. *Help Is on the Way for Math Skills.*

Mango, Karin N. *Codes, Ciphers, and Other Secrets.*

Tobias, Sheila. *Succeed with Math.*

Test

Find the value of each expression.

1. $9 \cdot 4 + 6 \cdot 7$

2. $6 + 3(9 + 11)$

3. $54 \div 9 - 3$

4. $36 \div 9 \times 4$

5. $4[6 \cdot (72 - 65) \div 3]$

6. $5[5 \times (41 - 36) - (8 + 14)]$

Evaluate each expression if $a = 7$, $b = 6$, and $c = 5$.

7. $3ac - 2ab + 7$

8. $7b + 15 \div c$

Restate each expression using the distributive property.

9. $a(3 + b)$

10. $2 \cdot x + 2 \cdot y$

11. $s \cdot 5 + 4 \cdot 5$

Simplify each expression.

12. $(x + 12) + 15$

13. $2(5a)$

14. $4(y \cdot 3)$

15. $27r + 14r$

16. $16xy - 9xy$

17. $7(m + 13) + 5(2m + 3)$

Solve each equation.

18. $y + 24 = 85$

19. $17 = x - 13$

20. $a + 5 = 5$

21. $7b = 784$

22. $\frac{k}{32} = 5$

23. $48 = 12z$

Solve.

24. Bill is reading a 216-page book. He needs to read twice as many pages as he has already read. Use the guess-and-check strategy to determine how many pages he has read.

25. Traffic Engineering In many cities, the number of cars on the road triples during rush hour. If the number of cars on the road during *non*rush hours is represented by c, then write an algebraic expression for how many cars there are *during* rush hour.

BONUS

Describe the solution of each equation.

a. $y = y$

b. $x + 4 = 2$

c. $x + 2 = x + 3$

d. $\frac{x}{x} = 1$

Academic Skills Test

1. If $a = 5$ and $b = 20$, what is the value of $2a + b$?

   A 50
   B 45
   C 40
   D 30

2. $6 \times (10 + 5) =$

   A $(6 + 10) + 5$
   B $(6 + 10) \times 5$
   C $(6 \times 10) + (6 \times 5)$
   D $(6 + 10) \times (6 + 5)$

3. If $a + 15 = 24$, what is the value of a?

   A 9
   B 11
   C 24
   D 39

4. Which shows using inverse operations to solve $x + 14 = 35$?

   A $2 + x = 5$
   B $x = 35 + 14$
   C $14x = 35$
   D $x = 35 - 14$

5. Dan must earn 250 points to win a prize. He has 176 points. Which sentence could be used to find p, the number of points he still needs to earn?

   A $176 + p = 250$
   B $176 - p = 250$
   C $176 + 250 = p$
   D $p = 250 + 176$

6. Which expression is equivalent to $12a + 8b - 6a$?

   A $6a + 8b$
   B $14ab$
   C $8b - 6a$
   D $12a - 2b$

7. If $12y = 60$, what is the value of y?

   A 5
   B 12
   C 44
   D 192

8. The product of a number and 4 is 104. What is the number ?

   A 416
   B 108
   C 100
   D 26

9. If $8x = 112$, what is the value of x?

   A 14
   B 104
   C 120
   D 896

10. Jose worked 20 hours last week. He earned $4.00 per hour. Which equation can be used to find his total earnings?

    A $x = 20 \div 4$
    B $4x = 20$
    C $20 \times 4 = x$
    D $20x = 4.00$

CHAPTER 2

CHAPTER OBJECTIVES

In this chapter you will learn to:

- graph integers on a number line
- compare and order integers
- compute with integers
- solve problems with integers
- use logic to find the negation of a statement

Integers

What rotates at 300 miles per hour, uproots large trees, overturns loaded railroad cars, and carries entire houses through the air? A tornado! This fiercest of storms can move forward at 30 to 40 miles per hour. That's so fast that a spotter on the ground can give people only about two minutes' warning. Luckily, Doppler radar can detect tornadoes 20 minutes before they touch down. Doppler radar can be a real lifesaver in tornado country.

Doppler radar shows a tornado's position, strength, wind speed, and direction. On the radar screen, bright green shows winds coming toward the radar, and red shows winds moving away. Where red and bright green appear close together, the winds are rotating—and that means a tornado.

Can you find the tornado on the radar screen shown here?

Science Connection

Class Project

Do research to find how many tornadoes touched down in each state for a recent year. Decide on a way to display the data on a class bulletin board.

Discuss tornado safety procedures with your classmates.

2-1 Integers and Absolute Value

Objectives:
Graph integers on a number line.
Find absolute value.

Key Terms:
integers
absolute value
coordinate

Which integer is neither positive nor negative?

Have you ever watched the launch of a space shuttle on TV? In the final seconds before liftoff, you might have heard "T minus 3, 2, 1, ignition, liftoff."

The number of seconds before the liftoff is indicated by negative numbers. In this case, the numbers are -3, -2, and -1. The number of seconds after liftoff is indicated by positive numbers like +1, +2, and +3. Ignition time is zero.

Negative and positive numbers are often used to show opposite situations. Zero is considered to be the starting point. These numbers are often shown on a number line.

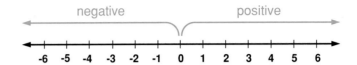

The numbers -1, -2, -3, . . . are called **negative integers.** The symbol for negative three is written -3.

The numbers 1, 2, 3, . . . are called **positive integers.** The symbol for positive four is written +4 or 4.

The set of all **integers** can be written { . . . -3, -2, -1, 0, 1, 2, 3 . . . } where . . . means *continues without end.*

To graph a particular set of integers, locate the integer points on a number line. The number that corresponds to a point on the number line is called the **coordinate** of the point.

History Connection
In the 15th century, flour merchants used integers to tell whether flour barrels were overweight or underweight. For example, +5 meant that the barrel was 5 pounds overweight; −5 meant that it was 5 pounds underweight.

Examples

1 Name the coordinates of *A, C,* and *D.*

The coordinate of *A* is -4, *C* is 3, and *D* is -2.

2 Graph { -1, 3, 5} on a number line.

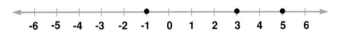

Looking at the number line shown below, you can see that 4 and -4 are different numbers. However, they are the same distance from the zero point. They both have the same **absolute value.**

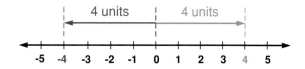

Absolute Value	The absolute value of a number is the distance the number is from the zero point on the number line.

What is the symbol for absolute value?

Say: *The absolute value of 4 is 4.*
Write: $|4| = 4$

Say: *The absolute value of -4 is 4.*
Write: $|-4| = 4$

Example

3 **Simplify $|-8| + |3|$.**

$$|-8| + |3| = 8 + 3 \qquad \text{The absolute value of -8 is 8.}$$
$$= 11 \qquad \text{The absolute value of 3 is 3.}$$

Checking for Understanding

Communicating Algebra

Replace each ■ with an integer that describes the situation.

1. 25° above zero, 25
   25° below zero, ■

2. stock value down $3, -3
   stock value unchanged, ■

3. 10 pound gain, 10
   8 pound loss, ■

4. 40 m below sea level, -40
   15 m above sea level, ■

True or false. **Explain your answer.**

5. All whole numbers are integers.

6. All integers are whole numbers.

Guided Practice

Name the coordinates of the points that are graphed.

7.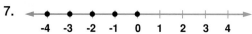

8.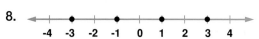

Find each absolute value.

9. $|-3|$

10. $|10|$

11. $|-16|$

12. $|25|$

Exercises

Graph each set of numbers on a number line.

13. {0, 2, 4}   **14.** {-2, -4, -6}   **15.** {-3, 0, 2}

Name the absolute value of each integer.

16. +6   **17.** -15   **18.** 13   **19.** 0

Simplify.

20. |-8| **21.** |-3| + |2| **22.** |15| − |-3| **23.** |-20| + |-19|

24. Evaluate the expression $5 \cdot 3 + 8 \div 2$.   (Lesson 1-1)

25. Solve mentally: $k + 4 = 11$.   (Lesson 1-5)

26. Using inverse operations, solve $16n = 48$.   (Lesson 1-7)

27. Write an equation and solve: Alicia spent $41 on a new outfit. The pants cost $23. How much was the blouse?   (Lesson 1-9)

If the number described below exists, name it.

28. the greatest positive integer

29. the least positive integer

30. the least negative integer

31. the greatest negative integer

32. an integer that is neither positive nor negative

33. a number that is not an integer

34. Geography   Death Valley, California, has the lowest altitude in the United States. Its elevation is 282 feet below sea level. What integer could be used to provide this information?

35. Space Exploration   About six seconds before liftoff, the three main shuttle engines start. About 120 seconds after liftoff, the solid rocket boosters burn out. Use integers to describe these events.

36. Find two values of x that make $|x| = 3$ a true statement.

  37. Research   Find three examples of negative integers used in games.

Team Problem Solving

Luisa threw five darts at the dartboard, and all five hit the board. Which of the following numbers could have been her score?

5, 18, 28, 31, 37

2-2 Comparing and Ordering

Objective:
Compare and order integers.

Brush fires occur when the weather has been hot and dry for a period of time. Forest rangers use signs like the one at the right to inform people of the fire hazard. As the fire hazard increases, the indicator is moved to the right.

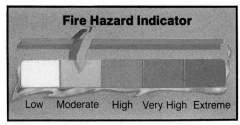

Fire Hazard Indicator

Low   Moderate   High   Very High   Extreme

Similarly on a number line, values increase as you move to the right. This makes it easy to determine which of two numbers is greater. On the number line shown below, 5 is to the right of -3.

Say: *5 is greater than -3.*
Write: $5 > -3$

-4 -3 -2 -1 0 1 2 3 4 5 6

You might also conclude that -3 is to the left of 5.

Say: *-3 is less than 5.*   Remember, the symbol points
Write: $-3 < 5$ to the lesser number.

Remember that any mathematical sentence containing $<$ or $>$ is called an inequality. In inequalities, numbers are compared.

Example

1 **Use the integers graphed on the number line to write two inequalities.**

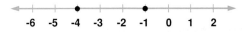

-6 -5 -4 -3 -2 -1 0 1 2

Since -4 is to the left of -1, write $-4 < -1$.
Since -1 is to the right of -4, write $-1 > -4$.

Checking for Understanding

Communicating Algebra

Use the numbers in each sentence to write an inequality. Use < or >.

1. -3 is less than 10.
2. 8 is greater than 7.
3. -9°C is colder than -8°C.
4. 65 mph is faster than 55 mph.
5. $19.95 is more than $15.
6. 6 ft is taller than 4 ft

Guided Practice

Write an inequality for each of the following.

7.
-7 -6 -5 -4 -3 -2 -1 0

8.
-3 -2 -1 0 1 2 3 4

Exercises

Replace each ● with <, >, or =.

9. 12 ● 14   **10.** -2 ● -3   **11.** 6 ● -6   **12.** 12 ● 0

13. 10 ● -3   **14.** -9 ● -15   **15.** -11 ● 5   **16.** |-4| ● 4

Order the numbers in each set from least to greatest.

17. {10, 0, -3}   **18.** {12, -9, -1}   **19.** {0, -4, -8, -3}

Use the numbers to write an inequality. Use < or >.

20. Water boils at 212°F. It freezes at 32°F.

21. Walking for 45 minutes burns 144 calories. Playing volleyball for the same amount of time burns 135 calories.

22. Yesterday's high temperature was 65°F. The low temperature was 34°F.

23. Evaluate the expression $ac - ab$ if $a = 8$, $b = 3$, and $c = 5$.   (Lesson 1-2)

24. Graph the numbers {-1, 1, 4} on a number line.   (Lesson 2-1)

25. Simplify the expression $|-5| + |3|$.   (Lesson 2-1)

26. Geology   Geologists often record important dates in Earth's history on a time line like the one shown below. The abbreviation M. Y. B. P. means *million years before present.*

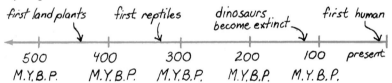

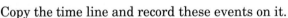

Copy the time line and record these events on it.

Event	M.Y.B.P.
Rocky Mountains rise	80
First dinosaurs	225
Beginning of ice age	1

27. Find all integer values for x that make $|x| < 3$ a true statement. Then graph the solution on a number line.

28. Measure your height and a friend's height. Write an inequality that compares them.

Adding Integers

Materials: counters, mat

In this Exploration, you will use counters to model addition.

▶ Remember that 2 + 3 means *combine a set of two items with a set of three items.* The integer addition +2 + (+3) means something similar. It tells you to combine a set of two positive items with a set of three positive items.

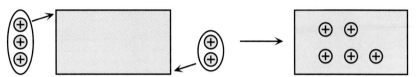

Therefore, +2 + (+3) = +5.

▶ Consider -2 + (-3). This tells you to combine a set of two negative items with a set of three negative items.

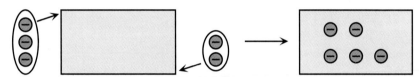

Therefore, -2 + (-3) = -5.

▶ Finally, consider -2 + (+3). This tells you to combine a set of two negative items with a set of three positive items. In this case it is possible to pair a positive counter with a negative counter. This is called a *zero pair.* You can remove as many zero pairs as possible, because removing zero does not change the value of the set.

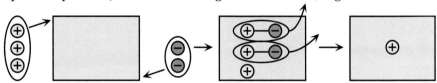

Therefore, -2 + (+3) = +1.

Your Turn: **What does +2 + (-3) mean? Model this operation.**

Model each problem.

1. 5 + 2   **2.** 5 + (-2)   **3.** -5 + 2   **4.** -5 + (-2)

Analysis

5. Suppose you add a positive integer and a negative integer. Explain how you use a zero pair to find the sum. Write your answer in paragraph form.

2-3 Adding Integers

Objective:
Add integers.

At night the average temperature on the surface of the planet Saturn is -150° Celsius. During the day the temperature rises 27°C. What is the average temperature on the planet's surface during the day?

Using counters to solve a problem like this would be inconvenient. Study the patterns in the following examples so that you can learn a rule for adding integers.

Example

If you know that
-3 + (-4) = -7, what
property allows you
to say that
-4 + (-3) = -7?

1 Find -3 + (-4).

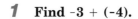

Start at zero. Move 3 units to the left. From there, move 4 more units to the left. The sum is -7.

-3 + (-4) = -7

This example suggests the following rule.

> **Adding Integers with the Same Signs**
>
> To add integers with the same sign, add their absolute values. Give the result the same sign as the integers.

Study the following examples to see if you can discover a rule for adding integers that have different signs.

Examples

2 Find -5 + 2.

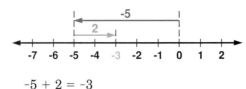

Start at zero. Move 5 units to the left. From there, move 2 units to the right. The sum is -3.

-5 + 2 = -3

3 Find 7 + (-3).

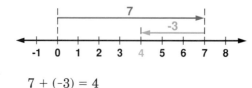

Start at zero. Move 7 units to the right. From there, move 3 units to the left. The sum is 4.

7 + (-3) = 4

Examples 2 and 3 suggest the following rule.

| **Adding Integers with Different Signs** | To add integers with different signs, subtract their absolute values. Give the result the same sign as the integer with the greater absolute value. |

Example

4 Find 9 + (-2).

$$9 + (-2) = +(|9| - |-2|)$$

Subtract absolute values. The result is

$$= +(9 - 2)$$

positive because the integer with the

$$= +7 \text{ or } 7$$

greater absolute value, 9, is positive.

Checking for Understanding

Communicating Algebra

Write an addition sentence for each diagram.

1.

2.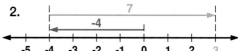

3.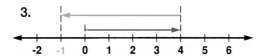

4.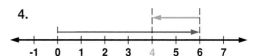

Guided Practice

State whether each sum is positive or negative.

5. $-5 + (-11)$ 6. $-14 + 13$ 7. $-9 + 15$ 8. $-5 + (-4)$

Find each sum or difference.

9. $|-13| - |9|$ 10. $|15| - |-3|$ 11. $|-12| + |-3|$ 12. $|-31| + |-12|$

Exercises

Independent Practice

Add.

13. $+9 + (+7)$ 14. $18 + 21$ 15. $-5 + (-13)$ 16. $-18 + (-3)$

17. $-2 + (+17)$ 18. $15 + (-9)$ 19. $5 + 21$ 20. $-10 + 3$

21. $-11 + (-18)$ 22. $-27 + (-43)$ 23. $88 + (-72)$ 24. $47 + (-63)$

25. $33 + (-48)$ 26. $-27 + 45$ 27. $67 + (-43)$ 28. $-22 + 51$

Evaluate each expression if $a = -3$, $k = 5$, and $m = -6$.

29. $8 + a$ 30. $13 + m$ 31. $-4 + k$ 32. $a + 12$

33. $-31 + k$ 34. $-18 + a$ 35. $m + (-21)$ 36. $k + (-16)$

Write an addition sentence for each situation. Then find the sum.

37. You withdraw $40 from your savings account. Then you withdraw $23.95 more.

38. In Saturday's football game, the Jackson Terriors lost 3 yards on one play. They gained 5 yards on the next.

39. The temperature was 16°F. The wind chill made it seem 25° colder.

40. Your paycheck is $124.95. You spend $15.50.

41. You jog 150 meters east. Then you turn around and jog 275 meters west.

Logical
Reasoning

Replace each with *always, sometimes,* or *never.*

42. The sum of two negative integers is ● negative.

43. The sum of two positive integers is ● negative.

44. The sum of a positive and a negative integer is ● positive.

Mixed Review

45. Simplify the expression $14y + 5(6 + 2y)$.   (Lesson 1-4)

46. Write a sentence that explains the meaning of $|-3|$.   (Lesson 2-1)

47. Write the inequality for *3 feet is shorter than 8 feet.*   (Lesson 2-2)

48. Write an inequality that compares the number of days in November to the number of days in January.   (Lesson 2-2)

49. Write the algebraic expression for the phrase *the number of pencils divided by three.*   (Lesson 1-8)

Applications

50. **Oceanography**   A submarine at 1300 m below sea level descends an additional 1150 m. How far below sea level is the submarine now?

51. **Astronomy**   Using the information at the top of page 60, find the average surface temperature on Saturn's surface during the day.

Critical
Thinking

52. **Make Up a Problem**   Write a problem that can be solved using the addition sentence $-15 + 25 = c$.

Wrap-Up

53. Draw a number line to show the addition sentence $-3 + (-2) = -5$.

2-4 More on Adding Integers

A weather balloon rises 200 feet from the ground,
drops 150 feet, and then rises 300 feet. What is the
new height of the balloon? One way to solve this
problem is by adding the integers 200, -150, and 300.

rise of		drop of		rise of		final height
200 feet	$+$	150 feet	$+$	300 feet	$=$	above the ground

$$200 \;+\; (-150) \;+\; 300 \;= h$$

$$[200 + (-150)] + 300 = h \qquad \text{Associative property, addition}$$

$$50 + 300 = h$$

$$350 = h$$

The final height of the balloon is 350 feet above the ground.

Sometimes you can add integers mentally. Study the following example
to see how the associative and commutative properties are used to find
the sum.

Example

*How can you check your
addition?*

1 **Solve $x = 10 + 22 + (-7) + (-31)$.**

$$x = 10 + 22 + (-7) + (-31)$$

$$= 10 + (-7) + 22 + (-31) \qquad \text{Think: } 10 + (-7) = 3$$
$$22 + (-31) = -9$$

$$= \qquad 3 \quad + \quad (-9)$$

$$= \qquad -6$$

CONNECTION TO ALGEBRA

You can use the distributive property and the rules for adding integers
to combine like terms. Notice that the distributive property can be
applied to more than two terms.

Example

2 **Simplify $5y + (-12y) + 6y$.**

$$5y + (-12y) + 6y = [5 + (-12) + 6]y \qquad \text{Distributive property}$$
$$= -1y \qquad\qquad\qquad \text{Add 5, -12, and 6.}$$
$$= -y \qquad\qquad\qquad \text{-1}y \text{ can be written as } -y.$$

Checking for Understanding

Communicating Algebra

1. What property allows you to add in any order?
2. Tell two different ways to solve $x = 8 + 24 + (-5) + (-25)$.

Guided Practice

State whether each sum is positive or negative.

3. $-3 + 4 + 7$
4. $-10 + 9 + 8$
5. $-8 + 7 + (-3)$
6. $-5 + (-6) + (-9)$
7. $-11 + 13 + 2$
8. $-15 + 9 + (-10)$

Exercises

Independent Practice

Solve each equation.

9. $x = 4 + (-12) + (-18)$
10. $y = 7 + (-11) + 32$
11. $h = 8 + (-15) + 13$
12. $5 + (-7) + 20 = a$
13. $-18 + (-23) + 10 = c$
14. $-12 + 14 + 8 = d$
15. $f = -17 + 36 + (-45)$
16. $g = -31 + (-9) + 62$
17. $r = 47 + 32 + (-16)$
18. $83 + (-19) + 16 = m$

Simplify each expression.

19. $5x + (-21x)$
20. $-3z + (-17z)$
21. $-12y + 5y$
22. $-3a + 12a + (-14a)$
23. $16d + (-9d) + (-27d)$
24. $9m + 43m + (-16m)$
25. $-4f + (-6f) + (-19f)$
26. $14b + (-21b) + 37b$

Mental Math

Solve each equation mentally.

27. $x = -11 + 10 + (-7) + 9$
28. $-14 + (-15) + 26 + 7 = y$
29. $-35 + 15 + 25 + (-10) = a$
30. $c = 98 + (-102) + 102 + (-98)$
31. $m = 28 + (-56) + 32 + (-75)$
32. $-69 + 33 + (-8) + (-15) = y$

Mixed Review

33. *True* or *false:* $30 - 18 \div 9 + 3 = 1$.   (Lesson 1-1)
34. Find two values for m that make $3m - 7 \leq 0$ true.   (Lesson 1-10)
35. Simplify $|19| - |-4|$.   (Lesson 2-1)
36. State whether the sum of -5 and -8 is positive or negative.   (Lesson 2-3)

Application

37. **Consumer Awareness**   Karen opened a checking account at a local bank on May 7. A record of her checks and deposits is shown below. What is the balance in her account on May 22?

DATE	DESCRIPTION	CHECK	DEPOSIT
5/7	Opening Deposit		$100.00
5/9	Jim's Record Mart	$15.95	
5/14	Babysitting		$18.00
5/22	Diller's Department Store	$52.87	

Critical Thinking

38. Roderico bought two shirts for $38. He paid $4 more for one shirt than for the other. How much did he pay for each shirt?

Wrap-Up

39. Give a real-world example that involves finding sums of many integers.

Subtracting Integers

Materials: counters, mat

In this Exploration, you will use counters to model subtraction.

▶ Consider +8 − (+3). To model this operation, start with a set of eight positive counters and remove three.

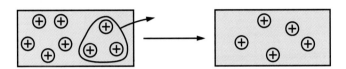

Therefore +8 − (+3) = +5.

▶ Consider -8 − (-3). Start with a set of eight negative counters and remove three.

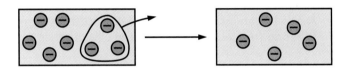

Therefore -8 − (-3) = -5.

▶ Now consider +8 − (-3). This means start with a set of eight *positive* counters and remove three *negative* counters. If you add three zero pairs to the set, the value of the set does not change. Now you can remove three negative counters.

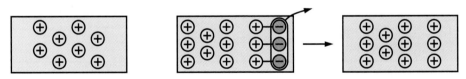

Therefore +8 − (-3) = 11.

Your turn: **What does -8 − (+3) mean? Model this operation. Explain how zero pairs are used in this subtraction.**

Model each problem.

1. 5 − 2   **2.** -5 − (-2)   **3.** 5 − (-2)   **4.** -5 − 2

5. 3 − 7   **6.** 3 − (-7)   **7.** -3 − 7   **8.** -3 − (-7)

Analysis

9. Notice the relationship between 5 − 2 and 5 + (-2). How are these two the same? How are they different? Model both operations.

2-5 Subtracting Integers

Objective:
Subtract integers.

Key Terms:
opposite
additive inverse

Lightning occurs when there are an unequal number of positive and negative electrical charges in the atmosphere. Positive charges and negative charges are opposites of each other.

Similarly, every positive integer can be paired with a negative integer. These pairs are called **opposites.** For example, the opposite of +3 is -3.

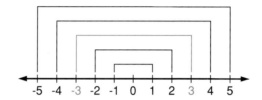

Notice that +3 and -3 are on opposite sides of 0. Both +3 and -3 are 3 units away from 0.

A number and its opposite are called **additive inverses** of each other. What do you notice about the sum of a number and its additive inverse?

$$3 + (-3) = 0 \qquad -41 + 41 = 0 \qquad 236 + (-236) = 0$$

Additive Inverse Property

In words: The sum of any number and its additive inverse is zero, the additive identity.
In symbols: For any number a, $a + (-a) = 0$.

Additive inverses can be used to define subtraction for integers. Compare the following addition and subtraction sentences.

Subtraction   **Addition**

additive inverses

$$8 - 3 = 5 \qquad\qquad 8 + (-3) = 5$$

same result

Subtracting Integers

In words: To subtract an integer, add its additive inverse.
In symbols: For any integers a and b, $a - b = a + (-b)$.

The subtraction rule allows you to rewrite any subtraction expression as an addition expression. This is useful because you already know how to add integers.

Examples

Solve each equation.

1 $a = 6 - 11$
   $a = 6 + (-11)$
   $a = -5$

2 $5 - (-8) = x$
   $5 + 8 = x$
   $13 = x$

3 $c = -4 - (-7)$
   $c = -4 + 7$
   $c = 3$

4 $-8 - 19 = d$
   $-8 + (-19) = d$
   $-27 = d$

CONNECTION TO ALGEBRA

The rule for subtraction of integers and the distributive property can be used to combine like terms.

Example

5 Simplify $3y - 5y$.

Why is -5y the additive inverse of 5y?

$3y - 5y = 3y + (-5y)$ Subtract 5y by adding its additive inverse, -5y.

$= [3 + (-5)]y$ Use the distributive property.

$= -2y$

Checking for Understanding

Communicating Algebra

Replace each ■ with a word or phrase to make a true sentence.

1. The opposite of a positive integer is ■ .
2. The opposite of a negative integer is ■ .
3. The opposite of the opposite of a positive integer is ■ .
4. If $n = 3$, then $-n =$ ■ .

Guided Practice

State the additive inverse of each number.

5. +6
6. -6
7. -13
8. 0
9. b
10. -b
11. cd
12. $-ef$
13. $-5x$
14. $9cd$

Restate each subtraction sentence as an addition sentence.

15. $7 - 13 = x$
16. $-8 - 5 = a$
17. $-17 - 9 = b$
18. $9 - (-2) = c$
19. $-18 - (-16) = y$
20. $-24 - (-23) = x$

Exercises

Solve each equation.

21. $x = -5 - 3$

22. $y = -6 - (-8)$

23. $a = 7 - 13$

24. $-11 - 9 = c$

25. $-13 - (-18) = d$

26. $-24 - 19 = h$

27. $7 - 25 = f$

28. $-17 - (-2) = g$

29. $j = -15 - 23$

30. $29 - 32 = a$

31. $b = -34 - (-19)$

32. $c = 51 - (-11)$

33. $-42 - 38 = q$

34. $n = -29 - 36$

35. $-18 - (-18) = m$

Evaluate each expression.

36. $y - 5$, if $y = -8$

37. $a - (-7)$, if $a = 19$

38. $b - (-5)$, if $b = -13$

39. $h - (-13)$, if $h = -18$

40. $k - (-12)$, if $k = 27$

41. $w - 37$, if $w = -18$

42. $9 - n$, if $n = -7$

43. $11 - m$, if $m = 5$

Simplify each expression.

44. $3x - 18x$

45. $11a - 21a$

46. $-2b - (-3b)$

47. $-16ab - 13ab$

48. $24cd - (-24cd)$

49. $-28d - 17d$

50. $41x - (-29x)$

51. $30y - (-11y)$

52. $-39ad - 15ad$

53. Which number, 4, 16, or 32, is a solution of the equation $8 = \frac{x}{4}$?
    (Lesson 1-5)

54. Name the absolute value of -21.   (Lesson 2-1)

55. Find the sum of 84 and -56.   (Lesson 2-3)

56. State whether the sum of -6, -3, and 10 is positive or negative.
    (Lesson 2-4)

57. Solve the equation $h = 1 + (-14) + 19$.   (Lesson 2-4)

58. Simplify the expression $-4x + 6x + (-13x)$.   (Lesson 2-4)

The $\boxed{+/-}$ key on a calculator is called the change-sign key. When it is pressed, the calculator changes the sign of the number in the display. For each of the following, tell what number is in the display.

59. Enter 8, press $\boxed{+/-}$ once.

60. Enter 10, press $\boxed{+/-}$ twice.

61. Enter 3, press $\boxed{+/-}$ 25 times.

62. Enter 5, press $\boxed{+/-}$ 100 times.

63. **Business**   The formula $P = I - E$ is used to find the profit (P) when income (I) and expenses (E) are known. Find P if $I = \$15,525$ and $E = \$17,000$.

64. **Meteorology**   The temperature outside was 20°F. The wind chill made it feel like -15°F. Find the difference between the real temperature and the apparent temperature.

65. *True* or *false:* $-n$ names a negative number.

66. **Make Up a Problem**   Write a subtraction problem in which the difference is a negative integer.

Algebra in Action-Geography

Time Zones

The prime meridian is a semicircle passing through Greenwich, England. Lines of longitude indicate degrees west or east of the prime meridian. The prime meridian is at 0°.

A city's longitude determines its time zone. The zone labeled zero is centered about the prime meridian. The time zones to the east of the prime meridian are named by negative numbers. The time zones to the west are named by positive numbers.

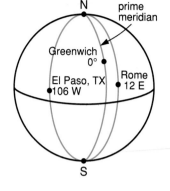

How does this differ from the integer number line?

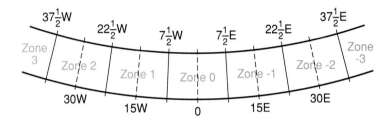

Various cities and their time zones are given below.

City	Time Zone	City	Time Zone
Athens, Greece	-2	Los Angeles, USA	8
Bombay, India	-5	Moscow, USSR	-3
Chicago, USA	6	New York, USA	5
El Paso, USA	7	Paris, France	0
Honolulu, USA	11	Rome, Italy	-1

To find the time in another city, subtract the time zone of the other city from the time zone of your own location. For example, if you are in El Paso and wish to know the time in Athens, subtract.

El Paso time zone − Athens time zone = $7 - (-2) = 9$

Since 9 is positive, the time in Athens is 9 hours *ahead* of El Paso. Thus, if it is 1:00 P.M. in El Paso, it is 10:00 P.M. in Athens.

1. If the time in Chicago is 4:00 P.M., find the time in Rome.

2. If the time in El Paso is 11:00 A.M., find the time in Honolulu.

3. If the time in Honolulu is 5:00 P.M., find the time in New York.

4. The person you wish to reach by phone lives in Paris and is available from 9:00 A.M. to 5:00 P.M., Paris time. During what hours should you call if you live in Chicago?

2-6 Statements and Negations

Objective:
Determine if a statement and its negation are true or false.

Key Terms:
statement
logic
negation

Which of the following sentences are true? Which are false?

> Mammals are warm-blooded.
> The word *pizza* is a verb.
> Indianapolis is the capital of Indiana.
> $1 + 2 \times 5 = 15$

Note that each of the previous sentences is either true or false. A **statement** is any sentence that is either true or false, but not both. **Logic,** the study of formal reasoning, is based on statements.

You can refer to a specific statement by representing it with a letter such as p or q. For example, let p represent the statement *Dublin is a city in Ireland.* This statement is true. To find the **negation** of statement p, write $\sim p$, which is read "not p." This represents the statement *Dublin is not a city in Ireland.* This statement is false.

The example above illustrates the principle that the negation of a true statement is false and the negation of a false statement is true.

Example

1 Let p represent the statement $9 + 3 = 11$. Let q represent the statement *A spider does not have six legs.* State whether each statement is true or false. Then state whether its negation is true or false.

a. p $9 + 3 = 11$ This statement is false.

   $\sim p$ $9 + 3 \neq 11$ This statement is true.

b. q A spider does not have six legs.   This statement is true.

   $\sim q$ A spider has six legs.   This statement is false.

What is the negation of the statement *No roses are yellow?* You might think it is *All roses are yellow.* But then the original statement and its negation are both false. It is true, though, that some roses are yellow. So the negation is *Some roses are yellow.*

Checking for Understanding

Communicating Algebra

Complete each sentence.

1. A _____?_____ is a sentence that is either true or false.

2. $\sim p$ refers to the _____?_____ of statement p.

3. Choose the statement that is the negation of the statement *All integers are even.*
   a. All integers are odd.   **b.** No integers are even.
   c. Some integers are even.   **d.** Some integers are odd.

Guided Practice

State whether each statement is *true* or *false*.

4. Water freezes at 0°C.   5. $9 \le 9$

State the negation of each statement. Then state whether the statement and its negation are *true* or *false*.

6. Corn is a vegetable.   7. $3(11 - 5) \ne 18$

Exercises

Independent Practice

State whether each statement is *true* or *false*.

8. Missouri is west of the Mississippi River.

9. Clarinets are not members of the woodwind family.

10. $2x = 6$ is an equation.

11. $2n = 4$ when $n = 1$

12. $\frac{1}{2} + \frac{3}{4} = \frac{3}{2} + \frac{1}{4}$

13. $0.1 + 0.11 = 0.21$

Write the negation of each statement. Then state whether the statement and its negation are true or false.

14. All years have 365 days.   15. Mozart was not a poet.

16. The solution to $5x = 5$ is 1.   17. $3y - 2y = 1$

18. Some cars have four doors.   19. $45 > 2(16 - 9)$

Using Logic

20. Ben hollered from his room, "Mom, I don't have no clean socks." His mom replied, "Please don't use double negatives, Ben!" What did she mean? Could Ben's statement be interpreted as "I have clean socks?" Why or why not?

Mixed Review

21. Solve the equation $-34 + 9 = h$.   (Lesson 2-3)

22. If $x = 4$, then $-x =$ ■ .   (Lesson 2-5)

Challenge

23. If p represents a true statement, is $\sim p$ true or false? $\sim(\sim p)$? $\sim(\sim(\sim p))$?

Critical Thinking

24. The gasoline tank of Mr. Steiner's car is one-third full. After he adds 2.5 gallons to it, the tank is one-half full. What is the capacity of his car's tank?

Wrap-Up

25. Write an example of a statement. State whether it is true or false. Then write its negation and state whether it is true or false.

2-8 Dividing Integers

Objective:
Divide integers.

The Bears football team was penalized the same amount on 3 consecutive plays. The total of the 3 penalties was 45 yards. The number of yards of each penalty can be expressed by the following sentence.

$$-45 \div 3 = \blacksquare$$

-45 represents a loss of 45 yards.
3 represents the number of plays.

To divide, think of a related multiplication sentence.

$$\blacksquare \times 3 = -45$$
$$-15 \times 3 = -45 \qquad \text{So, } -45 \div 3 = -15.$$

Why is the missing factor negative?

In the example above, the dividend is negative. Let's try a case where the divisor is negative.

$$27 \div (-3) = \blacksquare \quad \longrightarrow \quad \blacksquare \times (-3) = 27$$
$$-9 \times (-3) = 27 \quad \text{So, } 27 \div (-3) = -9.$$

These and other similar examples suggest the following rule.

Dividing Integers with Different Signs	The quotient of two integers with different signs is negative.

Examples

1 Find -38 ÷ 2.

$$-38 \div 2 = -19$$

2 Solve $y = \frac{54}{-9}$.

$$y = \frac{54}{-9}$$
$$y = -6$$

You know the quotient of two positive numbers is positive.

Both the dividend and divisor are positive. $48 \div 8 = 6$ The quotient is positive.

What do you think will be true about the quotient of two negative numbers? Check your prediction.

$$-32 \div (-4) = \blacksquare \longrightarrow \blacksquare \times (-4) = -32$$

For the product to be negative, one factor must be positive and the other negative.

$$-32 \div (-4) = 8$$

So, the missing factor is positive 8.

These and other similar examples suggest the following rule.

Dividing Integers with the Same Sign	The quotient of two integers with the same sign is positive.

Examples

3 Find **-90 ÷ (-15).**

4 Solve $\frac{72}{9} = x$.

$$\frac{72}{9} = x$$

$$-90 \div (-15) = 6 \qquad 8 = x$$

Checking for Understanding

Communicating Algebra

1. An elevator is moving up at a rate of 3 feet each second. The elevator is now at the street level. Write a division sentence using integers to represent the following questions. Answer each question.

    a. When was the elevator 15 feet below street level?

    b. When will the elevator be 30 feet above street level?

Guided Practice

State whether each quotient is positive or negative.

2. $-63 \div 7$ **3.** $-54 \div 9$ **4.** $16 \div 2$

5. $-36 \div (-6)$ **6.** $-48 \div (-16)$ **7.** $64 \div (-4)$

8. $42 \div (-7)$ **9.** $63 \div 21$ **10.** $\frac{-570}{19}$

11. $\frac{-804}{67}$ **12.** $\frac{68}{-17}$ **13.** $\frac{-48}{32}$

Exercises

Independent Practice

Divide.

14. $48 \div (-3)$ **15.** $-56 \div (-8)$ **16.** $-72 \div 9$

17. $84 \div (-7)$ **18.** $-52 \div (-4)$ **19.** $-72 \div (-12)$

20. $-91 \div 13$ **21.** $-51 \div (-17)$ **22.** $64 \div (-16)$

23. $98 \div (-14)$ **24.** $80 \div (-16)$ **25.** $-343 \div (-7)$

Solve each equation.

26. $d = \frac{-240}{-6}$ **27.** $\frac{-96}{24} = k$ **28.** $\frac{-105}{-15} = m$ **29.** $\frac{-450}{-45} = y$

30. $x = \frac{-120}{-15}$ **31.** $a = \frac{-144}{-36}$ **32.** $b = \frac{-175}{-25}$ **33.** $\frac{-288}{24} = k$

Evaluate each expression.

34. $\frac{y}{5}$, if $y = -50$

35. $\frac{32}{m}$, if $m = -8$

36. $\frac{b}{-7}$, if $b = -98$

37. $\frac{42}{z}$, if $z = -14$

38. $\frac{x}{-5}$, if $x = -65$

39. $\frac{-66}{w}$, if $w = -33$

Mixed Review

40. Write an equation and solve: Glen is reading a book that has 342 pages. If he is on page 176, how many pages does he have left to read?   (Lesson 1-9)

41. Name the absolute value of zero.   (Lesson 2-1)

42. *True* or *false:* The sum of two negative integers is always negative. (Lesson 2-3)

43. State whether the product of -9 and 18 is positive or negative.   (Lesson 2-7)

44. Multiply: $2x(-6)(3)$.   (Lesson 2-7)

Connection

45. Statistics   Five years ago, East Street High School had 250 more students than it does now. What was the average change in school population each year?

Application

46. Business   Stock in ABC Company has a change of -2 dollars each day for 4 days. What is the overall change?

Critical Thinking

47. Find two integers a and b so that $a + b = -5$ and $ab = 6$.

Wrap-Up

48. If you forget the rules for dividing integers, explain how you can use the rules for multiplying integers to decide whether the quotient is positive or negative.

History

The Egyptian Number System

About 5000 years ago, ancient Egyptians began to keep records by carving pictures into stone. These *hieroglyphics* contain a well organized number system based on groups of ten. In this system, a vertical stroke represents one unit. Other symbols are shown below.

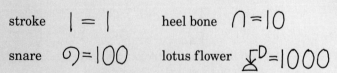

Using these hieroglyphics, the number 436 would be written

By studying carvings like the one pictured here, scholars have concluded that the ancient Egyptians were very accurate in counting and measuring.

2-9   Using Integers

Objective:
Solve problems using integers.

Suppose you started for school at 7:30 A.M. and found that the outside temperature was -4°F. You realized you had forgotten your homework and went back inside. When you stepped back outside at 7:32 A.M., the temperature was 45°F! This dramatic temperature change was recorded on January 22, 1943 in Spearfish, South Dakota. How much did the temperature rise in 2 minutes?

Explore

Read the problem carefully. Select the important data.

Estimate: Was the change greater than 45° or less than 45°?

- The beginning temperature was -4°F.
- The final temperature was 45°F.

What is asked?

- You need to find the temperature change.
- Let c = the temperature change.

Plan

The temperature rose during the time period. Will the amount of change be represented by a positive or a negative number?

You can compare the temperatures by using subtraction. Write an equation.

$$\underbrace{\text{change in temperature}}_{c} = \underbrace{\text{final temperature}}_{45} - \underbrace{\text{beginning temperature}}_{(-4)}$$

Solve

$c = 45 + 4$ Subtract -4 by adding its inverse, 4.

$c = 49$

In two minutes, the temperature rose 49°F!

FYI

A chinook wind is a warm dry wind that descends the eastern slopes of the Rocky Mountains. Such winds occasionally cause the temperature to rise 40° to 50°F in a few minutes.

Examine

Is your answer reasonable? Think about the scale on a thermometer. The temperature started at 4° *below* zero and ended at 45° *above* zero. The temperatures are 49° apart.

Compare your answer to the estimate. The answer is reasonable.

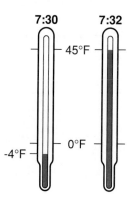

Example

1 In Fairfield, Montana, on December 24, 1924, the temperature fell from 63°F at noon to -21°F 12 hours later. What was the average temperature change each hour?

The temperature fell during the time period. Will the change be represented by a positive or a negative number?

Explore You need to find the average temperature change. You know the temperature at noon and 12 hours later.

Plan Find the total temperature change. Then divide the total by the number of hours.

Solve

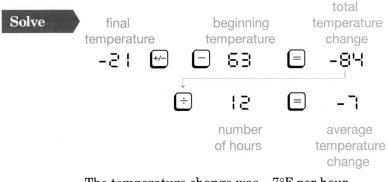

The temperature change was -7°F per hour.

Examine You can check the solution as follows:

$$7 \;\fbox{+/-}\; \fbox{×}\; 12 \;\fbox{=}\; -84 \qquad \text{change in 12 hours}$$

$$63 \;\fbox{+}\; 84 \;\fbox{+/-}\; \fbox{=}\; -21 \qquad \text{final temperature}$$

The solution checks.

Calculator Hint

Use the $\fbox{+/-}$ key on the calculator to enter a negative integer, or to change the sign of any integer. Notice that you press $\fbox{+/-}$ *after* you enter the number.

Checking for Understanding ──────────────────

For each problem, answer the related questions.

1. A traffic helicopter descended 160 meters to observe road conditions. It leveled off at 225 meters. What was its original altitude?
   a. What is asked?
   b. What was the final altitude of the helicopter?
   c. How far did the helicopter descend?
   d. Make a drawing to represent the problem.
   e. Write an equation that represents the original altitude.

2. The product of two integers is -36. One of the integers is 9. What is the other integer?
   a. What is asked?
   b. What is the product of the integers?
   c. What is one of the integers?
   d. How can you find the second integer?
   e. Write an equation that represents the second integer.

Exercises

Independent Practice

ESTIMATION
MENTAL MATH
CALCULATOR
PAPER/PENCIL

3. The record high temperature in Florida is 109°F. The record low temperature is -2°F. What is the difference in temperature between the record high and the record low?

4. A scuba diver descended to a depth of 75 meters below sea level. He then rose 30 meters to an underwater living quarters. How far beneath the surface were the living quarters?

5. Three subtracted from some number is equal to -7. What is the number?

6. Ray's score at the end of a game was 80. Halfway through the game his score was -35. How many points did he score during the last half of the game?

7. From the top of Mt. McKinley to the floor of Death Valley is 6280 meters. If Death Valley is 86 meters below sea level, what is the height of Mt. McKinley?

8. The temperature in Upper Scioto rose 2°F every minute for 7 minutes. The low temperature was -4°F. What was the temperature after 7 minutes?

9. The World Trade Center is 1350 feet tall. Each floor is about 12 feet high. Wei-Min rides the elevator 60 floors down from the top. About how far is she above the ground floor?

10. The quotient of two integers is -32. The divisor is -8. What is the other integer?

11. An elevator at the first floor takes one passenger up 10 floors and a second passenger up 12 more floors. If a third passenger goes down 14 floors, on what floor is the elevator?

Decision Making

12. Suppose your class is presenting a play to raise money. You pay a royalty of $45, scenery and costumes cost $77, and the programs and other expenses amount to $36. You make $128 selling advertisements. If you can sell 150 tickets, how much should you charge per ticket to make $300?

Mixed Review

13. State whether the quotient $72 \div 8$ is positive or negative.   (Lesson 2-8)

14. Solve the equation $x = -\frac{152}{4}$.   (Lesson 2-8)

Critical Thinking

15. Here is a famous problem. A snail at the bottom of a 10-foot hole crawls up 3 feet each day, but slips back 2 feet each night. How many days will it take the snail to reach the top of the hole and escape?

Wrap-Up

16. Choose one problem from among Exercises 3-12. Write the steps you used to solve the problem.

Review

Language and Concepts

Choose the letter of the correct word to complete each sentence.

1. __?__ is neither positive nor negative.

2. When comparing integers, the greater integer is found to the __?__ of the lesser integer on a number line.

3. If two addends are negative, their sum is __?__ negative.

4. To subtract an integer, add its __?__ .

5. The product of two positive integers is __?__ negative.

6. The quotient of two integers with the same sign is __?__ positive.

a. always
b. never
c. sometimes
d. left
e. right
f. opposite
g. zero

Skills

Replace each ■ with an integer that describes the situation. (Lesson 2-1)

7. a golf score of 3 under par, –3
   a golf score of 2 over par, ■

8. 780 feet above sea level, +780
   5 feet below sea level, ■

Write the absolute value of each integer. (Lesson 2-1)

9. –3

10. –5

11. 59

12. 8

Replace each ● with >, <, or = to make a true sentence. (Lesson 2-2)

13. 8 ● 5

14. –4 ● -2

15. –5 ● 5

16. 3 ● |-6|

Add. (Lessons 2-3, 2-4)

17. $-2 + (-6)$

18. $-7 + 5$

19. $-8 + 0$

20. $51 + (-33)$

21. $-3 + (-7) + 5$

22. $-15 + 25 + 20 + (-10)$

Solve each equation. (Lesson 2-5)

23. $a = 5 - (-2)$

24. $-10 - 4 = x$

25. $y = 15 - 20$

26. $-9 - 2 = b$

27. $x = 14 - 21$

28. $-10 - (-2) = z$

Simplify each expression. (Lessons 2-4, 2-5)

29. $-3a + 4a + (-5a)$

30. $7x - 9x$

31. $-13c + (-11c) + 8c$

Solve each equation. (Lessons 2-7, 2-8)

32. $s = -13(10)$

33. $-17(-8) = y$

34. $-5(-8)(3) = r$

35. $a = \frac{-15}{3}$

36. $\frac{-12}{-6} = b$

37. $c = \frac{18}{-3}$

Evaluate each expression. (Lessons 2-3, 2-4, 2-5, 2-7, 2-8)

38. $5x$, if $x = -3$

39. $z + (-6)$, if $z = 8$

40. $b - (-23)$, if $b = 20$

41. $a + b + (-4)$, if $a = -2, b = 7$

42. $\dfrac{-16}{m + n}$, if $m = -2, n = 6$

43. $8bn$, if $b = -3, n = -2$

Applications and Logic

Write the negation of each statement. (Lesson 2-6)

44. $4 + 3 \cdot 5 = 35$

45. Violins are members of the string family.

Solve. (Lesson 2-9)

46. Personal Finance Sarah Bly's account at the bank is overdrawn by $25. How much does she need to deposit to have a balance of $100?

47. Meteorology The temperature at 8:00 A.M. is 19°F. It rises 5° and then drops 7° by nightfall. What is temperature at nightfall?

ortfolio Suggestion

Select one of the assignments from this chapter that you found especially challenging and place it in your portfolio.

Curriculum Connection

- **Science** Investigate what is meant by *windchill*. How are integers used in determining windchill?

- **Social Studies** Research the effect weather has on economic issues.

- **Language Arts** Write a paragraph about a personal experience you have had during some type of severe weather.

Read More About It

Adler, Irving. *Integers: Positive and Negative*

Asimov, Isaac. *Realm of Algebra*

Stwertka, Albert. *Recent Revolutions in Mathematics*

Thomas, David A. *Math Projects For Young Scientists*

Test

Replace each ■ with an integer that describes the situation.

1. gain 4 kg, + 4
   lose 6 kg, ■

2. 8 inches below ground level, –8
   5 inches above ground level, ■

Replace each ● with >, <, or = to make a true sentence.

3. 0 ● 4

4. –3 ● 8

5. –6 ● |–9|

6. |–3| ● 3

Solve each equation.

7. $-4 + (-5) = a$

8. $-3 + 5 = b$

9. $y = 35 + (-19)$

10. $-9 + (-15) + 4 = q$

11. $-12 + 15 + (-3) = x$

12. $-8 - 3 = a$

13. $r = -15 - (-13)$

14. $12 - 25 = d$

15. $c = -33(3)$

16. $-14 \cdot (-7) = d$

17. $-15[7 \cdot (-2)] = y$

18. $-143 \div 11 = g$

19. $\frac{-270}{-90} = x$

20. $268 \div (-67) = b$

21. $c = \frac{-650}{-5}$

Simplify each expression.

22. $-21x + (-15x)$

23. $8x + (-9x) + 6x + 7x$

24. $-15y - (-10y)$

Evaluate each expression.

25. $-150 \div x$, if $x = -3$

26. $-16 + b$, if $b = -9$

27. $-6cd$, if $c = -5, d = -3$

28. $a \div b$, if $a = -12, b = -2$

29. $-25 - c$, if $c = -4$

30. $d + x$, if $d = -7, x = -15$

Solve.

31. Write the negation of the statement
    All dogs are dalmatians.

32. Maria's game scores were 40, –20, and 10.
    What was her total score?

33. Denver, Colorado, has an altitude of
    5280 feet above sea level. New Orleans,
    Louisiana, has an altitude of 5 feet
    below sea level. Find the difference in
    their altitudes.

BONUS

In your own words, explain why the product of seven negative integers is negative.

Academic Skills Test
Cumulative, Chapters 1–2

1. Which is equivalent to
   $3 \times 8 - 6 \div 2$?

   A 3
   B 9
   C 13
   D 21

2. The perimeter of a rectangle is $2(\ell + w)$
   where ℓ is the length and w is the
   width. What is the perimeter of a
   rectangle with length 15 cm and width
   5 cm?

   A 40 cm
   B 35 cm
   C 30 cm
   D 20 cm

3. If $36 - b = 20$, what is the value of b?

   A 56
   B 26
   C 16
   D 6

4. Ken has two more brothers than
   sisters. If Ken has n brothers, which
   sentence could be used to find s, the
   number of sisters he has?

   A $n = s - 2$
   B $n - 2 = s$
   C $s = n + 2$
   D $s = 2n$

5. Which set of numbers is in order from
   greatest to least?

   A {15, -8, -1}
   B {10, 0, -5}
   C {-5, -3, 1, 14}
   D {-1, -6, 8, 11}

6. You are riding in a hot air balloon. The
   balloon rises 100 m, drops 55 m, and
   then rises 50 m. How far from the
   ground are you now?

   A 5 m
   B 50 m
   C 95 m
   D 195 m

7. At 2:00 P.M. the temperature was 43°F.
   By 7:00 P.M. the temperature had fallen
   19°F. What was the temperature at
   7:00 P.M.?

   A 19°F
   B 22°F
   C 24°F
   D 62°F

8. If a diver descends at a rate of 5 m per
   minute, at what depth will she be after
   12 minutes?

   A +60 m
   B -17 m
   C -50 m
   D -60 m

9. Tom bought 80 shares of stock and later
   sold them for a loss of $240. Which
   integer represents the loss per share
   of stock?

   A 30
   B 3
   C -3
   D -30

10. What is the value of $-3k$ if $k = -33$?

    A -99
    B -11
    C 11
    D 99

CHAPTER 3

CHAPTER OBJECTIVES

In this chapter you will learn to:

- [ ] solve equations using the properties of equality
- [ ] apply equation-solving techniques to solve problems involving formulas
- [ ] solve inequalities
- [ ] solve problems involving equations and inequalities
- [ ] use logic to determine the truth value of compound statements

Solving One-Step Equations

How healthy are you?

Your answer may be "I'm almost never sick, so I'm very healthy." Or you might say, "I have a cold today, so I'm not feeling very healthy." But, did you know that there is much more to being healthy than how you feel physically?

Besides physical health, you need to consider your mental health and social health. These three parts of your health are connected, like the three sides of a triangle or the two sides of an equation. Each side affects the other one or two. To be considered really healthy, you need to have a *balanced* health triangle. You can accomplish this by working to keep each side of your health triangle healthy.

Now answer the question, "How healthy are you?"

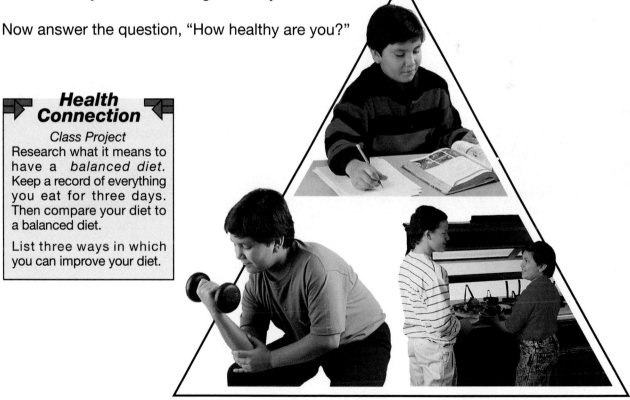

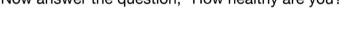

Health Connection

Class Project
Research what it means to have a *balanced diet*. Keep a record of everything you eat for three days. Then compare your diet to a balanced diet.

List three ways in which you can improve your diet.

Solving Equations

Materials: cups, counters, mats

In this Exploration, you will use cups and counters as models for building and solving equations.

▶ Recall from Chapter 1 that cups and counters are models for building an expression. For example, the expression $x + 2$ is shown below at the left. In the expression $x + 2$, x could be any value. But if there is a restriction on the expression, like $x + 2$ must have the same value as 5, you have an equation, $x + 2 = 5$.

What are the possible values of x so that $x + 2 = 5$ is a true equation? In other words, how many counters are in the cup? If you pair each counter on the left side with one on the right side, then you can see how many remain to fill the cup. In this case, the cup must contain 3 positive counters. So, $x = 3$.

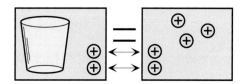

▶ Now consider $x + (-2) = 5$. In this example, there are no negative counters on the right side. Therefore, add enough zero pairs to match up the negative counters.

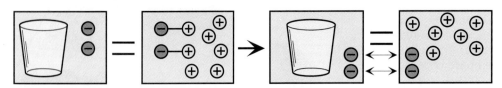

The cup contains 7 positive counters. So, $x = 7$.

Your Turn: Model $x + 2 = {}^-5$.

Model each equation and solve.

1. $s + 4 = 5$ **2.** $z + (-3) = -1$ **3.** $y + 7 = -4$ **4.** $t + (-6) = 7$

Analysis

5. Consider the equation $x - 2 = 5$. Model this equation. Describe the model and the procedure for solving this equation.

3-1   Solving Equations: $x + a = b$

Objective:
Solve equations by using the subtraction property of equality.

Key Term:
equivalent equation

Science Connection

The observations about action-reaction pairs are summarized in Newton's third law of motion. Isaac Newton also made important discoveries about gravitation and the spectrum of light.

Have you ever heard the expression *for every action there is an equal and opposite reaction?* In physical science, this means that forces always occur in pairs. The photo shows two teams having a tug-of-war. If the force of the red team equals the force of the blue team, the rope does not move. The teams are *balanced.* You can use this tug-of-war as a model of an equation.

Consider the equation $x + 8 = 3$. If you subtract 8 from the left side of the equation, you must also subtract 8 from the right side of the equation. This keeps the equation *balanced.*

Why was 8 subtracted from the left side?

$$x + 8 = 3$$

$$x + 8 - 8 = 3 - 8 \qquad \text{Subtract 8 from each side.}$$

$$x + 0 = \text{-}5 \qquad 8 - 8 = 0, 3 - 8 = \text{-}5$$

$$x = \text{-}5 \qquad x + 0 = x$$

All of the equations shown above are **equivalent equations** because they have the same solution, -5.

The property that you used to subtract 8 from each side of the equation is called the subtraction property of equality.

Subtraction Property of Equality	In words:	If you subtract the same number from each side of an equation, the two sides remain equal.
	In symbols:	For any numbers a, b, and c, if $a = b$ then $a - c = b - c$.

Example

1 Solve $m + 37 = -4$. Check your solution.

$$m + 37 = -4$$

$m + 37 - 37 = -4 - 37$ Subtract 37 from each side.

$$m = -41$$

Check: $m + 37 = -4$

$-41 + 37 \overset{?}{=} -4$ Replace m with -41.

$-4 = -4$ ✔ The solution is -41.

Subtracting a number is the same as adding its additive inverse. Therefore, another way to solve this type of equation is to use the additive inverse property, $a + (-a) = 0$.

Examples

Solve each equation. Check your solution.

2 $y + 21 = -7$

$$y + 21 = -7$$

$y + 21 + (-21) = -7 + (-21)$

$$y + 0 = -28$$

$$y = -28$$

Add -21 to each side because $21 + (-21) = 0$.

-21 is the additive inverse of 21.

Check: $y + 21 = -7$

$-28 + 21 \overset{?}{=} -7$ Replace y with -28.

$-7 = -7$ ✔ The solution is -28.

3 $-23 = k + (-5)$

$$-23 = k + (-5)$$

$-23 + 5 = k + (-5) + 5$ The additive inverse of -5 is 5.

$$-18 = k + 0$$

$$-18 = k$$ The solution is -18.

Checking for Understanding

Communicating Algebra

1. What is the name of the property that enables you to subtract the same number from each side of an equation?

2. Are $x + 4 = 10$ and $x = 5$ equivalent equations? Explain why or why not.

3. Explain how to solve $y + 12 = -7$.

4. Explain how to solve $z + (-13) = 12$.

Guided Practice

Solve each equation. Check your solution.

5. $k + 17 = 9$ **6.** $4 + b = -13$ **7.** $-11 = y + 27$

8. $-33 = x + 16$ **9.** $m + 10 = 7$ **10.** $9 = x + 13$

Exercises

Independent Practice

Solve each equation. Check your solution.

11. $18 + m = -57$
12. $11 + c = -5$
13. $w + 42 = -51$
14. $y + 23 = 15$
15. $44 = b + 63$
16. $67 = h + 38$
17. $-22 = z + 13$
18. $a + (-7) = 8$
19. $g + (-19) = 24$
20. $-13 + b = 14$
21. $-17 + z = 5$
22. $43 = r + (-15)$
23. $-12 + k = -37$
24. $k + (-11) = -21$
25. $p + (-8) = -21$
26. $15 = x + 42$
27. $x + (-21) = -59$
28. $24 = m + 37$
29. $q + (-3) = 17$
30. $b + 68 = 59$
31. $m + 37 = 14$

Language Skills

Write an equation for each situation. Then solve the equation.

32. Lone Star Supply Company bought a computer system with a color monitor for $1598. If the color monitor cost $699, how much did the rest of the system cost?

33. The best price Dawn found on a new compact car with a sports option package was $10,935. If the option package cost $850, what was the price of the car?

Mixed Review

34. Evaluate the expression $35 - c - b$ if $b = 13$ and $c = 7$. (Lesson 1-2)

35. Simplify the expression $21k + 3(k + 1)$. (Lesson 1-4)

36. Order the numbers $-2, -7, 9$ from least to greatest. (Lesson 2-2)

37. Solve the equation $x = -15 + 4 + (-9)$. (Lesson 2-4)

38. State whether the quotient of -42 and -7 is positive or negative. (Lesson 2-8)

Application

39. **Meteorology**   A maximum and minimum thermometer records both the high and low temperatures of the day. If the difference between one day's high and low was 49 degrees and the high temperature was 67 degrees, what was the low temperature?

Critical Thinking

40. Write two equations that are equivalent. Then write two equations that are not equivalent.

Wrap-Up

41. In this lesson you solved an equation by removing the same quantity from each side of the equation. Give an example of an everyday experience where both sides are treated equally.

Team Problem Solving

Suppose that one pyramid balances two cubes and one cylinder balances three cubes. Determine whether each statement below is *true* or *false*.

a. One pyramid and one cube balance one cylinder.

b. One cylinder and one pyramid balance four cubes.

c. One pyramid and one cube balance three cubes.

d. One pyramid and two cubes balance one cylinder.

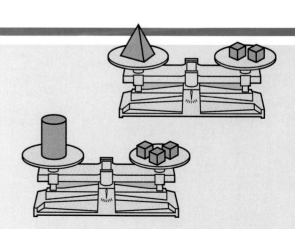

3-2 Solving Equations: $x - b = c$

Objective:
Solve equations by using the addition property of equality.

Mrs. Neish is a pharmacist at North Penn Hospital. She weighs various medicines using a scale like the one shown at the right. Suppose the scale is balanced. Next, she places a weight on one side of the scale. Then she must add an equal amount of the medicine to the other side to keep the scale balanced.

If a 12-gram weight were placed on the right side of the scale, then 12 grams of the medicine must be added to the left side to balance the scale.

In mathematics, some equations can be solved by adding the same number to each side of the equation. The equation $t - 12 = 42$ can be solved by adding 12 to each side.

Example

1 Solve $t - 12 = 42$. **Check your solution.**

$$t - 12 = 42$$
$$t - 12 + 12 = 42 + 12 \quad \text{Add 12 to each side to undo subtraction}$$
$$t = 54 \quad \text{of 12.}$$

Check: $t - 12 = 42$
$$54 - 12 \stackrel{?}{=} 42 \quad \text{Replace } t \text{ with 54.}$$
$$42 = 42 \quad ✓$$

Addition Property of Equality	In words:	If you add the same number to each side of an equation, the two sides remain equal.
	In symbols:	For any numbers a, b, and c, if $a = b$, then $a + c = b + c$.

Recall from Chapter 2 that to subtract an integer, you add its inverse.

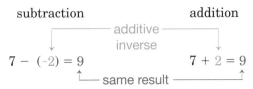

When solving equations, it is usually helpful to eliminate double signs first.

Examples

2 **Solve** $-17 = y - (-9)$**. Check your solution.**

$$-17 = y - (-9)$$
$$-17 = y + 9 \qquad \text{Rewrite as an addition sentence.}$$
$$-17 - 9 = y + 9 - 9 \qquad \text{Subtract 9 from each side.}$$
$$-26 = y$$

Check: $-17 = y - (-9)$
$$-17 \stackrel{?}{=} -26 - (-9) \qquad \text{Replace } y \text{ with } -26.$$
$$-17 = -17 \; \checkmark$$

The solution is -26.

Estimation Hint

THINK: $y - 20 = 30$
The solution should
be about 50.

Always check your
solution against your
estimate to see if it
is reasonable.

3 **Solve** $y + (-17) = 32$**.**

$$y + (-17) = 32$$
$$y - 17 = 32 \qquad \text{Rewrite as a subtraction sentence.}$$
$$y - 17 + 17 = 32 + 17 \qquad \text{Add 17 to each side.}$$
$$y = 49 \qquad \text{Check this solution.}$$

The solution is 49.

Checking for Understanding

Communicating Algebra

1. What is the name of the property that enables you to add the same number to each side of an equation?

2. Explain how to solve $b - 7 = 3$.

3. Explain how to solve $y - (-8) = 4$.

4. Write an equation in the form $x - b = c$ where the solution is -12.

Guided Practice

Rewrite each sentence as a corresponding addition or subtraction sentence. Then solve.

5. $x + (-9) = 15$
6. $y - (-16) = 24$
7. $p - (-7) = -18$
8. $-5 = r + (-12)$
9. $18 = t + (-4)$
10. $k - (-36) = 30$
11. $p - (-17) = 2$
12. $-23 = n + (-5)$
13. $m + (-8) = -15$

Exercises

Independent Practice

Solve each equation. Check your solution.

14. $a - 16 = 33$
15. $k - 36 = -37$
16. $y - 8 = -22$
17. $r - 21 = 58$
18. $y + (-7) = 19$
19. $y + (-8) = -31$
20. $b + (-14) = 6$
21. $k + (-13) = 21$
22. $z - (-7) = -19$
23. $x - (-18) = 14$
24. $t - (-34) = 66$
25. $s - (-47) = -27$
26. $42 = y - (-47)$
27. $59 = r - (-95)$
28. $-23 = q - 81$
29. $-17 = p - 93$
30. $x - (-33) = 14$
31. $d - 27 = -63$

Language Skill **Write an equation for each situation. Then solve the equation.**

32. Murielle had $175 when she started shopping for clothes. After buying a skirt, a pair of shoes, and a hat in three different stores, she had $41 left. How much did she spend?

33. Jennifer made a deposit of $150 for soccer camp. Her unpaid balance was $300. What was the fee for soccer camp?

Mixed Review

34. Write the algebraic expression $3x + 12$ using words.   (Lesson 1-8)

35. Name the absolute value of 131.   (Lesson 2-1)

36. Add the integers -27 and 5.   (Lesson 2-3)

37. Is -16 a solution of $-28 = z + 12$?   (Lesson 3-1)

38. Solve the equation $g + (-3) = 16$.   (Lesson 3-1)

39. Write an equation and solve: The temperature rose 34°F in one day. The high temperature was 21°. What was the low temperature?   (Lesson 3-1)

Challenge **Solve each equation. Check each solution.**

40. $[b + (-3)] + 2 = 4$

41. $(d + 5) + (-2) = 6$

42. $-10 = [n + (-4)] + 2$

43. $14 = [g - (-3)] + (-12)$

Applications

44. **Personal Finance**   After writing a check for $65 to pay her electric bill, Darlene has $139 left in her checking account. What was the original amount in her account?

45. **Recreation**   Phil is playing Jeopardy. After answering a 200-point question correctly, his score is -500. What was his score before he answered the question?

Critical Thinking

46. Place the digits 1, 2, 3, 5, and 8 in the boxes to make a true sentence. Use each digit exactly once. ▉▉ + -▉ = ▉▉

Wrap-Up

47. John owes Bill $7, that is, John has $-$7. Suppose Bill forgives $2 of that debt. In other words, he takes away $-$2 of John's debt. How much does John still owe? Is this similar to saying that subtracting a negative integer is the same thing as adding a positive integer? Give another example in everyday life that also illustrates this principle.

Writing Connection

The problem below is missing information. Write a sentence that would provide the necessary information to solve the problem.

(You do not have to solve the problem.)
In the first 4 days of a 5-day cold spell, the lowest temperature was -8°C. What was the highest temperature during the 4 days?

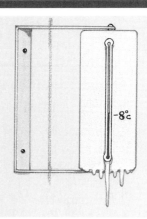

3-3 Solving Equations: $ax = c$

Objective:
Solve equations using the division property of equality.

As a forest ranger, Michael Volrath must check the size of trees in his forest district to see if they are growing at the expected rate. He records the distance around standing trees. He knows that the distance around a tree is about three times the thickness of the tree. Michael translates this into the equation $d = 3 \cdot t$, or $d = 3t$.

If Michael measures a tree that is 51 inches around, he must solve the equation $51 = 3t$ to estimate the thickness of the tree.

In mathematics, some equations can be solved by dividing each side of the equation by the same number. The equation $51 = 3t$ can be solved by dividing each side by 3.

How can you estimate the solution?

Example

1 **Solve $51 = 3t$. Check your solution.**

$51 = 3t$

$\dfrac{51}{3} = \dfrac{3t}{3}$ Divide each side by 3 to undo the multiplication $3 \cdot t$.

$17 = t$

Check: $51 = 3t$
$51 \overset{?}{=} 3(17)$ Replace t with 17.
$51 = 51$ ✔

The solution is 17. The tree is about 17 inches thick.

Division Property of Equality	In words: If you divide each side of an equation by the same nonzero number, the two sides remain equal.
	In symbols: For numbers a, b, and c, where $c \neq 0$, if $a = b$, then $\dfrac{a}{c} = \dfrac{b}{c}$.

FYI

One way to estimate the age, in years, of a sugar maple is to multiply the diameter, in inches, of the tree by 6.

2 **Solve $9y = -36$.**

$9y = -36$

$\dfrac{9y}{9} = \dfrac{-36}{9}$ Divide each side by 9.

$y = -4$ Check this solution.   The solution is -4.

Example

3 Solve $-1675 = -25c$. Check your solution.

$$-1675 = -25c$$

$$\frac{-1675}{-25} = \frac{-25c}{-25} \quad \text{Divide each side by } -25.$$

$$1675 \; \boxed{+/-} \; \boxed{\div} \; 25 \; \boxed{+/-} \; \boxed{=} \; 67$$

$$67 = c$$

Check: $-1675 = -25c$

$-1675 \stackrel{?}{=} -25 \cdot 67 \qquad$ Replace c with 67.

$-1675 = -1675 \quad \checkmark$

The solution is 67.

Checking for Understanding

Communicating Algebra

1. Explain how to check your solution to an equation.

2. What is the name of the property that enables you to divide each side of an equation by the same number?

3. Explain how to solve $7x = 91$.

4. Write an equation in the form $ax = c$ where the solution is -5

Guided Practice

Solve each equation. Check your solution.

5. $\dfrac{20}{-5} = y$

6. $\dfrac{-38}{2} = q$

7. $r = \dfrac{-56}{-7}$

8. $t = \dfrac{-84}{-4}$

9. $k = \dfrac{27}{-3}$

10. $\dfrac{-48}{16} = m$

11. $\dfrac{-24}{-6} = k$

12. $s = \dfrac{63}{9}$

Exercises

Independent Practice

Solve each equation. Check your solution.

13. $-7x = 56$

14. $4y = -52$

15. $6k = -78$

16. $-9m = 99$

17. $-104 = 8r$

18. $42 = -3y$

19. $-2p = -38$

20. $-13a = -39$

21. $-68 = 17c$

22. $165 = -11d$

23. $-13a = 52$

24. $-60 = -15h$

25. $-27q = 81$

26. $-13z = -65$

27. $-85 = 17r$

28. $-14x = -98$

29. $-180 = 12f$

30. $-14g = 406$

Write an equation for each situation. Then solve the equation.

Language Skills

31. Ben and his brother Jason are having their annual medical checkups. The nurse records their heights and weights. She tells Ben that he is 5 feet 7 inches tall and weighs 3 times as much as Jason. If Ben weighs 144 pounds, how much does Jason weigh?

32. Fax machines transmit printed copy by telephone lines but at different speeds. Model F transmits at twice the rate of Model G. If Model F transmits 9600 bits per second, how many bits per second does Model G transmit?

33. State whether the inequality $0 < 4k - 5$ is true for $k = 1$.   (Lesson 1-10)

34. If $y = -9$, then $-y = \underline{?}$.   (Lesson 2-5)

35. Solve the equation $x = -4(13)$.   (Lesson 2-7)

36. Solve the equation $b + (-17) = -4$.   (Lesson 3-1)

37. Rewrite the addition sentence $y + (-19) = 14$ as a subtraction sentence. (Lesson 3-2)

38. Solve the equation $d - (-14) = 5$.   (Lesson 3-2)

Connections

39. **Geometry**   The perimeter of any square is 4 times the length of one of its sides. If the perimeter of a square is 56 inches, what is the length of each side of the square?

40. **Statistics**   For many years, immigrants to the United States arrived at Ellis Island in New York. The flow of immigrants peaked in 1907 with more than 1 million. During that year, 37,807 people arrived from Germany. Estimate the average number of Germans who arrived daily at Ellis Island in 1907.

Application

41. **Sports**   Millicent is a hot air balloonist. The distance she can travel depends on the wind velocity. If the wind is blowing at a steady rate of 8 miles per hour, how long will it take her to fly 24 miles in her hot air balloon?

Critical Thinking

42. If a number is multiplied by itself and the answer is doubled, the result is 32. What is the number?

Wrap-Up

43. Twenty-three people shared the cost of lottery tickets. They split the $11,500,000 prize money. Write an equation that represents how to find each person's share of the prize money. How much was each person's share of the prize money?

Career

Medical Laboratory Technician

A medical laboratory technician is an important part of the health profession. Using a microscope and an automatic analyzer, a laboratory technician can measure the amount of different substances in your blood to determine if they are within normal ranges.

For example, a laboratory technician can measure the amount of cholesterol in a sample of a person's blood. The normal range is less than 200 milligrams/deciliter (mg/dL). If a blood test reveals a cholesterol level that is higher than 200 mg/dL, a doctor will prescribe a course of action to lower this level.

3-4 Solving Equations $\frac{x}{a} = b$

Objective:
Solve equations by using the multiplication property of equality.

Craig is training for the local Special Olympics. He practices by running around a $\frac{1}{4}$-mile track. Suppose he can run 1 lap in 2 minutes. At this rate, how many minutes does it take him to run 1 mile? Let t represent the number of minutes.

How do you know he runs 4 laps?

minutes to run 1 mile	divided by	laps	equals	minutes to run 1 lap
t	$\div$	4	$=$	2

In mathematics, some equations can be solved by multiplying each side of the equation by the same number. The equation $\frac{t}{4} = 2$ can be solved by multiplying each side by 4.

Example

1 **Solve $\frac{t}{4} = 2$. Check your solution.**

$$\frac{t}{4} = 2$$

$$\frac{t}{4} \cdot 4 = 2 \cdot 4 \qquad \text{Multiply each side by 4 to undo the division in } \frac{t}{4}.$$

$$t = 8$$

Check: $\quad \frac{t}{4} = 2$

$$\frac{8}{4} \stackrel{?}{=} 2 \qquad \text{Replace } t \text{ with 8.}$$

$$2 = 2 \quad \text{✔} \qquad \text{The solution is 8.}$$

Craig can run 1 mile in 8 minutes.

FYI

The Special Olympics includes over 1 million participants from 36 countries. Local games take place in more than 10,000 communities worldwide.

Multiplication Property of Equality	In words:	If you multiply each side of an equation by the same number, the two sides remain equal.
	In symbols:	For numbers a, b, and c, if $a = b$, then $ac = bc$.

Example

2 Solve $-13 = \dfrac{k}{-5}$. **Check the solution.**

$$-13 = \dfrac{k}{-5}$$

$$-13\,(-5) = \dfrac{k}{-5}\,(-5) \qquad \text{Multiply each side by -5.}$$

$$65 = k$$

Check: $\quad -13 = \dfrac{k}{-5}$

$$-13 \stackrel{?}{=} \dfrac{65}{-5} \qquad \text{Replace } k \text{ with 65.}$$

$$-13 = -13 \quad \checkmark \qquad \text{The solution is 65.}$$

Checking for Understanding

Communicating Algebra

1. What is the name of the property that enables you to multiply each side of an equation by the same number?

2. Explain how to solve $3 = \dfrac{z}{12}$.

3. Explain how to solve $\dfrac{p}{-7} = 23$.

4. Write an equation in the form $\dfrac{x}{a} = b$ where the solution is -14.

5. In an equation of the form $\dfrac{x}{a} = b$, explain why a cannot be 0.

6. Is 88 a solution of $8 = \dfrac{f}{-11}$? Explain why or why not.

Guided Practice

Name the number to multiply each side by to solve each equation. Then solve.

7. $3 = \dfrac{z}{12}$

8. $\dfrac{m}{13} = -9$

9. $4 = \dfrac{r}{-16}$

10. $-13 = \dfrac{m}{-7}$

11. $\dfrac{p}{-7} = 23$

12. $\dfrac{x}{9} = 40$

13. $-16 = \dfrac{q}{13}$

14. $\dfrac{h}{-5} = -14$

Exercises

Independent Practice

Solve each equation. Check your solution.

15. $14 = \dfrac{a}{-7}$

16. $\dfrac{b}{8} = -24$

17. $\dfrac{k}{-12} = 13$

18. $-10 = \dfrac{m}{11}$

19. $\dfrac{r}{-21} = -6$

20. $\dfrac{v}{-17} = -9$

21. $21 = \dfrac{x}{7}$

22. $18 = \dfrac{f}{-6}$

23. $-15 = \dfrac{f}{14}$

24. $\dfrac{h}{16} = 28$

25. $-32 = \dfrac{c}{22}$

26. $-26 = \dfrac{d}{47}$

27. $\dfrac{x}{-3} = 136$

28. $\dfrac{y}{8} = -117$

29. $-321 = \dfrac{t}{9}$

30. $5 = \dfrac{s}{-264}$

31. $\dfrac{a}{39} = -65$

32. $\dfrac{b}{-46} = 216$

33. $-71 = \dfrac{x}{24}$

34. $-42 = \dfrac{y}{33}$

 35. $-171 = \dfrac{x}{-124}$

36. $273 = \dfrac{p}{-784}$

37. $\dfrac{m}{319} = -467$

38. $\dfrac{z}{-639} = -408$

39. Name the operation you would do first in the expression $2 \cdot (3 + 4) - 9$. (Lesson 2-2)

40. Write an inequality for the phrase *45 mph is slower than 55 mph.* (Lesson 2-2)

41. Write an equation and solve: A skydiver jumps out of a plane at a height of 5140 feet. He descends 2900 feet before opening his parachute. What height is he at when he opens the parachute? (Lesson 2-9)

42. Solve the equation $r - 8 = -5$. (Lesson 3-2)

43. Name the number to divide each side by in the equation $84 = -7n$. (Lesson 3-3)

44. Solve the equation $15g = -75$. (Lesson 3-3)

Applications

45. Ecology On June 8, 1990 the tanker *Mega Borg* released millions of gallons of oil into the waters of the Gulf of Mexico. It is estimated that the spilled oil would fill 4222 tanker trucks, each capable of holding 9000 gallons. Estimate the total amount of oil spilled in gallons. If a barrel holds 42 gallons, about how many barrels were spilled?

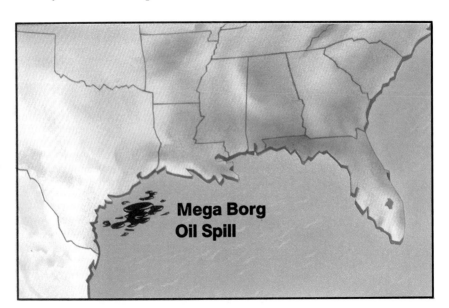

Mega Borg Oil Spill

46. Consumer Awareness Judy keeps a record of the gasoline she uses in her car. Last month she used 115 gallons. If her car averages 22 miles per gallon, how far did she drive last month?

47. Sports Randy is training for a triathalon. He runs an average of 27 miles per week. Since he began training, he has run 91 miles. How many weeks has he been training?

Critical Thinking

48. Alan took a weekend ski trip to the mountains. He drove his car an average of 60 miles per hour for the first two hours and, because of snow-covered roads, had to slow down to 40 miles per hour for the last three hours. What was his overall average speed for the trip?

Wrap-Up

49. Make Up a Problem Write a problem that can be solved using the division sentence $\frac{x}{3} = -2$.

3-5 Formulas

Objective:
Solve problems by applying formulas.

Key Term:
formula

In league bowling, the team that knocks down the most pins does not always win. Some teams have more skillful bowlers than others. To make sure that each team has an equal chance of winning, a handicap is given to each team based on the averages of the individual members. The final score is determined by adding the handicap to the actual score.

A **formula** shows the relationship among certain quantities. The formula below can be used to compute a league bowler's final handicap score.

$$\underbrace{\text{handicap score}}_{s} = \underbrace{\text{game score}}_{g} + \underbrace{\text{handicap}}_{h}$$

Example

What could be done instead of subtracting 14 from each side?

1 **Susan had a handicap bowling score of 186. Her handicap was 14. What was her game score?**

$$s = g + h$$
$$186 = g + 14 \qquad \text{Replace } s \text{ with 186 and } h \text{ with 14.}$$
$$186 - 14 = g + 14 - 14 \qquad \text{Subtract 14 from each side of the equation.}$$
$$172 = g \qquad \text{Susan's game score was 172.}$$

Formulas are often used in science. For example, the formula for the relationship among the current (I), voltage (V), and resistance (R) in an electrical circuit is $I = \dfrac{V}{R}$. This formula is known as Ohm's law.

Science Connection

Ohm's law is named after George Ohm (1789–1854). Ohm worked on electrical conduction, but when the Berlin Academy of Science refused to publish his work, he retired from scientific life in discouragement.

Example

2 **Suppose I is 24 amperes and R is 5 ohms. Find the voltage (V).**

$$I = \frac{V}{R} \qquad \text{Write the formula.}$$
$$24 = \frac{V}{5} \qquad \text{Replace } I \text{ with 24 and } R \text{ with 5.}$$
$$24 \times 5 = \frac{V}{5} \times 5 \qquad \text{Multiply each side by 5.}$$
$$120 = V \qquad \text{The voltage is 120 volts.}$$

Another important formula relates distance, rate, and time.

distance = rate · time
$$d = r \cdot t \quad \text{or } d = rt$$

Example

3 The Sanderson family plans to drive about 960 miles to visit relatives in Indianapolis. If the average rate of speed is 45 miles per hour, what will their driving time be?

$d = rt$ Write the formula.

$960 = 45t$

$\dfrac{960}{45} = \dfrac{45t}{45}$ Divide each side by 45.

$$960 \; \div \; 45 \; = \; 21.333333$$

The Sandersons' driving time will be about 21 hours.

Estimation Hint

THINK:
$1000 = 50t$
The solution should be about 20.

Checking for Understanding

Communicating Algebra

1. Write a formula with which you are familiar.

Translate each sentence into a formula.

2. The sale price (s) of an item is equal to the list price (l) less the discount (d).

3. The diameter (d) of a circle is twice the length of the radius (r).

Guided Practice

Solve. Use the correct formula.

4. Find the game score if the handicap score is 175 and the handicap is 12.

5. Find the distance you travel if your train goes 60 miles per hour for 4 hours.

6. Find the average number of amperes if $V = 60$ volts and $R = 5$ ohms.

7. Find the speed if your plane flies 2000 miles to Hawaii in 5 hours.

Exercises

Independent Practice

Solve mentally by substituting the variables with the given values.

8. $f = t - h$, if $t = 125$ and $h = 25$

9. $s = g + c$, if $g = 200$ and $c = 21$

10. $P = s + s + s$, if $s = 12$

11. $s = l - d$, if $l = \$50$ and $d = \$1$

12. $d = rt$, if $r = 50$ and $t = 6$

13. $d = rt$, if $r = 40$, and $t = 4$

14. $I = \dfrac{V}{R}$, if $V = 110$ and $R = 10$

15. $I = \dfrac{V}{R}$, if $V = 60$ and $R = 2$

16. Solve the equation $-13a = -78$. (Lesson 3-3)

17. Name the number to multiply each side by in the equation $\frac{r}{-6} = 53$. (Lesson 3-4)

18. Solve the equation $-5 = \frac{x}{-14}$. (Lesson 3-4)

Applications

Solve. Use the correct formula.

19. **Travel** The formula for finding gas mileage is $m = d \div g$ where m is miles per gallon, d is the distance traveled, and g is the number of gallons of gasoline used. Glen's car gets 34 miles per gallon. How many gallons of gasoline does he need to travel 289 miles?

**ESTIMATION
MENTAL MATH
CALCULATOR
PAPER/PENCIL**

20. **Sports** Angela's final handicap score in bowling was 186. Her handicap was 17. What was her game score?

21. **Travel** Mai Li travels 371 miles in 7 hours on a bus trip to her aunt's house. What rate of speed does she travel?

22. **Electronics** Automobiles have 12-volt systems. What is the current in a 12-volt circuit that has a resistance of 3 ohms?

23. **Meteorology** How far away is the lightning if you hear the sound of the thunder 4 seconds after you see the flash of lightning? Use 344 meters per second as the speed of sound.

Critical Thinking

24. The area (A) of a triangle is equal to the product of the base (b) and the height (h) divided by two. What are some strategies you could use to find the height of a triangle that has an area of 24 square inches and a base of 8 inches?

Wrap-Up

25. **Research** Find an example of a formula that is used in geometry.

Mid-Chapter Quiz

Solve each equation. (Lessons 3-1, 3-2, 3-3, 3-4)

1. $m + 59 = -123$
2. $27 = n + 64$
3. $278 = x + (-365)$
4. $w - 9 = -24$
5. $73 = y - (-81)$
6. $-19 = b - 76$
7. $-64 = -16r$
8. $-52 = 4m$
9. $-17x = 595$
10. $\frac{y}{11} = 13$
11. $-27 = \frac{x}{15}$
12. $\frac{a}{-7} = -35$

Which equation would you use to solve the problem below? (Lessons 3-1, 3-2)

13. Vickie removes 33.5 gallons of water from her baby sister's wading pool so she can add hot water to raise the temperature. The pool now contains 45 gallons of water. How many gallons did it contain before?

   a. $33.5 - x = 45$ b. $45 - 33.5 = x$ c. $x - 33.5 = 45$ d. $45 + x = 33.5$

Algebra in Action–Physics

Acceleration

When a skateboard accelerates, it changes speed. **Acceleration** is the rate at which speed is changing with respect to time.

To find the acceleration, first find the change in speed by subtracting the starting speed (s) from the final speed (f). Then divide by the time it took to make the change. The formula is shown below.

$$a = \frac{f - s}{t}$$

A race car goes from 44 meters per second (m/s) to 77 m/s in 11 seconds. Find the acceleration.

$a = \frac{f - s}{t}$

$a = \frac{77 - 44}{11}$ Replace f with 77, s with 44, and t with 11.

$a = \frac{33}{11}$ or 3

The car accelerates 3 meters per second each second.

Suppose a race car *decreases* its speed from 80 m/s to 20 m/s in 15 seconds. The same equation can be used. Subtracting the starting speed from the final speed gives a negative value. This means that the acceleration is negative. Negative acceleration is called *deceleration*.

$a = \frac{f - s}{t}$

$a = \frac{20 - 80}{15}$ Replace f with 20, s with 80, and t with 15.

$a = \frac{-60}{15}$ or -4   The acceleration is -4 meters per second each second.

Find the acceleration.

1. A motorcycle goes from 2 m/s to 14 m/s in 6 seconds.

2. A skateboard goes from 6 m/s to 0 m/s in 3 seconds.

3. A car starts from a standstill. It accelerates to 40 mph in 10 seconds.

4. A jet plane decreased its speed from 500 km/h to 350 km/h in 30 seconds.

5. A skateboard goes from 0 feet per second to 22 feet per second in 2 seconds.

6. A 110-car coal train goes from 1.2 mph to 6.8 mph in 4 minutes.

3-6 Perimeter and Area

Objective:
Find the perimeter and area of rectangles and squares.

Key Terms:
perimeter
area

Mr. O'Neill is the groundskeeper at Kennedy Jr. High. He wants to put a fence around the practice football field to keep people off the new grass he planted. How much fencing does he need?

160 ft

360 ft **360 ft**

160 ft

The distance around a geometric figure is called the **perimeter.** To find the perimeter of the field, Mr. O'Neill adds the measure of the sides.

$$160 + 360 + 160 + 360 = 160 + 160 + 360 + 360$$
$$= 320 + 720$$
$$= 1040$$

The perimeter of the football field is 1040 feet. So, Mr. O'Neill needs 1040 feet of fencing.

Suppose P represents the measure of the perimeter of a rectangle.
Let ℓ be the measure of the length.
Let w be the measure of the width.

$P = \ell + w + \ell + w$ Definition of perimeter

$P = \ell + \ell + w + w$ Commutative property of addition

$P = 2\ell + 2w$

$P = 2(\ell + w)$ Distributive property

Perimeter of a Rectangle	In words: If a rectangle has a length of ℓ units and a width of w units, then the perimeter is twice the sum of the length and width.
	In symbols: $P = 2(\ell + w)$.

Examples

Find the perimeter of each rectangle.

1

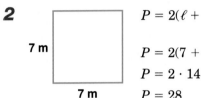

3 in.

8 in.

$P = 2(\ell + w)$ Write the formula for the
perimeter of a rectangle.

$P = 2(8 + 3)$ Replace ℓ with 8 and w with 3.

$P = 2 \cdot 11$ Add 8 and 3.

$P = 22$

The perimeter is 22 inches.

*Are all squares
rectangles? Explain.*

2

7 m

7 m

$P = 2(\ell + w)$ Write the formula for the
perimeter of a rectangle.

$P = 2(7 + 7)$ Replace ℓ with 7 and w with 7.

$P = 2 \cdot 14$

$P = 28$

The perimeter is 28 meters.

The previous examples all deal with the distance *around*
a geometric figure. Another measurement to consider is
the surface *enclosed by* a geometric figure.

The measure of the surface enclosed by
a geometric figure is called the **area.**
By counting the number of 1 yard by 1
yard squares, you can find that the area
of the rectangle shown at the right is 20
square yards. This area can also be
found by multiplying the measures of
the length and the width. So, the area is
$4 \cdot 5$ or 20 square yards.

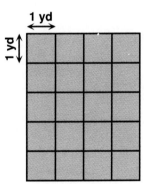

1 yd

1 yd

| **Area of a Rectangle** | In words: If a rectangle has a length of ℓ units and a width of w units, then the area is $\ell \cdot w$ square units. |
| | In symbols: $A = \ell w$ |

Examples

3 Find the area of a rectangle with length 15 m and width 4 m.

$A = \ell w$ Write the formula for the area of a rectangle.

$A = 15 \cdot 4$ Replace ℓ with 15 and w with 4.

$A = 60$ The area is 60 square meters.

The unit of measure of the sides is meter, so the area is stated in
square meters.

4 **Find the width of a rectangle with area 540 square feet and length 30 ft.**

$A = \ell w$ Write the formula for the area of a rectangle.

$540 = 30 \cdot w$ Replace A with 540 and ℓ with 30.

$\dfrac{540}{30} = \dfrac{30w}{30}$ Divide each side by 30.

$$ 540 \; \boxed{\div} \; 30 \; \boxed{=} \; 18 $$

$18 = w$

Check: $540 = 30w$

$540 \stackrel{?}{=} 30 \cdot 18$ Replace w with 18.

$540 = 540$ ✔

The width is 18 feet.

Checking for Understanding

Communicating Algebra

1. Draw and label a rectangle that has a length of 2 inches and a width of 1 inch.

2. Explain how to find the perimeter and area of a rectangle that is 9 cm long and 5 cm wide.

3. In your own words, write a definition of perimeter.

4. Explain how perimeter and area are different.

Guided Practice

Find the perimeter and area of each rectangle.

5.
2 m
2 m

6.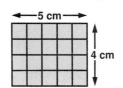
5 cm
4 cm

7.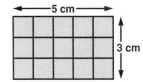
5 cm
3 cm

8.
8 in.
8 in.

9.
11 ft
3 ft

10.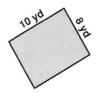
10 yd
8 yd

Exercises

Independent Practice

Find the perimeter and area of each rectangle.

	Length	Width
11.	9 m	4 m
12.	31 km	9 km
13.	129 yd	32 yd

Given each area, find the missing length or width of each rectangle.

	Area	Length	Width
14.	168 ft^2	24 ft	?
15.	195 cm^2	?	13 cm
16.	144 yd^2	16 yd	?

17. Simplify the expression $13rs + (-8rs) + 5rs$. (Lesson 2-4)

18. Solve the equation $\frac{m}{-38} = 14$. (Lesson 3-4)

19. Write a formula for the sentence *The radius (r) of a circle is equal to the diameter (d) divided by two.* (Lesson 3-5)

20. Alice travels 330 miles at a rate of 55 mph. How many hours was she driving? (Lesson 3-5)

Challenge

Find the area of the red part of each rectangle.

21.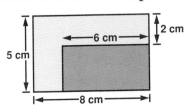

22.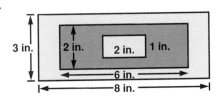

Applications

23. Construction Dan Block added a square deck to the back of his house. Draw and label a diagram of his deck if the perimeter of the deck is 84 feet.

24. Home Economics Barb cut some material to make a skirt. She cut a rectangular-shaped piece that is 28 in. wide and 36 in. long. Find the area of the cut material.

25. Landscaping Steve Kocher used old railroad ties to enclose a rectangular-shaped garden that was 30 ft long and 12 ft wide. What was the perimeter of the garden?

Connection

26. Geometry The outer triangle has an area of 56 square inches. This is 4 times as great as each inner triangle. Find the area of each inner triangle.

Critical Thinking

27. A square is a rectangle in which the length is equal to the width. Write a formula that can be used to find the area of a square.

Wrap-Up

28. Research Find the dimensions of the courts or playing fields of five different sports. Rank them from greatest to least with respect to area and then with respect to perimeter.

Area and Perimeter

Materials: small tiles or grid paper

In this Exploration, you will investigate the relationship between perimeter and area by building regions of given areas.

▶ The figures below show several figures with an area of 4 square units.

▶ Find the perimeter of each figure shown above by counting the units on the outside of the figure.

P = 10 units   P = 8 units   P = 10 units   P = 10 units   P = 10 units

▶ It appears that, for an area of 4 square units, the greatest possible perimeter is 10 units.

Your Turn: **Repeat the activities described above for areas of 1, 2, 3, 5, and 6 square units. Find the greatest possible perimeter for each area. Record your results in a table.**

For each given area, find the greatest possible perimeter. Draw a figure to show the greatest possible perimeter.

1. 8 square units   **2.** 10 square units   **3.** 15 square units

Analysis

4. A figure has an area of 50 square units. Predict the greatest possible perimeter.

5. Explain why the two figures shown at the right have the same perimeter. Draw another figure with the same perimeter.

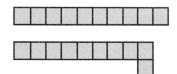

6. Suppose a figure has an area of x square units. Write an expression for the greatest possible perimeter.

3-7   Solving Inequalities: Adding or Subtracting

Objective:
Solve inequalities by using the addition and subtraction properties.

Beth had $99 in her savings account and Jim had $82 in his account. They each received $25 for mowing lawns and deposited the money in their accounts. Whose account has more money?

	Beth		Jim	
	99	>	82	
	$99 + 25$	>	$82 + 25$	Add 25 to each side.
	124	>	107	

Beth had more money in her account at the beginning, and she had more money in her account at the end.

Recall from Chapter 1 that an inequality is a mathematical sentence with $<$ or $>$. The sentences $99 > 82$ and $124 > 107$ are inequalities. Adding the same number to each side of the inequality $99 > 82$ did not change the truth of the inequality.

Suppose Jim and Beth each withdrew, or subtracted, $10 from their accounts.

	Beth		Jim	
	124	>	107	
	$124 - 10$	>	$107 - 10$	Subtract 10 from each side.
	114	>	97	

Beth still has more money than Jim. Subtracting the same number from each side of an inequality did not change the truth of the inequality.

These and other examples suggest the following properties.

Addition and Subtraction Properties of Inequalities	In words: Adding or subtracting the same number from each side of an inequality does not change the truth of the inequality.
	In symbols: For all numbers a, b, and c:
	1. If $a > b$, then $a + c > b + c$ and $a - c > b - c$.
	2. If $a < b$, then $a + c < b + c$ and $a - c < b - c$.

Examples

1 **Solve $m + 8 > 3$. Check your solution.**

$$m + 8 > 3$$
$$m + 8 - 8 > 3 - 8 \qquad \text{Subtract 8 from each side.}$$
$$m > \text{-}5$$

Check: Try -4, a number greater than -5.
$$\text{-}4 + 8 \overset{?}{>} 3$$
$$4 > 3 \quad \vee$$

The solution is $m > \text{-}5$, all numbers greater than -5.

2 **Solve $\text{-}26 \leq r - 16$. Check your solution.**

$$\text{-}26 \leq r - 16$$
$$\text{-}26 + 16 \leq r - 16 + 16 \qquad \text{Add 16 to each side.}$$
$$\text{-}10 \leq r$$

How can $\text{-}10 \leq r$ also be written using the $\geq$ symbol?

Check: Try -10 and 0, a number greater than -10.
$$\text{-}26 \overset{?}{\leq} \text{-}10 - 16 \qquad\qquad \text{-}26 \overset{?}{\leq} 0 - 16$$
$$\text{-}26 \leq \text{-}26 \quad \vee \qquad\qquad \text{-}26 \leq \text{-}16 \quad \vee$$

The solution is $r \geq \text{-}10$, all numbers greater than or equal to -10.

Checking for Understanding

Communicating Algebra

1. What is the name of the property that enables you to add the same number to each side of an inequality?
2. Explain how to solve $y - 9 > 3$.
3. Explain how to solve $\text{-}11 < b + 6$.
4. Write an inequality involving addition where the solution is $k \leq 18$.

Guided Practice

Solve each inequality. Check your solution.

5. $m + 6 > 5$
6. $r - 12 < \text{-}6$
7. $7 < t - 9$
8. $\text{-}8 > p - 1$
9. $k + 10 \geq \text{-}5$
10. $16 + s \leq \text{-}1$

Exercises

Independent Practice

Solve each inequality. Check your solution.

11. $y - 7 < 10$
12. $x - 2 < \text{-}14$
13. $m + 13 > 8$
14. $k + 17 > 36$
15. $5 + z > 27$
16. $12 + a > \text{-}9$
17. $\text{-}7 + b < \text{-}5$
18. $\text{-}9 + d < 10$
19. $w + (\text{-}3) < \text{-}7$
20. $\text{-}42 + k > 18$
21. $4 < y - 23$
22. $f + (\text{-}8) > \text{-}12$
23. $20 > z + (\text{-}19)$
24. $\text{-}31 < p - 7$
25. $t - (\text{-}5) > \text{-}6$
26. $10 < s - (\text{-}3)$
27. $\text{-}41 > r - (\text{-}8)$
28. $24 + m > 10$
29. $72 + k < 56$
30. $22 < m - (\text{-}16)$
31. $\text{-}30 \leq x + (\text{-}5)$
32. $\text{-}15 + t \leq 12$
33. $\text{-}67 + p \geq \text{-}48$
34. $11 + c \geq \text{-}29$

Write an inequality for each sentence.

35. Ellen spent more than $15 at the amusement park.

36. The temperature is less than or equal to negative seven degrees.

37. The number of players on the soccer team was greater than or equal to 13.

38. George plans to spend no more than $48 for clothes.

39. State the property shown by the statement $7 \cdot 1 = 7$.   (Lesson 1-3)

40. Write the negation of the statement *All integers are even.*   (Lesson 2-6)

41. Evaluate the expression $\dfrac{c}{-4}$ if $c = 28$.   (Lesson 2-8)

42. Use the formula $f = t - h$ to find the actual time (t) if the actual boat race score (f) is 93 hours and the handicap (h) is 12.   (Lesson 3-5)

43. Find the perimeter of the rectangle shown at the right.   (Lesson 3-6)

6 mi

2 mi

44. Find the area of a rectangle whose length is 18 ft and whose width is 4 ft.   (Lesson 3-6)

Write an inequality for each situation. Then solve the inequality.

45. **Sports**   The graph at the right shows the states with the most little league teams in 1990. How many teams could Florida add and still have fewer than Pennsylvania?

States with the most Little League teams

7,687   6,909   6,710

23,269   14,187

46. **Hobbies**   Emilio wants to spend at most $25 to buy new stamps for his collection. He has selected a new stamp that costs $4. What is the most he can spend on other stamps?

47. How can you use unmarked 7-quart and 4-quart containers to measure out exactly one quart? two quarts? three quarts? five quarts?

48. Measure your height and a friend's height. Write an inequality that compares your heights. Suppose that each of you were to grow 3 more inches. Now write an inequality that compares your new heights.

3-8 Solving Inequalities: Multiplying or Dividing

Objective:
Solve inequalities by using the multiplication and division properties.

When a pilot prepares for take-off, the engine power is increased, or multiplied, until there is enough thrust to push the airplane forward and get it airborne.

Upon landing, the airplane is slowed by again *multiplying* the power. This increase in power is made negative by using a thrust-reverser brake, which changes the direction of the thrust.

Similar characteristics exist when working with inequalities.

Consider multiplying or dividing each side of the inequality $4 < 6$ by a *positive* integer.

$4 < 6$	$4 < 6$
$4 \cdot 2 < 6 \cdot 2$ Multiply each side by 2.	$4 \div 2 < 6 \div 2$ Divide each side by 2.
$8 < 12$	$2 < 3$

The inequalities $8 < 12$ and $2 < 3$ are true. These and other examples suggest the following properties.

Multiplication and Division Properties of Inequalities	In words: When you multiply or divide each side of a true inequality by a *positive* integer, the result remains true. In symbols: For all integers a, b, and c, where $c > 0$, $\quad$ if $a > b$, then $a \cdot c > b \cdot c$ and $\frac{a}{c} > \frac{b}{c}$.

The rule is similar for $a < b$, $a \geq b$, and $a \leq b$.

Consider multiplying or dividing each side of $2 < 4$ by a *negative* integer.

$2 < 4$	$2 < 4$
$2\,(-1) < 4\,(-1)$ Multiply each side by -1.	$2 \div (-2) < 4 \div (-2)$ Divide each side by -2.
$-2 < -4$ False	$-1 < -2$ False

The inequalities $-2 < -4$ and $-1 < -2$ are both false. However, notice that both inequalities would be true if you reverse the order symbol. That is, change $<$ to $>$.

$$-2 > -4 \text{ and } -1 > -2 \text{ are both true.}$$

Multiplication and Division Properties of Inequalities	In words: When you multiply or divide each side of an inequality by a *negative* integer, you must *reverse the order symbol.*

In symbols: For all integers a, b, and c, where $c < 0$,

$$\text{if } a > b, \text{ then } a \cdot c < b \cdot c \text{ and } \frac{a}{c} < \frac{b}{c}.$$

The rule is similar for $a < b$, $a \le b$, and $a \ge b$.

Examples

1 Solve $\frac{x}{-5} > 2$. **Check your solution.**

$$\frac{x}{-5} > 2$$

$$\frac{x}{-5} \cdot (-5) < 2 \cdot (-5) \qquad \text{Multiply each side by } -5. \text{ Reverse the order symbol.}$$

$$x < -10$$

Check: Try -15, a number less than -10.

$$\frac{-15}{-5} \overset{?}{>} 2$$

$$3 > 2 \quad ✔$$

The solution is $x < (-10)$, all numbers less than -10.

2 **Solve $6x < -18$.**

$$6x < -18$$

$$\frac{6x}{6} < \frac{-18}{6} \qquad \text{Divide each side by 6.}$$

$$x < -3$$

Why was the order symbol not reversed?

The solution is $x < -3$, all numbers less than -3.

3 **Solve $-4k \le -20$.**

$$-4k \le -20$$

$$\frac{-4k}{-4} \ge \frac{-20}{-4} \qquad \text{Divide each side by } -4. \text{ Reverse the order symbol.}$$

$$k \ge 5$$

Why was the order symbol reversed?

The solution is $k \ge 5$, all numbers greater than or equal to 5.

Checking for Understanding

Communicating Algebra

1. Write an inequality that is always true.
2. Write an inequality that is always false.
3. Explain how to solve $-7m \ge -49$.
4. Write an inequality involving division where the solution is $x < 9$.

Guided Practice

State the number to multiply or divide each side by to solve each inequality. Then tell whether the order symbol should be reversed.

5. $-24 < 6a$ 6. $3y < 21$ 7. $-4x \ge 8$ 8. $-6z \le 18$

9. $\frac{d}{9} \le 12$ 10. $\frac{r}{3} \ge -4$ 11. $\frac{s}{-6} > -11$ 12. $-35 \le \frac{g}{-7}$

Exercises

Independent
Practice

Solve each inequality. Check your solution.

13. $-3y < -39$ **14.** $4x \geq -16$ **15.** $-5z > -65$ **16.** $-9a \leq 108$

17. $4r \geq -72$ **18.** $-12p < -108$ **19.** $8 \leq 2x$ **20.** $-48 > -6y$

21. $\frac{w}{-2} \leq 73$ **22.** $\frac{s}{-9} \geq 13$ **23.** $\frac{m}{8} > 21$ **24.** $\frac{k}{6} \leq -8$

25. $\frac{p}{-14} \leq 12$ **26.** $\frac{n}{-15} > -26$ **27.** $-20 < \frac{h}{-18}$ **28.** $-17 \geq \frac{f}{15}$

29. $21r \leq -378$ **30.** $-27x < -459$ **31.** $-3n \leq 51$ **32.** $-143 < -11y$

33. $-25 \leq \frac{a}{-16}$ **34.** $50 \leq \frac{k}{-32}$ **35.** $\frac{x}{34} \leq -3$ **36.** $\frac{g}{-21} \geq 33$

Language
Skill

Write an inequality for each situation. Then solve the inequality.

37. Seven times an integer is greater than -56.

38. The product of negative three and an integer is less than 21.

Mixed Review

39. Simplify the expression $|-12| - |9|$. (Lesson 2-1)

40. Solve the equation $15x = 150$. (Lesson 3-3)

41. Find the area of a rectangle whose length is 15 yards and whose width is 9 yards. (Lesson 3-6)

42. Solve the inequality $-23 < p - 17$. (Lesson 3-7)

43. Write the inequality for the sentence *Mr. Luez will spend no more than $19 for a shovel.* (Lesson 3-7)

Applications

44. Economics The *Norfolk Ledger* pays 8 cents per paper to the delivery person. How many papers must Kristen deliver to earn $4 or more per day?

45. Automobile Repair Felix Virtullo is a repair service estimator. The owner of a vehicle describes the problem and Felix makes an estimate of the cost of repair. When Mr. Virtullo gives a customer an estimate, he guarantees that the cost of the repair will be no more than $30 over the estimate. Suppose Mr. Virtullo thinks that the work is going to cost $300. Write an inequality to represent his estimate.

Critical
Thinking

46. The product of an integer and negative four is greater than -24. Find the greatest integer that meets this condition.

Wrap-Up

47. In your own words, explain why when you multiply or divide an inequality by a negative integer you must change the order of the inequality.

3-9 Compound Statements

Objective:
Determine the truth value of compound statements.

Key Terms:
compound statement
conjunction
disjunction
counterexample

For their summer vacation, the Elliott family drove to Yellowstone National Park. On the way there, they noticed on several of the interstate highways that the minimum speed was 45 mph and the maximum speed was 65 mph.

For the Elliotts to be driving at a legal speed, they had to drive:

> at least 45 mph
>
> *and*
>
> no faster than 65 mph.

The word *and* is used to connect the statements.

Two statements connected by the word *and* form a **compound statement.** Such a statement is called a **conjunction.** For a conjunction to be true, *both* statements must be true. For example, suppose the Elliotts drove at 68 mph. Then the first statement would be true, but the second statement would be false. So they would not be driving at a legal speed.

You can use inequalities to describe the speed at which the Elliotts should drive. Let s represent their speed.

They should drive at least 45 mph.	$s \geq 45$
They should drive no faster than 65 mph.	$s \leq 65$

These inequalities can then be combined in a compound inequality.

$$45 \leq s \leq 65$$

This is read:
"45 is less than or equal to s is less than or equal to 65."

Examples

1 **Determine whether the following conjunction is *true* or *false*. A triangle has three sides, and a pentagon has four sides.**

For a conjunction to be true, both parts of it must be true. Since triangles have three sides, the first part is true. However, pentagons have five sides, not four, so the conjunction is false.

2 **Write the inequalities $x > -2$ and $x < 4$ as a compound inequality.**

$x > -2$ is the same as $-2 < x$.
So, $x > -2$ and $x < 4$ can be written as $-2 < x < 4$.

FYI

There are 42,500 miles of interstate highways in the U.S. If they were connected end to end, they would go around the equator almost two times.

Every year, Steve Herr and his sister Suzi plan an outing to King's Point Amusement Park with their 12 cousins, parents, and grandparents. Admission is free if you are under 2 years of age or if you are a senior citizen (65 or older).

So, for anyone to get in free, he or she must be:

Steve's cousin Jordan is 2 years old. Should he get in free?

less than 2 years old

or

65 years or older.

The word *or* is used to connect the statements.

Two statements connected by the word *or* form another type of compound statement called a **disjunction.** For a disjunction to be true, *only one* of the statements must be true. For example, suppose Steve's cousin Kelsey is 10 months old. Then she gets into the park free because she is less than 2 years old.

Obviously, it is not possible for anyone to be less than 2 years old and also 65 years or older. However, in logic it is possible for both statements of a disjunction to be true. For example, if we say that it may rain Friday or it may rain Saturday, we mean that it could rain one or both days. So, a disjunction is true if at least one of its statements is true. It is false only if both of its statements are false.

Example

3 **Determine whether the following disjunction is *true* or *false*. The solution to the equation $x - 14 = 76$ is 62, or the sum of two odd numbers is always an odd number.**

Solve $x - 14 = 76$ to see whether the first statement is true or false.

$$x - 14 = 76$$
$$x - 14 + 14 = 76 + 14$$
$$x = 90$$

The first statement is false.

Check the sums of several pairs of odd numbers to see if the second statement is true or false. You only need one example to show that it is false. The example is called a **counterexample.**

$9 + 7 = 16$ even   The statement is false.

Since both statements are false, the disjunction is false.

Checking for Understanding

1. Name the two types of compound sentences. State how they are alike and how they are different.

2. The compound statement *Alan may go to the football game, or Mieko might go to the mall* is an example of a __?__ .

Determine whether each compound statement is *true* or *false*.

3. A hexagon has six sides, and an octagon has nine sides.

4. The product of two odd numbers is even, or $-17 + 4 = -13$.

5. $3 > 4$ or $7 < 6$

6. $7 \geq 6$ and $7 > 4$

Write each pair of inequalities as a compound inequality.

7. $0 \leq n$ and $n \leq 10$

8. $r > -3$ and $r < 8$

Exercises

Determine whether each compound statement is *true* or *false*.

9. Austin is the capital of Texas, and Cincinnati is the capital of Ohio.

10. Abraham Lincoln was a U.S. president, or Dolly Madison was the wife of President Madison.

11. The sum of $4n$ and $2n$ is $6n$, and 5 is a solution to the equation $3 + c = 8$.

12. $13 > 9$ and $13 > 12$

13. $9 > 0$ and $12 < 17$

14. $5 > -3$ and $-5 > -1$

15. $13 > 6$ or $0 < -2$

Write each pair of inequalities as a compound inequality.

16. $0 < y$ and $y \leq 12$

17. $p > 3$ and $p \leq 11$

18. $z > -4$ and $z < 0$

19. $m > -13$ and $m < -6$

20. *True* or *false*: All open sentences are equations. (Lesson 1-5)

21. Simplify the expression $-17ab - (-5ab)$. (Lesson 2-5)

22. State whether the order symbol should be reversed to solve the inequality $\frac{r}{4} > -12$. (Lesson 3-8)

23. **Consumer Awareness** The label on a can of paint reads *For proper results, do not apply if the temperature is below 30°F or above 85°F.* Write a compound inequality that describes the range of acceptable temperatures for applying the paint.

24. Suppose you are to determine if a conjunction is true. You determine that the first statement of the conjunction is false. Is it necessary to determine if the second statement is true or false? Why or why not?

25. Describe a situation relating to water where the inequality $32 < t < 212$ might be used.

3-10   Using Equations and Inequalities

Objective:
Solve verbal problems
by translating them
into equations
and inequalities.

Pedro rides his 10-speed bike for exercise. The last time he rode his bike he had trouble changing gears. Pedro took his bike to the Lewis Bike Shop to have it repaired. At the shop Pedro was told the total bill for labor and parts would be at least $48. The cost of the parts was $33. How much could Pedro expect to pay for labor?

In the first example below, the problem is solved using an inequality.

Example

1 Find the amount Pedro could expect to pay for labor.

Explore

Skim the problem for the general idea. You need to find the expected cost of labor. Let c = cost of labor.

Select the important data.
• The total bill will be at least $48.
• The cost for parts was $33.

Since c is the cost of labor, the total cost of parts and labor is $33 + c$.

Plan

At least $48 means $48 or more. That is, the total cost will be more than or equal to $48.

Solve

Total cost	is greater than or equal to	$48.
$33 + c$	$\geq$	48

$$33 - 33 + c \geq 48 - 33$$
$$c \geq 15$$

Pedro will pay $15 or more for labor.

Examine

Check the answer against your estimate. Then, check the answer in the words of the problem. $33 plus $15 is at least $48.

Estimation Hint

THINK:
$50 - 30 = 20$
The answer should
be close to $20.

Example

2 The last-place runner in a 10-kilometer race took 88 minutes to finish the race. This was twice the winner's time. What was the first-place time?

Explore You need to find the time of the first-place runner. You know the last runner took 88 minutes, twice the winner's time. Let t be the winner's time.

Plan Translate the words into an equation using the variable. *Twice* suggests multiplication.

How can you estimate the answer?

Last-place time	is	twice the winning time.
88	=	$2t$

Solve

$$\frac{88}{2} = \frac{2t}{2}$$

$$44 = t$$

The first-place time was 44 minutes.

Examine Check the answer against your estimate and in the words of the problem. 88 is two times 44.

Checking for Understanding

Communicating Algebra **For each problem, answer the related questions.**

1. Jimmy wants to spend at most $25 for new baseball cards for his collection. He has selected a new card that costs $9. How much can he spend on other cards?

   a. What is asked?

   b. What does the phrase *at most $25* mean?

   c. Write an equation or inequality that describes the problem.

   d. Would you reverse the inequality symbol to solve this problem?

2. A florist shop makes a profit of $5 on each special flower arrangement sold. How many of these arrangements must be sold in order to make a profit of at least $150?

   a. What is asked?

   b. What is the profit on one arrangement?

   c. What does the phrase *at least $150* mean?

   d. Write an equation or an inequality that describes the problem.

Guided Practice **Define a variable and translate each situation into an equation or an inequality.**

3. A number increased by 4 is 16.

4. Five times some number equals –85.

5. The result of dividing a number by 7 is 49.

6. Seven dollars less than the cost of the compact disc is $12.

7. The product of 21 and a number is at most –84.

8. The quotient when dividing a number by –7 is less than 112.

Exercises

Define a variable, write an equation or an inequality, and then solve.

9. Four times a number is greater than 76. What is the number?

10. Janice has averaged 17 points per game as a forward for the Northern Potter High School girls varsity basketball team. She has scored 221 points this season. In how many games has she played?

11. The quotient of a number and 8 is greater than 72. Find the number.

12. Room 308 at Darien High School sold more class play tickets than any other room, more than $1350 worth. If the tickets cost $3 each, how many tickets did Room 308 sell?

13. Maria bought 3 pairs of jeans for $75. If each pair cost the same amount, how much did each pair cost?

14. Tracy works for a hotel taking reservations over the phone. She earns $6 per hour. Last week she made $72. How many hours did she work?

15. Three friends went to dinner together to celebrate one of them getting a job promotion. The total of the check was less than $36. If the cost was shared evenly, how much did each person pay for dinner?

16. Marty saves $8 each week from money he earns from his paper route. How many weeks will he have to save before he can buy a camera that sells for $192?

17. Girl Scouts who sell at least 75 boxes of cookies earn a sweatshirt. Matilda needs to sell 17 more boxes to qualify for the sweatshirt. Each box costs $2. How many boxes has she sold so far?

Mixed Review

18. Solve the inequality $\frac{f}{3} \geq -9$.   (Lesson 3-8)

19. Write the inequalities $s > -4$ and $s < -2$ as a compound inequality. (Lesson 3-9)

Decision Making

20. Carlos needs to buy at least seven pairs of socks to use the week he goes to basketball camp. The socks he wants are sold separately for $2 a pair or in packages of three pair for $5. List all of the different combinations of packages that Carlos could buy. In what combination should Carlos buy the socks? Explain your answer.

Critical Thinking

21. In the figures at the right, the area of rectangle C is 6 times the area of rectangle A. If the difference between the areas of rectangle B and rectangle A is 12 square units, what is the area of rectangle C?

Wrap-Up

22. **Make Up a Problem**   Write a problem using the following information: Let w = Noel's weight in pounds. Use the equation $w - 4 = 139$.

Review

Language and Concepts

Choose the letter that best matches each phrase.

1. a mathematical sentence that shows the relationship between certain quantities

2. *x is less than or equal to 2* written as a mathematical sentence

3. an example of an equation

4. a measure of the space enclosed by a rectangle

5. the number added to each side of the equation $n - 42 = 9$ to solve it

6. the number you would multiply each side of the equation $\frac{y}{-42} = 9$ by to solve it

7. the distance around a rectangle

a. 9
b. 42
c. -42
d. x + 3 = 10
e. x + 3
f. x ≥ 2
g. x ≤ 2
h. formula
i. variable
j. perimeter
k. area

Skills

Solve each equation. Check your solution. (Lessons 3-1, 3-2, 3-3, 3-4)

8. $x + 28 = 42$

9. $-7 = -12 + g$

10. $72 = p + (-56)$

11. $k - 2 = -6$

12. $a - 5 = 4$

13. $r - (-7) = -19$

14. $-156 = -12f$

15. $180 = -15y$

16. $9k = -117$

17. $\frac{x}{11} = 10$

18. $\frac{m}{25} = -4$

19. $-18 = \frac{r}{-16}$

Solve. Use the formula $d = rt$. (Lesson 3-5)

20. Find d if $r = 40$ mph and $t = 9$ h.

21. Find r if $d = 15$ miles and $t = 5$ h.

22. Find t if $d = 144$ miles and $r = 48$ mph.

Find the perimeter of each rectangle. (Lesson 3-6)

23.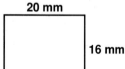
20 mm
16 mm

24.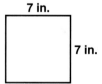
7 in.
7 in.

25.
15 ft
25 ft

Find the area of each rectangle. (Lesson 3-6)

26.
48 m
20 m

27.
100 yd
75 yd

28.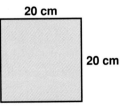
20 cm
20 cm

Solve each inequality. (Lessons 3-7, 3-8)

29. $a - 4 > -2$

30. $-18 \leq x - 10$

31. $56 > m + 16$

32. $b + 33 < -2$

33. $-5c > 30$

34. $\frac{x}{-11} \geq 8$

35. $-14 < \frac{d}{-6}$

36. $3y \leq -123$

37. $\frac{-k}{4} \leq -21$

Applications and Logic

Determine whether each compound statement is *true* or *false*. (Lesson 3-9)

38. A heptagon has 6 sides or $2 + 3 = 5$.

39. A square has four equal sides and $10 - 4 > 6$.

40. $7 > 7$ and $6 > 4$

41. $0 > -4$ or $2 \leq -2$

Define a variable, write an equation or an inequality, and then solve. (Lesson 3-10)

42. Alberto buys 2 dozen eggs. After making omelets he has 9 eggs left. How many eggs did he use in the omelets?

43. Sally earns $5 for mowing the lawn and $4 for weeding the garden. How many times must she mow the lawn to earn $60?

44. A number increased by 13 is at least -39. What is the number?

45. Five friends share the cost of a lunch equally. Each person pays $4. What is the total of the lunch?

ortfolio Suggestion

Select an item from your work in this chapter that shows your creativity and place it in your portfolio.

Curriculum Connection

• **Health** Research the chemical formulas for vitamins. Include a description of how the body uses each vitamin and natural sources of each vitamin.

• **Science** Write a brief paragraph about Isaac Newton's contributions to science.

Read More About It

Bitter, Gary G. *Exploring with Pocket Calculators.*

Burns, Marilyn. *Math for Smarty Pants.*

Ross, Frank X. *The Metric System — Measures for all Mankind.*

Test

Solve each equation. Check your solution.

1. $11 + m = 19$

2. $-15 + z = 3$

3. $8 = 15 + b$

4. $x + 32 = 58$

5. $-9 = -12 + h$

6. $95 = p + (-38)$

7. $k - 5 = -8$

8. $a - 6 = 5$

9. $r - (-4) = -19$

10. $34 = w - 55$

11. $-13 = z - 21$

12. $94 = t - (-76)$

13. $-8y = 72$

14. $84x = 252$

15. $-12y = -84$

16. $-180 = -12d$

17. $165 = -15a$

18. $9r = -144$

19. $\frac{w}{15} = 10$

20. $\frac{n}{30} = -6$

21. $-14 = \frac{r}{-18}$

Solve. Use the formula $d = rt$.

22. Find r if $t = 3$ hours and $d = 114$ miles.

23. Find t if $d = 177$ kilometers and $r = 59$ kilometers/hour.

Find the area and perimeter of each rectangle described below.

24. length, 8 ft; width, 9 ft

25. length, 30 in.; width, 17 in.

Solve each inequality. Check your solution.

26. $11 \leq r + 28$

27. $z - 25 > -30$

28. $-7s \leq 63$

29. $7 > \frac{x}{-8}$

Determine whether each compound statement is *true* or *false*.

30. $6 > 0$ and $-6 < 0$

31. $2 = 0$ or $-2 > 3$

Define a variable, write an equation or an inequality, then solve.

32. The YMCA had registration for a summer soccer league. The number of players was divided by 12 to form 11 teams. How many players signed up?

33. Allie and her father spent at least $110 while shopping. Allie spent $47. How much did her father spend?

BONUS

The length of a rectangular picture frame is 3 inches longer than twice its width. The perimeter is 78 inches. Make a labeled drawing of the picture frame and find its dimensions.

Academic Skills Test

Cumulative, Chapters 1-3

1. If $m = 12$ and $p = 10$, what is the value of $m + m \cdot p$?

   A 120   **C** 240
   B 132   **D** 1440

2. Which equation is equivalent to $x + 5 = -4$?

   A $x = -4$
   B $x + 5 - 5 = -4 - 5$
   C $x + 10 = -8$
   D $x + 5 - 5 = -4 + 5$

3. At dawn the temperature was -5°C. By noon, the temperature had risen 8°. What was the temperature at noon?

   A 13°C
   B 3°C
   C -3°C
   D -13°C

4. Which equation is equivalent to $\frac{c}{5} = 2.5$?

   A $\frac{c}{5} = \frac{2.5}{5}$ **C** $\frac{c}{5} \cdot 5 = 2.5 \cdot 5$

   B $\frac{c}{5} = 2.5 \cdot 5$ **D** $\frac{c}{5} = 2.5 \div 5$

5. The distance traveled is given by the formula $d = rt$, where r is the rate of speed and t is the time spent traveling. How long would it take to ride a bike 9 miles at a rate of 6 miles per hour?

   A 1.5 hours
   B 15 minutes
   C 54 hours
   D 54 miles

6. If $15 = \frac{c}{6}$, what is the value of c?

   A 90   **C** 9
   B 21   **D** $2\frac{1}{2}$

7. If $8g = (-32)$, what is the value of g?

   A -256   **C** -24
   B -40   **D** -4

8. Georgia wants to lose 2 pounds a week for 6 weeks. What will the change in her weight be?

   A -12 lb
   B -2 lb
   C +8 lb
   D +12 lb

9. An airplane descended 250 feet in 5 minutes. What was its average rate of descent?

   A 50 ft/min
   B 250 ft/min
   C 1250 ft/min
   D Not here

10. Brent knows that to have an average of more than 85 in Biology, his two exam scores must total more than 170. If his first grade was 82, which inequality could he use to find g, his second grade needed for an average greater than 85?

    A $82 + g > 170$

    B $82 > 170 + g$

    C $g + 82 < 170$

    D $\frac{82 + g}{2} > 170$

CHAPTER 4

CHAPTER OBJECTIVES

In this chapter you will learn to:

- find factors of numbers and monomials
- use prime factorization to find the GCF and LCM
- use the LCM to write equivalent fractions
- multiply and divide powers
- solve problems by drawing a diagram

Factors and Fractions

You probably think of music as being sounds or notes. But did you know that music can also be represented mathematically?

Take two notes commonly played together: C and E. We find these two notes harmonious. That's because each note vibrates at a frequency that harmonizes with the other. Even if you can't tell harmony from noise, you can figure whether the frequencies can be expressed as a simple fraction made up of small whole numbers or not. If the fraction can be simplified into something like $\frac{8}{5}$ or $\frac{2}{3}$, you have harmony. If you end up with a fraction like $\frac{247}{264}$, then you have noise.

From the chart below you can see that C is to E as 264 is to 330. In simplest form $\frac{264}{330}$ is equal to $\frac{4}{5}$.

Can you find other harmonious combinations of notes?

Music Connection

Class Project

A famous sequence of numbers is the Fibonacci sequence, 1, 1, 2, 3, 5, 8, ... The sequence begins with 1 and each number that follows is the sum of the previous two numbers. Find examples of Fibonacci numbers in music.

Frequency (hertz)							
264	294	330	349	392	440	494	528
C	D	E	F	G	A	B	C

4-1 Factors and Monomials

Objective:
Determine whether one number is a factor of another.

Key Terms:
factor
monomial
divisible

The Beechcroft Junior High School Marching Band has 78 members. Can they be arranged in six equal rows to march in the Holiday Parade?

This is another way of asking if 6 is a **factor** of 78. You can divide to find factors. The factors of a whole number divide that number with a remainder of 0.

What are the other factors of 78?

Example

1 **Determine whether 6 is a factor of 78.**

$$78 \boxed{\div} 6 \boxed{=} 13$$

Since the quotient is a whole number, 6 is a factor of 78. You can also say that 78 is **divisible** by 6. Therefore, the band can march in 6 equal rows.

Sometimes you can test for divisibility mentally. The following rules will help you determine whether a number is divisible by 2, 3, 5, 6, or 10.

> A number is divisible by:
> - 2 if the ones digit is divisible by 2.
> - 3 if the sum of its digits is divisible by 3.
> - 5 if the ones digit is 0 or 5.
> - 6 if the number is divisible by 2 *and* 3.
> - 10 if the ones digit is 0.

Example

What are the factors of 135?

2 **Determine whether 135 is divisible by 2, 3, 5, 6, or 10.**

- The ones digit is *not* divisible by 2, so 135 is *not* divisible by 2.
- The sum of the digits, $1 + 3 + 5 = 9$, is divisible by 3, so 135 is divisible by 3.
- The ones digit is 5, so 135 is divisible by 5.
- The number is divisible by 3 but *not* by 2, so 135 is *not* divisible by 6.
- The ones digit is *not* 0, so 135 is *not* divisible by 10.

Therefore, 135 is divisible by 3 and 5, but not 2, 6, or 10.

CONNECTION TO ALGEBRA

You know that 6 and 13 are factors of 78 because $6 \cdot 13 = 78$. Consider the algebraic expression $3x$. Since this notation means $3 \cdot x$, it follows that 3 and x are factors of $3x$.

What makes 2x + 3 and m − 6 different from the monomials listed?

An expression such as $3x$ is called a **monomial.** A monomial is an integer, a variable, or a product of integers or variables. Other examples of monomials are y, 15, ab, and $4mk$. Expressions like $2x + 3$ and $m - 6$ are not monomials.

Examples

Determine whether each expression is a monomial.

3 $2(\ell + w)$

This expression is not a monomial because it involves addition.

4 $-12abc$

This expression is a monomial because it is the product of integers and variables.

Checking for Understanding

Communicating Algebra

1. In your own words, explain why 7 is a factor of 63.
2. Is 8 a factor of 68? Explain why or why not.
3. Discuss why 1 is a factor of every number.
4. Can 0 be a factor of any number?

Guided Practice

Determine whether each expression is a monomial. Explain why or why not.

5. $5m$
6. $2x + 1$
7. $-6xy$
8. -175

Using divisibility rules, state whether each number is divisible by 2, 3, 5, 6, or 10.

9. 38
10. 117
11. 576
12. 1630

Exercises

Independent Practice

Use divisibility rules to determine if the first number is divisible by the second number.

13. 1075; 5
14. 999; 3
15. 285; 6
16. 705; 10
17. 117; 3
18. 11,112; 2
19. 3241; 3
20. 1002; 6

Use mental math skills, paper and pencil, or a calculator to find a number that satisfies the given conditions.

21. a four-digit number divisible by 3
22. a five-digit number divisible by both 2 and 5
23. a three-digit number *not* divisible by 6
24. a three-digit number divisible by 2 but *not* 4
25. a three-digit number divisible by 2, 3, and 5.
26. a three-digit number *not* divisible by 2, 3, or 5

Logical Thinking

27. What number is a factor of every nonzero number?
28. Is every nonzero number a factor of itself?
29. Name two factors for every nonzero number.

Mixed Review

30. Evaluate the expression $k + 14$ if $k = 8$. (Lesson 1-2)
31. If $n = -5$, then $-n = $ ■ . (Lesson 2-5)
32. Solve the equation $x = (6)(-4)(-3)$. (Lesson 2-7)
33. *True* or *false:* If $y + c = d$, then $y = d - c$. (Lesson 3-1)
34. Solve the equation $63 = -9y$. (Lesson 3-3)
35. **Physics**   Lorena rides her bike 21 miles in 3 hours. At what rate of speed does she travel? (Lesson 3-5)
36. Write an inequality for the sentence: *The temperature is greater than sixty degrees.* (Lesson 3-7)

Applications

37. **Measurement**   A leap year occurs when the year is divisible by 4. A century year is a leap year if it is divisible by 400. Name the last and the next century leap years.

38. **Arts and Crafts**   A rectangular quilt has 24 squares. Draw all the shapes in which the quilt can be arranged.

Critical Thinking

39. What is the greatest 3-digit number that is *not* divisible by 2, 3, or 5?

Wrap-Up

40. If a number is divisible by 2 and 10, is it always divisible by 20? If not, give a counterexample.

Enrichment

Perfect Numbers

A **perfect number** is equal to the sum of its factors, except itself. There are two perfect numbers less than 30. The third perfect number is 496.

$$496 = 1 + 2 + 4 + 8 + 16 + 31 + 62 + 124 + 248$$

1. Find a perfect number between 1 and 10.
2. Find a perfect number between 20 and 30.
3. **Research**   Find other perfect numbers.

4-2 Powers and Exponents

Objective:
Use powers and exponents in expressions and equations.

Key Terms:
standard form
base
exponent
power

There are over 10,000 private airports in the United States. The number 10,000 is in **standard form.** You can also express 10,000 as the product $10 \cdot 10 \cdot 10 \cdot 10$.

A shorter way to express $10 \cdot 10 \cdot 10 \cdot 10$ is by using exponents. An **exponent** tells how many times a number, called the **base,** is used as a factor.

The expression $10 \cdot 10 \cdot 10 \cdot 10$ can be written 10^4.

$$base \rightarrow \mathbf{10^4} \leftarrow exponent$$

The number 10 is the **base.**
The number 4 is the **exponent.**

Numbers that are expressed using exponents are called **powers.** The *powers* 5^2, 9^3, and 8^4 are read as follows.

5^2 five to the *second* power or five *squared*

9^3 nine to the *third* power or nine *cubed*

8^4 eight to the *fourth* power

Using exponents, how would you write x to the fifth power?

The value of any number to the first power is the number. For example, $6^1 = 6$

Examples

Write each power as a product of the same factor.

1 2^4

The base is 2. The exponent 4 means 2 is a factor 4 times.

$2^4 = 2 \cdot 2 \cdot 2 \cdot 2$

2 b^3

The base is b. The exponent 3 means b is a factor 3 times.

$b^3 = b \cdot b \cdot b$

Write each product using exponents.

3 $6 \cdot 6$

The base is 6. Because 6 is a factor 2 times, the exponent is 2.

$6 \cdot 6 = 6^2$

4 $x \cdot x \cdot x \cdot x \cdot x$

The base is x. Because x is a factor 5 times, the exponent is 5.

$x \cdot x \cdot x \cdot x \cdot x = x^5$

Calculator Hint

Many calculators have a key labeled y^x. This key allows you to compute exponents very quickly. Suppose you want to find 16^5.

16 $\boxed{y^x}$ 5 $\boxed{=}$

1048576

The result appears immediately.

Since powers are a short form of repeated multiplication, they need to be included in the rules for order of operations.

Order of Operations	1. Do all operations within grouping symbols first; start with the innermost grouping symbols. 2. *Evaluate all powers from left to right.* 3. Next do all multiplications and divisions from left to right. 4. Then do all additions and subtractions from left to right.

Examples

Evaluate each expression.

5 n^2 if $n = 5$

$n^2 = 5^2$ Replace n with 5.

$= 5 \cdot 5$

$= 25$

6 $8r^3$ if $r = 3$

$8r^3 = 8 \cdot 3^3$ Replace r with 3.

$= 8 \cdot 27$ Find the power

$= 216$ first.

Checking for Understanding

Communicating Algebra

1. To evaluate the expression five to the seventh power, __?__ would be used as a factor __?__ times.

2. In the power 3^5, the base is __?__ and the exponent is __?__ .

3. What is another way to say *four to the second power?*

4. What is another way to say *x to the third power?*

5. Explain how you can find 1^{405} very quickly.

Guided Practice

Write each power as the product of the same factor.

6. 2^4 7. m^2 8. 5^3 9. a^5 10. 6^4

Write each product using exponents.

11. $2 \cdot 2 \cdot 2$ 12. $8 \cdot 8$ 13. $1 \cdot 1 \cdot 1 \cdot 1$ 14. $9 \cdot 9$ 15. $6 \cdot 6 \cdot 6$

Exercises

Independent Practice

Write each product using exponents.

16. $4 \cdot 4 \cdot 4 \cdot 4$ 17. $t \cdot t \cdot t \cdot t \cdot t$ 18. $(p \cdot p)(p \cdot p)$ 19. $a(a \cdot a)$

 Use a calculator to determine whether each sentence is *true* or *false*.

20. $2^{10} > 10^2$ 21. $9^8 > 8^9$ 22. $2^4 = 4^2$ 23. $2^3 \neq 3^2$

Evaluate each expression.

24. x^3 if $x = 1$ 25. m^2 if $m = 9$ 26. r^4 if $r = 3$

27. $3y^4$ if $y = 2$ 28. $4r^3$ if $r = 3$ 29. $2m^3$ if $m = 5$

Language Skills

Evaluate.

30. nineteen cubed   31. eight squared   32. three to the sixth power

33. eleven squared   34. sixteen cubed   35. five to the seventh power

36. the sum of four squared and six cubed

37. seven cubed divided by seven squared

38. Solve the equation $4 - r = -15$.   (Lesson 3-2)

39. Write the inequalities $b < 1.6$ and $b \geq 0.3$ as a compound inequality. (Lesson 3-9)

40. *True* or *false:* Every number is a factor of itself.   (Lesson 4-1)

41. Is the expression $-17cd$ a monomial?   (Lesson 4-1)

42. Use divisibility rules to determine if 3 is a factor of 2345.   (Lesson 4-1)

Challenge

Evaluate.

43. $a^2 - b^2$ if $a = 3$ and $b = 2$

44. $(a + b)^2 - 2b$ if $a = 3$ and $b = 2$

45. $a^2 - b^2$ if $a = 2$ and $b = 4$

46. $(a^2 - b)^2$ if $a = 2$ and $b = 4$

Connection

47. Geometry   To find the measure of the volume of a cube, you find the product of its length, width, and height. Use exponents to write an expression that represents the volume of a cube that measures s units on each side.

Application

48. Computer Science   Computers use the binary, or base two, number system. Each place-value position is a power of two. For example, $101011_2 = (1 \times 2^5) + (0 \times 2^4) + (1 \times 2^3) + (0 \times 2^2) + (1 \times 2^1) + (1 \times 1) = 43$. Find the value of 11010_2.

Critical Thinking

49. Each whole number from 1 to 15 can be expressed using the numerals 1, 2, 3, and 4 exactly once and any of the operation signs. For example, $1 = (1 + 4) \div (2 + 3)$. Write an expression for each of the others.

Wrap-Up

50. Explain why $2^3 \neq 3^2$.

Reading Algebra

Powers

Remember that in a power the exponent indicates the number of times the base is used as a factor. Suppose you are to write each of the following in symbols.

Words	Symbols
three times x squared	$3x^2$
three times x, the quantity squared	$(3x)^2$

Suppose $x = 2$. Then
$3x^2 = 3 \cdot 2^2$ or 12, and
$(3x)^2 = (3 \cdot 2)^2$ or 36.
The values are not the same.

In the second expression, the parentheses are used to show that the expression $3x$ is used as a factor twice.

$$(3x)^2 = (3x)(3x)$$

The phrase *the quantity* is used to indicate parentheses when reading expressions.

State how to read each expression.

1. 3^3 **2.** $(12r)^5$ **3.** $(x + 2y)^2$ **4.** $4m^2n^4$ **5.** $a - b^3$ **6.** $(2a)^4 + b$

Decide if the expressions are equivalent. Write *yes* or *no*.

7. $12 \cdot x^3$ and $12 \cdot x \cdot x \cdot x$ **8.** $3xy^5$ and $3(xy)^5$ **9.** $(2a)^3$ and $8a^3$

10. $(ab)^2$ and $a^2 \cdot b^2$ **11.** xy^3 and x^3y^3 **12.** $4(x^3)^2$ and $4x^6$

4-3　Strategy: Draw a Diagram

Objective:
Solve problems by
drawing diagrams.

Seth is planning a chess tournament for his
class. Eight students will participate in a
single elimination format; that is, only the
winners continue to play. Seth needs to
know how many games will be played so he
can contact the teachers that are to be at
each game. How many games will be played
during this tournament?

Explore　　There are 8 students in a single elimination tournament.
The problem asks how many games will be played.

Plan　　In planning the tournament, Seth makes a diagram to
show how the winners advance through the tournament.
Such a diagram will help to count the number of games.

Solve

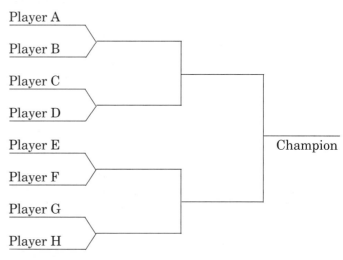

First Round　**Second Round**　**Finals**

Player A
Player B
Player C
Player D
Player E
Player F
Player G
Player H

Champion

Notice that 4 games are played in the first round, 2
games are played in the second round, and 1 final game
is played for the championship.

$$4 + 2 + 1 = 7$$

There will be 7 games played in the tournament.

Examine　　Every player loses exactly once except the champion.
Therefore, $8 - 1$ or 7 games are played. The answer
is correct.

Diagrams are useful tools in solving some problems. These diagrams do
not need to be elaborate, but can be very simple illustrations.

Checking for Understanding

Communicating Algebra

1. Why is drawing a diagram a good strategy for solving problems?

2. Is drawing a diagram a good strategy for solving all problems? Why or why not?

Guided Practice

3. Draw a diagram to determine how many games have to be played to determine a champion from 32 teams in a single elimination softball tournament.

4. Look at the staircase at the right. How many cubes would be needed for a staircase with seven steps?

Exercises

Independent Practice

Solve. Use any strategy.

5. Sam made a list of his mother's parents, grandparents, great grandparents, and great-great grandparents. If Sam did not list any stepgrandparents, how many people are listed?

6. Sybil's score on her math test is 89. Some of the questions are worth 5 points and the rest of the questions are worth 2 points. The teacher gave no partial credit for any question. If 37 problems are correct, how many 5-point questions did Sybil answer correctly?

7. A volleyball team has 6 members. Suppose each member shakes hands with every other member. How many handshakes take place?

8. During the first minute of a game of tug-of-war, team A pulled team B forward 2 feet. During the second minute, team B pulled team A forward 1 foot. Then, during the third minute, team A pulled team B forward 2 feet. If this pattern continues, how long will it take team A to pull team B forward a total of 10 feet to win the game?

9. Replace each ● with an operation symbol to make the equation true. Add parentheses if necessary.

$$4 ● 3 ● 6 ● 3 = 45$$

10. A ball is dropped from 10 meters above the ground. It hits the ground and bounces up $\frac{1}{2}$ as high as it fell. If each bounce is $\frac{1}{2}$ the height of the last bounce, how far has the ball traveled down on the last bounce when it hits the ground the fifth time?

Critical Thinking

11. Willis is cutting a round pizza. What is the greatest number of pieces he can get from 1 pizza with 4 straight cuts?

Wrap-Up

12. **Make Up a Problem** Write a problem where a diagram would be helpful in solving the problem.

4-4 Prime Factorization

Objectives:
Identify prime and composite numbers. Write the prime factorization of a composite number.

Key Terms:
prime
composite
factor tree

How are divisibility rules helpful in finding prime factors?

The Food Basket has 30 boxes of a new flavored pretzel to put on display. The display is being set up by two stock clerks.

$30 = 2 \cdot 15$ Each of the 2 clerks sets up 15 boxes. $30 \div 2 = 15$

$30 = 2 \cdot (3 \cdot 5)$ Suppose each clerk is responsible for setting up 3 rows with 5 boxes per row. $15 \div 3 = 5$

The work cannot be divided into any smaller whole number parts.

Notice that the only factors of 2 are 1 and 2. Also notice that the only factors of 3 and 5 are 1 and the original number. A whole number *greater than one* that has *exactly* two factors, 1 and itself, is called a **prime number.**

Any whole number, besides 0 and 1, that is not prime can be written as a product of prime numbers. A whole number *greater than one* that has more than two factors is called a **composite number.**

The numbers 1 and 0 are *neither* prime *nor* composite. Zero has an endless number of factors. The number 1 has only one factor, itself. However, every number greater than one is either prime or composite.

Sometimes a number can be factored in several ways. The diagrams below show how three different students found all the prime factors of 36. These diagrams are called **factor trees.**

$$36 = \underset{\wedge}{4} \cdot \underset{\wedge}{9}$$
$$= 2 \cdot 2 \cdot 3 \cdot 3$$

$$36 = 2 \cdot \underset{\wedge}{18}$$
$$= 2 \cdot 2 \cdot 9$$
$$= 2 \cdot 2 \cdot 3 \cdot 3$$

$$36 = 3 \cdot \underset{\wedge}{12}$$
$$= 3 \cdot 2 \cdot 6$$
$$= 3 \cdot 2 \cdot 2 \cdot 3$$

Although the factors are in different order, the result is the same. The order in which you factor does not matter. The factoring process stops when all factors are prime. When a number is expressed as a product of factors that are all prime, the expression is called the **prime factorization** of the number.

A calculator can also be used to find the prime factorization of a number.

Example

1 **Find the prime factorization of 180.** Divide by prime factors until the quotient is prime.

$$180 = 2 \cdot 2 \cdot 3 \cdot 3 \cdot 5$$

Sometimes you may need to factor a negative integer. Any negative integer may be written as the product of -1 and a whole number.

Example

2 **Factor -273 completely.**

$$-273 = -1 \cdot 273$$
$$= -1 \cdot 3 \cdot 91$$
$$= -1 \cdot 3 \cdot 7 \cdot 13$$

Now find the prime factorization of 273. Will 2 divide 273 evenly? No, try 3. 91 is not divisible by 2, 3, or 5. Try 7, the next greater prime.

$$-273 = -1 \cdot 3 \cdot 7 \cdot 13$$ 13 is a prime number.

CONNECTION TO ALGEBRA

A monomial can also be written in factored form as a product of prime numbers, -1, and variables with no exponent greater than 1.

Example

3 **Factor $18xy^2$.**

$$18xy^2 = 2 \cdot 9 \cdot x \cdot y^2$$
$$= 2 \cdot 3 \cdot 3 \cdot x \cdot y \cdot y$$

Checking for Understanding

Communicating Algebra

1. In your own words, explain the difference between prime and composite numbers.

2. Draw a factor tree to find the prime factorization of 144.

Guided Practice

Determine whether each number is *prime* or *composite*.

3. 33   4. 57   5. 13   6. 19

Find the least prime number that is a factor of each number.

7. 36   8. 55   9. 77   10. 88

Find the prime factorization of each number.

11. 63   12. 51   13. 98   14. 110

Exercises

Factor each number or monomial completely.

15. 30   16. 45   17. 49   18. 81

ESTIMATION
MENTAL MATH
CALCULATOR
PAPER/PENCIL

19. 68   20. 48   21. -26   22. -63

23. 80   24. 128   25. -112   26. -95

27. $28x^2y$ 28. $400a^2b^3$ 29. $-72ab^3$ 30. $560x^4y^2$

31. $42xy^2$ 32. $75m^2k$ 33. $210ab^3$ 34. $-65r^2s$

Mixed Review

35. Order the numbers in the set {3, -6, 0, -2} from least to greatest. (Lesson 2-2)

36. Solve the equation $\frac{b}{-11} = -6$.   (Lesson 3-4)

37. Write the expression *eight cubed* using exponents.   (Lesson 4-2)

38. Evaluate the expression $(x - y^2)^2$ if $x = 11$ and $y = 3$.   (Lesson 4-2)

Challenge

39. Primes that differ by 2 are called *twin primes*. One such pair is 3 and 5. Find all pairs of twin primes less than 100. What can you say about the numbers between twin primes?

Connection

40. **Geometry**   Mr. Rathburn's back yard measures 10 yards wide by 10 yards long. He wants to construct a rectangular garden that has an area of 90 square feet. What are the whole number dimensions that are possible for the garden?

Critical Thinking

41. A number is divisible by 2 if its last digit is divisible by 2, and a number is divisible by 4 if the number formed by its last two digits is divisible by 4. Continue the pattern and determine the rule for divisibility by 8.

Wrap-Up   *Journal Entry*   42. In your own words, write an explanation of how to use a calculator to find the prime factorization of a number.

History

The Mayan Number System

The Mayan civilization of Central America was highly advanced in astronomy, the calendar, architecture, and commerce. It was at its height from A.D. 300 to A.D. 900, and mathematics was an important part of its civilization.

One notable accomplishment was the development of a base-twenty number system with a special symbol for zero. The symbol for zero, ⬭ resembles a shell, but more probably it is a front view of a closed fist.

The Mayans also developed an amazingly accurate calendar. The Mayan calendar consisted of eighteen months of 20 days each, plus an additional "month" of 5 days, for a total of 365 days. In fact, the Mayans modified their base-twenty system to correspond to their calendar. Instead of the third position being 20^2, or 400, it was 18×20 or 360.

Factor Patterns

Materials: computer

You already know that prime numbers have exactly two factors. What numbers have exactly three factors? or four factors? In this Exploration, you will investigate factor patterns.

The numbers 2 through 18 and their factors have been placed in the chart according to the number of factors.

Exactly 2 Factors	Exactly 3 Factors	Exactly 4 Factors	Exactly 5 Factors	Exactly 6 Factors
2: 1, 2 3: 1, 3 5: 1, 5 7: 1, 7 11: 1, 11 13: 1, 13 17: 1, 17	4: 1, 2, 4 9: 1, 3, 9	6: 1, 2, 3, 6 8: 1, 2, 4, 8 10: 1, 2, 5, 10 14: 1, 2, 7, 14 15: 1, 3, 5, 15	16: 1, 2, 4, 8, 16	12: 1, 2, 3, 4, 6, 12 18: 1, 2, 3, 6, 9, 18

Your Turn: **Copy and complete the chart for the numbers 19 through 30.**

Work at a computer.

1. The BASIC program at the right will print the factors of a given number. Using the program, find the factors of the numbers 31 through 50. Then place these numbers in the correct column of the chart.

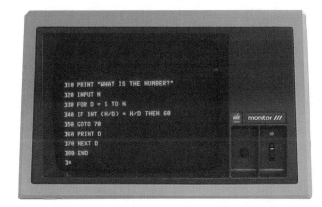

```
310 PRINT "WHAT IS THE NUMBER?"
320 INPUT N
330 FOR D = 1 TO N
340 IF INT (N/D) = N/D THEN 60
350 GOTO 70
360 PRINT D
370 NEXT D
380 END
]▪
```

Analysis

2. Predict a number from 50 through 150 that can be placed in each column. Check your prediction with the computer.

3. Write a paragraph that describes the pattern in each column.

4-5 Greatest Common Factor (GCF)

Objective:
Find the greatest common factor for two or more integers or monomials.

Key Terms:
greatest common factor
GCF

How are divisibility rules helpful in finding the GCF?

For his project in home economics, Adita wants to create a giant American flag. The widest white material he could find was 78 inches wide, while the widest red material he could find was 91 inches wide. The stripes are to be the same width. If he does not want to waste any material, what is the widest the stripes on his flag can be?

factors of 78: 1, 2, 3, 6, 13, 26, 39, 78

factors of 91: 1, 7, 13, 91

The *common factors* of 78 and 91, shown in blue, are 1 and 13. The greatest of the factors common to two or more numbers is called the **greatest common factor (GCF)** of each number. The greatest common factor of 78 and 91 is 13.

Each stripe should be 13 inches wide. Adita cuts the 78-inch material into 6 pieces of 13 inches each, and he cuts the 91-inch material into 7 pieces of 13 inches each.

Prime factorization can be used to find the GCF of numbers as follows. Find the prime factorization of each number. Then find the product of their common factors.

Examples

1 **Use prime factorization to find the GCF of 56 and 84.**

First, find the prime factorization of each number.

$$56 = 2 \cdot 28 \qquad\qquad 84 = 2 \cdot 42$$
$$= 2 \cdot 2 \cdot 14 \qquad\qquad = 2 \cdot 2 \cdot 21$$
$$= 2 \cdot 2 \cdot 2 \cdot 7 \qquad\qquad = 2 \cdot 2 \cdot 3 \cdot 7$$

Then find the common factors.

$$56 = \boxed{2} \cdot \boxed{2} \cdot 2 \cdot \boxed{7}$$
$$84 = \boxed{2} \cdot \boxed{2} \cdot 3 \cdot \boxed{7}$$

The loops indicate each common factor. They are 2, 2, and 7.

The greatest common factor of 56 and 84 is $2 \cdot 2 \cdot 7$ or 28.

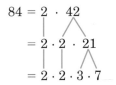

FYI

The largest American flag ever made was first displayed in Evansville, Indiana, on March 22, 1980. It weighs 7.7 tons and is 411 feet long by 210 feet wide.

2 Use prime factorization to find the GCF of 60, 20, and 180.

Write each number as a product of prime factors.

$$60 = 2 \cdot 2 \cdot 3 \cdot 5$$
$$20 = 2 \cdot 2 \cdot 5$$
$$180 = 2 \cdot 2 \cdot 3 \cdot 3 \cdot 5$$

The common factors are 2, 2, and 5.

Thus, the GCF is $2 \cdot 2 \cdot 5$ or 20.

CONNECTION TO ALGEBRA

The GCF of two or more monomials is also the product of their common factors.

Examples

3 Find the GCF of $24x^2$ and $18x$.

$$24x^2 = 2 \cdot 2 \cdot 2 \cdot 3 \cdot x \cdot x$$
$$18x = 2 \cdot 3 \cdot 3 \cdot x$$

The common factors are 2, 3, and x.

The GCF of $24x^2$ and $18x$ is $2 \cdot 3 \cdot x$ or $6x$.

4 Find the GCF of $20m^2k$, $30mk^2$, and $45mk$.

How do you know just by looking that a number will have 5 as a factor?

$$20m^2k = 2 \cdot 2 \cdot 5 \cdot m \cdot m \cdot k$$
$$30mk^2 = 2 \cdot 3 \cdot 5 \cdot m \cdot k \cdot k$$
$$45mk = 3 \cdot 3 \cdot 5 \cdot m \cdot k$$

The common factors are 5, m, and k.

The GCF of $20m^2k$, $30mk^2$, and $45mk$ is $5mk$.

Checking for Understanding

Communicating Algebra

1. In your own words, explain how to find the GCF of two or more numbers.

2. Use factor trees to find the prime factorization of 135 and 315. Then circle the common factors.

3. Name two numbers whose GCF is 12.

Guided Practice

Find the GCF of each pair of numbers or monomials.

4. $6 = 2 \cdot 3$
   $8 = 2 \cdot 2 \cdot 2$

5. $20 = 2 \cdot 2 \cdot 5$
   $12 = 2 \cdot 2 \cdot 3$

6. $135 = 3 \cdot 3 \cdot 3 \cdot 5$
   $315 = 3 \cdot 3 \cdot 5 \cdot 7$

7. $20 = 2 \cdot 2 \cdot 5$
   $30 = 2 \cdot 3 \cdot 5$

8. $15r^2s = 3 \cdot 5 \cdot r \cdot r \cdot s$
   $27rs^2 = 3 \cdot 3 \cdot 3 \cdot r \cdot s \cdot s$

9. $24a^2b = 2 \cdot 2 \cdot 2 \cdot 3 \cdot a \cdot a \cdot b$
   $36ab^2 = 2 \cdot 2 \cdot 3 \cdot 3 \cdot a \cdot b \cdot b$

10. $3, 12$

11. $18ab, 27a$

12. $12x, 40x^2$

13. $-5, -7$

Exercises

Find the GCF for each set of numbers or monomials.

14. 40, 24 **15.** 24, 36 **16.** 93, 69 **17.** 42, 56

18. 56, 16 **19.** 48, 84 **20.** 6, 8, 12 **21.** 10, 15, 25

22. 108, 144 **23.** 72, 216 **24.** 12, 18, 30 **25.** 9, 15, 24

26. $7m^2, 35m$ **27.** $14k, 42k^2$ **28.** $33y^2, 44y$

29. $-26m^2, 52m^3$ **30.** $-25ab, 35b$ **31.** $-18, 45xy$

32. $24x^2, -60x$ **33.** $16y, 40y^2$ **34.** $14m, -56m^3$

35. $18a, 30ab, 42b$ **36.** $15x^2, 35y^2, 70xy$ **37.** $18mk, 6m^2, 42m^2k$

Mixed Review

38. Simplify the expression $-18xy + 12xy + (-13xy)$. (Lesson 2-4)

39. Find the perimeter of a rectangle whose sides are 12 cm and 3 cm. (Lesson 3-6)

40. Solve the inequality $-26 < \frac{m}{-3}$. (Lesson 3-8)

41. Use divisibility rules to determine if 231 is divisible by 6. (Lesson 4-1)

42. Evaluate the expression $4k^3$ if $k = 3$. (Lesson 4-2)

Applications

43. Engineering A bicycle path being planned along State Route 80 will run 42 miles between the first two cities. The path will continue on to a third city 70 miles farther away. The planners wish to make rest-stops at each city and at equal distances along the entire course. If the distances between stops is as great as possible, how far apart should the rest-stops be?

44. Carpentry The industrial technology class is building a storage shelf. The class wants to make the best use of a piece of plywood that measures 48 in. by 72 in. How many shelves measuring 12 in. by 16 in. could be cut from the plywood with no waste?

Connection

45. Algebra Alberto is 14 years old. The GCF of his age and his younger sister's age is 7. How old is his sister?

Critical Thinking

46. Numbers that have a GCF of 1 are said to be **relatively prime.** Find the least two composite numbers that are relatively prime.

Wrap-Up

47. Write three numbers that have a GCF of 8.

4-6 Simplifying Fractions

Objective:
Simplify fractions by using the GCF.

Key Terms:
ratio
simplest form
algebraic fraction

A pizza was cut into 4 equal pieces. There are three pieces left. The ratio of pieces left to total pieces is 3 to 4.

A **ratio** is a comparison of two numbers by division. The ratio above can be expressed in the following ways.

$$3 \text{ to } 4 \qquad 3{:}4 \qquad \frac{3}{4} \qquad 3 \div 4$$

A ratio is most commonly expressed as a fraction in **simplest form.** A fraction is in simplest form when the GCF of the numerator and denominator is 1.

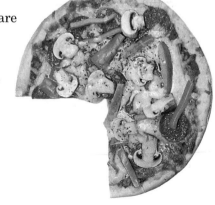

Example

1 Write $\frac{24}{40}$ in simplest form.

$$24 = \boxed{2} \cdot \boxed{2} \cdot \boxed{2} \cdot 3$$
$$40 = \boxed{2} \cdot \boxed{2} \cdot \boxed{2} \cdot 5$$

The GCF of 24 and 40 is $2 \cdot 2 \cdot 2$ or 8.

$$\frac{24}{40} \overset{\div 8}{\underset{\div 8}{=}} \frac{3}{5}$$

Since the GCF of 3 and 5 is 1, the fraction $\frac{3}{5}$ is in simplest form.

The division in the example above can be represented in another way.

$$\frac{24}{40} = \frac{\cancel{2} \cdot \cancel{2} \cdot \cancel{2} \cdot 3}{\cancel{2} \cdot \cancel{2} \cdot \cancel{2} \cdot 5} = \frac{3}{5}$$

The slashes indicate that the numerator and denominator are divided by $2 \cdot 2 \cdot 2$, the GCF.

Example

2 Write $\frac{20}{36}$ in simplest form.

$$\frac{20}{36} = \frac{5 \cdot \cancel{2} \cdot \cancel{2}}{3 \cdot 3 \cdot \cancel{2} \cdot \cancel{2}}$$

Divide both the numerator and the denominator by $2 \cdot 2$, the GCF.

$$= \frac{5}{9}$$

CONNECTION TO ALGEBRA

Fractions with variables in the numerator or denominator are called **algebraic fractions.** Algebraic fractions can also be written in simplest form.

Example

3 Simplify $\frac{8a^2b}{12ab}$. ——— Both a and b must not be zero.

What is the GCF of $8a^2b$ and $12ab$?

$$\frac{8a^2b}{12ab} = \frac{\cancel{2} \cdot \cancel{2} \cdot 2 \cdot \cancel{a} \cdot a \cdot \cancel{b}}{\cancel{2} \cdot \cancel{2} \cdot 3 \cdot \cancel{a} \cdot \cancel{b}}$$
Divide both the numerator and the denominator by $2 \cdot 2 \cdot a \cdot b$.

$$= \frac{2a}{3}$$

Checking for Understanding _____

Communicating Algebra

1. Explain how you would write $\frac{24}{26}$ in simplest form.

2. Explain how you know whether a fraction is in simplest form.

Guided Practice

Express each ratio as a fraction.

3. $7 \div 8$ **4.** 5 to 8   **5.** 9:10   **6.** three to six

State whether each fraction is in simplest form. If it is not in simplest form, write it in simplest form.

7. $\frac{3}{5}$ **8.** $\frac{3}{6}$ **9.** $\frac{15}{21}$ **10.** $\frac{30}{37}$ **11.** $\frac{51}{60}$

12. $\frac{8ab}{15cd}$ **13.** $\frac{7x}{14xy}$ **14.** $\frac{11m}{13t}$ **15.** $\frac{8z^2}{16z}$ **16.** $\frac{12cd}{20}$

Exercises _____

Independent Practice

Write each fraction in simplest form.

17. $\frac{2}{10}$ **18.** $\frac{8}{12}$ **19.** $\frac{12}{16}$ **20.** $\frac{2}{14}$ **21.** $\frac{9}{15}$

22. $\frac{4}{18}$ **23.** $\frac{7}{21}$ **24.** $\frac{15}{30}$ **25.** $\frac{16}{24}$ **26.** $\frac{10}{22}$

27. $\frac{20}{28}$ **28.** $\frac{24}{32}$ **29.** $\frac{18}{36}$ **30.** $\frac{27}{33}$ **31.** $\frac{30}{35}$

32. $\frac{12}{40}$ **33.** $\frac{25}{40}$ **34.** $\frac{40}{42}$ **35.** $\frac{18}{44}$ **36.** $\frac{21}{45}$

37. $\frac{12m^2k}{15mk}$ **38.** $\frac{36a^3b}{48a^2b}$ **39.** $\frac{15rs^2}{50rs}$ **40.** $\frac{3x^2y^2}{51xy}$

41. $\frac{28rsy}{52rsy}$ **42.** $\frac{40p^2q}{52pq^2}$ **43.** $\frac{17k^2z}{51z}$ **44.** $\frac{48m^2r^2}{56m^2r}$

Mixed Review	**45.** Write the negation of the statement *The greatest common factor of 10 and 15 is 5.* (Lesson 2-6)

45. Write the negation of the statement *The greatest common factor of 10 and 15 is 5.*   (Lesson 2-6)

46. Solve the equation $a + 9 = -13$.   (Lesson 3-1)

47. Sports   The Brandon Bobcats need at least 10 more points to win the football game. If the opponent's score is 26, what is the Bobcats' score?   (Lesson 3-10)

48. Is the number 51 a *prime* or *composite?*   (Lesson 4-4)

49. State the least prime number that is a factor of the number 48.   (Lesson 4-4)

Applications

50. Conservation   The fish commission is stocking fish in Beechwood Lake. Suppose 2800 brown trout and 2100 rainbow trout are put in the lake. Express the ratio of rainbow trout to the total fish stocked as a fraction in simplest form.

51. Sports   Karen made 8 of 14 free throws in her last basketball game. How would you describe her success as a fraction in simplest form?

Connection

52. Statistics   In a school district census, 2800 students are female and 2400 students are male. Express the ratio of males to females as a fraction in simplest form.

Critical Thinking

53. Any odd number that is greater than 5 can be written as the sum of three prime numbers, $(7 = 3 + 2 + 2)$. Show that this statement is true for odd numbers between 50 and 60.

Wrap-Up

54. Describe a situation where a fraction in simplest form would be more easily understood than the original equivalent fraction.

Mid-Chapter Quiz

Use divisibility rules to determine whether the first number is a factor of the second number. Write *yes* or *no*.   (Lesson 4-1)

1. 6; 236   **2.** 3; 342   **3.** 2; 838   **4.** 5; 1024

Evaluate each expression.   (Lesson 4-2)

5. $42 - 3 \cdot 9$ **6.** $\dfrac{4^2 - 2}{5 + 3^2}$ **7.** $\dfrac{2^2 \cdot 3^2 + 5}{2^2(3^2 + 5)}$ **8.** $\dfrac{4 \cdot 9 - 2^2}{4 - 3 \cdot 4}$

Solve.   (Lesson 4-3)

9. Eight softball teams are to play in a round-robin tournament; that is, they play each other once. For how many games will umpires be needed?

Factor each number or monomial completely.   (Lesson 4-4)

10. 40   **11.** 75   **12.** -180   **13.** $154a^2bc^2$

Find the GCF.   (Lesson 4-5)

14. 54, 72   **15.** $25x^2$, $30xy$ **16.** 36, 120   **17.** $24xy^2$, $36x^2y$, $48xy$

4-7 Least Common Multiple (LCM)

Objective:
You will learn how to find the least common multiple of two or more numbers.

Key Terms:
multiple
least common multiple
LCM

Joe and Rosa are training for the Boston Marathon. Joe can go around the park in his wheelchair in 3 minutes; Rosa can go the same distance in 4 minutes. If they start at the same time, when will they be side-by-side again?

Multiples can be used to answer this question. A **multiple** of a number is the product of that number and any whole number.
Recall that whole numbers are 0, 1, 2, 3, . . .

multiples of 3: 0, 3, 6, 9, 12, 15, 18, 21, 24, . . .

multiples of 4: 0, 4, 8, 12, 16, 20, 24, 28, 32, . . .

Common multiples of 3 and 4 are 0, 12, 24, 36, ... The least of the nonzero common multiples of two or more numbers is called the **least common multiple (LCM)** of the numbers. The least common multiple of 3 and 4 is 12. So, Joe and Rosa will be side-by-side again in 12 minutes.

Examples

1 List the first five multiples of 4.

$$0 \cdot 4 \quad 1 \cdot 4 \quad 2 \cdot 4 \quad 3 \cdot 4 \quad 4 \cdot 4$$

multiples of 4   0,   4,   8,   12,   16

What can you say about the first multiple of any number?

2 List the first five multiples of *n*.

$$0 \cdot n \quad 1 \cdot n \quad 2 \cdot n \quad 3 \cdot n \quad 4 \cdot n$$

multiples of *n*   0,   *n*,   2*n*,   3*n*,   4*n*

Calculator Hint
You can use a calculator to find the LCM. Divide multiples of the greater number by the lesser number until you get a whole number quotient.

Find the LCM for each pair of numbers.

3 3, 5

multiples of 3:
  0, 3, 6, 9, 12, 15, . . .
multiples of 5:
  0, 5, 10, 15, 20, . . .

The LCM is 15.

4 4, 6

multiples of 4:
  0, 4, 8, 12, 16, . . .
multiples of 6:
  0, 6, 12, 18, . . .

The LCM is 12.

Prime factorization can be used to find the LCM of a set of numbers. A common multiple contains *all* of the prime factors of each number. The LCM contains *each* factor the greatest number of times it appears for any of the numbers.

Examples

Use prime factorization to find the LCM for each set of numbers.

5 **12, 16**

$12 = \boxed{2} \cdot \boxed{2} \cdot 3$ or $2^2 \cdot 3$
$16 = \boxed{2} \cdot \boxed{2} \cdot 2 \cdot 2$ or 2^4

Find the common factors. Multiply all the factors, using the common factors only once.

So, the LCM of 12 and 16 is $2 \cdot 2 \cdot 2 \cdot 2 \cdot 3$ or 48.

6 **15, 18, 12**

$15 = ③ \cdot 5$
$18 = \boxed{2} \cdot ③ \cdot 3$ or $2 \cdot 3^2$
$12 = \boxed{2} \cdot 2 \cdot ③$ or $2^2 \cdot 3$

Use the common factors, which are circled, only once.

The greatest power of 2 is 2^2.
The greatest power of 3 is 3^2.
The greatest power of 5 is 5^1.

The LCM is $2^2 \cdot 3^2 \cdot 5$ or 180.

CONNECTION TO ALGEBRA

The LCM of two or more monomials is also the product of the greatest power of each factor.

Example

7 **Find the LCM of $5x$ and $10x^2$.**

$5x = \qquad \boxed{5} \cdot \boxed{x}$
$10x^2 = 2 \cdot \boxed{5} \cdot \boxed{x} \cdot x$

The greatest power of 2 is 2^1.
The greatest power of 5 is 5^1.
The greatest power of x is x^2.

The LCM is $2 \cdot 5 \cdot x^2$ or $10x^2$.

Checking for Understanding

Communicating Algebra

1. Explain how to find the first five multiples of 6.

2. In your own words, describe how to find the LCM of a set of numbers if you have the prime factorization of each number.

Guided Practice

List the first four multiples of each of the following.

3. 12   4. 7   5. 30   6. k

Find the LCM.

7. multiples of 2: 0, 2, 4, 6, 8, 10, 12, ...
   multiples of 5: 0, 5, 10, 15, 20, 25, ...

8. multiples of 6: 0, 6, 12, 18, 24, ...
   multiples of 9: 0, 9, 18, 27, 36, ...

Use prime factorization to find the LCM for each set of numbers.

9. 8, 9   10. 10, 14   11. 20, 12   12. 45, 30, 35

Exercises

Independent Practice

Determine whether the first number is a multiple of the second number.

13. 28; 7   14. 32; 5   15. 24; 3   16. 42; 6

17. 52; 13   18. 144; 9   19. 108; 8   20. 116; 7

21. 156; 12   22. 162; 18   23. 4500; 18   24. 8664; 36

Find the LCM for each set of numbers or monomials.

25. 6, 10   26. 4, 10   27. 8, 12   28. 2, 9

29. 15, 18   30. 6, 15   31. 5, 7   32. 8, 14

33. 21, 28   34. 14, 21   35. 18, 32   36. 15, 75

37. $7y, 12y$ 38. $20cd, 50d$ 39. $16x, 24y$ 40. $12x, 35x^2$

41. $4k, 8k, 12k$ 42. 9, 12, 15   43. 7, 21, 84   44. $3c, 5c^2, 7$

Mixed Review

45. Solve the equation $d = \frac{225}{-5}$.   (Lesson 2-8)

46. *True* or *false:* To solve the equation $-4m = 56$, divide each side of the equation by 56.   (Lesson 3-3)

47. Using the formula $I = \frac{V}{R}$, find the amperes if the voltage is 88 volts and the resistance is 8 ohms.   (Lesson 3-5)

48. Factor the monomial $45a^2b^3$.   (Lesson 4-4)

49. Find the GCF of 39 and 65.   (Lesson 4-5)

Applications

50. **Biology**   Cicadas are sometimes called 17-year locusts because they emerge from the ground every 17 years. The number of one type of caterpillar peaks every 5 years. If the peak cycles of the caterpillars and cicadas coincided in 1990, what will be the next year in which they coincide?

51. **Interior Decorating**   In a pattern of floor tiles laid in rows, one row has tiles 4 inches long, another has tiles 5 inches long, and a third row has tiles 6 inches long. In how many inches will the ends of all three rows be even and the pattern start to repeat?

Critical Thinking

52. The LCM of two numbers is $2^2 \cdot 3 \cdot 5^2$. The GCF of the same numbers is $2 \cdot 5$. If one of the numbers is $2 \cdot 3 \cdot 5$, what is the other number?

Wrap-Up

53. Name a pair of numbers that have 24 as their LCM.

4-8 Comparing Fractions

Objective:
Compare fractions with different denominators and find the least common denominator for algebraic fractions.

Key Terms:
least common denominator
LCD

Jenny and her friends went to the Pizza Parlor. The pepperoni pizza was cut into 8 equal pieces while the Hawaiian pizza was cut into 7 equal pieces. After eating 3 pieces from each pizza, $\frac{5}{8}$ remained of the pepperoni pizza and $\frac{4}{7}$ remained of the Hawaiian pizza. Which pizza has more remaining?

One way to compare $\frac{5}{8}$ and $\frac{4}{7}$, is to write them as equivalent fractions with the *same* denominator. Any common denominator could be used. However, the computation may be easier if the least common denominator is used.

The **least common denominator (LCD)** is the least common multiple of the denominators. The LCM of 8 and 7 is 56.

Multiplying the numerator and denominator of a fraction by the same number is like multiplying the fraction by what number?

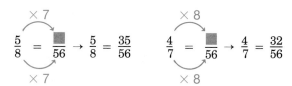

Comparing $\frac{35}{56}$ to $\frac{32}{56}$, the greater fraction has the greater numerator.

Thus, $\frac{5}{8} > \frac{4}{7}$. So, there is more pepperoni pizza remaining.

Pizza is often thought of as junk food. However, the vitamins thiamin and niacin are in the flour used to make the crust. Tomato sauce is high in vitamins A and C, while cheese contains calcium and riboflavin. Toppings such as peppers and mushrooms also contain several vitamins and minerals.

Example

1 **Which is greater, $\frac{3}{8}$ or $\frac{5}{12}$?**

Write $\frac{3}{8}$ and $\frac{5}{12}$ as equivalent fractions with the same denominators.
The LCM of 8 and 12 is 24.

Find equivalent fractions with the LCM as the denominator.

$$\frac{3}{8} \xrightarrow[\times 3]{\times 3} \frac{\blacksquare}{24} \rightarrow \frac{3}{8} = \frac{9}{24} \qquad \frac{5}{12} \xrightarrow[\times 2]{\times 2} \frac{\blacksquare}{24} \rightarrow \frac{5}{12} = \frac{10}{24}$$

Since $\frac{10}{24} > \frac{9}{24}$, $\frac{5}{12} > \frac{3}{8}$.

CONNECTION TO ALGEBRA

Prime factorization can be used to find the LCD for algebraic fractions that have different denominators.

Example

2 Find the LCD for $\frac{5}{3b^2}$ and $\frac{4}{2b}$.

$3b^2 = 3 \cdot b \cdot b$ First, find the LCM of $3b^2$ and $2b$.

$2b = 2 \cdot b$

The LCM of $3b^2$ and $2b$ is $3 \cdot 2 \cdot b \cdot b$ or $6b^2$.

The LCD of $\frac{5}{3b^2}$ and $\frac{4}{2b}$ is $6b^2$.

Checking for Understanding

Communicating Algebra

1. Write the steps you would take to compare $\frac{3}{8}$ and $\frac{1}{3}$.

2. Explain how the LCM and LCD are alike and how they are different.

Guided Practice

Find the LCD for each pair of fractions.

3. $\frac{1}{2}, \frac{3}{5}$

4. $\frac{7}{9}, \frac{1}{4}$

5. $\frac{1}{4}, \frac{3}{8}$

6. $\frac{2}{9}, \frac{5}{12}$

7. $\frac{7}{a}, \frac{8}{a^2}$

8. $\frac{1}{2m}, \frac{3}{5m^2}$

9. $\frac{4}{5a}, \frac{7}{25a^2}$

10. $\frac{5}{6am}, \frac{7}{9a^2m}$

Replace each ● with < or > to make a true statement.

11. $\frac{2}{3}$ ● $\frac{3}{4}$

12. $\frac{7}{8}$ ● $\frac{5}{6}$

13. $\frac{9}{10}$ ● $\frac{10}{11}$

14. $\frac{6}{5}$ ● $\frac{7}{6}$

Exercises

Independent Practice

Find the LCD for each pair of fractions.

15. $\frac{1}{2}, \frac{1}{4}$

16. $\frac{1}{3}, \frac{1}{8}$

17. $\frac{1}{5}, \frac{2}{7}$

18. $\frac{5}{6}, \frac{6}{9}$

19. $\frac{7}{8a}, \frac{2}{15a}$

20. $\frac{2}{9m}, \frac{3}{25m^2}$

21. $\frac{5}{6k^2}, \frac{4}{8km}$

22. $\frac{7}{9x^2}, \frac{5}{8x}$

Replace each ● with < or > to make a true statement.

23. $\frac{1}{3}$ ● $\frac{3}{12}$

24. $\frac{3}{8}$ ● $\frac{5}{16}$

25. $\frac{11}{12}$ ● $\frac{50}{60}$

26. $\frac{3}{4}$ ● $\frac{4}{5}$

27. $\frac{10}{12}$ ● $\frac{9}{11}$

28. $\frac{5}{7}$ ● $\frac{7}{9}$

29. $\frac{7}{11}$ ● $\frac{8}{13}$

30. $\frac{7}{9}$ ● $\frac{8}{10}$

Mixed Review

31. Graph the set of numbers {0, -1, 3} on a number line.   (Lesson 2-1)

32. Solve the inequality $-10 < x + 23$.   (Lesson 3-7)

33. *True* or *false:* All squares are rectangles and all rectangles are squares. (Lesson 3-9)

34. Find the GCF for 9, -63, 108.   (Lesson 4-5)

35. Is the fraction $\frac{9}{28}$ in simplest form?   (Lesson 4-6)

36. Write the fraction $\frac{14r^2}{2rs^2}$ in simplest form.   (Lesson 4-6)

Applications

37. Ecology   The graph at the right gives statistics for the amount of ticker tape produced in parades in New York City. Use equivalent fractions to *estimate* a simple fraction that compares the amount of paper in the Mets' victory parade of 1969 to the parade for the return of the U.S. hostages in 1981.

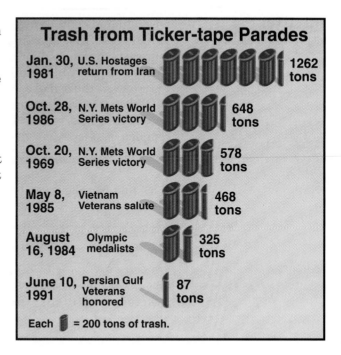

Trash from Ticker-tape Parades

Jan. 30, 1981   U.S. Hostages return from Iran   1262 tons

Oct. 28, 1986   N.Y. Mets World Series victory   648 tons

Oct. 20, 1969   N.Y. Mets World Series victory   578 tons

May 8, 1985   Vietnam Veterans salute   468 tons

August 16, 1984   Olympic medalists   325 tons

June 10, 1991   Persian Gulf Veterans honored   87 tons

Each = 200 tons of trash.

38. Auto Mechanics   Phil is tuning up a foreign car and left his metric wrenches at home. He needs to remove a 12 mm bolt. He measures the bolt with a tape measure and finds 12 mm to be about $\frac{15}{32}$ in. Will his $\frac{9}{16}$ in. wrench or $\frac{1}{2}$ in. wrench be a closer fit?

Critical Thinking

39. I am a fraction less than one and in simplest form. My numerator and denominator are both prime numbers between 25 and 36. Who am I?

Wrap-Up

40. Make Up a Problem   Write a problem that involves comparing fractions.

Team Problem Solving

Work in small groups.
In the diagram at the right, find the total number of squares. Then tell how many of them contain a number of dots that is a multiple of 3. (Hint: There are more than 10 squares.)

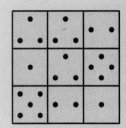

4-9 Multiplying Powers

Objective:
Multiply monomials.

Key Term:
product of powers

One hundred eighth-graders at Colestock Junior High School participated in a relay marathon over the Labor Day weekend to raise money for Muscular Dystrophy research. At the end of each lap of the 1000-meter course the runner relayed the baton to the next runner. If each student completed 10 laps, how many meters were run?

Each number above can be expressed as a power of 10.

8th-graders	laps per student	meters per lap
$100 = 10^2$	$10 = 10^1$	$1000 = 10^3$

To answer the question, you need to find the product of 10^2, 10^1, and 10^3.

Example

1 Find $10^2 \cdot 10^1 \cdot 10^3$.

Do you see a pattern with the exponents?

$$10^2 \cdot 10^1 \cdot 10^3 = (10 \cdot 10) \cdot 10 \cdot (10 \cdot 10 \cdot 10)$$
$$= 10^6 \quad \text{10 is used as a factor 6 times.}$$

 10 $\boxed{y^x}$ 6 $\boxed{=}$ 1000000

One million meters were run in the relay marathon.

FYI

A googolplex is defined as 100^{100}. It is the largest number to have its own name. The term *googol* was invented in 1938 by Milton Sirotta. He was 9 years old at the time!

Study the pattern of products shown below. Each factor and each product have been replaced with a power of 2.

$2^1 = 2$
$2^2 = 4$
$2^3 = 8$
$2^4 = 16$
$2^5 = 32$
$2^6 = 64$
$2^7 = 128$
$2^8 = 256$
$2^9 = 512$

$2 \cdot 4 = 8 \qquad 8 \cdot 4 = 32 \qquad 16 \cdot 32 = 512$

$\downarrow \quad \downarrow \quad \downarrow \qquad \downarrow \quad \downarrow \quad \downarrow \qquad \downarrow \quad \downarrow \quad \downarrow$

$2^1 \cdot 2^2 = 2^3 \qquad 2^3 \cdot 2^2 = 2^5 \qquad 2^4 \cdot 2^5 = 2^9$

Look at the powers only. Do you see a pattern with the exponents? In Example 2 you will see if your pattern holds for $32 \cdot 8 = 256$.

Example

2 Find $32 \cdot 8$ or $2^5 \cdot 2^3$.

$$2^5 \cdot 2^3 = (2 \cdot 2 \cdot 2 \cdot 2 \cdot 2) \cdot (2 \cdot 2 \cdot 2)$$
$$= 2 \cdot 2 \cdot 2 \cdot 2 \cdot 2 \cdot 2 \cdot 2 \cdot 2$$
$$= 2^8$$
$$2^5 \cdot 2^3 = 2^{5+3} \text{ or } 2^8$$

$2 \boxed{y^x} 8 \boxed{=} 256$

CONNECTION TO ALGEBRA

Monomials such as m^4 and m^3 can also be multiplied using the pattern shown in Example 2.

Example

3 Find $m^4 \cdot m^3$.

$$m^4 \cdot m^3 = (m \cdot m \cdot m \cdot m) \cdot (m \cdot m \cdot m) \quad \text{4 factors} \cdot \text{3 factors}$$
$$= m \cdot m \cdot m \cdot m \cdot m \cdot m \cdot m \quad \text{7 factors}$$
$$= m^7$$
$$m^4 \cdot m^3 = m^{4+3} \text{ or } m^7$$

These and other examples suggest the following rule.

Product of Powers	In words: You can multiply powers *that have the same base* by adding their exponents. In symbols: For any number a and positive integers m and n, $a^m \cdot a^n = a^{m+n}$.

Example

4 Find $(-7y^2)(3y^3)$.

$$(-7y^2)(3y^3) = (-7 \cdot 3)(y^2 \cdot y^3) \quad \text{Commutative and associative properties}$$
$$= -21 \cdot y^{2+3} \quad \text{Product of powers}$$
$$= -21 y^5$$

Checking for Understanding

Communicating Algebra

1. Explain how you would find the product $3^5 \cdot 3^2$.

2. How would you show that $7^2 \cdot 7 = 7^{2+1}$?

3. Make up a product of powers problem that has a solution of a^6.

Guided Practice

Verify each product by multiplication.

4. $3^1 \cdot 3^2 = 3^3$ 5. $2^3 \cdot 2^2 = 2^5$ 6. $4^2 \cdot 4^2 = 4^4$

7. $2^5 \cdot 2^2 = 2^7$ 8. $5^2 \cdot 5 = 5^3$ 9. $3^2 \cdot 3^4 = 3^6$

Find each product.

10. $x^3 \cdot x^5$ 11. $y^7 \cdot y^2$ 12. $a^4 \cdot a^4$

Exercises

Independent Practice

Find each product.

13. $3^4 \cdot 3^6$

14. $10^5 \cdot 10^5$

15. $8^2 \cdot 8^3 \cdot 8$

16. $b^5 \cdot b^2$

17. $m \cdot m^3$

18. $d \cdot d^6$

19. $(3a^2)(4a^3)$

20. $(5x^3)(4x^4)$

21. $(3x^4)(-5x^2)$

22. $(5n^3)(-6n^2)$

23. $(-10x^3y)(2x^2)$

24. $(-3y^3z)(7y^4)$

25. $(a^2b)(ab^3)$

26. $(y^3z^4)(yz^5)$

27. $m^4(m^3b^2)$

Language Skill

Write a mathematical expression for each phrase.

28. the sum of three to the eighth power and six

29. seven to the fourth power times four to the fifth power

30. four times three to the second power plus four times five squared

Mixed Review

31. Solve the equation $17 = a - (-4)$.   (Lesson 3-2)

32. By which number should you multiply to solve the equation $-9 = \dfrac{m}{11}$?
(Lesson 3-4)

33. Write the fraction $\dfrac{16a^3b^2}{2a^2b^2}$ in simplest form.   (Lesson 4-6)

34. *True* or *false*: 3 is the LCM of 18.   (Lesson 4-7)

35. Find the LCM of 9 and 21.   (Lesson 4-7)

Application

36. **Meteorology**   The intensity of an earthquake is measured on the Richter scale. Each increase of one on the Richter scale means a ten-times increase in intensity. In 1906 an earthquake in San Francisco measured 8.3 on the Richter scale. Another earthquake that measured 7.1 shook the same area in October of 1989. Estimate how many times more intense the 1906 earthquake was.

Critical Thinking

37. Complete this problem using only two different digits.

■ , ■ ■ ■ × ■ ■ ■ = ■ ■ ■ , ■ ■ ■

Wrap-Up

38. In your own words, describe or write the rule for multiplying powers.

4-10　Dividing Powers

Objective:
Divide monomials and simplify expressions containing negative exponents.

Key Term:
quotient of powers

The population of the People's Republic of China exceeds one billion people. The land area of this country is about ten million square kilometers. What is the average number of people per square kilometer?

Each number above can be expressed as a power of ten.

one billion $(1,000,000,000) = 10^9$

ten million $(10,000,000) = 10^7$

To answer the question, you need to find the quotient of one billion and ten million.

Do you see a pattern with the exponents?

Example

1　Find $\dfrac{10^9}{10^7}$.

Use a calculator.

$10\ \boxed{y^x}\ 9\ \boxed{\div}\ 10\ \boxed{y^x}\ 7\ \boxed{=}\ 100$ or 10^2

There are about 100 people per square kilometer.

Social Studies Connection

Sociology is the study of human society. A sociologist who specializes in **demography** studies the size, composition, and distribution of human populations within a specific geographical area. This could involve the use of very large numbers as seen at the beginning of this lesson.

Study the pattern of products below. Each dividend, divisor, and quotient has been replaced with a power of 2.

$8 \div 2 = 4$　　　$32 \div 8 = 4$　　　$512 \div 16 = 32$
$\downarrow \quad \downarrow \quad \downarrow$　　$\downarrow \quad \downarrow \quad \downarrow$　　$\downarrow \quad \downarrow \quad \downarrow$
$2^3 \div 2^1 = 2^2$　$2^5 \div 2^3 = 2^2$　$2^9 \div 2^4 = 2^5$

Look at the powers only. Do you see a pattern with the exponents? In Example 2 you will see if your pattern is true for $256 \div 8 = 32$.

$2^1 = 2$
$2^2 = 4$
$2^3 = 8$
$2^4 = 16$
$2^5 = 32$
$2^6 = 64$
$2^7 = 128$
$2^8 = 256$
$2^9 = 512$

Example

2 Find $256 \div 8$ or $\dfrac{2^8}{2^3}$.

$$\dfrac{2^8}{2^3} = \dfrac{2 \cdot 2 \cdot 2 \cdot 2 \cdot 2 \cdot 2 \cdot 2 \cdot 2}{2 \cdot 2 \cdot 2}$$ Divide the numerator and denominator by $2 \cdot 2 \cdot 2$.

$$= 2 \cdot 2 \cdot 2 \cdot 2 \cdot 2$$

$$= 2^5$$

| 2 | y^x | 5 | = | 32 |

$$2^8 \div 2^3 = 2^{8-3} \text{ or } 2^5$$

CONNECTION TO ALGEBRA

Monomials such as a^6 and a^2 can also be divided using the pattern shown in Example 2.

Example

*What value of **a** is not allowed? Why?*

3 Find $\dfrac{a^6}{a^2}$.

$$\dfrac{a^6}{a^2} = \dfrac{a \cdot a \cdot a \cdot a \cdot a \cdot a}{a \cdot a}$$ ← 6 factors   Notice that $\dfrac{a \cdot a}{a \cdot a} = 1$.
← 2 factors

$$= a \cdot a \cdot a \cdot a$$ The quotient has 4 factors.

$$= a^4$$

$$\dfrac{a^6}{a^2} = a^{6-2} \text{ or } a^4$$

| **Quotient of Powers** | In words: You can divide powers *that have the same base* by subtracting their exponents.
 In symbols: For any whole numbers m and n, and nonzero number a, $$\dfrac{a^m}{a^n} = a^{m-n}.$$ |

Examples

4 Simplify $\dfrac{7^4}{7^2}$.

$$\dfrac{7^4}{7^2} = 7^{4-2}$$ The bases are the same.

$$= 7^2$$ Subtract the exponents.

5 Simplify $\dfrac{x^7}{x^3}$.

$$\dfrac{x^7}{x^3} = x^{7-3}$$

$$= x^4$$

What happens if $m = n$ in the rule above? Try $\dfrac{a^2}{a^2}$.

$$\dfrac{a^2}{a^2} = \dfrac{a \cdot a}{a \cdot a} \qquad \text{or} \qquad \dfrac{a^2}{a^2} = a^{2-2}$$ Use the rule for dividing powers.

$$= 1 \qquad\qquad\qquad\qquad = a^0$$

Since $\dfrac{a^2}{a^2}$ cannot have two different values, you can conclude that $a^0 = 1$.

You can use the quotient of powers rule to simplify $\dfrac{a^2}{a^6}$.

$$\dfrac{a^2}{a^6} = a^{2-6} \quad \text{Use the rule for} \qquad \text{or} \qquad \dfrac{a^2}{a^6} = \dfrac{\cancel{a} \cdot \cancel{a}}{\cancel{a} \cdot \cancel{a} \cdot a \cdot a \cdot a \cdot a}$$

$$\phantom{\dfrac{a^2}{a^6}} \text{dividing powers.}$$

$$= a^{-4} \quad 2 - 6 = -4 \qquad\qquad\qquad = \dfrac{1}{a \cdot a \cdot a \cdot a}$$

$$\qquad\qquad\qquad\qquad\qquad\qquad = \dfrac{1}{a^4}$$

Since $\dfrac{a^2}{a^6}$ cannot have two different values, you can conclude that a^{-4} is equal to $\dfrac{1}{a^4}$. In general, any power such as a^{-n} can be written $\dfrac{1}{a^n}$.

Example

6 **Express 5^{-3} with positive exponents.**

$$5^{-3} = \dfrac{1}{5^3}$$

Checking for Understanding

Communicating Algebra

1. In your own words, explain the quotient of powers rule.

2. How would you show that $\dfrac{7^2}{7} = 7^{2-1}$ or 7?

3. Explain how to express a^{-2} with a positive exponent.

Guided Practice

Verify each quotient by replacing the powers with their values.

4. $\dfrac{3^5}{3^2} = 3^3$ 5. $\dfrac{4^3}{4} = 4^2$ 6. $\dfrac{2^4}{2^6} = 2^{-2}$ 7. $\dfrac{5^4}{5^4} = 5^0$

Find each quotient.

8. $\dfrac{x^9}{x}$ 9. $\dfrac{y^8}{y^6}$ 10. $\dfrac{a^4}{a^4}$ 11. $\dfrac{x^2}{x^3}$

Exercises

Independent Practice

Find each quotient.

12. $\dfrac{8^4}{8^3}$ 13. $\dfrac{(-2)^6}{(-2)^5}$ 14. $\dfrac{10^{10}}{10^3}$ 15. $\dfrac{(-7)^3}{(-7)}$

16. $\dfrac{y^5}{y^2}$ 17. $\dfrac{a^{10}}{a^6}$ 18. $\dfrac{(-x)^4}{(-x)^3}$ 19. $\dfrac{c^8}{c}$

20. $\dfrac{z^3}{z}$ 21. $\dfrac{m^{15}}{m^{11}}$ 22. $\dfrac{f^{20}}{f^8}$ 23. $\dfrac{y^{100}}{y^{100}}$

Simplify. Express each result with positive exponents. Assume no denominator is equal to zero.

24. 5^0 **25.** $(-3)^0$ **26.** 4^{-1} **27.** 10^{-2}

28. 3^{-2} **29.** x^{-1} **30.** n^{-3} **31.** $s^{-2}t^3$

Mixed Review

32. *True* or *false:* The sum of a positive number and a negative number is always negative. (Lesson 2-3)

33. Find the area of a rectangle whose sides are 8 ft and 6 ft. (Lesson 3-6)

34. Write an inequality for the sentence: *The quotient of an integer and four is greater than or equal to nine.* (Lesson 3-8)

35. Write the fraction $\frac{18}{126}$ in simplest form. (Lesson 4-6)

36. Find the LCM of 4, 6, and 18. (Lesson 4-7)

37. Which is greater, $\frac{1}{4}$ or $\frac{3}{11}$? (Lesson 4-8)

38. Find the LCD of $\frac{1}{9r}$ and $\frac{3}{4r^2}$. (Lesson 4-8)

Logical Thinking

Determine which column contains the expression that does not belong. Be prepared to give the reason for your answer.

	A	B	C	D
39.	$\frac{21}{82}$	$\frac{5}{20}$	$\frac{10}{40}$	$\frac{15}{60}$
40.	$\frac{9}{12}$	$\frac{36}{49}$	$\frac{12}{14}$	$\frac{26}{52}$
41.	10^3	10^{-2}	10^8	10^5
42.	31	21 ·	11	37

Applications

43. Microbiology A human blood cell is about 2^{-17} meters in diameter. The length of a DNA strand is about 2^{-18} meters long. How many DNA strands could be laid end to end across the diameter of a human blood cell?

44. Computer Science Large computers can perform operations in nanoseconds. A nanosecond is one billionth of a second or 10^{-9} seconds. If a computer takes 10 seconds to compute a problem, how many nanoseconds of time elapsed?

Connection

45. Measurement A millimeter is $\frac{1}{1000}$ of a meter or 10^{-3} meters. A kilometer is 1000 meters. How many millimeters are in a kilometer?

Critical Thinking

46. Simplify $x^4 \cdot x^{-2} \cdot x^8 \cdot x^{-10}$.

Wrap-Up

47. Write in your own words the relationship between the rules for multiplying powers and dividing powers.

Algebra in Action-Physics

Measuring Sound

Scientists use a unit called the decibel to measure the intensity level of sound. A decibel equals one-tenth of a *bel,* a unit named after the Scottish-born inventor and scientist Alexander Graham Bell. Decibel measurements use numbers in a different way. For instance, 20 decibels does not mean the sound is 20 times the intensity of one decibel. A measurement of 20 decibels means ten used as a factor two times ($10 \cdot 10 = 10^2$) or 100 times the intensity of one decibel.

The chart lists the intensity of various sounds and the algebraic expression for each measurement.

Sound	Decibels	Expression
barely heard	0	10^0
breathing	10	10^1
whispering	30	10^3
normal conversation	50	10^5
noisy office	60	10^6
telephone bell	70	10^7
vacuum cleaner	80	10^8
subway train	100	10^{10}
motorcycle	110	10^{11}
rock concert	130	10^{13}
jet airplane	150	10^{15}

Sounds of 140 decibels or more produce pain in the ear and could damage the tissues of the inner ear.

How many times more intense is the sound of a motorcycle than the sound of a telephone bell?

motorcycle → 110 decibels → 10^{11} or ten used as a factor 11 times

bell → 70 decibels → 10^7 or ten used as a factor 7 times

Notice that $11 - 7 = 4$. This means that the sound of the motorcycle is 10^4 or 10,000 times more intense than the sound of a telephone bell.

1. Is it safer for your ears if you run a vacuum cleaner or attend a rock concert?
2. What is the expression for the decibel level that results in pain?
3. How many more times intense is the sound of a subway than someone whispering?
4. **Research** Find the decibel levels of other common activities such as using a power mower.

Review

Language and Concepts

State whether each sentence is *true* or *false*. If false, replace the underlined word or number to make a true sentence.

1. The least common multiple of 10 and 15 is <u>30</u>.

2. The number 27 is an example of a <u>prime</u> number.

3. $\frac{8}{12}$ is the simplest form of $\frac{16}{24}$.

4. The fraction $\frac{3}{4}$ is equivalent to the fraction $\frac{12}{16}$.

5. 14 is the <u>GCF</u> of 28 and 42.

6. <u>5m</u> is a factor of $675mn$.

7. Numbers such as $2, 3, 5,$ and 7 are examples of <u>odd</u> numbers.

8. In the expression 10^5, the <u>base</u> is 5.

9. The expression 10^{-3} is equivalent to $\frac{1}{10^3}$.

10. $3^4 \cdot 3^5 = \underline{9}^{20}$.

Skills

Determine whether each expression is a monomial. (Lesson 4-1)

11. $2x$

12. $\frac{m}{3}$

13. amq

14. $p - 3$

Determine whether each number is divisible by 2, 3, 5, 6, or 10. (Lesson 4-1)

15. 84

16. 45

17. 420

18. 51

Write each product using exponents. (Lesson 4-2)

19. $4 \cdot 4$

20. $7 \cdot 7$

21. $3 \cdot 3 \cdot 3$

22. $y \cdot y \cdot y \cdot y$

Evaluate each expression. (Lesson 4-2)

23. $3t^4$ if $t = 2$

24. a^2 if $a = 12$

25. $2y^3$ if $y = 5$

Factor each number or monomial completely. (Lesson 4-4)

26. 28

27. -50

28. 124

29. $200a^2b$

Find the GCF for each set of numbers or monomials. (Lesson 4-5)

30. 24, 120

31. 64, 48

32. $-16a^2, 30a^3$

33. $24xy^3, 36z^2$

34. $12xy^3, 72x^2y, 24x^3y$

35. $60a^3b, 150a^2b^2, 36a^2b$

Write each fraction in simplest form. (Lesson 4-6)

36. $\frac{10}{30}$

37. $\frac{13a^2}{52a}$

38. $\frac{12}{60}$

39. $\frac{27x^2y^2}{15xy^3}$

Find the LCM for each pair of numbers or monomials. (Lesson 4-7)

40. 5, 8

41. 6, 14

42. $4x, 15x^2$

43. $12a, 16b$

Replace each ● with < or > to make a true statement. (Lesson 4-8)

44. $\frac{1}{3}$ ● $\frac{1}{4}$

45. $\frac{1}{7}$ ● $\frac{1}{5}$

46. $\frac{15}{28}$ ● $\frac{19}{42}$

47. $\frac{7}{18}$ ● $\frac{11}{24}$

Find each product. (Lesson 4-9)

48. $a^5 \cdot a^2$

49. $(2x^2)(3x^3)$

50. $(2ab)(-5a^2b)$

51. $(x^2y^3)(x^4y)$

Simplify. Express each result with positive exponents. Assume that no denominator is equal to zero. (Lesson 4-10)

52. $\frac{x^8}{x^2}$

53. $\frac{y^5}{y^5}$

54. 3^{-2}

55. $\frac{4^3}{4^5}$

Applications and Problem Solving

56. Lance has two pieces of cloth. One is 68 inches wide and the other is 84 inches wide. Find the width of the widest strips that can be cut from both pieces with nothing left over. (Lesson 4-7)

57. Beverly and Pam are painting a fence. For every 3 sections Pam paints, Beverly paints 5. If Beverly paints 20 sections in 2 hours, how long will it take Pam to paint 18 sections? (Lesson 4-3)

58. Is $\frac{5}{6}$ yard of material enough for a pattern calling for $\frac{3}{4}$ yard? (Lesson 4-8)

Portfolio Suggestion

Select an item from this chapter that you feel shows your best work and place it in your portfolio.

Curriculum Connection

- **Music** Research the mathematics involved in the relationship between the notes on a musical scale.

- **Automotive Industry** Check the owner's manual of several types and makes of vehicles to find instances where fractions are used.

Read More About It

Asimov, Isaac. *Quick and Easy Math.*

Collins, A. Frederick. *Rapid Math without a Calculator.*

Luce, Marnie. *Primes are Builders.*

Test

Write a number that satisfies each condition.

1. a three-digit number divisible by 2, but not 6

2. a four-digit number divisible by either 3 or 5

Evaluate each expression.

3. y^2 if $y = 4$

4. t^2 if $t = 11$

5. $2m^4$ if $m = 5$

Solve. Use a diagram.

6. Five teams are to play each other in a round robin basketball tournament, that is, each team will play all four of the other teams. How many games will be played?

Factor each number or monomial completely.

7. 36

8. -54

9. $130x^3yz^2$

Find the GCF for each set of numbers or monomials.

10. 40, 64

11. $28a^2b$, $70a^2b^3c$

12. 27, 75, 15

Write each fraction in simplest form.

13. $\frac{9}{27}$

14. $\frac{4x^3y^5}{16x^2y}$

15. $\frac{48}{60}$

Find the LCM for each set of numbers or monomials.

16. 6, 7

17. 5, 12

18. $4c$, $6cd$, $10d^2$

Replace each ■ with the term that will make a true statement.

19. $\frac{3}{15} = \frac{■}{45}$

20. $\frac{5}{6} = \frac{■}{18}$

21. $\frac{8}{16x} = \frac{■}{48x^2}$

Find each product.

22. $x^4 \cdot x^5$

23. $(a^3b)(a^5b^2)$

Simplify. Write each result with a positive exponent. Assume that no denominator is equal to zero.

24. $\frac{(-2)^4}{(-2)}$

25. $\frac{x^2}{x^9}$

BONUS

True or *false:* Every whole number, greater than 1, is either prime or can be expressed as the sum of primes in exactly one way. Write a sentence that supports your answer.

Academic Skills Test

Cumulative, Chapters 1-4

1. How is the product $5 \times 5 \times 5$ expressed in exponential notation?

   A 3×5
   B 5^3
   C 3^5
   D 5^5

2. Which is equivalent to 4^{-3}?

   A $3 \cdot 3 \cdot 3 \cdot 3 \cdot 3$
   B $4 \cdot 4 \cdot 4$
   C $\dfrac{1}{3 \cdot 3 \cdot 3 \cdot 3}$
   D $\dfrac{1}{4 \cdot 4 \cdot 4}$

3. Which fraction is in simplest form?

   A $\dfrac{12a^2}{15ab}$
   B $\dfrac{15s}{50}$
   C $\dfrac{5b^2 c}{7bh^2}$
   D $\dfrac{12d^2}{25ab}$

4. Which sentence is true?

   A $-8 > -12$
   B $-11 > 5$
   C $-4 = |4|$
   D $6 < -6$

5. Which equation is equivalent to $32 + x = 193$?

   A $32 = x + 193$
   B $x + 32 = 193$
   C $193 = 32 - x$
   D $x - 32 = 193$

6. Which is equivalent to $5^3 \cdot 5^2$?

   A 5^5 **C** 25
   B 5^6 **D** 25^5

7. Which is equivalent to $\dfrac{3^6}{3^2}$?

   A 3^3 **C** 3^8
   B 3^4 **D** 3^{12}

8. To paint a room with four walls, each 8 feet tall and 12 feet wide, you would need to buy a can of paint that covers at least—

   A 80 square feet
   B 96 square feet
   C 160 square feet
   D 384 square feet

9. Marcus bought and sold 50 shares of stock for a loss of $2 per share. Which integer represents the total loss?

   A 100 dollars
   B 25 dollars
   C -25 dollars
   D -100 dollars

10. The Sanchez's added a 6 foot by 15 foot rectangular deck to the back of their house. How much edging do they need? (The deck is attached to the house on one of the long sides.)

    A 27 feet
    B 36 feet
    C 42 feet
    D 90 feet

CHAPTER 5

CHAPTER OBJECTIVES

In this chapter you will learn to:

- add and subtract rational numbers
- solve equations and inequalities with rational numbers
- find terms of an arithmetic sequence
- convert within the customary system
- solve problems by looking for a pattern

Rationals: Adding and Subtracting Patterns

What do volleyball, hockey, and tennis have in common? Yes, they all are team sports, but that's not the answer. Here's another hint. This particular characteristic is not shared by football, basketball, and tennis.

Give up? In volleyball, every time your team wins a point, the score increases by one. The same is true when you score a goal in hockey or win a set in tennis. What about basketball? Some baskets are worth one, two, or three points. The scoring does not increase in a regular pattern.

The scoring in volleyball, hockey, and tennis is an example of an arithmetic sequence. Can you think of another example of an arithmetic sequence?

Health Connection

Chapter Project

Different sports offer different fitness benefits. Some are better for increasing flexibility, while others build muscle strength or endurance.

Do research to find out about the fitness benefits of an exercise or sport you like.

5-1 Rational Numbers

Objective:
Identify and compare rational numbers.

Key Term:
rational number

One skill that you will need to use throughout your life is the ability to organize. The loose-leaf binder that you carry to school is probably organized by subject. You may have noticed that numbers can be organized into sets. One set of numbers is *whole numbers*.

Whole Numbers
This set includes 0, 1, 2, 3, ... It also includes any number, such as $\frac{4}{4}$, that can be written as a whole number.

Whole Numbers

Another set of numbers is *integers*.

Name an integer that is not a whole number.

Integers
This set includes . . . -2, -1, 0, 1, 2, ... Notice that all whole numbers are included in the set of integers. Integers also include any number, such as $\frac{-3}{1}$, that can be written as an integer.

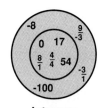
Integers

Of course, you have used numbers that are not whole numbers or integers. Consider this situation.

Suppose you and your friend form a small lawn-care company. You agree to share the profit or loss equally. During the first month the company loses $15. If *a* represents the amount due each person, then the equation $2a = -15$ can be used to find each person's loss. To solve the equation, divide each side by 2.

$$2a = -15$$

$$a = \frac{-15}{2} \text{ or } -7\frac{1}{2} \text{ or } -7.5 \qquad \text{Each person loses \$7.50.}$$

Both 2 and -15 are integers, but $\frac{-15}{2}$ is not an integer. The number system must be extended to include numbers like $\frac{-15}{2}$. The number $\frac{-15}{2}$ is called a **rational number.**

Definition of a Rational Number	Any number that can be expressed in the form $\frac{a}{b}$, where a and b are integers and $b \neq 0$, is called a rational number.

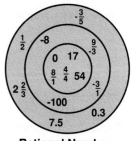
Rational Numbers
This includes common fractions, such as $\frac{1}{2}$. It also includes mixed numbers and decimals. All integers are included because any integer, such as -5, can be written in the form $\frac{-5}{1}$.

Examples

Name the set of numbers to which each number belongs.

1 15   15 is a whole number and an integer. Since it can be written as $\frac{15}{1}$, it is also a rational number.

2 -9   -9 is an integer and a rational number.

3 $-2\frac{1}{2}$ Since $-2\frac{1}{2}$ can be written as $\frac{-5}{2}$, it is a rational number.

Recall from Chapter 2 that $\frac{-4}{2} = \frac{4}{-2} = -2$. Similarly, any rational number such as $\frac{-2}{5}$ can also be written as $\frac{2}{-5}$ or $-\frac{2}{5}$. This will be useful when you graph rational numbers on a number line. You can also use a number line to compare rational numbers.

Examples

4 **Graph $\frac{-2}{3}$ on a number line. Then give two other names for $\frac{-2}{3}$.**

$\frac{-2}{3}$ can be written as $-\frac{2}{3}$ or $\frac{2}{-3}$.

5 **Use the number line to determine which is greater, $-\frac{3}{4}$ or $-\frac{1}{2}$.**

$-\frac{1}{2} = -\frac{2}{4}$

On the number line shown above, $-\frac{1}{2}$ is graphed to the right of $-\frac{3}{4}$. Therefore, $-\frac{1}{2} > -\frac{3}{4}$.

Checking for Understanding

Communicating Algebra

1. In your own words, what is a rational number?

2. Graph $-1\frac{2}{5}$ on a number line.

3. Give an example of a rational number that is not an integer.

Guided Practice

Name the set of numbers to which each number belongs.

4. -6 5. $-7\frac{2}{3}$ 6. 0 7. $\frac{20}{7}$

Replace each ⬤ with <, >, or = to make a true sentence. Use a number line if necessary.

8. $\frac{2}{3}$ ⬤ $-\frac{2}{3}$ 9. -5 ⬤ $\frac{-10}{2}$ 10. $\frac{-4}{3}$ ⬤ 0

Exercises

Independent Practice

Name the set of numbers to which each number belongs.

11. -1 12. -379 13. 9.0 14. $\frac{1}{100}$

Graph each number on a number line. Then give two other names for each number.

15. 2 16. $-\frac{1}{2}$ 17. $\frac{6}{-2}$ 18. $-1\frac{3}{4}$

Replace each ⬤ with <, >, or = to make a true sentence.

19. $\frac{3}{4}$ ⬤ $-\frac{3}{4}$ 20. $-5\frac{1}{2}$ ⬤ $\frac{-11}{2}$ 21. $\frac{2}{5}$ ⬤ $-\frac{2}{3}$

22. $3\frac{1}{3}$ ⬤ $\frac{10}{3}$ 23. $-\frac{3}{8}$ ⬤ $-\frac{1}{8}$ 24. $\frac{10}{12}$ ⬤ $\frac{5}{6}$

Mixed Review

25. Solve $y + 18 = -5$.   (Lesson 3-1)

26. **Personal Finance**   In order to open a special savings account, Jan needs a minimum of $100. She currently has $82. How many hours would she have to baby-sit at a rate of $3 per hour to have enough money to open the account? (Lesson 3-10)

27. State whether 37 is prime or composite.   (Lesson 4-4)

28. Simplify $(3m^3)(-7m^2)$.   (Lesson 4-9)

Applications

29. **Measurement**   Tina has a complete set of wrenches measured in 16th inches, 8ths, and 4ths. Her $\frac{5}{8}$-inch wrench is too small for a job. What is the next size larger that Tina has?

30. **Personal Finance**   On Monday Mr. Brown received a check for $47.58 and paid a bill for $59.23. At the end of the day, does he have more or less money than he had in the morning? How do you know?

Critical Thinking

31. Does $\frac{6}{2.4}$ name a rational number? If so, what one? If not, why not?

Wrap-Up

32. Write a sentence explaining why rational numbers are needed.

5-2 Writing Decimals as Fractions

Objective:
Rename decimals as fractions.

Key Terms:
terminating
repeating
bar notation

History Connection

The decimal system was named for the Latin word for tenth, *decimus*. Decimeter means one tenth of a meter. In the old Roman calendar, December was the tenth month.

Florence Griffith-Joyner set a world record by running 100 meters in just 10.49 seconds. While you might often hear this read as 10 *point* 49 seconds, 10.49 means 10 and $\frac{49}{100}$. The decimal 10.49 is an example of a *terminating* decimal. The chart below shows examples of other terminating decimals.

Decimal	Words	Fraction
9.3	nine and three tenths	$9\frac{3}{10}$
-0.7	negative seven tenths	$-\frac{7}{10}$
65.12	sixty-five and twelve hundredths	$65\frac{12}{100}$
0.13	thirteen hundredths	$\frac{13}{100}$
0.004	four thousandths	$\frac{4}{1000}$

Every **terminating decimal** can be written as a fraction with a denominator of 10, 100, 1000, and so on.

Examples

Express each decimal as a fraction or mixed number in simplest form.

1   0.08   ⬛▶ $0.08 = \frac{8}{100}$

$= \frac{2}{25}$ Simplify. The GCF of 8 and 100 is 4.

2   2.25   ⬛▶ $2.25 = 2\frac{25}{100}$

$= 2\frac{1}{4}$ Simplify.

Decimals either terminate or they go on forever. Decimals like 0.3333333 . . . are called **repeating decimals.** Because it is inconvenient to write all of these digits, you can use the **bar notation** $0.\overline{3}$ to indicate that the 3 repeats. Other examples follow.

5.19191919 . . . $= 5.\overline{19}$ The digits 19 repeat.

2.4123123123 . . . $= 2.4\overline{123}$ The digits 123 repeat.

Here are some commonly used decimal-fraction equivalences. It is helpful to know them by memory.

$$0.5 = \frac{1}{2}$$

$$0.\overline{3} = \frac{1}{3}$$

$$0.25 = \frac{1}{4}$$

$$0.2 = \frac{1}{5}$$

$$0.125 = \frac{1}{8}$$

Examples 3 and 4 explain how to rename repeating decimals as fractions.

Examples

3 **Express $0.\overline{6}$ as a fraction.**

Let $N = 0.666\ldots$. Then $10N = 6.666\ldots$.

Multiply N by 10, because 1 digit repeats.

Subtract $N = 0.666$ to eliminate the repeating part, $0.666\ldots$.

$$10N = 6.666\ldots$$
$$-\ N = 0.666\ldots$$
$$9N = 6$$

$$N = \frac{6}{9} \text{ or } \frac{2}{3}$$

Recall that $10N - 1N = 9N$.

So, $0.\overline{6} = \frac{2}{3}$.

4 **Express $2.272727\ldots$ as a fraction.**

Let $N = 2.272727\ldots$. Then $100N = 227.2727\ldots$.

Multiply N by 100, because 2 digits repeat.

$$100N = 227.2727\ldots$$
$$-\ N = \quad\ 2.2727\ldots$$
$$99N = 225$$

$$N = \frac{225}{99} \text{ or } 2\frac{3}{11}$$

Subtracting eliminates the repeating part, $0.2727\ldots$.

So, $2.272727\ldots = \frac{225}{99}$ or $2\frac{3}{11}$.

Which multiplier should you choose to express $0.8\overline{17}$ as a fraction?

These and other examples suggest that you can rename any repeating decimal as a fraction. Therefore, repeating decimals are included in the set of rational numbers.

Checking for Understanding

Communicating Algebra

1. In your own words, explain why 0.125 is a rational number.

2. Explain how you would choose a multiplier to express $6.\overline{12}$ as a fraction.

Write each expression as a decimal and then as a fraction.

3. five hundredths
4. seven tenths
5. six thousandths
6. ninety-two hundredths
7. two and six tenths
8. one and three thousandths
9. sixteen and twenty-two hundredths

Guided Practice

Complete.

10. $0.9 = \frac{\blacksquare}{10}$

11. $0.7 = \frac{\blacksquare}{10}$

12. $0.25 = \frac{\blacksquare}{100} = \frac{\blacksquare}{4}$

13. $2.53 = 2\frac{\blacksquare}{100}$

14. $3.05 = 3\frac{\blacksquare}{100} = 3\frac{\blacksquare}{20}$

15. $0.408 = \frac{\blacksquare}{1000} = \frac{\blacksquare}{125}$

What multiplier, 10 or 100, would you use to express each decimal as a fraction?

16. $9.222\ldots$

17. $0.181818\ldots$

18. $5.\overline{29}$

Exercises

Independent Practice

Express each decimal as a fraction or mixed number in simplest form.

19. 0.5 **20.** 0.8 **21.** 0.32 **22.** 0.75 **23.** 0.84

24. 0.05 **25.** 0.98 **26.** 0.52 **27.** -0.66 **28.** 0.57

29. -2.26 **30.** 3.54 **31.** -9.64 **32.** -1.38 **33.** 5.31

34. 0.125 **35.** -0.744 **36.** 2.75 **37.** 1.51 **38.** 0.101

39. -0.562 **40.** 0.303 **41.** 0.486 **42.** 4.309 **43.** 9.626

44. $0.\overline{2}$ **45.** $0.\overline{3}$ **46.** $-0.\overline{5}$ **47.** $1.\overline{4}$ **48.** $0.\overline{8}$

49. $0.\overline{12}$ **50.** $1.\overline{13}$ **51.** $0.\overline{15}$ **52.** $-0.\overline{81}$ **53.** $2.\overline{45}$

Challenge

54. Express $0.\overline{9}$ as a fraction.

55. Express $0.1\overline{6}$ as a fraction.

Mixed Review

56. Translate this phrase into an expression: *the sum of 8 and x.*   (Lesson 1-8)

57. Translate this sentence into a formula: *The unit price (u) of an item equals the total price (t) divided by the weight (w).*   (Lesson 3-5)

58. Find the LCD for the fractions $\frac{2}{7}$ and $\frac{3}{4}$.   (Lesson 4-8)

59. Rename $\frac{-4}{5}$ as a fraction in two different ways.   (Lesson 5-1)

60. *True* or *false:* $-2\frac{1}{2} < -3$.   (Lesson 5-1)

Applications

61. Measurement   A gauge measured the thickness of a piece of metal as 0.023 inches. What fraction of an inch is this?

62. Measurement   A trash bag has a thickness of 1.75 mils. This is 0.00175 inches. What fraction of an inch is this?

Critical Thinking

63. A machinist made a stainless steel peg 2.37 inches in diameter for a $2\frac{3}{8}$-inch diameter hole. Will the peg fit? How do you know?

Wrap-Up

64. Write a sentence explaining why terminating and repeating decimals are rational numbers.

5-3 Estimating Sums and Differences

Objectives:
Round decimals.
Estimate sums and differences of decimals.

The freshman class needs to earn $250 for their annual trip to a waterpark. They made $49.50 at a car wash, $37.83 at a bake sale, $121.93 at a concert, and $52.65 at a game booth. The class treasurer made the following estimate.

$$
\begin{array}{r}
\$\ 49.50 \\
37.83 \\
121.93 \\
+\ 52.65 \\
\end{array}
\quad\longrightarrow\quad
\begin{array}{r}
\$\ 50 \\
40 \\
120 \\
+\ 50 \\
\hline
\$260 \\
\end{array}
$$

Round to the nearest ten.

The class has earned *about* $260 and should have enough for the trip.

One way to estimate is to use rounding. A review of the rules for rounding is given in Examples 1 and 2.

Examples

1 Round 2.6 to the nearest whole number.

Look at the digit to the right of the ones place. Since $6 > 5$, round the digit in the ones place up.

To the nearest whole number, 2.6 is 3.

2 Round 9.328 to the nearest tenth.

Look at the digit to the right of the tenths place. Since $2 < 5$, the digit in the tenths place remains the same.

To the nearest tenth, 9.328 is 9.3.

Estimation is often used to provide a quick and easy answer when an exact answer is not necessary. It is also an excellent way to check the reasonableness of answers.

Calculator Hint

Always estimate before using a calculator to solve problems. It is a good way to check that you have entered the numbers correctly.

Example

3 Suppose you choose three items at the grocery store that cost $1.83, $0.95, and $2.10. The cashier tells you that the total is $6.88. Is this reasonable?

Round each item to the nearest dollar amount.

$1.83 rounds to $2. $0.95 rounds to $1. $2.10 rounds to $2.

The total should be about $2 + $1 + $2 or $5. Therefore, a total of $6.88 is *not* reasonable.

Checking for Understanding

Communicating Algebra

1. State two reasons for using estimation.

2. Give two examples of numbers that round to the nearest whole number. One example should round up; the other should not.

Guided Practice

Round to the nearest whole number or dollar.

3. 7.058
4. 6.92
5. 12.461
6. 902.496

7. 40.47
8. $5.69
9. $45.04
10. $0.88

Estimate. Then state whether the answer shown in color is reasonable.

11. $7.6 + 12.4 + 13.2$ 43.2
12. $24.62 + 15.31 + 12.76$ 52.69
13. $195.7 - 156.8$ 38.9
14. $7810 - 5670$ 1140

Exercises

Independent Practice

Round to the nearest tenth.

15. 6.028
16. 1.84
17. 24.692
18. 35.95

19. 95.499
20. 35.506
21. 49.573
22. 24.692

Estimate. Use the data at the right. Be prepared to explain your answers.

23. cost of small popcorn and large drink

24. cost of nachos with cheese and medium drink

25. change from $10 for 2 orders of nachos

26. cost of a T-shirt and poster

27. Is $3.50 enough to buy a large popcorn, apple, and medium drink?

28. Is $16.23 the right change from $20 for a large ice cream and medium drink?

29. What items would you buy if you had $10 to spend?

Popcorn .... sm.	$1.38
.... lg.	$2.29
Nachos w/cheese	$1.88
Soft drinks .... sm.	$0.85
.... med.	$1.05
.... lg.	$1.75
Apples	$0.65
Ice cream .... sm.	$1.29
.... lg.	$1.72

T-shirts	$5.25
Posters	$3.75

Mixed Review

30. Write an expression for the phrase *the product of nine to the third power and two.* (Lesson 4-2)

31. Graph $\dfrac{-8}{4}$ on a number line. (Lesson 5-1)

32. Express -0.34 as a fraction in simplest form. (Lesson 5-2)

Application

33. **Personal Finance** Martin had $10 when he went to the store. He bought toothpaste, $1.89, suntan lotion, $4.39, paper, $1.27, and soap, $2.04. He estimated his total to see whether he had enough money. What is *your* estimate? Did he have enough money?

Critical Thinking

34. The sum of 11, 29, and 61 is 101. Find three 3-digit numbers that have a sum of 1001. Use the digits 1, 1, 2, 9, 6, 1, and any three other digits.

Wrap-Up

35. Tell about a real-life situation that does not require an exact answer, only an estimate.

5-4 Rationals: Adding and Subtracting Decimals

Objective:
Add and subtract decimals.

Key Terms:
closure
commutative
associative
identity
inverse

The Mustang Band Boosters ledger shows these entries for the week.

Item	Credit	Debit
○ Beginning Balance	$148.50	
Rummage Sale	129.38	
Bus Trip		$258.30
Dues Collected	155.00	
○ Refund on Sales Tax	43.28	
Postage and Mailing		74.80

Credits represent positive rational numbers and debits represent negative rational numbers. In order to find the amount of money that the band boosters have at the end of the week, you will need to add rational numbers. All the properties of addition of whole numbers and integers also apply to rational numbers.

Is subtraction of whole numbers closed? If not, give a counterexample.

Properties of Addition	Examples
Closure Property For any rational numbers a and b, a + b is a rational number.	3 + 4 = 7 1 + (−5) = −4 3.7 + 4.2 = 7.9
Commutative Property For any rational numbers a and b, a + b = b + a.	−3 + 4 = 4 + (−3) −5.9 + 6.3 = 6.3 + (−5.9)
Associative Property For any rational numbers a, b, and c, (a + b) + c = a + (b + c).	(−8.2 + 3.7) + 6.5 = −8.2 + (3.7 + 6.5)
Identity Property For every rational number a, a + 0 = a and 0 + a = a.	−2 + 0 = −2 −9.3 + 0 = −9.3
Inverse Property For every rational number a, a + (−a) = 0.	5 + (−5) = 0 17.5 + (−17.5) = 0

You will use these properties and the rules you learned for adding and subtracting integers to add and subtract all rational numbers.
Estimating will help you determine the reasonableness of your answers.

Examples

Solve each equation.

1 $d = 12.4 + 13.7$ Estimate: $12 + 14 = 26$

$$\begin{array}{r} 12.4 \\ + \ 13.7 \\ \hline 26.1 \end{array}$$ or $12.4 \ \boxplus \ 13.7 \ \boxminus \ 26.1$

$d = 26.1$ Compare with the estimate. The answer is reasonable.

2 $f = 119 - 105.7$ Estimate: $120 - 110 = 10$

$$\begin{array}{r} 119.0 \\ - \ 105.7 \\ \hline 13.3 \end{array}$$ or $119 \ \boxminus \ 105.7 \ \boxminus \ 13.3$

$f = 13.3$ Compare with the estimate.

3 $m = 7.5 - 9.8$ Estimate: $8 - 10 = -2$

$= 7.5 + (-9.8)$ Subtraction rule

$$\begin{array}{r} 9.8 \\ - \ 7.5 \\ \hline 2.3 \end{array}$$

$m = -2.3$ The difference is negative because $|-9.8| > |7.5|$.

4 Simplify $3.9n + 5.3n - 12.4n$.   Estimate: $4 + 5 - 12 = -3$

$3.9n + 5.3n - 12.4n = (3.9 + 5.3 - 12.4)n$ Distributive property

$= (9.2 - 12.4)n$

$= -3.2n$ The answer is reasonable.

Calculator Hint

When you use a calculator to compute subtraction problems like the one in Example 3, it is not necessary to write it as an addition problem.

$7.5 \ \boxminus \ 9.8 \ \boxminus$

The display shows -2.3.

Checking for Understanding

Communicating Algebra

1. Using rational numbers, not all integers, give an example of the commutative property of addition.

2. Give an example of a rational number and its additive inverse.

3. In your own words, tell how to subtract rational numbers. Give examples of several different cases.

Guided Practice

State where the decimal point should be placed in each sum or difference.

4. $3.5 + 2.7 = 62$ 5. $5.92 - 3.61 = 231$ 6. $4.05 + 6.7 = 1075$

7. $9.32 - 5.10 = 422$ 8. $7.08 + 5.932 = 13012$ 9. $80.5 + 8.05 = 8855$

Exercises

Name the property shown by each sentence.

10. $9.27 + 0 = 9.27$

11. $3.2 + (-4.5) = -4.5 + 3.2$

12. $5.6 + (-5.6) = 0$

13. $3.47 + 0.16$ is a rational number.

Solve each equation.

14. $x = 3.5 + 7.3$

15. $t = 90.47 + 14.3$

16. $19.2 - 17.8 = c$

17. $47.7 - 39.09 = p$

18. $53.72 - 7.093 = a$

19. $r = 0.593 - 0.3879$

20. $-2.3 + 5.7 = x$

21. $5.3 - 2.6 = c$

22. $s = -4.9 + (-3.2)$

23. $p = -9.31 - (-0.5)$

24. $-21.4 - 16.72 = w$

25. $b = -11.28 - (-4.8)$

Simplify each expression.

26. $4n + 2.1n$

27. $5.3s - 4 + 2s$

28. $9.62 + (2.1 + 5.06)t$

29. $35.2w - 20.91w$

30. $0.2k + 3.1r + 1.3k$

31. $19y - 5.6y - 10.32y$

Evaluate each expression if $a = 2.1$, $b = 6.08$, $x = 12.52$, and $y = 0.9$.

32. $a + x$

33. $b - y$

34. $x - (a + b)$

35. $a + b + y$

36. $b + y - a$

37. $(a - y) + x$

38. Solve the inequality $k + (-3) \leq 8$.   (Lesson 3-7)

39. Evaluate the expression $2d^3$ if $d = 4$.   (Lesson 4-2)

40. Express the decimal 3.2 as a mixed number.   (Lesson 5-2)

41. Round 24.561 to the nearest tenth.   (Lesson 5-3)

42. Mr. Tsou drove 11.3 miles to the dentist, 7.5 miles to the grocery, and 4.4 miles back home. Estimate how far he drove.   (Lesson 5-3)

43. Utilities   In a recent year, the average cost of one kilowatt-hour of electricity was $0.074. The following year, it had risen to $0.113. What was the amount of increase to the nearest cent?

44. Accounting   See the ledger on page 174. Write an equation that represents the Mustang Band Boosters weekly ledger and find the balance at the end of the week.

45. Personal Finance   Charley has $475.29 in his checking account. He receives his paycheck for $125.90. The same day, he receives bills for $397.28 and $225.40. If he pays both bills, what is his financial condition at the end of the day?

46. Sue has 45 ft of chicken wire for a fence. She wants to build a triangular pen that is 11.2 ft on one side and 9.35 ft on a second side. How much wire will she have left for the third side? Can she use all of it? Why or why not?

47. Write a paragraph explaining how estimation helps in solving addition and subtraction problems with decimals.

5-5 Rationals: Adding and Subtracting Like Fractions

Objective:
Add and subtract fractions with like denominators.

Key Terms:
mixed number

In music two eighth notes are played for the same time as one quarter note. One way to illustrate this is to use fractions.

$$\frac{1}{8} + \frac{1}{8} = \frac{2}{8}$$ THINK: One-eighth plus one-eighth is two-eighths.

$$= \frac{1}{4}$$ Write the fraction in simplest form.

The example shown above suggests the following rule.

Adding Like Fractions	In words:	To add fractions with like denominators, add the numerators.
	In symbols:	For fractions $\frac{a}{c}$ and $\frac{b}{c}$, where $c \neq 0$, $\frac{a}{c} + \frac{b}{c} = \frac{a+b}{c}$.

What is the sum of $\frac{a}{2}$ and $\frac{b}{2}$?

Sometimes the sum of two fractions is greater than 1. When this happens, we usually write the sum as a mixed number in simplest form. A **mixed number** is the sum of a whole number and a fraction.

Examples

Solve each equation.

1 $\frac{7}{12} + \frac{11}{12} = a$ Since the denominators are the same, add the numerators.

$$\frac{18}{12} = a$$

$$1\frac{6}{12} = a$$ Rename $\frac{18}{12}$ as a mixed number, $1\frac{6}{12}$.

$$1\frac{1}{2} = a$$ Write the mixed number in simplest form.

2 $n = 2\frac{1}{4} + 3\frac{3}{4}$

$$n = \left(2 + \frac{1}{4}\right) + \left(3 + \frac{3}{4}\right)$$

$$n = (2 + 3) + \left(\frac{1}{4} + \frac{3}{4}\right)$$ Associative and Commutative properties

$$n = 5 + 1$$ Add the numerators and simplify.

$$n = 6$$

Mental Math Hint

THINK: $\frac{1}{4} + \frac{3}{4} = 1$

$2 + 3 + 1 = 6$

Subtracting fractions with like denominators is similar to adding.

| Subtracting Like Fractions | For fractions $\frac{a}{c}$ and $\frac{b}{c}$, where $c \neq 0$, $\frac{a}{c} - \frac{b}{c} = \frac{a-b}{c}$. |

Examples

3 Solve $x = \frac{5}{16} - \frac{25}{16}$.

$x = \frac{5}{16} - \frac{25}{16}$ Since the denominators are the same, subtract the numerators.

$x = \frac{-20}{16}$

$x = -1\frac{4}{16}$ or $-1\frac{1}{4}$ Rename as a mixed number and simplify.

4 Simplify $2\frac{1}{3}n + \frac{2}{3}n - 1\frac{2}{3}n$.

$2\frac{1}{3}n + \frac{2}{3}n - 1\frac{2}{3}n = \left(2\frac{1}{3} + \frac{2}{3} - 1\frac{2}{3}\right)n$ Distributive property

$= \left(2\frac{3}{3} - 1\frac{2}{3}\right)n$

$= 1\frac{1}{3}n$

Checking for Understanding

Communicating Algebra

1. Define like fractions and give several examples of them.

2. In your own words, state a simple rule for adding and subtracting like fractions.

3. What properties allow you to add the whole number and fraction parts of mixed numbers separately?

4. Express $\frac{22}{3}$ as a mixed number.

Guided Practice

Solve each equation. Write each solution in simplest form.

5. $\frac{5}{8} + \frac{2}{8} = a$

6. $\frac{25}{13} + \frac{12}{13} = x$

7. $\frac{10}{17} + \frac{-3}{17} = y$

8. $\frac{15}{18} - \frac{14}{18} = c$

9. $\frac{19}{25} - \frac{7}{25} = n$

10. $\frac{-24}{40} + \frac{15}{40} = z$

Exercises

Independent Practice

Solve each equation. Write each solution in simplest form.

11. $\frac{7}{9} + \frac{1}{9} = a$

12. $c = \frac{9}{15} + \frac{1}{15}$

13. $b = \frac{12}{13} - \frac{10}{13}$

14. $\frac{17}{20} - \frac{2}{20} = g$

15. $\frac{30}{12} - \frac{14}{12} = x$

16. $\frac{42}{16} + \frac{12}{16} = j$

17. $b = \frac{32}{18} - \frac{7}{18}$

18. $n = \frac{25}{19} + \frac{13}{19}$

19. $\frac{36}{21} - \frac{8}{21} = y$

20. $\frac{18}{26} + \frac{-15}{26} = m$

21. $h = \frac{25}{30} + \frac{-4}{30}$

22. $\frac{56}{32} - \frac{16}{32} = s$

Simplify each expression.

23. $5\frac{1}{2}n + \frac{1}{2}n - 2\frac{1}{2}n$

24. $-3\frac{1}{2}r + \left(-2\frac{1}{2}r\right) + 5r$

25. $7\frac{1}{5}x + \left(-3\frac{2}{5}x\right) + 5\frac{1}{5}x$

26. $-3y - 7\frac{1}{3}y + \left(-5\frac{1}{3}y\right)$

27. $\frac{9}{5}t + \left(-2\frac{1}{5}t\right) - \frac{3}{5}t$

28. $2\frac{5}{8}b - 3\frac{1}{8}b + \left(-1\frac{7}{8}b\right)$

Mixed Review

29. *True* or *false:* Quebec is a city in South America or Peru is a country in North America. (Lesson 3-9)

30. Find the GCF of 48 and 104. (Lesson 4-5)

31. Round 99.55 to the nearest whole number. (Lesson 5-3)

32. Solve the equation $y = -5.3 - 18.1$. (Lesson 5-4)

33. Simplify the expression $8.3d + 2.2d + 7.91d$. (Lesson 5-4)

Applications

34. **Home Economics** Andre used $2\frac{1}{2}$ yards of material for a shirt and $3\frac{1}{2}$ yards for a jacket. How many yards did he use altogether?

35. **Carpentry** A carpenter needs a $14\frac{3}{4}$-inch piece of molding. How much remains after she cut off the end of a 36-inch piece of molding?

Critical Thinking

36. On a balance scale a brick balances with a 3-pound weight and one-half of a brick. How much does the brick weigh?

Wrap-Up

37. Write a few sentences that would explain to a younger student how to add fractions with like denominators.

Mid-Chapter Quiz

Replace each ⬤ with <, >, or = to make each statement true. (Lesson 5-1)

1. $-\frac{15}{5}$ ⬤ -3

2. $-\frac{3}{4}$ ⬤ 0

3. 0.24 ⬤ -0.24

4. 2.61 ⬤ 2.065

Express each decimal as a fraction or mixed number in simplest form. (Lesson 5-2)

5. 0.6

6. 0.35

7. -2.4

8. $0.\overline{6}$

9. $1.\overline{5}$

Estimate each sum or difference. (Lesson 5-3)

10. $\$12.91 + \$6.52$

11. $58.9 - 28.76$

12. $8.1 + 7.5 + 6.2 + 9.9$

Solve each equation. (Lessons 5-4, 5-5)

13. $b = 5.3 + 8.1$

14. $24 - 5.6 = x$

15. $c = -3.2 + 2.5$

16. $n = \frac{9}{11} + \frac{13}{11}$

17. $\frac{7}{8} - \frac{3}{8} = m$

18. $2\frac{1}{3} + 1\frac{2}{3} = d$

5-6 Rationals: Adding and Subtracting Unlike Fractions

Objective:
Add and subtract fractions with unlike denominators.

The diagram at the right is typical of the kind of drawing an architect or builder uses. This diagram shows a cross-section of an outside wall. Notice that the measurements are given as fractions or mixed numbers. How thick is the entire wall?

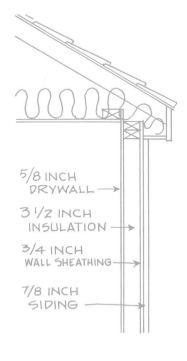

5/8 INCH DRYWALL →

3 1/2 INCH INSULATION →

3/4 INCH WALL SHEATHING →

7/8 INCH SIDING →

To find the thickness of the wall, add $\frac{5}{8}$, $3\frac{1}{2}$, $\frac{3}{4}$, and $\frac{7}{8}$. The fractions do not have the same denominator, but they can be renamed with a common denominator of 8.

$$\frac{5}{8} = \frac{5}{8}$$

$$3\frac{1}{2} = 3\frac{4}{8} \qquad \text{Rename } \frac{1}{2} \text{ as } \frac{4}{8}.$$

$$\frac{3}{4} = \frac{6}{8} \qquad \text{Rename } \frac{3}{4} \text{ as } \frac{6}{8}.$$

$$+\frac{7}{8} = \frac{7}{8}$$

$$\overline{}$$

$$3\frac{22}{8} = 5\frac{6}{8} \text{ or } 5\frac{3}{4} \qquad \text{The wall is } 5\frac{3}{4} \text{ inches thick.}$$

> **Adding and Subtracting Unlike Fractions**
>
> To find the sum or difference of two fractions with unlike denominators, rename the fractions with a common denominator. Then add or subtract and simplify.

In Chapter 4, you learned that the least common denominator is the least common multiple of the denominators. The LCD is helpful in renaming fractions for adding and subtracting.

Examples

1 Solve $a = \frac{7}{12} + \frac{8}{15}$. **Write the solution in simplest form.**

$$a = \frac{7}{12} + \frac{8}{15} \qquad \begin{array}{l} 12 = 2^2 \cdot 3 \text{ and } 15 = 3 \cdot 5 \\ \text{The LCM of 12 and 15 is } 2^2 \cdot 3 \cdot 5 \text{ or } 60. \end{array}$$

$$a = \frac{35}{60} + \frac{32}{60} \qquad \frac{7 \cdot 5}{12 \cdot 5} = \frac{35}{60} \text{ and } \frac{8 \cdot 4}{15 \cdot 4} = \frac{32}{60}$$

$$a = \frac{67}{60} \text{ or } 1\frac{7}{60} \qquad \text{Rename } \frac{67}{60} \text{ as } 1\frac{7}{60}.$$

2 **Solve $\frac{1}{3} - \frac{3}{4} = c$.**

$$\frac{1}{3} - \frac{3}{4} = c$$

$$\frac{4}{12} - \frac{9}{12} = c \qquad \text{The LCM of 3 and 4 is 12.}$$

$$\frac{4}{12} + \left(-\frac{9}{12}\right) = c \qquad \text{Subtract } \frac{9}{12} \text{ by adding its inverse, } -\frac{9}{12}.$$

$$-\frac{5}{12} = c$$

Likewise, to add or subtract mixed numbers with unlike denominators, first rename the fractions with a common denominator.

Examples

3 **Solve $b = 2\frac{5}{8} + 3\frac{11}{24}$.** Estimate: $3 + 3 = 6$

$$b = 2\frac{15}{24} + 3\frac{11}{24} \qquad \text{Use the LCM of 8 and 24 to rename } \frac{5}{8} \text{ as } \frac{15}{24}.$$

$$b = 5\frac{26}{24}$$

$$b = 6\frac{2}{24} \text{ or } 6\frac{1}{12} \qquad \text{Simplify.}$$

In Example 4, why is it necessary to rename $8\frac{10}{36}$ as $7\frac{46}{36}$?

4 **Solve $8\frac{5}{18} - 6\frac{7}{12} = y$.** Estimate: $8 - 7 = 1$

$$8\frac{5}{18} \quad \longrightarrow \quad 8\frac{10}{36} \quad \longrightarrow \quad 7\frac{46}{36} \qquad \text{Rename } 8\frac{10}{36} \text{ as } 7\frac{46}{36}.$$

$$-6\frac{7}{12} \qquad\qquad -6\frac{21}{36} \qquad\qquad -6\frac{21}{36}$$

$$\overline{\phantom{-6\frac{7}{12}}} \qquad\qquad \overline{\phantom{-6\frac{21}{36}}} \qquad\qquad \overline{1\frac{25}{36}}$$

$$1\frac{25}{36} = y$$

5 **Solve $d = 3\frac{2}{3} - 5\frac{1}{6}$.** Estimate: $4 - 5 = -1$

$$d = 3\frac{2}{3} - 5\frac{1}{6}$$

$$d = 3\frac{4}{6} - 5\frac{1}{6} \qquad \text{Find a common denominator.}$$

$$d = 3\frac{4}{6} + \left(-5\frac{1}{6}\right) \qquad \text{Subtract } 5\frac{1}{6} \text{ by adding its inverse, } -5\frac{1}{6}.$$

$$d = 3\frac{4}{6} + \left(-4\frac{7}{6}\right) \qquad -5\frac{1}{6} = -4\frac{7}{6}$$

$$d = -1\frac{3}{6} \text{ or } -1\frac{1}{2}$$

Checking for Understanding

Communicating
Algebra

1. What is the first step in adding or subtracting fractions with unlike denominators?

2. State the rules for finding the sign of sums of positive and negative numbers. (Look in Lesson 2-3 to review these rules.)

Guided
Practice

Find the LCD. Then find each sum or difference.

3. $\frac{1}{4} + \frac{3}{8}$

4. $\frac{5}{6} - \frac{1}{3}$

5. $\frac{2}{7} + \frac{2}{14}$

6. $\frac{7}{8} - \frac{2}{5}$

7. $\frac{5}{8} - \frac{1}{2}$

8. $\frac{9}{10} + \frac{5}{6}$

9. $\frac{8}{12} - \frac{1}{3}$

10. $\frac{4}{5} + \frac{11}{15}$

Complete.

11. $7\frac{2}{5} = 6\frac{\blacksquare}{5}$

12. $2\frac{1}{9} = 1\frac{\blacksquare}{9}$

13. $5\frac{8}{30} = 4\frac{\blacksquare}{30}$

14. $6\frac{4}{12} = 5\frac{\blacksquare}{12}$

Exercises

Independent
Practice

Solve each equation. Write each solution in simplest form.

15. $\frac{6}{7} + \frac{11}{14} = c$

16. $\frac{7}{8} - \frac{3}{10} = x$

17. $\frac{11}{10} - \frac{3}{5} = k$

18. $a = \frac{7}{6} + \frac{5}{18}$

19. $-5 - 3\frac{1}{2} = j$

20. $-2 - (-4) = m$

21. $a = -7 - \left(-2\frac{3}{4}\right)$

22. $p = -\frac{3}{4} - \left(-\frac{1}{4}\right)$

23. $3\frac{5}{7} - \left(-1\frac{1}{7}\right) = x$

24. $b = 3\frac{1}{3} - 2\frac{1}{6}$

25. $s = 6\frac{3}{4} - 3\frac{1}{2}$

26. $4\frac{1}{3} - 2\frac{1}{2} = t$

27. $c = -\frac{9}{5} - \left(-\frac{3}{5}\right)$

28. $r = -\frac{3}{11} - \frac{7}{11}$

29. $8\frac{1}{5} - 2\frac{1}{4} = v$

Mixed Review

Evaluate each expression if $c = 4.3$ and $d = 17.04$. (Lesson 5-4)

30. $c + d$

31. $c - d$

Solve each equation. Write each solution in simplest form. (Lesson 5-5)

32. $n = \frac{23}{14} - \frac{5}{14}$

33. $\frac{33}{12} + \frac{9}{12} = g$

34. $x = \frac{4}{10} + \left(-\frac{3}{10}\right)$

Application

35. **Publishing** The length of a page in a yearbook is 10 inches. The top margin is $\frac{1}{2}$-inch and the bottom margin is $\frac{3}{4}$-inch. What is the length of the page inside the margins?

Critical
Thinking

36. A rope is cut in half and one-half is used. Then one-fifth of the remaining rope is cut off and used. The piece left is 12 feet long. How long was the rope originally?

Wrap-Up

37. **Make Up a Problem** Write a problem in which you would add the mixed numbers $12\frac{1}{2}$ and $18\frac{3}{4}$.

Algebra in Action-Business

Stock Market Reports

Companies often are funded through the sale of shares of stock. The stock is purchased by individuals or groups in the hope that the price of the stock will rise and that the company will be profitable and pay dividends. Stock prices change daily. Rational numbers are used to show the increase or decrease in the price of a share of stock.

The newspaper clipping at the right shows part of the stock market page. Listed are

- the name of the company (usually abbreviated);
- the high and low price for the past 52 weeks;
- the high, low, and closing price, in dollars, for the last business day;
- the change, in dollars, from the previous day's closing price.

A listing of $23\frac{1}{2}$ is read as $23\frac{1}{2}$ points and means \$23.50. (A point is a dollar, so $\frac{1}{2}$ point is \$0.50.)

52-Week High	Low	Stock - Div	High	Low	Last	Chg.
$24\frac{7}{8}$	17	ChWstes .16	$23\frac{1}{2}$	23	23	$-\frac{1}{4}$
$23\frac{7}{8}$	$17\frac{7}{8}$	Chspk .72	$19\frac{1}{8}$	$18\frac{1}{2}$	$18\frac{5}{8}$	$-\frac{3}{4}$
$73\frac{1}{2}$	$52\frac{3}{4}$	Chevrn 2.80	70	$68\frac{3}{4}$	$68\frac{7}{8}$	$-\frac{7}{8}$
149	$136\frac{3}{4}$	ChiMlw 6.00r	$138\frac{1}{4}$	$138\frac{1}{4}$	$138\frac{1}{4}$	$-\frac{1}{4}$
$22\frac{1}{2}$	13	Chile n .34e	$17\frac{7}{8}$	$17\frac{5}{8}$	$17\frac{5}{8}$	$-\frac{3}{8}$
$38\frac{1}{2}$	$27\frac{5}{8}$	Chilis s	$37\frac{1}{8}$	37	37	$-\frac{3}{8}$
$26\frac{3}{4}$	$13\frac{1}{8}$	Chiqula .40	$26\frac{7}{8}$	$25\frac{3}{8}$	$25\frac{1}{2}$	$-\frac{7}{8}$
$8\frac{3}{4}$	$4\frac{7}{8}$	ChkFull .31t	$7\frac{7}{8}$	$7\frac{3}{4}$	$7\frac{3}{4}$	....
$42\frac{7}{8}$	$30\frac{1}{8}$	ChrisCr 1.61t	$32\frac{1}{4}$	$31\frac{1}{8}$	$31\frac{3}{8}$	$-\frac{5}{8}$
16	$13\frac{1}{4}$	ChCft pf 1.00	$14\frac{1}{4}$	$14\frac{1}{4}$	$14\frac{1}{4}$	$+\frac{1}{4}$
$13\frac{1}{2}$	$8\frac{1}{4}$	Christn	$12\frac{3}{4}$	$12\frac{5}{8}$	$12\frac{5}{8}$	$-\frac{1}{4}$
$27\frac{1}{8}$	$14\frac{1}{2}$	Chryslr 1.20	$16\frac{5}{8}$	16	$16\frac{1}{8}$	$-\frac{1}{4}$
$51\frac{3}{8}$	$34\frac{1}{8}$	Chubbs 1.32	46	$44\frac{7}{8}$	$45\frac{1}{8}$	$-\frac{5}{8}$
$4\frac{5}{8}$	$1\frac{3}{8}$	Chyron	2	$1\frac{7}{8}$	2	....
$39\frac{3}{8}$	$31\frac{3}{4}$	Cilcorp 2.46	$33\frac{3}{4}$	$33\frac{1}{2}$	$33\frac{1}{2}$	$-\frac{3}{8}$
35	22	CinnBel .76	$24\frac{1}{4}$	24	$24\frac{1}{8}$	....
$32\frac{3}{8}$	$27\frac{1}{8}$	CinGE 2.40	$30\frac{3}{8}$	30	30	$-\frac{1}{2}$
46	40	CinG pf 4.00	45	45	45	+1
99	93	CinG pf 9.30	94	$93\frac{1}{2}$	94	$+\frac{1}{2}$
81	$74\frac{1}{4}$	CinG pf 7.44	$77\frac{5}{8}$	$77\frac{5}{8}$	$77\frac{5}{8}$	$-\frac{7}{8}$
98	$91\frac{1}{2}$	CinG pf 9.28	$93\frac{1}{2}$	$93\frac{1}{2}$	$93\frac{1}{2}$	....
$100\frac{1}{2}$	95	CinG pf 9.52	97	97	97	+1
$21\frac{7}{8}$	15	CinMil .72	$17\frac{3}{4}$	$17\frac{5}{8}$	$17\frac{3}{4}$	$+\frac{1}{8}$
$13\frac{5}{8}$	$4\frac{1}{2}$	CineOd	$5\frac{1}{8}$	5	$5\frac{1}{8}$	....
$13\frac{1}{8}$	1	viCircK	$1\frac{1}{4}$	1	$1\frac{1}{8}$	....
$28\frac{7}{8}$	$18\frac{5}{8}$	CirClys .10	$25\frac{5}{8}$	$25\frac{3}{8}$	$25\frac{3}{8}$	$-\frac{1}{2}$
$66\frac{1}{4}$	40	Circus	65	63	64	$-\frac{5}{8}$
$35\frac{1}{2}$	$21\frac{3}{4}$	Citicorp 1.78	$22\frac{5}{8}$	22	$22\frac{3}{8}$	....
70	59	Citcp pf 6.00e	$61\frac{1}{2}$	$61\frac{1}{2}$	$61\frac{1}{2}$	$-\frac{1}{4}$
$81\frac{5}{8}$	$67\frac{5}{8}$	Citcp pf A 7.00e	$69\frac{1}{8}$	69	69	$-\frac{1}{4}$
$25\frac{1}{2}$	$22\frac{1}{4}$	Citcp pfC 2.28	$23\frac{7}{8}$	$23\frac{1}{4}$	$23\frac{1}{4}$	$-\frac{1}{4}$
36	23	CtzSCp 1.40	$24\frac{1}{8}$	$23\frac{3}{8}$	$23\frac{1}{2}$	$-\frac{5}{8}$
28	20	ClyNCs .64	$20\frac{5}{8}$	$20\frac{1}{8}$	$20\frac{1}{8}$	$-\frac{3}{8}$

Look at the shaded line, the stock of Chrysler (Chryslr). Its high for the year is $27\frac{1}{8}$ and its low is $14\frac{1}{2}$. It had a high of $16\frac{5}{8}$, a low of 16, and closed at $16\frac{1}{8}$ yesterday. It closed down $\frac{1}{4}\left(-\frac{1}{4}\right)$ from the day before. What did it close at the day before?

Answer each question for Chevron (Chevrn).

1. What is its high for the year? What is its low? How many points difference is there between the high and low?

2. What did it close at yesterday? What did it close at the day before?

Complete the following.

3. Jean bought some shares of Fidelity Trust stock at $96\frac{1}{2}$. Fidelity Trust closed yesterday at $87\frac{5}{8}$. Write and solve an equation that shows what happened to Jean's investment.

4. **Research** Choose a stock from a newspaper listing to follow for several days. Make a graph and then write a few sentences about your stock's activity.

5-7 Solving Equations and Inequalities

Objective:

Solve equations and inequalities with rational numbers.

In 1989 the top two oil-producing states were Texas and Alaska. Together they produced 1372.2 million barrels of oil. How many barrels of oil were produced in Texas?

You know that the two states produced 1372.2 million barrels. From the graph, you know that Alaska produced 684.0 million barrels.

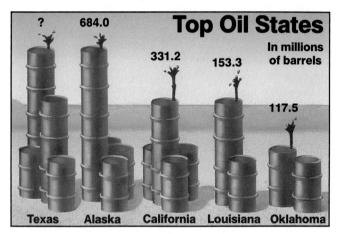

Top Oil States

In millions of barrels

? 684.0 331.2 153.3 117.5

Texas   Alaska   California   Louisiana   Oklahoma

Let b represent the number of barrels, in millions, produced in Texas. Then the equation $b + 684.0 = 1372.2$ represents this problem.

$$b + 684.0 = 1372.2$$
$$b + 684.0 - 684.0 = 1372.2 - 684.0$$

$$1372.2 \; \boxed{-} \; 684.0 \; \boxed{=} \; 688.2$$

Estimate:
1400 − 700 = 700

$$b = 688.2$$

Compare with the estimate. Is the answer reasonable?

Texas produced 688.2 million barrels of oil.

You can solve rational number equations using the same skills you used to solve equations in Chapters 1 and 3.

Examples

Solve each equation. Check your solution.

1
$$x - 4.7 = 1.5$$
$$x - 4.7 + 4.7 = 1.5 + 4.7 \qquad \text{Add 4.7 to each side.}$$
$$x = 6.2$$

Check: $x - 4.7 = 1.5$
$$6.2 - 4.7 \stackrel{?}{=} 1.5$$
$$1.5 = 1.5 \quad ✔$$

FYI

The United States has 590 million barrels of government-owned crude oil in reserve for an emergency. This is the largest reserve in the world.

Estimation Hint

Since $\frac{2}{3}$ is close to 1, and $\frac{5}{2}$ is $2\frac{1}{2}$, x must be about $1\frac{1}{2}$.

2

$$x + \frac{2}{3} = \frac{5}{2}$$

$$x + \frac{2}{3} - \frac{2}{3} = \frac{5}{2} - \frac{2}{3} \qquad \text{Subtract } \frac{2}{3} \text{ from each side.}$$

$$x = \frac{15}{6} - \frac{4}{6} \qquad \text{The LCD of 2 and 3 is 6.}$$

$$x = \frac{11}{6}$$

$$x = 1\frac{5}{6}$$

The solution is $1\frac{5}{6}$.

Check: $x + \frac{2}{3} = \frac{5}{2}$

$$1\frac{5}{6} + \frac{2}{3} \stackrel{?}{=} \frac{5}{2}$$

$$\frac{11}{6} + \frac{4}{6} \stackrel{?}{=} \frac{5}{2}$$

$$\frac{15}{6} \stackrel{?}{=} \frac{5}{2}$$

$$\frac{5}{2} = \frac{5}{2} \quad \checkmark$$

Likewise, rational number inequalities are solved using the same skills used for solving whole number and integer inequalities.

Examples

Solve each inequality.

3

$$x + \frac{1}{4} > 3$$

$$x + \frac{1}{4} - \frac{1}{4} > 3 - \frac{1}{4} \qquad \text{Subtract } \frac{1}{4} \text{ from each side.}$$

$$x > 2\frac{3}{4}$$

Check: Try 3, a number greater than $2\frac{3}{4}$.

$$x + \frac{1}{4} > 3$$

$$3 + \frac{1}{4} \stackrel{?}{>} 3$$

$$3\frac{1}{4} > 3 \quad \checkmark$$

Is $2\frac{3}{4}$ a solution? Why or why not?

The solution is $x > 2\frac{3}{4}$, all numbers greater than $2\frac{3}{4}$.

4 $14.29 + m \leq 8.73$

$$m \leq 8.73 - 14.29 \qquad \text{Subtract 14.29 from each side.}$$

$$m \leq 8.73 + (-14.29) \qquad \text{Subtract by adding the inverse.}$$

$$m \leq -5.56 \qquad \text{Check this solution.}$$

Is -5.56 a solution? Why or why not?

The solution is $m \leq -5.56$, all numbers less than or equal to -5.56.

Checking for Understanding

Communicating Algebra

1. Estimate the solution of the equation $n + 8.7 = 25.2$.

2. Explain how you should solve the equation $y + 7\frac{1}{2} = 3\frac{3}{4}$.

3. Explain how you should solve the inequality $y - 3\frac{1}{2} \geq -5\frac{1}{2}$.

Guided Practice

Explain how to solve each equation. Then solve.

4. $x + (-7) = 10$

5. $y - 1.4 = -9.3$

6. $14 = x + 7\frac{1}{2}$

7. $-14.2 = t - 5$

8. $r + 17\frac{1}{2} = 12\frac{3}{4}$

9. $m - \left(-7\frac{1}{2}\right) = -7\frac{1}{2}$

Explain how to solve each inequality. Then solve.

10. $x + \frac{1}{2} > 5$

11. $a - 1.6 < 3.4$

12. $y - 3 \leq \frac{5}{2}$

Exercises

Independent Practice

Solve each equation. Check your solution.

13. $t - 2.6 = 5.8$

14. $s - 1.3 = 4.7$

15. $x + 5 = \frac{1}{2}$

16. $b + 3\frac{1}{4} = 5$

17. $y - 2.6 = 11.4$

18. $p + 1.7 = -3.4$

19. $b + 16.7 = -4.3$

20. $a + \frac{3}{8} = 2\frac{1}{2}$

21. $z - \frac{7}{8} = 3\frac{1}{4}$

22. $m + \frac{3}{2} = \frac{7}{2}$

23. $t - 2\frac{1}{8} = 3\frac{3}{8}$

24. $n + 21.6 = 16.8$

Solve each inequality. Check your solution.

25. $a + \frac{1}{2} > 4$

26. $d - \frac{2}{3} < \frac{7}{3}$

27. $n - 5.7 < -10$

28. $b - 1.6 \leq 4.3$

29. $f + 2.8 \geq -7.3$

30. $n + \frac{1}{4} \leq -2$

31. $-3.5 + p > -7$

32. $y - 6 \leq -8.5$

33. $a - 7\frac{1}{2} \geq -8$

34. $-5 + d > -\frac{11}{2}$

35. $f - 2 \geq 6\frac{1}{4}$

36. $-2\frac{1}{3} \leq a - \frac{5}{6}$

Language Skill

Write using symbols.

37. x is less than or equal to -4.2.

38. The value of x is greater than 4.5.

39. The value of b is 3.78.

Mixed Review

40. Evaluate the expression $10 + y$ if $y = -15$.   (Lesson 2-3)

41. Find the width of a rectangle with area 240 square feet and length 30 feet. (Lesson 3-6)

42. Evaluate $2x^3$ if $x = -2$.   (Lesson 4-2)

Solve each equation.   (Lessons 5-4, 5-5, 5-6)

43. $-3\frac{1}{5} + \frac{3}{5} = s$

44. $1\frac{3}{8} - 4\frac{1}{8} = t$

45. $3.67 - 0.74 = p$

46. $a = \frac{1}{3} - \left(-\frac{2}{5}\right)$

47. $y = 4\frac{3}{4} - 5\frac{1}{6}$

48. $c = -\frac{2}{7} + \frac{11}{8}$

Applications

Write and solve an equation or inequality for each application.

49. **Landscaping**   Tom needed at least $5\frac{1}{3}$ pounds of grass seed for his lawn. He has $2\frac{1}{2}$ pounds. How much more does he need?

50. **Cooking**   Serafina owes her friend Maria $2\frac{1}{2}$ cups of sugar and she needs $2\frac{1}{4}$ cups to bake cookies. How much sugar does she need to buy to repay Maria and bake the cookies?

51. **Driver's Education**   The odometer on the car read 269.5 miles at the beginning of class. After class the odometer read 298.2 miles. How far was the car driven during class?

52. **Meteorology**   At 2:00 P.M. the barometric pressure was 29.85 inches. It had dropped 0.28 inches from the previous reading. What was the previous reading?

Connection

53. **Geometry**   The sum of the measures of the angles in a triangle is 180°. If two of the angles measure 75.5° and 60.3°, what is the measure of the third angle?

Critical Thinking

54. The sign $\geq$ can also mean *not less than*. What can the sign $\leq$ mean?

Wrap-Up

55. Explain how the solutions of $x + 7 = -5$ and $x + 7 > -5$ differ.

Writing Connection

Write one or more complete sentences to answer each question.

1. How do you know that 3.16 is greater than 3.158?

2. How do you know that $\frac{7}{16}$ is equal to $\frac{14}{32}$?

3. How do you know that -5 is a rational number?

4. How do you know that 3.5 is not the solution of $x + 1.5 = 6.0$?

5-8 Strategy: Look for a Pattern

Objective:
Solve problems by looking for a pattern.

Marguerite has part of a bus schedule. She wishes to take the bus to visit her grandmother but she cannot leave until after 1:00 P.M. What is the earliest time Marguerite can catch the bus?

Bus Schedule

Departures

8:35 AM
9:23 AM
10:11 AM
10:59 AM

Explore
The schedule shows several morning departure times. Marguerite needs to know the earliest departure time after 1:00 P.M.

Plan
Since bus schedules often follow patterns, look for a pattern. Once you find a pattern, you can extend the schedule beyond 1:00 P.M.

Solve
Notice that there are 48 minutes between each of the departure times.

8:35 +48 min. → 9:23 +48 min. → 10:11 +48 min. → 10:59

According to this pattern, the next scheduled departure would be 48 minutes after 10:59, or 11:47 P.M. The schedule would continue as follows.

11:47 +48 min. → 12:35 +48 min. → 1:23

Marguerite can catch the bus at 1:23 P.M.

What patterns can you observe in your classroom?

Examine
You can estimate to check the reasonableness of your answer. Since there are about 50 minutes between departures, and the last known time is about 11:00, the next times should be about 11:50, 12:40, and 1:30. Therefore, the answer 1:23 seems reasonable.

Looking for, and then extending, patterns is a good problem-solving strategy. Sometimes the patterns are easy to see and sometimes they may be more hidden.

Checking for Understanding

Communicating Algebra

1. Use the pattern at the right to find 9999 × 5 and 9999 × 6. Explain your thinking.

9999 × 1 = 9999
9999 × 2 = 19,998
9999 × 3 = 29,997
9999 × 4 = 39,996

2. Look at the pattern of numbers shown below.

    1   2   4   7

Is the next number found by multiplying by 2? Why or why not?

Exercises

Solve. Look for a pattern.

3. Use the pattern at the right to find $11{,}111 \times 11{,}111$ and $111{,}111 \times 111{,}111$.

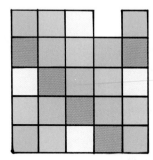

$$
\begin{array}{ll}
1 \times 1 & = 1 \\
11 \times 11 & = 121 \\
111 \times 111 & = 12{,}321 \\
1{,}111 \times 1{,}111 & = 1{,}234{,}321
\end{array}
$$

4. At Fairview High School the bell rings at 8:05, 8:51, 8:55, 9:41, and 9:45 each weekday morning. When do the next three bells ring?

5. Look at the quilt at the right. What color is the piece needed to complete the quilt?

ESTIMATION
MENTAL MATH
CALCULATOR
PAPER/PENCIL

Solve. Use any strategy.

6. Two peaches and one pear cost $1.15. Three peaches and three pears cost $2.40. What is the cost of one peach?

7. Willa is conditioning for the swim team. On the first day, she did 5 push-ups. The second day she did 6 push-ups. The third day she did 8 and on the fourth day she did 11. To continue this pattern, how many push-ups should she do on the fifth day, the sixth day, and the seventh day? On which day will she reach her goal of 40 push-ups?

8. Willa swam 1 lap on the first day. She swam 2 laps on the second day, 4 laps on the third day, and 8 laps on the fourth day. To continue this pattern, how many laps should she swim on the seventh day?

9. Ross travels south on his bicycle riding 8 miles per hour. One hour later, his friend Mason starts riding his bicycle from the same location. He travels south trying to catch up with Ross. If Mason rides 10 miles per hour for 3 hours, will he catch up with Ross?

10. What is the sum of the first 50 odd numbers?

11. The Bulldogs have lost $\frac{2}{3}$ of their games. If they win their last four games, they will have won as many games as they lost. How many games have the Bulldogs won so far?

Critical Thinking

12. Jon is stacking oranges to form a triangular pyramid. Each orange touches three oranges below it as shown at the right. In order to have a pyramid with 6 layers, how many oranges should Jon put on the bottom layer?

Wrap-Up

13. Suppose another student is having difficulty finding a pattern in a list of numbers. Write some hints you could give about how to find a pattern.

5-9 Arithmetic Sequences

Objectives:
Find terms of
arithmetic sequences.
Represent a
sequence algebraically.

Key Terms:
sequence
term
arithmetic sequence
common difference

In 1993, the first class postage rate was 29¢ for the
first ounce and 23¢ for each additional ounce. A chart
showing the postage for weights up to 5 ounces is
shown below. What was the cost for a 9-oz letter?

Weight	1 oz	2 oz	3 oz	4 oz	5 oz	6 oz	7 oz	8 oz	9 oz
Postage	$0.29	$0.52	$0.75	$0.98	$1.21				

Look for a pattern in the chart. After the first ounce, each additional
ounce costs an additional 23¢. By extending the chart, you can see that
the cost for a 9-oz letter was $2.13.

A list of numbers in a certain order, such as 1, 2, 3, 4, ... or 0.29, 0.52,
0.75, 0.98, ... , is called a **sequence.** Each number is called a **term** of
the sequence. When the difference between any two consecutive terms is
the same, the sequence is called an **arithmetic sequence.** The
difference is called the **common difference.**

Arithmetic Sequence	An arithmetic sequence is a sequence in which the difference between any two consecutive terms is the same.

Examples

**State whether each sequence is arithmetic. Then write the next
three terms of each sequence.**

1 5, 6.5, 8, 9.5, ...

Since the difference between any two consecutive terms is 1.5, the
sequence is arithmetic.

$$5 \quad\quad 6.5 \quad\quad 8 \quad\quad 9.5$$
$$+ 1.5 \quad\quad + 1.5 \quad\quad + 1.5$$

The next three terms are 11, 12.5, and 14.

2 1, 3, 6, 10, 15, ...

Since there is no common difference, the sequence is *not* arithmetic.

$$1 \quad\quad 3 \quad\quad 6 \quad\quad 10 \quad\quad 15$$
$$+ 2 \quad\quad + 3 \quad\quad + 4 \quad\quad + 5$$

The next three terms are 21, 28, and 36.

FYI

The United States
Postal Service has
the world's largest
vehicle fleet, 158,758
cars and trucks. The
United States also
ranks first in the
average number of
letters that each
person mails, 635
letters per year.

Given the sequence o, t, t, f, f, s, s, . . . find the next three terms.

3 7, 4, 1, -2, -5, . . .

Notice the pattern of common differences, -3. This sequence is arithmetic.

7   4   1   -2   -5

-3   -3   -3   -3

The next three terms are -8, -11, and -14.

CONNECTION TO ALGEBRA

If we know any term of an arithmetic sequence and the common difference, we can list the terms of the sequence. Consider the postage sequence at the beginning of the lesson: 29, 52, 75, 98, ...

Here's how to write an expression that represents a term in the sequence. The first term is 29; call the first term a. The difference between terms is 23; call the common difference d. Study this pattern.

1st term: a 29
2nd term: $a + d$ 29 + 23 = 52
3rd term: $a + d + d$ or $a + 2d$ 52 + 23 = 75
4th term: $a + d + d + d$ or $a + 3d$ 75 + 23 = 98
⋮   ⋮
nth term: $a + (n - 1)d$

Because n represents any term, you can use this expression to find any term in the sequence.

Example

What is the advantage of using the algebraic expression over extending the sequence to find a particular term?

4 **Use the expression $a + (n - 1)d$ to find the 9th term of the postage sequence 29, 52, 75, 98, ...**

The first term, a, is 29. The common difference, d, is 23. n is 9.

$a + (n - 1)d = 29 + (9 - 1)23$ Replace a with 29, d with 23, and n with 9.
$\quad\quad\quad\quad = 29 + (8)23$
$\quad\quad\quad\quad = 29 + 184$
$\quad\quad\quad\quad = 213$ The ninth term is 213.

A nine-ounce letter would cost $2.13.   Compare this answer to the one in the postage table.

Checking for Understanding

Communicating Algebra

1. Give an example of an arithmetic sequence.

2. In your own words, define an arithmetic sequence.

3. Explain how you can find any term of an arithmetic sequence if you know the first term and the common difference.

Guided Practice

Write the next three terms of each arithmetic sequence.

4. 1, 2, 3, 4, 5, . . .

5. 1, 3, 5, 7, 9, . . .

6. 20, 19, 18, 17, . . .

7. 109, 104, 99, 94, . . .

8. 15, 30, 45, . . .

9. 90, 86, 82, . . .

10. 0.5, 1, 1.5, . . .

11. 7, 7.15, 7.3, . . .

State whether each sequence is an arithmetic sequence.

12. 1, 3, 9, 18, . . .

13. 1, 2, 4, 8, . . .

14. 100, 200, 300, 400, . . .

15. 8, 8, 8, 8, . . .

Exercises

Independent Practice

Write the next three terms of each sequence.

16. 84, 78, 72, 66, . . .

17. 3, 7, 11, 15, . . .

18. 12, 19, 26, 33, . . .

19. 0, 4, 8, 12, . . .

ESTIMATION
MENTAL MATH
CALCULATOR
PAPER/PENCIL

20. 0.8, 0.6, 0.4, 0.2, . . .

21. 4.53, 5.65, 6.77, 7.89, . . .

22. 1, 4, 9, 16, 25, . . .

23. 1, 2, 5, 10, 17, . . .

24. 1, 2, 4, 8, 16, . . .

25. 1, 8, 27, 64, . . .

Language Skill

26. Write the first five terms in an arithmetic sequence with a constant difference of 5. The first term is 2.5.

27. Name the first term and common difference of the sequence 0, 3, 6, 9, . . .

28. Find the sixth term in the sequence 4, 8, 12, 16, . . .

29. Find the twelfth term in the sequence 100, 98, 96, 94, . . .

30. The sixth term in a sequence is 5. The constant difference is -2. Find the first five terms.

Mixed Review

31. Solve the equation $y - (-13) = 5$.   (Lesson 3-2)

32. Find the LCM of 9 and 24.   (Lesson 4-7)

33. Solve the inequality $z - \frac{2}{5} \geq -2$.   (Lesson 5-7)

34. **Home Economics**   Cecilia needs $4\frac{1}{4}$ yards of fabric to make curtains for her bedroom. She needs an additional $1\frac{1}{2}$ yards to make matching pillows. How much fabric does she need?
(Lesson 5-7)

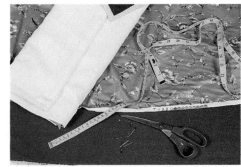

35. Consumer Awareness An apartment rents for $750 a month. The monthly rent is expected to increase $15 each year. What will the monthly rent be at the end of five years?

36. Business A stock analyst predicts that a certain stock will lose an average of $\frac{1}{8}$ point a month for the year. If the stock is listed at $9\frac{7}{8}$ on January 1 what is it predicted to be at the end of the year?

Critical Thinking

37. Find the seventh term of the sequence 2, 3, 5, 9, 17, . . . Is this an arithmetic sequence?

Wrap-Up

38. In the sequence $1, 1 + b, 1 + 2b, 1 + 3b, \ldots$, what is the constant difference? What is the fifth term?

Biography

Sonya Kovalevsky

Sonya Kovalevsky, born in Moscow in 1850, was a mathematical genius. At fifteen she grasped the ideas of differential calculus with such speed that she astonished her teachers. At that time, higher education was not open to women in Russia, so at 18 she had to find a way to travel to Europe to study. However, it was unacceptable for an unmarried woman of 18 to travel. Sonya tricked her father into letting her marry Vladimer Kovalevsky and they traveled to Germany to study. After her husband's death, Sonya went to live in Stockholm, where she lectured on mathematics at the University.

In the course of her study and lecturing, Sonya addressed number sequences. The following sequence represents a basic foundation of Sonya's work with infinite series and calculus. In the figures below, each fraction is represented by shading a part of a square. Notice that the shaded part approaches, but never quite reaches, the whole square.

This shading illustrates that the sum $\frac{1}{2} + \frac{1}{4} + \frac{1}{8} + \frac{1}{16} + \frac{1}{32} + \ldots$ approaches 1.

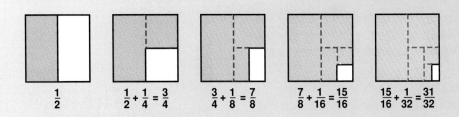

$$\frac{1}{2} \qquad \frac{1}{2} + \frac{1}{4} = \frac{3}{4} \qquad \frac{3}{4} + \frac{1}{8} = \frac{7}{8} \qquad \frac{7}{8} + \frac{1}{16} = \frac{15}{16} \qquad \frac{15}{16} + \frac{1}{32} = \frac{31}{32}$$

Fibonacci Sequence

Materials: calculator

In this Exploration, you will study a pattern of numbers known as the **Fibonacci sequence**. The first six terms in the Fibonacci sequence are:

$$1, 1, 2, 3, 5, 8, \ldots$$

The numbers in this sequence are called **Fibonacci numbers**.

▶ Although the Fibonacci sequence is not an arithmetic sequence, it is a well defined sequence. The sequence begins with 1 and each term is the sum of the previous two terms of the sequence.

		$1 + 1$	$1 + 2$	$2 + 3$	$3 + 5$
1,	1,	2,	3,	5,	8, ...

The next term is $5 + 8$ or 13.

Your turn: List the first 10 terms of the Fibonacci sequence.

▶ It may appear to you that this sequence is nothing more than random numbers. But there are many patterns imbedded in the sequence. Here's one.

- Every third Fibonacci number is divisible by 2.
- Every fourth Fibonacci number is divisible by 3.
- Every fifth Fibonacci number is divisible by 5.
- Every sixth Fibonacci number is divisible by 8.

Your turn: Complete this sentence: Every seventh Fibonacci number is divisible by __?__ . Give two examples.

Analysis

Answer each question using the Fibonacci sequence.

1. There is only one perfect square and one perfect cube, except 1, in the sequence. Name them.

2. If the Fibonacci numbers are squared and then adjacent squares are added together, a new sequence is formed. Describe the sequence.

3. **Research** Find out how the Fibonacci numbers are related to the pineapple and sunflower.

5-10 Adding and Subtracting Measures

Objectives:
Convert within the customary system. Add and subtract measures.

Key Terms:
customary units
length
weight
capacity

Megan enjoys riding her bicycle cross country. Last fall she rode her bicycle from her home to Atlanta, a distance of 645 miles.

The *mile* is a unit of **length** commonly used in the United States. Other **customary units** of length are *inches*, *feet*, and *yards*.

The most weight that Megan carries in her backpack when bicycling is 10.5 pounds. Customary units of **weight** are *ounces*, *pounds*, and *tons*.

Megan always carries a 1-quart canteen of water when bicycling. In the customary system, **capacity** is measured using *fluid ounces*, *cups*, *pints*, *quarts*, and *gallons*.

These relationships make it possible to convert measures within the customary system.

The Customary System
1 foot (ft) = 12 inches (in.)
1 yard (yd) = 3 feet (ft)
1 mile (mi) = 5,280 feet (ft)
1 pound (lb) = 16 ounces (oz)
1 ton = 2,000 pounds (lb)
1 cup (c) = 8 fluid ounces (fl oz)
1 pint (pt) = 2 cups (c)
1 quart (qt) = 2 pints (pt)
1 gallon (gal) = 4 quarts (qt)

Examples

1   27 in. = ■ ft.

$$\frac{27}{12} = 2\frac{3}{12} \text{ or } 2\frac{1}{4}$$

27 in. = $2\frac{1}{4}$ ft

You are changing from a smaller unit to a larger unit, so DIVIDE.
Since 12 in. = 1 ft, divide by 12.

2   10.5 lb = ■ oz

$$10.5 \;\boxed{\times}\; 16 \;\boxed{=}\; 168$$

10.5 lb = 168 oz

You are changing from a larger unit to a smaller unit, so MULTIPLY.

Since 1 lb = 16 oz, multiply by 16.

Customary units can be added or subtracted. Sometimes it is necessary to rename measurements as shown in the following examples.

Examples

3
$$
\begin{array}{r}
5 \text{ ft } 8 \text{ in.} \\
+\ 2 \text{ ft } 6 \text{ in.} \\
\hline
7 \text{ ft } 14 \text{ in.}
\end{array}
$$
    Add the units separately.

↑_____ 14 in. = 1 ft 2 in.

7 ft 14 in. = 8 ft 2 in.

4
$$
\begin{array}{r}
6 \text{ gal } 1 \text{ qt} \\
-\ 3 \text{ gal } 3 \text{ qt}
\end{array}
$$
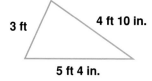
$$
\begin{array}{r}
\overset{5}{6} \text{ gal } \overset{5}{1} \text{ qt} \\
-\ 3 \text{ gal } 3 \text{ qt} \\
\hline
2 \text{ gal } 2 \text{ qt}
\end{array}
$$
Since 3 qt is greater than 1 qt, rename 6 gal 1 qt as 5 gal 5 qt. Then subtract.

Estimation Hint

THINK:
4 ft 10 in. is almost 5 ft; 5 ft 4 in. is a little more than 5 feet. The perimeter should be about 5 + 3 + 5 or 13 feet.

5 **Find the perimeter of the triangle shown at the right.**

To find the perimeter, add the measures of the sides.

$$
\begin{array}{r}
4 \text{ ft } 10 \text{ in.} \\
3 \text{ ft} \\
+\ 5 \text{ ft}\ \ \ 4 \text{ in.} \\
\hline
12 \text{ ft } 14 \text{ in.} = 13 \text{ ft } 2 \text{ in.}
\end{array}
$$

3 ft   **4 ft 10 in.**

5 ft 4 in.

The perimeter of the triangle is 13 ft 2 in.

Checking for Understanding _____

Communicating Algebra

1. Explain how to convert 20 inches to feet.
2. Explain how to convert 5 gallons to quarts.
3. State the formula for the perimeter of a rectangle.
4. If you know the measure of one side of a square, how would you find the perimeter?

Guided Practice

Complete.

5. 24 in. = ■ ft   6. 12 ft = ■ yd   7. 5 mi = ■ yd

8. 3 lb = ■ oz   9. 6 tons = ■ lb   10. 3 c = ■ fl oz

11. 0.5 lb = ■ oz   12. 96 yd = ■ ft   13. 1.25 mi = ■ ft

Add or subtract.

14.
$$
\begin{array}{r}
10 \text{ ft } 3 \text{ in.} \\
+\ 5 \text{ ft } 9 \text{ in.}
\end{array}
$$

15.
$$
\begin{array}{r}
3 \text{ yd } 2 \text{ ft} \\
+\ 8 \text{ yd } 1 \text{ ft}
\end{array}
$$

16.
$$
\begin{array}{r}
9 \text{ ft } 4 \text{ in.} \\
-\ 7 \text{ ft } 1 \text{ in.}
\end{array}
$$

Rational Numbers: Adding and Subtracting Patterns

Exercises

Add or subtract.

17. 6 lb 12 oz
    + 4 lb 8 oz

18. 10 ft
    − 4 ft 6 in.

19. 2 gal 3 qt
    + 4 gal 1 qt

20. 5 gal 3 qt
    − 2 gal 2 qt

21. 7 lb 8 oz
    − 3 lb 13 oz

22. 4 gal
    − 2 gal 3 qt

Find the perimeter.

23.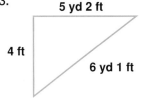
5 yd 2 ft
4 ft
6 yd 1 ft

24.
11 ft 9 in.
5 ft 5 in.

Applications

25. **Consumer Awareness**   A small car weighs 1.2 tons. Find the weight of the car in pounds.

26. **Home Economics**   A pitcher holds 3 quarts. How many 12-oz glasses can be filled from the pitcher?

Connection

27. **Geometry**   The perimeter of a triangle is 8 ft 10 in. The lengths of two sides are 3 ft 8 in. and 2 feet 9 in. Find the length of the third side.

Mixed Review

28. Solve the inequality $-3p > 105$.   (Lesson 3-8)

29. Write 3^{-2} using positive exponents.   (Lesson 4-10)

30. Solve the equation $b - 2\frac{3}{8} = 1\frac{1}{2}$.   (Lesson 5-7)

31. Write the next three terms of the sequence 5, 4, 2, -1, -5.   (Lesson 5-9)

Critical
Thinking

32. Mr. Hiroshi has an office that he wants to panel. The office is a square, 11 feet 6 inches on a side and 8 feet high. If panelling comes in sheets 8 feet high by 4 feet wide, how many sheets of panelling will he need to buy?

Wrap-Up

33. Explain in general how you would find the perimeter of any polygon.

Team Problem Solving

Mr. Cutter has a back yard that is 25 meters long and 10 meters wide. The path that Mr. Cutter's lawn mower cuts is 1 meter wide. If Mr. Cutter mows his lawn using the path shown, how far will he have walked when he is finished?

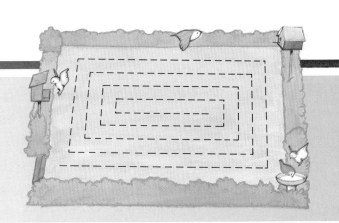

Review

Language and Concepts

Choose the letter of the best word or words to complete each sentence.

1. To subtract a rational number, add its __?__ .

2. To add fractions with __?__ , add the numerators.

3. The decimal 0.333 . . . is an example of a __?__ .

4. The closure property for addition states that the sum of any two rational numbers is a __?__ .

5. __?__ are five properties of addition that apply to rational numbers.

a. associative
b. closure
c. commutative
d. identity
e. inverse
f. like denominators
g. rational number
h. repeating decimal
i. terminating decimal
j. unlike denominators

Skills

Replace each ● with < , > , or = to make a true sentence.   (Lesson 5-1)

6. 4.6 ● 3.9

7. -5.5 ● 1.6

8. $\frac{4}{5}$ ● $-\frac{4}{5}$

9. $7\frac{1}{2}$ ● $\frac{15}{2}$

Express each decimal as a fraction or mixed number in simplest form.
(Lesson 5-2)

10. -0.6

11. 0.85

12. -0.25

13. 0.08

14. 1.4

15. -2.36

16. $0.333 \ldots$

17. $0.2727 \ldots$

18. $2.\overline{4}$

Round each decimal to the nearest whole number or dollar.   (Lesson 5-3)

19. 7.45

20. 9.621

21. 2.50

22. 109.92

Estimate.   (Lesson 5-3)

23. $10.03 + $5.84

24. $44.03 - 32.9$

25. $20.3 + 59.7 + 62.8$

Name the property shown by each sentence.   (Lesson 5-4)

26. $9.2 + (-9.2) = 0$

27. $(2 + 3.4) + (-5) = -5 + (2 + 3.4)$

28. $-1.5 + 2.3 = 2.3 + (-1.5)$

29. $8.3 + (-6.1)$ is a rational number.

Solve each equation.   (Lesson 5-4)

30. $p = 0.9 + 4.5$

31. $s = 3.05 - 1.94$

32. $w = 4.6 - 2.7$

33. $h = -3.6 + 9.4$

34. $5.8 - (-3.2) = q$

35. $m = -1.8 + (-3.7)$

Find the LCM for the denominators in each sum or difference. (Lesson 5-6)

36. $\frac{3}{4} + \frac{1}{8}$ **37.** $\frac{5}{8} - \frac{1}{6}$ **38.** $\frac{2}{5} - \frac{1}{6}$ **39.** $\frac{3}{7} + \frac{1}{2}$

Solve each equation. Write each solution in simplest form. (Lessons 5-5, 5-6)

40. $a = \frac{7}{9} + \frac{5}{9}$ **41.** $\frac{27}{18} - \frac{9}{18} = h$ **42.** $2\frac{3}{8} + 1\frac{5}{8} = p$ **43.** $6 - 3\frac{3}{4} = a$

44. $a = \frac{5}{9} - \frac{7}{18}$ **45.** $b = -\frac{1}{6} + \frac{1}{3}$ **46.** $-\frac{3}{5} - \frac{3}{10} = t$ **47.** $1\frac{3}{5} + 6\frac{7}{8} = m$

Explain how to solve each equation or inequality. Then solve. (Lesson 5-7)

48. $g - 1.4 = 8$ **49.** $-3.6 = x - 9$ **50.** $a + \frac{2}{3} < 2$

Write the next three terms of each sequence. Then state whether or not the sequence is arithmetic. Write *yes* or *no*. (Lesson 5-9)

51. 20, 23, 26, 29, . . .   **52.** 8, 9, 12, 17, . . .

Add or subtract. (Lesson 5-10)

53.   5 gal 2 qt
     $+$ 2 gal 3 qt

54.   10 lb 15 oz
     $+$ 8 lb 12 oz

55.   10 yd
     $-$ 2 yd 1 ft

Applications and Problem Solving

56. A volleyball team has 6 members. Suppose each member shakes hands with every other member. How many handshakes take place? (Lesson 5-8)

57. A rectangular field is 30 yd 1 ft wide and 75 yd 2 ft long. What is the perimeter of the field? (Lesson 5-10)

ortfolio Suggestion

Select some of your work from this chapter that shows how you used a calculator or computer. Place it in your portfolio.

Curriculum Connection

• **Language Arts** Research the Dewey Decimal System to find out how decimals are used to organize books in the library.

Read More About It

Burns, Marilyn. *The I Hate Math! Book.*

Luce, Marnie and A. B. Lerner. *Infinity, What is It?*

Wallace, G. David. *Money Basics.*

Test

Replace each ● with < , > , or = to make a true sentence.

1. $2\frac{3}{5}$ ● $2\frac{5}{8}$

2. 54.349 ● 54.36

3. 1.04 ● $1\frac{2}{50}$

Express each decimal as a fraction or mixed number in simplest form.

4. 0.72

5. 2.702

6. $0.\overline{7}$

7. $-1.\overline{6}$

Name the property shown by each sentence.

8. $7.9 + 3.6 = 3.6 + 7.9$

9. $1.49 + 0 = 1.49$

10. $\frac{4}{5} + \left(-\frac{4}{5}\right) = 0$

11. $1\frac{2}{7} \div (-5)$ is a rational number.

Solve each equation.

12. $6.8 + 0.7 = m$

13. $y = 15 - 8.7$

14. $2.13 - 0.95 = s$

15. $3.9 + 4.1 - 2.6 = h$

16. $a = 1.6 - (-3.4)$

17. $n = 4.5 - 8.6$

18. $\frac{9}{16} + \frac{9}{16} = a$

19. $2\frac{13}{15} - 1\frac{14}{15} = g$

20. $6\frac{1}{7} + \left(-2\frac{4}{7}\right) = h$

21. $\frac{4}{5} - \frac{5}{10} = g$

22. $j = \frac{2}{3} + 3\frac{4}{5}$

23. $-\frac{3}{5} - \frac{1}{10} = c$

Solve each equation or inequality. Write the solutions in simplest form.

24. $h - \frac{7}{2} = 2\frac{1}{2}$

25. $r - 6.8 = 10.2$

26. $1.2 + s > 3$

27. $a - 9 \leq 1.5$

Write the next three terms of each sequence. Then state whether the sequence is arithmetic. Write *yes* or *no*.

28. $1, 1.2, 1.6, 2.2, \ldots$

29. $7, 14, 21, 28, \ldots$

Solve.

30. Luisa plants $\frac{1}{4}$ of her garden in flowers, $\frac{1}{3}$ in tomatoes, and the rest in beans. How much of the garden is planted in beans?

31. Write the first 5 terms in an arithmetic sequence with a common difference of 4. The first term is 20.

32. Find the sum of the whole numbers from 1 through 100.

33. Estimate the perimeter of a triangle with sides 4.55 m, 6.81 m, and 9.27 m.

BONUS

True or *false*: $0.3 = \frac{1}{3}$. Write your answer in paragraph form.

Academic Skills Test

Cumulative, Chapters 1-5

1. If $x = 9$ and $y = 18$, what is the value of $9y - 18x$?

   A 729
   B 0
   C -9
   D -243

2. If $a + 2.5 = 5.1$, what is the value of a?

   A 2.6
   B 3.6
   C 5.1
   D 7.6

3. Which equation is equivalent to $-14d = 98$?

   A $-14d \cdot (-14) = 98$
   B $-14d \div (-14) = 98$
   C $-14d \div (-14) = 98 \cdot (-14)$
   D $-14d \div (-14) = 98 \div (-14)$

4. Which is equivalent to 3^4?

   A 12
   B 64
   C 81
   D 243

5. Which is equivalent to $\dfrac{a^2}{a^4}$?

   A a^{-2}
   B a^2
   C a^6
   D a^8

6. If $m + 2.1 = 8$, what is the value of m?

   A 5.9
   B 6.9
   C 8
   D 10.1

7. Which number should come next in this pattern?

   $$1, 3, 7, 13, 21, \ldots$$

   A 23
   B 29
   C 31
   D 35

8. A group of divers are at a depth of -20 m. If they descend 15 m more, at what depth will they be?

   A 5 m
   B -5 m
   C 35 m
   D -35 m

9. Sandy bowls in a league. Her handicap score is found using the formula:

   $$s = g + c$$

   where s is the handicap score, g is the game score, and c is the handicap. What was Sandy's game score if her handicap score is 186 and her handicap is 15?

   A 161
   B 171
   C 191
   D 201

10. Gladys bought grocery items for the following prices: $1.39, $2.89, 58¢, and $1.19. The best estimate of the total cost is—

    A $4
    B $5
    C $6
    D $9

CHAPTER OBJECTIVES

In this chapter you will learn to:

- compare and order rational numbers expressed as fractions and decimals
- multiply and divide rational numbers
- solve equations and inequalities with rational numbers
- find terms of a geometric sequence
- use deductive and inductive reasoning

Rationals: Multiplying and Dividing Patterns

Archaeologists can learn a lot studying animal fossils. One question they always have is, "When was this animal alive?" A good way to find out is to use radiocarbon dating.

All living things contain a radioactive isotope of carbon called carbon-14. When an animal dies, the carbon-14 in its body begins very slowly to change, or decay, into nitrogen-14. When we find the fossilized body many thousands of years later, some of the carbon-14 is still there. By comparing the amounts of carbon-14 and nitrogen-14 in the fossil, we can estimate how long ago the animal died. The graph shows the rate at which carbon-14 decays into nitrogen-14.

How long does it take for half of the carbon-14 to become nitrogen-14? (This is known as the half-life of carbon-14.)

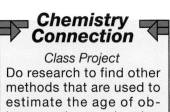

Chemistry Connection

Class Project
Do research to find other methods that are used to estimate the age of objects, such as rocks, that are not suitable for carbon-14 dating.

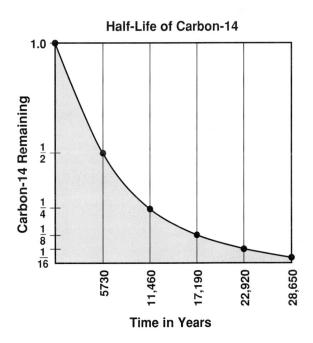

Half-Life of Carbon-14

6-1   Writing Fractions as Decimals

Objective:
Write fractions as terminating or repeating decimals.

Key Terms:
terminating
repeating

Which part accounts for the greatest heat loss?

Much of the energy consumed in the United States is used for heating buildings. Some of this energy is lost because buildings lose heat through their exteriors. The chart below shows the fraction of heat lost from different parts of a house.

Part of House	Fraction of Total Heat Loss
Exterior walls	$\frac{1}{8}$
Ceiling	$\frac{3}{8}$
Doors	$\frac{1}{20}$
Windows	$\frac{3}{25}$
Basement	$\frac{2}{25}$
Air leakage	$\frac{1}{4}$

It is sometimes more convenient to write numbers as decimals instead of fractions. One reason is that it is easier to compare with decimals. It may also be more convenient to use decimals when computing with a calculator.

Consider the fraction $\frac{1}{4}$. Remember that a fraction is another way of writing a division problem. So, $\frac{1}{4}$ means 1 ÷ 4. Divide using a calculator.

$$1 \;\boxed{\div}\; 4 \;\boxed{=}\; 0.25$$

Any fraction can be expressed as a decimal by dividing the numerator by the denominator.

$$\frac{1}{4} \implies \begin{array}{r} 0.25 \\ 4\overline{)1.00} \\ \underline{-8} \\ 20 \\ \underline{-20} \\ 0 \end{array}$$

Annex zeros to the numerator.   1 = 1.00

The fraction $\frac{1}{4}$ can be expressed as the decimal 0.25. Remember that a decimal like 0.25 is called a **terminating decimal** because the division ends or terminates when the remainder is zero.

FYI

A thermogram is a photograph-like image of a building that shows areas that are emitting large amounts of heat. The red areas show maximum heat loss. Usually these are areas where more insulation is needed.

204   *Rationals: Multiplying and Dividing Patterns*

Examples

1 Express $\frac{2}{5}$ as a decimal.

Use a calculator.

$$2 \;\boxed{\div}\; 5 \;\boxed{=}\; 0.4$$

$$\frac{2}{5} = 0.4$$

Use paper and pencil.

$$\frac{2}{5} \;\blacktriangleright\; 5\overline{)2.0}^{\,0.4}$$

2 Express $2\frac{3}{8}$ as a decimal.

Use a calculator.

$$2\frac{3}{8} = 2 + \frac{3}{8}$$

$$2 \;\boxed{+}\; 3 \;\boxed{\div}\; 8 \;\boxed{=}\; 2.375$$

$$2\frac{3}{8} = 2.375$$

Use paper and pencil.

Consider only $\frac{3}{8}$.

$$\frac{3}{8} \;\blacktriangleright\; \begin{array}{r} 0.375 \\ 8\overline{)3.000} \\ -24 \\ \hline 60 \\ -56 \\ \hline 40 \end{array}$$

$$2 + 0.375 = 2.375$$

An alternative strategy is to first write $2\frac{3}{8}$ as the improper fraction $\frac{19}{8}$. What decimal is equivalent to $\frac{19}{8}$?

Not all fractions can be expressed as terminating decimals. Consider the fraction $\frac{1}{3}$. How can $\frac{1}{3}$ be expressed as a decimal? Use a calculator to divide.

$$1 \;\boxed{\div}\; 3 \;\boxed{=}\; 0.3333333$$

The calculator displays only seven decimal places. What digit would be in the eighth place to the right of the decimal point? Check using paper and pencil as shown at the right.

$$\frac{1}{3} \;\blacktriangleright\; \begin{array}{r} 0.333 \\ 3\overline{)1.000} \\ -9 \\ \hline 10 \\ -9 \\ \hline 10 \\ -9 \\ \hline 1 \end{array}$$

Notice that the remainder after each step is 1. If you continue dividing, the pattern will repeat. Therefore, the digit 3 will be in the eighth place and will continue indefinitely.

Remember that a decimal like 0.333333 . . . , or $0.\overline{3}$, is called a **repeating decimal.**

Examples

3 **Express $\frac{5}{9}$ as a decimal.**

Use a calculator.

$5\ \boxed{\div}\ 9\ \boxed{=}\ 0.5555556$

The calculator rounds.

$5\ \boxed{\div}\ 9\ \boxed{=}\ 0.5555555$

The calculator truncates.

$\frac{5}{9} = 0.555\ldots$ or $0.\overline{5}$

Use paper and pencil.

$$\frac{5}{9} \implies \begin{array}{r} 0.555\ldots \\ 9\overline{)5.000\ldots} \\ \underline{-45} \\ 50 \\ \underline{-45} \\ 50 \\ \underline{-45} \\ 5 \end{array}$$

4 **Express $-5\frac{2}{3}$ as a decimal.**

Use a calculator.

$-5\frac{2}{3} = -(5 + \frac{2}{3})$

$5\ \boxed{+}\ 2\ \boxed{\div}\ 3\ \boxed{=}\ 5.6666667$

$\boxed{+/-}\ -5.6666667$

$-5\frac{2}{3} = -5.666\ldots$ or $-5.\overline{6}$

Use paper and pencil.

Consider only $\frac{2}{3}$.

$$\frac{2}{3} \implies \begin{array}{r} 0.666\ldots \\ 3\overline{)2.000\ldots} \\ \underline{-18} \\ 20 \\ \underline{-18} \\ 20 \\ \underline{-18} \\ 2 \end{array}$$

$-(5 + 0.666\ldots) = -5.666\ldots$

Checking for Understanding

Communicating Algebra

1. In your own words, explain how to express a fraction as a decimal.

2. Give two examples of fractions that can be expressed as terminating decimals. Give two examples that can be expressed as repeating decimals.

Express each decimal using bar notation.

3. $0.85858585\ldots$ 　　4. $0.0023232323\ldots$ 　　5. $0.833333333\ldots$

Write the first ten decimal places.

6. $0.1\overline{8}$ 　　7. $0.\overline{846}$ 　　8. $0.0\overline{528}$ 　　9. $0.0\overline{528}$

Guided Practice

Express each fraction as a decimal. Use a bar to show a repeating decimal.

10. $\frac{1}{2}$ 　　11. $-\frac{1}{3}$ 　　12. $-\frac{3}{4}$ 　　13. $\frac{1}{8}$ 　　14. $\frac{2}{3}$

15. $-\frac{1}{6}$ 　　16. $\frac{5}{8}$ 　　17. $-2\frac{5}{6}$ 　　18. $\frac{2}{9}$ 　　19. $1\frac{7}{10}$

Exercises

Independent Practice

Express each fraction as a decimal. Use a bar to show a repeating decimal.

20. $\frac{2}{5}$ 21. $\frac{7}{10}$ 22. $\frac{12}{25}$ 23. $-\frac{3}{8}$ 24. $\frac{11}{20}$

 25. $\frac{1}{9}$ 26. $\frac{3}{11}$ 27. $\frac{7}{18}$ 28. $7\frac{3}{4}$ 29. $\frac{7}{16}$

30. $\frac{2}{3}$ 31. $2\frac{5}{9}$ 32. $\frac{7}{9}$ 33. $\frac{5}{9}$ 34. $\frac{10}{33}$

35. $\frac{23}{45}$ 36. $\frac{28}{45}$ 37. $2\frac{9}{16}$ 38. $-4\frac{5}{16}$ 39. $\frac{31}{40}$

Calculator

Exercises 40-43 give several different calculators' displays for $1 \div 6$. What would each calculator display for $8 \div 9$?

40. .16666666 41. 0.1666666 42. 0.1666667 43. .17

Mixed Review

44. State whether $\frac{-5r}{s}$ is a monomial. (Lesson 4-1)

45. Find the product $(mn^3)(mp)(n^2p^2)$. (Lesson 4-9)

46. *True* or *false*: $7.43 < 7.5$. (Lesson 5-1)

47. Round 26.549 to the nearest tenth. (Lesson 5-3)

Solve each equation. (Lessons 5-5, 5-7)

48. $x = \frac{32}{21} - \frac{14}{21}$

49. $b - \frac{5}{8} < 3\frac{1}{2}$

Applications

50. **Consumer Awareness** Sarah bought $2\frac{1}{8}$ yards of fabric at \$4.95 per yard. Explain how you could find the total cost using your calculator.

 51. **Business** On Tuesday, XYZ stock fell $1\frac{5}{8}$ points. This means that the price of the stock dropped $\$1\frac{5}{8}$. Express this amount in dollars and cents.

52. **Biology** The fastest monarch butterfly can fly $\frac{1}{3}$ mile in one minute. Express $\frac{1}{3}$ as a decimal rounded to the nearest hundredth.

Critical Thinking

53. Find a terminating decimal that lies between $\frac{1}{3}$ and $\frac{5}{9}$ on a number line.

Wrap-Up

54. Express $\frac{1}{11}, \frac{2}{11}, \ldots, \frac{10}{11}$ as decimals. Explain the pattern you have found.

Enrichment

Are there decimals that neither terminate nor repeat? Consider the decimal 0.101001000100001000001 . . . . It is constructed by writing a 1 followed by a 0, then a 1 followed by two 0s. Certainly this decimal is well-defined in that you can determine the digit in any decimal place. Yet it is non-terminating, non-repeating. Can you write another decimal that is non-terminating, non-repeating?

Decimal Patterns

Materials: calculator

Every fraction can be expressed as either a terminating or repeating decimal. In this Exploration, you will look for a pattern that will help you identify whether the decimal for a given fraction terminates or repeats.

Explore: Use a calculator to express each fraction $\frac{1}{2}, \frac{1}{3}, \frac{1}{4}, \ldots, \frac{1}{20}$ as a decimal.

Here are some examples.

▶ $\frac{1}{8}$ → $1 \;\boxed{\div}\; 8 \;\boxed{=}\; 0.125$

The decimal terminates.

▶ $\frac{1}{12}$ → $1 \;\boxed{\div}\; 12 \;\boxed{=}\; 0.0833333$

The decimal repeats.

Your Turn: **Express each fraction $\frac{1}{2}, \frac{1}{3}, \ldots, \frac{1}{20}$ as a decimal. Identify each decimal as terminating or repeating. Keep track of your results in a chart.**

Analysis

1. Have you discovered a pattern? If so, write a sentence that describes it.

2. If you haven't discovered a pattern, try finding the prime factorization of each denominator. Look for a pattern in the prime factorizations. Write a sentence that describes the pattern.

3. Suppose one of your classmates predicts that $\frac{1}{30}$ can be expressed as a terminating decimal. Do you agree? Write a sentence that explains your reasoning.

6-2 Estimating Products and Quotients

Objective:
Estimate products and quotients of rational numbers.

Key Term:
compatible numbers

What mistake might Marta have made?

Marta solves this problem using a calculator.

$$56.3 \;\boxed{\times}\; 1.25 \;\boxed{\div}\; 9.1 \;\boxed{=}$$

The display reads 77.335165.
Is this the correct answer? You can use estimation to determine if an answer is reasonable.

$$
\begin{array}{ccl}
56.3 & \rightarrow & 60 \\
\times\,1.25 & \rightarrow & \underline{\times\; 1} \\
& & 60
\end{array}
\quad
\begin{array}{l}
\text{Round each factor to its}\\
\text{greatest place-value position.}
\end{array}
$$

$$
\downarrow
$$

$$60 \div 9.1 \quad \rightarrow \quad 60 \div 9 \approx 7$$

The answer should be close to 7.
Therefore, 77.335165 is incorrect.

Another estimation strategy is to use **compatible numbers.** In this strategy, numbers are rounded so that they *"fit together."* That is, it is easy to compute with them mentally. This strategy is effective when estimating the product of a fraction and a whole number. For example, estimate the product of $\frac{1}{3}$ and 11.

$$\frac{1}{3} \times 11 = \blacksquare \qquad
\begin{array}{l}
\text{The denominator tells how many equal parts.}\\
\text{THINK: What is } \frac{1}{3} \text{ of 11?}
\end{array}$$

$$\frac{1}{3} \times 12 = 4 \qquad
\begin{array}{l}
\text{The nearest multiple of 3 is 12.}\\
\text{3 and 12 are compatible numbers.}
\end{array}$$

The product of $\frac{1}{3}$ and 11 is about 4.

Examples

Estimate each product.

1 $\frac{2}{5} \times 49$

$\frac{2}{5} \times 49 \rightarrow \frac{2}{5} \times 50$

THINK: $\frac{1}{5}$ of 50 is 10.

$\frac{2}{5}$ of 50 is 20.

$\frac{2}{5} \times 49$ is about 20.

2 $\frac{25}{49} \times 180$

THINK: $\frac{25}{49}$ is about $\frac{1}{2}$.

$\frac{25}{49} \times 180 \rightarrow \frac{1}{2} \times 180$

THINK: $\frac{1}{2}$ of 180 is 90.

$\frac{25}{49} \times 180$ is about 90.

History Connection

One of the most common customary measurements, the mile, originated as an estimate. Historically, a mile was defined as the distance covered by a Roman soldier in 1000 double steps. Notice the similarity between the words *mile* and *mille*, meaning 1000.

Checking for Understanding

Communicating Algebra

1. Explain why estimation is an important step when computing answers with a calculator.

Guided Practice

Rewrite each problem as you would to estimate using rounding. Then estimate each product or quotient.

2. 4.61×1.9 **3.** $35 \div 7.3$ **4.** $5.8 \div 2.29$

Rewrite each problem as you would to estimate using compatible numbers. Then estimate each product.

5. $\frac{1}{3} \times 5$ **6.** $\frac{8}{23} \times 9$ **7.** $\frac{1}{2} \times 10\frac{1}{8}$

Exercises

Independent Practice

Estimate each product or quotient.

8. 15.93×9.8 **9.** 2.49×1.9 **10.** 59.1×2.09

11. $\$18.20 \div 2.7$ **12.** $27.26 \div 2.6$ **13.** $8.1 \div 2.2$

14. $\frac{1}{3} \times 5$ **15.** $\frac{1}{4} \times 13$ **16.** $\frac{1}{5} \times 24$

17. $\frac{5}{6} \times 10$ **18.** $\frac{2}{3} \times 16$ **19.** $\frac{5}{9} \times 20$

20. $\frac{9}{19} \times 120$ **21.** $\frac{10}{14} \times 30$ **22.** $\frac{31}{40} \times 200$

Mixed Review

23. State the least prime number that is a factor of 91.   (Lesson 4-4)

24. Find the twelfth term in the arithmetic sequence 99, 90, 81, 72, ... (Lesson 5-9)

25. Express $3\frac{3}{8}$ as a decimal.   (Lesson 6-1)

Application

26. Biology   A dolphin can swim at a speed of 37 mph. A human can swim about one-eighth as fast. About how fast can a human swim?

Connection

27. Geometry   The formula for the area of a rectangle is $A = \ell w$, where ℓ is the length and w is the width. If the length is 5.8 meters and the width is 3.1 meters, is the area less than or greater than 20 square meters?

Critical Thinking

28. Choose the correct phrase: The product of a whole number and a fraction less than one is always (less than, greater than, equal to) the whole number.

Wrap-Up

29. Write a problem involving decimals whose estimated answer is 3.

6-3 Rationals: Multiplying Fractions

Objective:
Multiply fractions.

About $\frac{1}{3}$ of Earth's land can be used for farming. About $\frac{2}{5}$ of this farmland is used to grow grain crops. What part of Earth's land is used to grow grain?

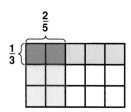

About $\frac{2}{5}$ of $\frac{1}{3}$, or $\frac{2}{15}$, of Earth's land is used to grow grain. You get the same result if you compute $\frac{2}{5} \cdot \frac{1}{3}$.

$$\frac{2}{5} \cdot \frac{1}{3} = \frac{2 \cdot 1}{5 \cdot 3} \qquad \text{Multiply the numerators.}$$
$$\text{Multiply the denominators.}$$
$$= \frac{2}{15}$$

Multiplying Fractions	In words: To multiply fractions, multiply the numerators and multiply the denominators.
	In symbols: For fractions $\frac{a}{b}$ and $\frac{c}{d}$, where $b \neq 0$ and $d \neq 0$, $\frac{a}{b} \cdot \frac{c}{d} = \frac{ac}{bd}$.

FYI

Grain is one of the best sources of energy for people and animals. The eight most common grains are wheat, rice, corn, barley, sorghum, oats, rye, and millet.

Example

1 Solve $x = \frac{5}{9} \cdot \frac{8}{15}$.

Method 1

$x = \frac{5}{9} \cdot \frac{8}{15}$

$x = \frac{5 \cdot 8}{9 \cdot 15}$ Multiply.

$x = \frac{40}{135}$ or $\frac{8}{27}$ Simplify.

Method 2

$x = \frac{5}{9} \cdot \frac{8}{15}$

$x = \frac{\overset{1}{5}}{9} \cdot \frac{8}{\underset{3}{15}}$ The GCF of 5 and 15 is 5. Divide 5 and 15 by 5.

$x = \frac{1 \cdot 8}{9 \cdot 3}$ or $\frac{8}{27}$ Multiply.

The multiplication skills that you have developed with integers and fractions can be used when you multiply negative fractions.

Examples

Solve each equation.

2 $3\frac{1}{2} \cdot \left(-4\frac{2}{3}\right) = c$ Estimate: $3(-5) = -15$

$\frac{7}{2} \cdot \left(\frac{-14}{3}\right) = c$ Rename $3\frac{1}{2}$ as $\frac{7}{2}$ and $-4\frac{2}{3}$ as $\frac{-14}{3}$.

$\frac{7}{\overset{2}{\underset{1}{\cancel{2}}}} \cdot \left(\frac{\overset{-7}{\cancel{-14}}}{3}\right) = c$

$\frac{-49}{3} = c$ The product of two rational numbers with different signs is negative.

$-16\frac{1}{3} = c$ Compare with the estimate.

3 $x = -\frac{3}{4} \cdot \left(-\frac{5}{7}\right)$

$x = \frac{-3}{4} \cdot \left(\frac{-5}{7}\right)$

$x = \frac{-3(-5)}{4 \cdot 7}$ The product of two rational numbers with the same sign is positive.

$x = \frac{15}{28}$

The rules for exponents that were stated for whole numbers in Lesson 4-2 also hold for fractions. For example, the expression $\left(\frac{3}{4}\right)^2$ means $\frac{3}{4} \cdot \frac{3}{4}$.

Examples

Solve each equation.

What is the difference between $\left(-\frac{3}{4}\right)^2$ and $-\left(\frac{3}{4}\right)^2$?

4 $x = \left(\frac{2}{3}\right)^2$

$x = \frac{2}{3} \cdot \frac{2}{3}$

$x = \frac{4}{9}$

5 $y = \left(-\frac{5}{6}\right)^2$

$y = \left(-\frac{5}{6}\right)\left(-\frac{5}{6}\right)$

$y = \frac{25}{36}$

Checking for Understanding

Communicating
Algebra

1. Explain how the model at the right shows the product of $\frac{2}{3}$ and $\frac{1}{4}$. What is the product?

2. An animal preserve is $\frac{3}{4}$ of a state park. About $\frac{1}{2}$ of the preserve is wooded. Draw a model that shows how to find what part of the park is wooded.

3. What should be your first step when computing $5\frac{1}{4} \cdot 3\frac{1}{6}$?

Guided Practice

State whether each product is positive or negative.

4. $-\frac{1}{2} \cdot 4$

5. $\frac{2}{3} \cdot \frac{3}{4}$

6. $2\frac{1}{2}(-3)$

7. $(-5)\left(-2\frac{1}{2}\right)$

Solve each equation. Write each solution in simplest form.

8. $\frac{1}{2} \cdot \frac{5}{6} = a$

9. $\frac{2}{3} \cdot \frac{9}{10} = b$

10. $\frac{8}{12} \cdot \frac{4}{6} = c$

11. $y = \frac{5}{6}(-2)$

12. $r = 1\frac{2}{3} \cdot 2$

13. $2\frac{3}{4}\left(-\frac{4}{5}\right) = f$

Exercises

Independent Practice

Solve each equation. Write each solution in simplest form.

14. $c = \frac{1}{2} \cdot \frac{2}{7}$

15. $r = -2\frac{1}{2}(-6)$

16. $k = -\frac{5}{6}\left(-\frac{2}{5}\right)$

17. $d = -4\left(\frac{3}{8}\right)$

18. $(-7)\left(-2\frac{1}{3}\right) = h$

19. $5\left(-3\frac{1}{2}\right) = x$

20. $m = \left(-1\frac{1}{3}\right)\left(-\frac{3}{4}\right)$

21. $\left(2\frac{1}{4}\right)\left(-\frac{4}{3}\right) = t$

22. $j = \left(\frac{3}{4}\right)\left(-\frac{4}{3}\right)$

23. $\left(-9\frac{3}{5}\right)\left(\frac{5}{12}\right) = y$

24. $f = (-16)\left(-\frac{3}{8}\right)$

25. $(-12)\left(-\frac{5}{6}\right) = c$

26. $(-7)\left(-8\frac{1}{2}\right) = x$

27. $\left(3\frac{1}{6}\right)\left(-2\frac{1}{3}\right) = m$

28. $\left(-8\frac{1}{3}\right)\left(2\frac{2}{5}\right) = k$

29. $x = \left(\frac{1}{2}\right)^2$

30. $d = \left(-\frac{3}{4}\right)^2$

31. $\left(\frac{5}{8}\right)^2 = n$

32. $a = \left(-\frac{2}{3}\right)^2$

33. $\left(-\frac{4}{5}\right)^2 = y$

34. $2 \cdot \left(\frac{5}{6}\right)^2 = m$

Evaluate each expression if $a = \frac{1}{2}$, $b = -\frac{2}{3}$, $x = 2\frac{1}{4}$, and $y = -1\frac{5}{6}$.

35. ax

36. $2y$

37. b^2

38. $by + \frac{1}{2}$

39. $3b - 4a$

40. $a^2(x + 5)$

41. $y - (a + b)$

42. $x - ay$

Mixed Review

43. Solve the equation $x = -15 + 5 + (-2)$.   (Lesson 2-4)

44. Write $\frac{4a^2b}{12b^3}$ in simplest form.   (Lesson 4-6)

45. Express -2.44 as a mixed number in simplest form.   (Lesson 5-2)

46. Simplify the expression $9.4d + 4.3d + 1.8d$.   (Lesson 5-4)

47. Express $-5\frac{7}{15}$ as a decimal.   (Lesson 6-1)

48. Estimate the product $\frac{1}{6} \times 23$.   (Lesson 6-2)

Applications

49. Home Economics   A recipe for 16 pancakes calls for $2\frac{1}{2}$ cups of flour. How much flour is needed to make one-half the recipe?

50. Carpentry   Four pieces of wood each $11\frac{3}{8}$ inches long are required to build a cabinet. If all four pieces are cut from one board and 1 inch is allowed for waste in cutting each piece, how long should the board be?

Critical Thinking

51. Which point shown on the number line could be the graph of the product of the numbers graphed at A and B?

Wrap-Up

52. In your own words, explain how to multiply fractions.

Enrichment

Algebraic Fractions

Just as you can multiply rational numbers expressed as fractions, you can multiply algebraic fractions.

Find $\frac{a}{b} \cdot \frac{b}{c}$, where $b \neq 0$ and $c \neq 0$.

$\frac{a}{b} \cdot \frac{b}{c} = \frac{a \cdot b}{b \cdot c}$ The GCF of the numerator and denominator is b. Divide the numerator and denominator by b.

$= \frac{a}{c}$

How should you find the product $\frac{a^2}{b} \cdot \frac{b}{a}$?

6-4 Rationals: Multiplying Decimals

Objective:

Multiply decimals.

Juan is a member of a springboard diving team. Recently his team participated in a diving competition. Each dive that Juan performs is scored by several judges. The individual scores are totaled and then multiplied by a number called the degree of difficulty. This number is between 1 and 4.

The judges' scores for Juan's first dive are shown below. This dive had a 1.5 degree of difficulty. We can find his score for this dive as follows:

$$6.5 + 7.0 + 6.5 + 5.5 + 7.0 = 32.5 \quad \text{Add the scores.}$$

$$32.5 \times 1.5 = \blacksquare \quad \text{Multiply by the degree of difficulty.}$$

Estimation is helpful in placing the decimal point in the product.

THINK: $32 \times 1 = 32$ Since 1.5 is between 1 and 2,
$32 \times 2 = 64$ 32×1.5 should be between 32 and 64.

$$
\begin{array}{r}
32.5 \\
\times\ 1.5 \\
\hline
1625 \\
325 \\
\hline
48.75
\end{array}
$$

Multiply as with whole numbers.

Place the decimal point using the estimate.

What is the rule for locating the decimal point in a product?

Juan scored 48.75 points on his first dive. Notice that 48.75 is between 32 and 64.

Check using a calculator.

32.5 ⊠ 1.5 ⊟ 48.75   ✔ The solution checks.

Examples

Solve each equation.

1 $y = (2.3)(3.5)$ Estimate: $2 \times 4 = 8$

$$
\begin{array}{r}
2.3 \\
\times\, 3.5 \\
\hline
115 \\
69 \\
\hline
8.05
\end{array}
$$

or   2.3 ⊠ 3.5 ⊟ 8.05

$y = 8.05$ Compare with the estimate.
The answer is reasonable.

2 $(-0.105)(0.03) = k$ Estimate: $-\dfrac{1}{10} \cdot \dfrac{3}{100} = -\dfrac{3}{1000}$

$$
\begin{array}{r}
-0.105 \\
\times\;\;\; 0.03 \\
\hline
-0.00315
\end{array}
$$

or   0.105 ⊞ ⊠ 0.03 ⊟ -0.00315

$-0.00315 = k$ Compare with the estimate.

3 $c = (100)(2.4)$ Estimate: $100 \cdot 2 = 200$

$$
\begin{array}{r}
100 \\
\times\, 2.4 \\
\hline
400 \\
200 \\
\hline
240.0
\end{array}
$$

or   100 ⊠ 2.4 ⊟ 240

$c = 240$ Compare with the estimate.

Mental Math Hint

When multiplying by powers of 10 like 10, 100, 1000, . . . , you can easily find the product mentally. The decimal point is moved to the right the same number of places as the number of zeros in the power of 10.

Checking for Understanding

Communicating Algebra

1. State the rule for placement of the decimal point in multiplication.

2. Explain why you should make an estimate when using a calculator to multiply.

3. In your own words, state how to find a product when one of the factors is 10, 100, 1000, ...

Use estimation to choose the correct product.

4. $(1.3)(0.02)$; 0.260 or 0.026

5. $(0.04)(0.01)$; 0.0004 or 0.4000

6. $(5.1)(0.8)$; 40.8 or 4.08

7. $(0.02)6$; 0.012 or 0.12

State where the decimal point should be placed in each product.

8. $16 \cdot 4.7 = 752$ **9.** $3.62 \cdot 30 = 10860$ **10.** $(3.2)(7.5) = 2400$

Find each product.

11. $(9.16)(39)$ **12.** $(0.118)(-4.8)$ **13.** $(12.8)(1.1)$

Exercises _____

**Independent
Practice**

Solve each equation.

14. $x = 6(3.2)$ **15.** $z = (11.8)(5)$ **16.** $w = -7(8.46)$

17. $p = 2.4 \cdot 8.7$ **18.** $r = (-5.03)(-6.1)$ **19.** $m = (40.6)(0.9)$

20. $(2.1)(0.6) = x$ **21.** $0.98(-0.6) = k$ **22.** $0.22 \cdot 0.008 = y$

Mental Math

Solve each equation mentally.

23. $(100)(11.6) = t$ **24.** $(1000)(7.2) = m$ **25.** $p = (12.79)(10)$

26. $(0.02)(10) = y$ **27.** $n = 10(-3.05)$ **28.** $x = (-100)(-3.6)$

Estimation

Estimate each product. Then compute with a calculator.

29. $(2.01)(1.5) = r$ **30.** $0.082 \cdot 6.6 = y$ **31.** $(35.3)4.9 = t$

32. $8.022(0.03) = s$ **33.** $(1.08)(4.75) = p$ **34.** $0.035 \cdot 4.1 = k$

35. $d = 5.92 \cdot 0.47$ **36.** $g = (75.9)(0.372)$ **37.** $x = 9.0006(9.01)$

38. $w = (7.2)(3.5)(0.05)$ **39.** $y = (0.23)(1.7)(0.04)$ **40.** $a = (81)(0.02)(1.5)$

Evaluate each expression.

41. $2.5x$ if $x = 8$ **42.** $3y^2$ if $y = 0.4$

43. ab if $a = -1.5, b = 10$ **44.** $r(s + t)$ if $r = 0.9, s = 4.2, t = 5.1$

45. $(x + y)^2$ if $x = 0.4, y = -0.2$ **46.** $(x + 4)(x - 2)$ if $x = 1.8$

Mixed Review

47. Find the LCD for $\frac{1}{3b^2}$ and $\frac{7}{18b}$.
(Lesson 4-8)

48. Solve the equation $s = 5\frac{2}{3} - 1\frac{5}{6}$.
(Lesson 5-6)

49. Social Studies Tuy builds a model of the Pentagon for his social studies project. The length of each side is 10.7 inches. What is the total perimeter? (Lesson 5-10)

50. Estimate $6.08 \times 3.99 \div 7.96$.
(Lesson 6-2)

Solve each equation. Write each solution in simplest form.   (Lesson 6-3)

51. $d = (-3)\left(2\frac{1}{7}\right)$

52. $f = \left(-2\frac{1}{4}\right)\left(-\frac{1}{3}\right)$

Applications

53. **Economics**   When Emilio's baseball team visited Mexico, $1 in American money could be exchanged for 2702.7 pesos. If Emilio exchanged $25 in American money, how many pesos did he receive?

54. **Home Economics**   Suppose you are in charge of baking a ham for your family picnic. The directions tell you to cook the ham for 22 minutes per pound at 325°. The ham weighs 7.45 pounds. To the nearest minute how long should you bake the ham?

55. **Consumer Awareness**   In a recent year, American consumers spent $0.8 billion for golf equipment. They spent four times more for athletic shoes than on golf equipment. How much did they spend for athletic shoes?

Critical Thinking

56. Name two decimals whose sum is 2.4 and whose product is 1.44.

Wrap-Up

57. Copy. Place a decimal point in each factor so that the product is correct.

$$493 \times 17 = 8.381$$

Reading Algebra

The Metric System

During the 1790s, a group of French scientists developed a system of measurement called the metric system. Some basic units are the **meter** (length), **liter** (capacity), and the **gram** (mass). Other units are named by using *prefixes* with the basic units. The chart at the right shows how the prefixes relate to the powers of ten.

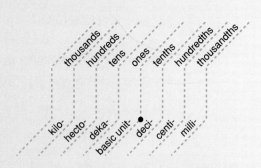

When the prefix *centi-* is added to *meter,* the result is *centimeter.* A *centimeter* is one-hundredth of a meter.

prefix ⌐   basic unit

centimeter   =   **0.01 meter**

Name the metric unit equivalent to each measurement.

1. 0.001 meter

2. 1000 liters

3. 100 grams

4. 0.1 meter

5. 10 liters

6. 0.01 gram

7. **Research**   Use a dictionary to find the meaning of the prefixes *mega-* and *micro-*.

6-5 Multiplicative Inverses

Objective:
Identify and use the properties of multiplication.

Key Terms:
multiplicative
inverse
reciprocal

Tim and Hiro want to buy a surfboard together. Since Tim will be gone part of the summer, Hiro suggests that Tim pay only $\frac{1}{3}$ of the cost instead of $\frac{1}{2}$. If Tim pays \$80, what is the total cost of the surfboard?

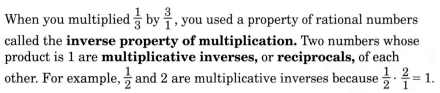

You can use an equation to solve this problem. Let c = total cost.

$$\tfrac{1}{3} \text{ of total cost} = 80$$

$$\frac{1}{3}c = 80$$

$$\frac{3}{1} \cdot \frac{1}{3}c = 80 \cdot \frac{3}{1} \qquad \text{Multiply each side of the equation by 3.}$$

$$1c = 240 \qquad \frac{3}{1} \cdot \frac{1}{3} = 1$$

$$c = 240 \qquad 1c = c$$

The surfboard costs \$240.

Are $-\frac{5}{8}$ and $\frac{8}{5}$ multiplicative inverses?

When you multiplied $\frac{1}{3}$ by $\frac{3}{1}$, you used a property of rational numbers called the **inverse property of multiplication.** Two numbers whose product is 1 are **multiplicative inverses,** or **reciprocals,** of each other. For example, $\frac{1}{2}$ and 2 are multiplicative inverses because $\frac{1}{2} \cdot \frac{2}{1} = 1$.

In the same way, $-\frac{3}{2}$ and $-\frac{2}{3}$ are multiplicative inverses because $-\frac{2}{3} \cdot -\frac{3}{2} = 1$.

Inverse Property of Multiplication

In words: The product of a number and its multiplicative inverse is 1.

In symbols: For every nonzero number $\frac{a}{b}$, where a, b ≠ 0, there is exactly one number $\frac{b}{a}$ such that $\frac{a}{b} \cdot \frac{b}{a} = 1$.

All the properties that were true for multiplication of integers are also true for multiplication of rationals. The properties for multiplication of rationals are summarized in the following chart.

Properties of Multiplication	Examples
Closure Property For all rational numbers x and y, $x \cdot y$ is a rational number.	$-3 \cdot 4 = -12 \quad 3.2 \cdot 0.5 = 1.6$ $-\dfrac{2}{5} \cdot \dfrac{3}{4} = -\dfrac{3}{10}$
Commutative Property For all rational numbers x and y, $x \cdot y = y \cdot x$.	$-3(-5) = -5(-3)$ $0.2(0.3) = 0.3(0.2)$ $\dfrac{2}{3} \cdot \dfrac{1}{2} = \dfrac{1}{2} \cdot \dfrac{2}{3}$
Associative Property For all rational numbers x, y, and z, $(x \cdot y) \cdot z = x \cdot (y \cdot z)$.	$(-2 \cdot 3) \cdot 4 = -2(3 \cdot 4)$ $\left(-\dfrac{1}{4} \cdot \dfrac{2}{3}\right) \cdot \dfrac{1}{2} = -\dfrac{1}{4} \cdot \left(\dfrac{2}{3} \cdot \dfrac{1}{2}\right)$ $(-6.5 \cdot 9.3) \cdot 2.1 = -6.5 \cdot (9.3 \cdot 2.1)$
Identity Property For every rational number x, $x \cdot 1 = x$ and $1 \cdot x = x$.	$-7 \cdot 1 = -7 \qquad 1 \cdot \dfrac{2}{3} = \dfrac{2}{3}$ $7.45 \cdot 1 = 7.45$
Inverse Property For every rational number $\dfrac{x}{y}$, $x, y \neq 0$, there is a unique rational number $\dfrac{y}{x}$ such that $\dfrac{x}{y} \cdot \dfrac{y}{x} = 1$.	$2 \cdot \dfrac{1}{2} = 1 \quad -\dfrac{2}{3}\left(-\dfrac{3}{2}\right) = 1$ $0.5\left(\dfrac{1}{0.5}\right) = 1$

What number has no multiplicative inverse?

The distributive property is the only property of rationals that involves two operations. The operations are multiplication and addition.

Distributive Property	For all rational numbers, x, y, and z, $x \cdot (y + z) = x \cdot y + x \cdot z.$

You can use the distributive property to compute products mentally.

Examples

1 **Compute $4 \times 5\frac{1}{2}$.**

$$4 \times 5\tfrac{1}{2} = 4\left(5 + \tfrac{1}{2}\right)$$

THINK: 4 groups of 5 is 20. 4 groups of $\frac{1}{2}$ is 2.

$$= 20 + 2 \text{ or } 22$$

2 **Compute $\frac{1}{3} \times 6\frac{3}{4}$.**

$$\tfrac{1}{3} \times 6\tfrac{3}{4} = \tfrac{1}{3}\left(6 + \tfrac{3}{4}\right)$$

THINK: $\frac{1}{3}$ of 6 is 2.

$\frac{1}{3}$ of $\frac{3}{4}$ is $\frac{1}{4}$.

$$= 2 + \tfrac{1}{4} \text{ or } 2\tfrac{1}{4}$$

Checking for Understanding

Communicating Algebra
1. Name two numbers that are multiplicative inverses of each other.

2. Write a number sentence that shows the commutative property of multiplication for rational numbers.

3. What property allows you to compute $9 \times 3\frac{2}{3}$ as $9(3) + 9\left(\frac{2}{3}\right)$?

Guided Practice

State which pairs of numbers are multiplicative inverses. Write *yes* or *no*.

4. $5, \frac{1}{5}$

5. $-\frac{3}{4}, 1\frac{1}{3}$

6. $3\frac{3}{4}, \frac{15}{4}$

7. $0.5, 2$

Exercises

Independent Practice

Name the multiplicative inverse of each rational number.

8. $\frac{2}{3}$

9. $-\frac{8}{9}$

10. 7

11. 0.8

12. $2\frac{4}{5}$

13. -1

14. $\frac{a}{b}$

15. $\frac{x}{y}$

Name the property shown by each statement.

16. $\left(2 \cdot \frac{5}{2}\right) \cdot \frac{3}{4} = \frac{3}{4}\left(2 \cdot \frac{5}{2}\right)$

17. $-5 \cdot \frac{4}{5}$ is a rational number.

18. $-1\frac{14}{17} \cdot 1 = -1\frac{14}{17}$

19. $\left(-\frac{5}{6}\right)\left(-\frac{6}{5}\right) = 1$

20. $(-4 \cdot 5) \cdot \frac{1}{4} = [5 \cdot (-4)] \cdot \frac{1}{4}$

21. $10 \times 3\frac{1}{5} = 10(3) + 10\left(\frac{1}{5}\right)$

Mental Math

Compute.

22. $100 \times 2\frac{1}{2}$

23. $6 \times 2\frac{1}{3}$

24. $3\frac{1}{4} \times 8$

25. $3\frac{2}{5} \times 20$

26. $\frac{1}{2} \times 8\frac{1}{2}$

27. $\frac{1}{4} \times 12\frac{1}{2}$

28. $\frac{2}{3} \times 6\frac{3}{5}$

29. $\frac{3}{4} \times 16\frac{2}{3}$

Mixed Review

30. *True* or *false*: $\frac{11}{3} \geq 3\frac{2}{3}$.   (Lesson 5-1)

31. Find the product $\left(-\frac{1}{2}\right)(8)$.   (Lesson 6-3)

32. Solve the equation $p = (-7.8)(0.04)$.   (Lesson 6-4)

33. Evaluate $a(b + c)$ if $a = 4.1$, $b = 0.8$, and $c = 9.3$.   (Lesson 6-4)

Application

34. **Personal Finance**   Marsha bought a dozen plates for $3.25 each and a dozen bowls for $1.25 each. Write two different expressions that Marsha can use to find the total cost.

Critical Thinking

35. What is the product of the first 10 terms of the sequence $\frac{1}{2}, \frac{2}{3}, \frac{3}{4}, \frac{4}{5}, \ldots$?

Wrap-Up

36. Explain how multiplicative inverses and additive inverses differ. Give an example for each using $-\frac{5}{9}$.

6-6　Rationals: Dividing Fractions

Objective:
Divide fractions.

LeRoy Freeman needs 3 cups of pineapple for a fruit salad. If every can of pineapple contains $\frac{3}{4}$ cup, how many cans should Mr. Freeman buy?

To decide how many cans are needed, divide 3 by $\frac{3}{4}$. Recall that dividing by a number is the same as multiplying by its multiplicative inverse.

Will Mr. Freeman need more than 3 or less than 3 cans of pineapple?

$3 \div \frac{3}{4} = 3 \cdot \frac{4}{3}$ Dividing by $\frac{3}{4}$ is the same as multiplying by $\frac{4}{3}$.

$= \frac{\overset{1}{\cancel{3}}}{1} \cdot \frac{4}{\underset{1}{\cancel{3}}}$

$= 4$

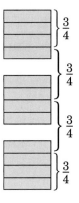

Mr. Freeman needs to buy 4 cans of pineapple. Compare with the model.

Division with Rational Numbers	In words:	To divide by a rational number, multiply by its multiplicative inverse.
	In symbols:	For rational numbers $\frac{a}{b}$ and $\frac{c}{d}$, where $b, c, d \neq 0$, $$\frac{a}{b} \div \frac{c}{d} = \frac{a}{b} \cdot \frac{d}{c}.$$

Example

1　Solve $y = \frac{4}{5} \div \frac{2}{3}$.

$y = \frac{4}{5} \div \frac{2}{3}$

$y = \frac{4}{5} \cdot \frac{3}{2}$ Dividing by $\frac{2}{3}$ is the same as multiplying by $\frac{3}{2}$.

$y = \frac{\overset{2}{\cancel{4}}}{5} \cdot \frac{3}{\underset{1}{\cancel{2}}}$ Divide 4 and 2 by 2.

$y = \frac{6}{5}$ or $1\frac{1}{5}$ Rename as a mixed number in simplest form.

The division skills you have developed with integers and fractions can be used to divide with negative fractions.

Examples

Solve each equation.

2 $y = 3 \div \left(-\frac{1}{2}\right)$

$y = \frac{3}{1} \cdot \frac{-2}{1}$ Dividing by $-\frac{1}{2}$ is the same as multiplying by -2.

$y = -6$

3 $a = -\frac{10}{7} \div \left(-\frac{5}{14}\right)$

$a = -\frac{10}{7} \cdot \left(-\frac{14}{5}\right)$

$a = -\frac{\overset{2}{\cancel{10}}}{\cancel{7}_{1}} \cdot \left(-\frac{\overset{2}{\cancel{14}}}{\cancel{5}_{1}}\right)$

$a = 4$

4 $b = -3\frac{1}{2} \div 1\frac{3}{4}$

$b = -\frac{7}{2} \div \frac{7}{4}$

$b = -\frac{7}{2} \cdot \frac{4}{7}$

$b = -\frac{\overset{1}{\cancel{7}}}{\cancel{2}_{1}} \cdot \frac{\overset{2}{\cancel{4}}}{\cancel{7}_{1}}$

$b = -2$

Checking for Understanding

Communicating Algebra

1. State another name for *reciprocal*.

2. Complete the sentence: Dividing by any number, except zero, is the same as ▇ .

3. Name the multiplicative inverse of -3.

Guided Practice

State a multiplication expression for each division expression. Then compute.

4. $\frac{2}{5} \div \frac{3}{4}$

5. $\frac{5}{6} \div \frac{10}{11}$

6. $-\frac{7}{9} \div \frac{2}{3}$

7. $\frac{13}{15} \div 6$

8. $5 \div \left(-1\frac{1}{3}\right)$

9. $2\frac{3}{5} \div 3\frac{6}{7}$

10. $5 \div 1\frac{9}{11}$

11. $7\frac{5}{9} \div (-8)$

Exercises

Independent Practice

Solve each equation. Write each solution in simplest form.

12. $a = 6 \div \left(-\frac{2}{3}\right)$

13. $c = -10 \div \left(\frac{5}{2}\right)$

14. $\frac{3}{4} \div \frac{2}{3} = t$

15. $h = 2\frac{1}{2} \div \frac{3}{4}$

16. $p = -3\frac{1}{5} \div 4\frac{2}{5}$

17. $6\frac{1}{8} \div 4\frac{2}{3} = v$

18. $x = 10 \div (-2)$

19. $-3 \div \left(\frac{2}{3}\right) = y$

20. $\frac{3}{4} \div \frac{1}{2} = g$

Solve each equation. Write each solution in simplest form.

21. $p = -\frac{5}{7} \div \frac{1}{14}$

22. $2\frac{3}{4} \div \left(-\frac{3}{4}\right) = a$

23. $-8 \div \left(-\frac{4}{3}\right) = h$

24. $f = -\frac{3}{8} \div (-3)$

25. $c = 5\frac{5}{6} \div 2\frac{1}{3}$

26. $4\frac{2}{3} \div \left(-\frac{6}{7}\right) = d$

27. $q = -2 \div \left(-\frac{1}{3}\right)$

28. $s = -3\frac{1}{4} \div 2\frac{1}{6}$

29. $-7\frac{1}{2} \div \left(1\frac{1}{5}\right) = n$

30. $m = -\frac{16}{7} \div \left(-\frac{12}{35}\right)$

31. $a = \frac{21}{30} \div \left(-\frac{7}{15}\right)$

32. $12\frac{1}{4} \div \left(-\frac{14}{3}\right) = j$

Evaluate each expression.

33. $a \div b$, if $a = \frac{2}{3}$ and $b = 1\frac{1}{3}$

34. $r \div s$, if $r = -\frac{8}{9}$ and $s = \frac{7}{18}$

35. $a^2 \div b^2$, if $a = -\frac{3}{4}$ and $b = 1\frac{1}{3}$

36. $m + n \div p$, if $m = \frac{2}{3}$, $n = 1\frac{1}{3}$, and $p = \frac{1}{9}$

Mixed Review

37. Evaluate $2d^3$ if $d = 3$.   (Lesson 4-2)

38. Round 44.005 to the nearest tenth.   (Lesson 5-3)

39. **Personal Finance**   Ana buys two pairs of shoes at a store that is having a "buy one pair, get the second for half price" sale. If the first pair of shoes costs $35.98, the second pair costs $29.98, and the tax is $3.73, how much can Ana expect to pay?   (Lesson 6-4)

40. Name the multiplicative inverse of $-3\frac{1}{4}$.   (Lesson 6-5)

41. Compute $\frac{1}{3} \times 1\frac{4}{5}$.   (Lesson 6-5)

Applications

42. **Home Economics**   Mrs. Blair needs $8\frac{1}{4}$ cups of raisins to make 2 fruitcakes. A 15-ounce box of raisins contains $2\frac{3}{4}$ cups. How many boxes should Mrs. Blair buy?

43. **Publishing**   A page of type is to be divided into three columns. If the page is $6\frac{3}{4}$ inches wide, how many inches wide is each column?

44. **Carpentry**   How many boards, each 2 feet 6 inches long, can be cut from a board 18 feet long?

Critical Thinking

45. What is the quotient if the eighth term of the sequence $1, \frac{1}{2}, \frac{1}{4}, \frac{1}{8}, \ldots$ is divided by the ninth term?

Wrap-Up   46. A number is both multiplied and divided by the same rational number n, where $0 < n < 1$. Which is greater, the product or the quotient? Explain your reasoning.

6-7 Rationals: Dividing Decimals

Objective:
Divide decimals.

The world outdoor sprint records as of 1992 are shown in the graph. Carl Lewis ran 100 meters in 9.86 seconds. How many meters, on the average, did he run in 1 second? To find this rate, divide 100 by 9.86.

World Sprint Records
MEN
100 meters—Carl Lewis (U.S.) 9.86
200 meters—Pietro Mennea (Italy)19.72
400 meters—Butch Reynolds (U.S.) 43.29
WOMEN
100 meters—Florence Griffith Joyner (U.S.) 10.49
200 meters—Florence Griffith Joyner (U.S.) 21.34
400 meters—Marita Koch (E. Germany) 47.60

$$100 \div 9.86 = 10.141988$$

To the nearest hundredth, Carl Lewis ran an average of 10.14 meters in 1 second.

Let's review how this division is done. The division can be completed using paper and pencil more easily if the divisor, 9.86, is a whole number. Think of the fraction $\frac{100}{9.86}$.

$$\frac{100}{9.86} \times \frac{100}{100} = \frac{10,000}{986}$$

You can multiply by 100 mentally.
$100 \times 100 = 10,000$
$9.86 \times 100 = 986$

Since $\frac{100}{100} = 1$, the value of the fraction is unchanged. This becomes a division problem involving whole numbers.

$$\frac{100}{9.86} \implies 9.86 \overline{)100.00} \implies 986 \overline{)10,000.000}$$

```
        10.                      10.14
  9.86 )100.00            986 )10,000.000
        986                     986
        140                     140
          0                       0
        140                     1400
                                 986
                                4140
                                3944
                                 196
```

To the nearest hundredth, the quotient is 10.14. This matches the calculator result.

Examples

Solve each equation.

1 $50 \div 2.5 = a$ Estimate: $50 \div 2 = 25$

$2.5\overline{)50.0}$ Multiply both 2.5 and   or 50 ÷ 2.5 = 20
50 by 10 to get a
whole number divisor.

$$\begin{array}{r} 20 \\ 25\overline{)500} \end{array}$$

$20 = a$ Compare with the estimate.

2 $-0.0078 \div 0.003 = y$ Estimate: $-8 \div 3 = -2\frac{2}{3}$

$0.003\overline{)0.0078}$ Multiply 0.003 and 0.0078 by 1000 to get
a whole number divisor.

$$\begin{array}{r} -2.6 \\ 3\overline{)-7.8} \end{array}$$ The quotient is negative.

or

 0.0078 +/- ÷ 0.003 = -2.6

$-2.6 = y$ Compare with the estimate.

Checking for Understanding ───────────────

Communicating Algebra

1. What is the first step in preparing to divide two decimals?

2. Explain why you should make an estimate when using a calculator to divide.

Use estimation to choose the correct quotient.

3. $36 \div 1.2$; 3 or 30

4. $4.5 \div 0.9$; 5 or 0.5

5. $1.6 \div 8$; 2 or 0.2

6. $1.6 \div 0.8$; 2 or 0.2

Guided Practice

State where the decimal point should be placed in each quotient.

7. $0.48 \div 0.2 = 24$

8. $25 \div 0.05 = 5$

9. $3.6 \div 0.04 = 9$

10. $0.012 \div 0.04 = 3$

11. $0.0018 \div 0.2 = 9$

12. $0.56 \div 0.007 = 8$

Find each quotient.

13. $24 \div 0.3$

14. $0.3 \div 6$

15. $15.5 \div (-0.5)$

Exercises

Solve each equation.

16. $a = 0.128 \div 8$ **17.** $r = 19.26 \div 6$ **18.** $0.335 \div 5 = d$

19. $y = 5.0 \div 0.25$ **20.** $z = 2.25 \div 2.5$ **21.** $-3.75 \div 0.25 = b$

22. $1.69 \div (-1.3) = x$ **23.** $-7.5 \div (-1.5) = m$ **24.** $4.156 \div 0.4 = k$

Estimation

Estimate each quotient. Then compute with a calculator. Round the quotient to the nearest tenth, if necessary.

25. $a = 5.98 \div 2.9$ **26.** $1.98 \div 0.71 = p$ **27.** $7.81 \div 5 = y$

28. $z = 0.597 \div 0.25$ **29.** $t = 93.702 \div 2.4$ **30.** $a = 17.3 \div 5.2$

Mixed Review

31. Solve the equation $-5x = -100$.   (Lesson 3-3)

32. Simplify the expression $-\frac{4}{5}y + 2\frac{3}{5}y + 7\frac{1}{5}y$.   (Lesson 5-5)

33. Name the property shown by the sentence $8 \times 2\frac{1}{4} = 8(2) + 8\left(\frac{1}{4}\right)$.
(Lesson 6-5)

Applications

34. Sports   Cheryl competes in the 100-meter freestyle event at a swim meet. Her times in the previous five races were 53.53 s, 54.24 s, 54.82 s, 54.69 s, and 54.60 s. Find Cheryl's average time.

35. Business   While taking inventory at The Builder's Store, a clerk measures a stack of table tops. If each top is 1.875 inches thick and the stack is 3 feet 9 inches high, how many table tops are in the stack?

**Critical
Thinking
Wrap-Up**

36. Name two decimals whose sum is 2.7 and whose quotient is 2.

37. Using the information in the chart at the beginning of the lesson, find the average number of meters Florence Griffith-Joyner ran in 1 second in the 100-meter event.

Mid-Chapter Quiz

1. Express $\frac{2}{3}$ as a decimal.   (Lesson 6-1)

2. Estimate $\frac{2}{3} \times 20$.   (Lesson 6-2)

3. Explain how to compute $8 \times 3\frac{1}{4}$ using the distributive property.   (Lesson 6-5)

Solve each equation.   (Lesson 6-3, 6-4, 6-6)

4. $x = -2\frac{1}{2}(-8)$ **5.** $0.008 \times 0.9 = y$ **6.** $a = -\frac{3}{8} \times 1\frac{1}{3}$

7. $b = \left(\frac{4}{9}\right)^2$ **8.** $p = -2.5 \times 1.8$ **9.** $\frac{3}{4} \div 1\frac{1}{3} = c$

6-8 Solving Equations and Inequalities

Objective:
Solve equations and inequalities containing rational numbers.

What is the product of a number and its multiplicative inverse?

You can apply the multiplication and division skills that you have learned for rational numbers to solve equations and inequalities containing rational numbers. Consider this problem.

On the first day Cherry Creek Limited stock was traded, it gained $4\frac{1}{8}$ points. The amount of gain was $\frac{1}{4}$ of its opening price. What was the stock price when it opened?

Let x = the opening price.

$$\frac{1}{4} \text{ of opening price} = 4\frac{1}{8}$$

$$\frac{1}{4}x = 4\frac{1}{8}$$

$$4 \cdot \frac{1}{4}x = 4 \cdot 4\frac{1}{8} \qquad \text{4 and } \frac{1}{4} \text{ are multiplicative inverses.}$$

$$x = \frac{4}{1} \cdot \frac{33}{8}$$

$$x = \frac{33}{2} \text{ or } 16\frac{1}{2}$$

Cherry Creek Limited stock opened at $16\frac{1}{2}$ points or \$16.50.

Examples

Solve each equation. Check your solution.

1 $\qquad \dfrac{x}{-1.5} = 4.5$

$(-1.5)\,\dfrac{x}{-1.5} = (-1.5)4.5$ Multiply each side by –1.5.

$\qquad x = -6.75$

Check: $\dfrac{x}{-1.5} = 4.5$

$\qquad \dfrac{-6.75}{-1.5} \overset{?}{=} 4.5$

$\qquad 4.5 = 4.5$ ✔

6.75 $\boxed{+/-}$ $\boxed{\div}$ 1.5 $\boxed{+/-}$ $\boxed{=}$ 4.5

FYI

Two hundred years ago, stock brokers in New York City met underneath a buttonwood tree on Wall Street to transact their business. Today, this site on Wall Street is the location of the New York Stock Exchange.

2 $2.3x = -9.2$

$\dfrac{2.3x}{2.3} = \dfrac{-9.2}{2.3}$ Divide each side by 2.3.

$x = -4$

Check:

$\boxed{2.3}\ \boxed{\times}\ \boxed{4}\ \boxed{+/-}\ \boxed{=}\ \text{-9.2}\ \ \checkmark$

$\boxed{9.2}\ \boxed{+/-}\ \boxed{\div}\ \boxed{2.3}\ \boxed{=}\ \text{-4}$

In Example 3, why should you multiply by $\frac{6}{5}$?

3 $\dfrac{5}{6}m = -\dfrac{2}{3}$

$\dfrac{6}{5} \cdot \dfrac{5}{6}m = \left(\dfrac{6}{5}\right)\left(-\dfrac{2}{3}\right)$ Multiply each side by $\frac{6}{5}$.

$m = -\dfrac{4}{5}$

Check: $\dfrac{5}{6}m = -\dfrac{2}{3}$

$\dfrac{5}{6}\left(-\dfrac{4}{5}\right) \overset{?}{=} -\dfrac{2}{3}$

$-\dfrac{2}{3} = -\dfrac{2}{3}\ \ \checkmark$

Inequalities with rational numbers are solved using the same skills you developed for solving inequalities with integers.

Examples

Solve each inequality.

How can you check the solution of an inequality?

4 $\dfrac{a}{2.6} > 4.5$

$(2.6)\dfrac{a}{2.6} > (2.6)4.5$ Multiply each side by 2.6.

$\boxed{2.6}\ \boxed{\times}\ \boxed{4.5}\ \boxed{=}\ \text{11.7}$

$a > 11.7$

Any number greater than 11.7 is a solution.

5 $\dfrac{c}{-2} \geq 3.6$

$(-2)\dfrac{c}{-2} \leq (-2)3.6$ Multiply each side by -2. Remember to reverse the order symbol.

$c \leq -7.2$

Any number less than or equal to -7.2 is a solution.

Checking for Understanding ⎯⎯⎯⎯⎯⎯⎯⎯⎯⎯⎯⎯

Communicating Algebra

1. Explain how you would use a multiplicative inverse to solve $\frac{2}{3}x = 8$.

2. Name the multiplicative inverse of $-\frac{4}{9}$.

Explain how to solve each equation or inequality. Then solve.

3. $0.5x = 10$

4. $1.1y = -2.2$

5. $3.6 < 0.9c$

6. $\frac{2}{3}x > 6$

7. $-\frac{5}{6}y = \frac{3}{5}$

8. $\frac{5}{8}t = 4$

9. $\frac{k}{1.5} = 1.1$

10. $\frac{z}{-7} \leq 12$

11. $-11t > 12\frac{1}{2}$

Exercises

Independent
Practice

Solve each equation or inequality. Check your solution.

12. $2a = -12$

13. $-3y = 1.5$

14. $\frac{x}{3} > -6$

15. $-\frac{1}{4}c = 3.4$

16. $\frac{x}{3.2} < -4.5$

17. $-1.6n = 0.48$

18. $\frac{n}{2} \geq -1.6$

19. $\frac{1}{3}d = -0.36$

20. $\frac{m}{2.3} = -5.2$

21. $-\frac{3}{5}r \leq \frac{2}{3}$

22. $-7\frac{1}{2}x = 5\frac{1}{4}$

23. $-2h > 4.6$

24. $\frac{4}{5}y = \frac{3}{8}$

25. $\frac{x}{-2.5} \leq 3.2$

26. $7x < 1\frac{4}{10}$

Mixed Review

27. Evaluate the expression $4[2(3 + 5) - 8]$. (Lesson 1-1)

28. Solve the inequality $f - 4.6 < -3.1$. (Lesson 5-7)

29. Evaluate the expression $x^2 \div y$ if $x = \frac{1}{2}$ and $y = -\frac{2}{3}$. (Lesson 6-6)

30. Solve the equation $y = 30.66 \div 7$. (Lesson 6-7)

31. **Measurement**   The distance from Jeff's house to school is about 2000 feet. What part of a mile is this?   (Lesson 6-7)

Applications

32. **Personal Finance**   Alma has $24 to spend on cassette tapes. The CD and Tape Mart is selling tapes for $6.99. How many tapes can Alma buy?

33. **Personal Finance**   Kioka earns 5¢ for every newspaper she delivers. She earns between $2.50 and $3.00 every day. What is the least number of newspapers Kioka delivers each day? What is the greatest number?

Critical
Thinking

34. **Make Up a Problem**   Write a problem that can be solved using the inequality in Exercise 6.

Wrap-Up

35. Write an equation that can be solved by multiplying by $\frac{3}{4}$.

Team Problem Solving

The Absent-Minded Professor has two glasses of Sparkling Soda. The first is half full. The second glass, twice the size of the first glass, is one-quarter full. The professor fills both glasses with water and mixes the contents into a third container. What part of the final mixture is Sparkling Soda?

6-9 Inductive and Deductive Reasoning

Objective:
Use deductive and inductive reasoning.

Key Terms:
inductive
deductive

FYI

Galileo (1564–1642) is sometimes called the father of modern experimental science. His father wanted him to be a medical doctor. However, his interest in mathematics led him to serve as professor of mathematics for 18 years at the University of Padua in Italy.

Carla noticed that for the last six Thursdays her pre-algebra teacher, Mrs. Hodges, has given a pop quiz. So, Carla assumed that the next Thursday there would be a pop quiz in pre-algebra.

Carla used **inductive reasoning**. Inductive reasoning *makes* a rule after seeing several examples. Carla made a conclusion based on what happened in the past.

Example

1 The Italian scientist Galileo discovered that there was a relationship between the time of the swing back and forth of a pendulum and its length.

Time of Swing	Length of Pendulum
1 second	1 unit
2 seconds	4 units
3 seconds	9 units
4 seconds	16 units

How long do you think a pendulum with a swing of 5 seconds is?

Look at the pattern in the table. The measure of the pendulum is the square of the measure of time.

Apply inductive reasoning. A pendulum with a swing of 5 seconds would have a length of 5^2 or 25 units.

Let's return to Carla's pop quiz situation. Suppose her teacher does not give a pop quiz the next Thursday. Then the method of inductive reasoning did not work. Since inductive reasoning is based on past evidence only, it may sometimes fail.

At the beginning of the year, Carla's science teacher, Mr. Steiner, said he would give a quiz every Friday. Carla knows that if it is a Friday, there will be a quiz in science.

This is an example of **deductive reasoning**. Deductive reasoning *uses* a rule to make a conclusion or a decision.

Example

2 All polygons with five sides are pentagons. The polygon shown at the right has five sides. What can you conclude?

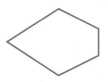

Use deductive reasoning to conclude that the polygon is a pentagon.

Checking for Understanding

1. In Example 1, suppose the time of the swing of the pendulum is 1.5 seconds. How long is the pendulum?

2. Which type of reasoning goes from a general rule to specific examples?

3. How is the problem-solving strategy *look for a pattern* like inductive reasoning?

Guided Practice

Determine whether each exercise is an example of *inductive* or *deductive* reasoning. Explain your reasoning.

4. It has snowed every New Year's Day for the past four years. Akiko says it will snow on New Year's Day this year.

5. The team that wins two more games will make the playoffs. Carter Junior High wins two more games, so they go to the playoffs.

6. If the angles of a triangle have the same measure, the triangle is an equilateral triangle. The triangle shown at the right is an equilateral triangle.

Exercises

Independent Practice

Determine whether each exercise is an example of *inductive* or *deductive* reasoning. Explain your reasoning.

7. Patty noticed the school cafeteria served pizza on the last five Mondays. Patty decides that the cafeteria always serves pizza on Monday.

8. Numbers whose last two digits are divisible by 4 are divisible by 4. So, 18,324 is divisible by 4.

9. Band members are admitted free to all football games. Rachel plays flute in the band. She gets into every football game free.

10. Every customer who came into Balmer's Clothing was wearing a raincoat. Mrs. Balmer decided it was raining.

Use inductive reasoning to determine the next two numbers in each list.

11. 5, 10, 15, __?__ , __?__

12. 39, 33, 27, __?__ , __?__

13. 1, 10, 100, __?__ , __?__

14. 6, 7, 9, 12, 16, __?__ , __?__

Critical Thinking

15. If you are not passing English, your parents will not let you go on the class field trip. You are not allowed to go on the field trip. Does this mean you did not pass English? Give reasons to support your answer.

Wrap-Up

16. Explain the difference between inductive and deductive reasoning. Then give an example of each.

6-10   Geometric Sequences

Objectives:
Recognize and extend geometric sequences. Represent a sequence algebraically.

Key Terms:
geometric sequence
common ratio

A certain golf ball, dropped from any height onto a hard surface will rebound 0.8 of the way back to its starting point. Suppose the ball is dropped from a height of 20 meters. The chart below shows its height after the first four bounces.

Bounce	0	1	2	3	4	5
Height (m)	20	16	12.8	10.24	8.192	?

$\times 0.8$ $\times 0.8$ $\times 0.8$ $\times 0.8$ $\times 0.8$

How high will it bounce on the fifth bounce? Use your calculator.

$$8.192 \;\boxtimes\; 0.8 \;\boxminus\; 6.5536$$

Can a geometric sequence contain zero as a term?

The successive heights form a sequence. When the terms of a sequence increase or decrease by a constant *factor,* the sequence is called a **geometric sequence**. The factor is called the **common ratio**. In the sequence shown above, the common ratio is 0.8.

Geometric Sequence	A geometric sequence is a sequence in which the ratio between any two successive terms is the same.

Examples

State whether each sequence is geometric. Then write the next three terms of each sequence.

1 $9, 3, 1, \frac{1}{3}, \ldots$

Since the common ratio is $\frac{1}{3}$, the sequence is geometric.

$$9 \quad 3 \quad 1 \quad \frac{1}{3}$$

$\times \frac{1}{3}$ $\times \frac{1}{3}$ $\times \frac{1}{3}$

The next three terms are $\frac{1}{9}$, $\frac{1}{27}$, and $\frac{1}{81}$.

2 $-1, -1, -2, -6, -24, \ldots$

Since there is no common ratio, the sequence is *not* geometric.

$$-1 \quad -1 \quad -2 \quad -6 \quad -24$$

$\times 1$ $\times 2$ $\times 3$ $\times 4$

Are you using inductive or deductive reasoning when you find the terms?

The next three terms are -120, -720, and -5040.

Example

State whether the sequence is geometric. Then write the next three terms.

3 8, -4, 2, -1, . . .

The common ratio is $-\frac{1}{2}$. The sequence is geometric.

$$8 \quad -4 \quad 2 \quad -1$$
$$\times\left(-\frac{1}{2}\right) \times\left(-\frac{1}{2}\right) \times\left(-\frac{1}{2}\right)$$

The next three terms are $\frac{1}{2}$, $-\frac{1}{4}$, and $\frac{1}{8}$.

CONNECTION TO ALGEBRA

If we know the first term of a geometric sequence and the common ratio, we can find any other term of the sequence. Consider the sequence 2, 6, 18, 54, ...

Here's how to write an expression that represents a term in the sequence. The first term is 2; call the first term a. The common ratio is 3; call the common ratio r. Study this pattern.

1st term: a
2nd term: $a \cdot r$
3rd term: $a \cdot r \cdot r$ or ar^2
4th term: $a \cdot r \cdot r \cdot r$ or ar^3
      .   .
      .   .
      .   .
nth term: $a \cdot r^{n-1}$ The exponent is one less than the number of the term.

Because n represents any term, you can use this expression to find any term in the sequence.

Examples

4 **Use the expression ar^{n-1} to find the sixth term of the sequence 2, 6, 18, 54, . . .**

The first term, a, is 2, the common ratio is 3, and n is 6.

$ar^{(n-1)} = 2 \cdot 3^{6-1}$ or $2 \cdot 3^5$ $a = 2, r = 3, n = 6$

$$2 \; \boxed{\times} \; 3 \; \boxed{y^x} \; 5 \; \boxed{=} \; 486$$

The sixth term is 486.   You can check this answer by extending the sequence.

5 **Find the height of the golf ball on page 233 after 10 bounces.**

$ar^{(n-1)} = 20(0.8)^{(10-1)}$ $a = 20, r = 0.8, n = 10$

$$20 \; \boxed{\times} \; 0.8 \; \boxed{y^x} \; 9 \; \boxed{=} \; 2.6843546$$

The height is about 2.7 meters.

Checking for Understanding

Communicating
Algebra

1. Give an example of a geometric sequence.

2. Explain the difference between an arithmetic sequence and a geometric sequence.

3. In the geometric sequence 8, -4, 2, -1, . . . , what is the *common ratio*?

Guided
Practice

State whether each sequence is a geometric sequence. If so, state the common ratio.

4. 2, 4, 8, 16, . . .   5. 2, 4, 6, 8, . . .   6. -5, 1, $-\frac{1}{5}$, $\frac{1}{25}$, . . .

7. 24, 12, 6, 3, . . .   8. 9, 3, -3, -9, . . .   9. $\frac{1}{2}$, $\frac{1}{4}$, $\frac{1}{8}$, $\frac{1}{16}$, . . .

Exercises

Independent
Practice

Write the next three terms of each sequence.

10. $\frac{1}{2}$, 1, 2, 4, . . .   11. 2, 3, 5, 8, 12, . . .

12. 8, 4, 2, 1, . . .   13. 3, 7, 11, 15, . . .

14. 7, -14, 28, -56, . . .   15. $4\frac{1}{2}$, $2\frac{1}{2}$, $\frac{1}{2}$, $-1\frac{1}{2}$, . . .

16. 18, -6, 2, $-\frac{2}{3}$, . . .   17. 5, $-\frac{5}{2}$, $\frac{5}{4}$, $-\frac{5}{8}$, . . .

Language
Skill

18. Write the first five terms in a geometric sequence with a common ratio of $\frac{1}{2}$. The first term is 2.

19. Find the fourth term of a geometric sequence if $a = -5$ and $r = -2$.

20. Use the expression $ar^{(n-1)}$ to find the eighth term of the geometric sequence 16, 8, 4, . . .

Mixed Review

21. Evaluate the expression $rs \div t$ if $r = 2.8$, $s = 0.5$, $t = 0.7$.   (Lesson 6-7)

Solve each equation or inequality.   (Lesson 6-8)

22. $\frac{k}{-1.3} = -4.7$ 23. $-\frac{7}{8}f > \frac{1}{3}$

Applications

24. **Physics**   A ball rebounds $\frac{2}{3}$ of its height after every fall. If it is dropped from a height of 48 feet, how high will it bounce at the end of the third bounce?

25. **Business**   A certain automobile loses $\frac{3}{10}$ of its value every year. If Tina bought an automobile for $8400, what is the value at the end of three years?

Critical
Thinking

26. Find the value of x so that 3, x, 12, . . . is a geometric sequence.

Wrap-Up

27. Write the first four terms of a geometric sequence that has a common ratio of $-\frac{1}{3}$.

Algebra in Action-Computers

Spreadsheets

An electronic **spreadsheet** program is a computer application that allows users to prepare tables easily. Using a spreadsheet program, it is possible to project results, make calculations, and print almost anything that can be arranged in a table.

Spreadsheets can also be used as a tool to simulate experiments when the variables can be described algebraically. Consider the bouncing golf ball from page 233. We can use a spreadsheet to record the height of the ball after each bounce. The computer automatically performs the calculations. A sample printout is shown at the right.

	BALL BOUNCE	
	A	**B**
1	Initial Ht (m) =	20
2	Common Ratio =	0.8
3	HIT NUMBER	RETURN HT
4	1	20.000
5	2	16.000
6	3	12.800
7	4	10.240
8	5	8.192
9	6	6.554
10	7	5.243

In the printout, the vertical columns and the horizontal rows are combined and named to form **cells**. Each cell is identified by a row and column. Cell B6 shows the height 12.800 meters.

This spreadsheet was built so that the initial height can be changed by altering the value in cell B1. The common ratio can be changed by altering the value in B2. The remaining cells are determined by formulas. The computer does the calculations.

A4 + 1 means add 1 to the value in cell A4.

	BALL BOUNCE	
	A	**B**
1	Initial Ht (m) =	20
2	Common Ratio =	0.8
3	HIT NUMBER	RETURN HT
4	1	=B1
5	=A4+1	=B4*B2
6	=A5+1	=B5*B2
7	=A6+1	=B6*B2

B4 * B2 means multiply the value in cell B4 by the value in cell B2.

1. Explain the meaning of the formula in cell B6.
2. A certain ball will return to a height that is 0.5 times the previous height. How would you modify the spreadsheet?
3. Suppose you wanted to investigate this question: If a ball is dropped from twice as high, will it bounce twice as many times? Describe how you could use a spreadsheet to answer this question.

6-11 Scientific Notation

Objective:
Write numbers in scientific notation.

If you have ever visited Luray Caverns in Virginia or Carlsbad Caverns in New Mexico, you have probably wondered about the icicle-like structures that hang from the roof of the caves. These stalactites are formed as water containing limestone drips through the roof of the cave, one drop at a time.

Some stalactites are as old as 40,000 years. If one drop of water drips through the roof of the cave every second for 40,000 years, there would be about 1,262,304,000,000 drops. If each drop contained just 0.00001 g of limestone, 12,623,040 g of limestone would have formed. This is 12,623 kg or nearly 28,000 pounds.

Large numbers such as 1,262,304,000,000 and small numbers such as 0.00001 are difficult to use. It is easy to make mistakes by omitting zeros or having the decimal point in the wrong place. People who deal regularly with such numbers use **scientific notation.**

Scientific Notation	Numbers expressed in scientific notation are expressed as the product of a factor and a power of 10. The factor must be greater than or equal to 1 and less than 10.

Examples

Express each number in standard form.

1 2.73×10^5

$2.73 \times 10^5 = 2.73 \times 100,000$ 2.73000
$ = 273,000$ five places

2 4.53×10^{-3}

$4.53 \times 10^{-3} = 4.53 \times \dfrac{1}{10^3}$ Remember from Lesson 4-10

$\phantom{4.53 \times 10^{-3}} = 4.53 \times \dfrac{1}{1000}$ that $10^{-3} = \dfrac{1}{10^3}$.

$\phantom{4.53 \times 10^{-3}} = 4.53 \times 0.001$ 004.53

$\phantom{4.53 \times 10^{-3}} = 0.00453$ three places

Mental Math Hint

When dividing by 10, 100, 1000, . . . , the decimal point is moved left as many places as the number of zeros. In Example 3, moving the decimal point 4 places left is the same as dividing the number by 10,000 or 10^4. You then multiply by 10^4 so that the number is not changed.

What is the relationship between the absolute value of the exponent and the number of places the decimal is moved?

Examples

Express each number in scientific notation.

3 20,000

$$20,000 = 2 \times 10^4$$

Move the decimal point 4 places to the left. Multiply by 10^4.

4 57,800,000

$$57,800,000 = 5.78 \times 10^7$$

Move the decimal point 7 places to the left. Multiply by 10^7.

5 -0.001

$$-0.001 = -1 \times 10^{-3}$$

Move the decimal point 3 places to the right. Multiply by 10^{-3}.

6 0.000623

$$0.000623 = 6.23 \times 10^{-4}$$

Move the decimal point 4 places to the right. Multiply by 10^{-4}.

Most scientific calculators allow you to compute with numbers in scientific notation. Example 7 shows how to enter a number in standard form and have the calculator express it in scientific notation. Example 8 shows how to enter a number that will not fit on the display screen in standard form.

Examples

7 **Use a calculator to express 0.0000089 in scientific notation.**

$$0.0000089 \quad \boxed{\text{EE}} \quad \boxed{=} \quad 8.9 - 06$$

exponent of the power of ten

$$0.0000089 = 8.9 \times 10^{-6}$$

8 **Enter 1,262,304,000,000 into a calculator.**

First write the number in scientific notation.

$$1,262,304,000,000 = 1.262304 \times 10^{12}$$

Then enter the number.

$$1.262304 \quad \boxed{\text{EE}} \quad 12$$

The display shows 1.2623   12.

Checking for Understanding

Communicating Algebra

1. Explain why 23.8×10^4 is not written in scientific notation.

2. Give an example of a number that is written in scientific notation.

Guided Practice

State where the decimal point should be placed in order to express each number in scientific notation. Then state the power of ten by which you should multiply.

3. 159   4. 46   5. 1800   6. 85,000

7. 0.85   8. 0.00246   9. 0.0000593   10. 0.0000124

Exercises

Independent Practice

Express each number in standard form.

11. 5.2×10^5 12. 3.9×10^3 13. -6.1×10^4

14. 1.23×10^{-2} 15. -5.765×10^{-3} 16. 7.02×10^0

17. the distance to the moon, 2.39×10^5 miles

18. the height of the Sears Tower, 1.454×10^3 feet

19. the volume of a drop of liquid, 5×10^{-5} liter

Express each number in scientific notation.

20. 820   21. 0.504   22. -6100

23. 97,000   24. -0.004976   25. 0.00623

26. 5,000,000   27. 0.00000892   28. 96

29. the diameter of a red blood cell, 0.0003 inch

30. the distance to the sun, 93,000,000 miles

Mixed Review

31. Find the quotient $\dfrac{(-2)^4}{(-2)^2}$. (Lesson 4-10)

32. State whether the sequence 2, 4, 8, 16, . . . is an arithmetic sequence. (Lesson 5-9)

33. Solve the inequality $\dfrac{t}{2.3} < -0.4$. (Lesson 6-8)

34. The people of Waynesburg, PA, celebrate July 26 as Rain Day, because it almost always rains on that day. Is this an example of inductive or deductive reasoning? (Lesson 6-9)

Applications

35. **Physics** The speed of light in a vacuum is 29,979,280,000 centimeters per second. Explain how to enter this number into a calculator.

36. **Chemistry** An oxygen atom has a mass of 2.66×10^{-23} grams. Explain how to enter this number into a calculator.

Critical Thinking

37. Use the associative and commutative properties of multiplication to compute $(2.72 \times 10^5)(3.25 \times 10^{-3})$.

Wrap-Up

38. Explain how the movement of the decimal point is related to expressing numbers in scientific notation.

6-12 Comparing Rational Numbers

Objective:
Compare rational numbers.

Sometimes you need to determine which of two numbers is greater. A number line is sometimes a useful tool in comparing numbers. However, there are times when a number line is not convenient. Consider this problem.

After playing 21 baseball games, the Pittsburgh Pirates had won 14 games. During the same time, the New York Mets won 15 games out of 23. Which team has the better record?

Why is it inconvenient to compare $\frac{14}{21}$ and $\frac{15}{23}$ on a number line?

To solve this problem, compare $\frac{14}{21}$ and $\frac{15}{23}$. One way to compare these fractions is to express them as decimals and then compare the decimals.

$\frac{14}{21}$ 0.6666667　　　$\frac{15}{23}$ ➡ 0.6521739

In the hundredths place, $6 > 5$.

Since $0.6666667 > 0.6521739$, $\frac{14}{21} > \frac{15}{23}$.
The Pirates have the better record.

Another way to compare two rational numbers is to express them as equivalent fractions with like denominators.

Mental Math Hint

You can also compare fractions by using cross products.

$\frac{4}{9}　\frac{5}{12}$

$48　　45$

Since $48 > 45$,
$\frac{4}{9} > \frac{5}{12}$.

Examples

1 Replace each ● with <, >, or = to make a true sentence.

$\frac{4}{9}$ ● $\frac{5}{12}$

$\frac{16}{36}$ ● $\frac{15}{36}$　　The common denominator is 36.

$\frac{16}{36} > \frac{15}{36}$　　Since $16 > 15$, $\frac{16}{36} > \frac{15}{36}$.

Therefore, $\frac{4}{9} > \frac{5}{12}$.

2 $\frac{2}{3}$ 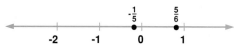 -0.5

> Any positive number is graphed to the right of any negative number on the number line.

> Therefore, $\frac{2}{3} > -0.5$.

3 -0.8 ● $-\frac{2}{3}$

-0.8 ● $-\frac{2}{3}$

-0.8 ● -0.6666 . . .   Express $-\frac{2}{3}$ as a decimal.

In the tenths place, -8 < -6.
Therefore, $-0.8 < -\frac{2}{3}$.

4 35.9 ● 3.59 × 10²

35.9 ● 3.59 × 10²

35.9 ● 359   Express 3.59 × 10² in standard form.

35.9 < 359

Therefore, $35.9 < 3.59 \times 10^2$.

Checking for Understanding ──────────────────

Communicating Algebra

1. Compare the numbers that are graphed on the number line shown below.

2. Using a number line, explain why -1.5 < -1.

3. In your own words, describe two different ways to compare $\frac{5}{8}$ and 0.4.

Guided Practice

State the greater number for each pair.

4. 25.23, 25.13

5. -5.049, -5.149

6. 63.024, -63.028

7. $1\frac{3}{5}$, $1\frac{3}{6}$

8. $-\frac{4}{5}$, $-\frac{9}{10}$

9. 1.2×10^3, 1.2×10^4

Exercises

Independent
Practice

ESTIMATION
MENTAL MATH
CALCULATOR
PAPER/PENCIL

Replace each ● with <, >, or = to make a true sentence.

10. 5.2 ● 5.18

11. 4.38 ● 4.48

12. -5.25 ● 5.2

13. -93.25 ● -93.35

14. -50.03 ● -50.13

15. 97.2 ● -96.23

16. $\frac{5}{9}$ ● $\frac{2}{3}$

17. $\frac{3}{8}$ ● $\frac{13}{25}$

18. $-\frac{2}{6}$ ● $\frac{7}{21}$

19. $-\frac{2}{3}$ ● $-\frac{1}{3}$

20. $\frac{25}{27}$ ● $\frac{17}{19}$

21. $-\frac{4}{5}$ ● $-\frac{5}{8}$

22. $1\frac{1}{8}$ ● 1.29

23. 107.58 ● $107\frac{58}{100}$

24. $-75\frac{3}{10}$ ● 75.29

25. $-2\frac{1}{2}$ ● -2.59

26. 1.27 ● 1.27×10^{1}

27. $5\frac{3}{5}$ ● 5.35×10^{-2}

Mixed Review

28. Find the LCD of $\frac{5}{7x^2}$ and $\frac{9}{21x}$. (Lesson 4-9)

29. Solve the equation $r = -8\frac{3}{4} + 9\frac{7}{8}$. (Lesson 5-6)

30. State whether the sequence 16, 10, 4, -2, . . . is a geometric sequence. (Lesson 6-10)

31. Express 0.024 in scientific notation. (Lesson 6-11)

Applications

32. Consumer Awareness Zina found tennis balls priced at three for $4.69. At a second store they were priced at seven for $8.50. Which is the better buy?

33. Sports Use the information in the problem presented on page 240. Suppose the Pirates win two of their next three games and the Mets win all of their next three games. Which team has the better record?

Critical Thinking

34. Find a fraction between $\frac{3}{4}$ and $\frac{4}{5}$. Explain how you found it.

Wrap-Up Journal Entry **35.** Explain how to order $\frac{3}{5}$, 0.63, and 6.02×10^{-1} from least to greatest. Graph the numbers on a number line to show your solution.

Enrichment

Density

How many rational numbers are there between $\frac{1}{3}$ and $\frac{1}{2}$? Would you guess *a few,*

many, or *none?* If you express $\frac{1}{3}$ as 0.333 . . . and $\frac{1}{2}$ as 0.5, you can see that 0.4 falls

between them. So does 0.41, 0.42, 0.43, and so on. In fact, there are infinitely many

rational numbers between $\frac{1}{3}$ and $\frac{1}{2}$. The **density property** states that between any

two rational numbers, no matter how close they may seem, there is at least one other

rational number.

6-13 Circles and Circumference

Objective:
Find the circumference of a circle.

Key Terms:
circle
center
radius
diameter
circumference

A favorite ride at amusement parks, the Ferris wheel is a power-driven, upright wheel with passenger cars mounted on the rim of the wheel. Each car is the same distance from the center. As the wheel turns, each car traces the path of a circle.

A **circle** is the set of all points in a plane that are the same distance from a given point in the plane. The given point is called the **center.** The distance from the center to any point on the circle is called the **radius (r).** The distance across the circle through its center is its **diameter (d).** The diameter of a circle is twice its radius, or $d = 2r$.

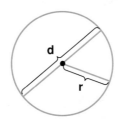

FYI

The Ferris wheel was invented by the American engineer G. W. Ferris for the 1893 World's Exposition in Chicago. The first Ferris wheel stood 250 feet tall and carried nearly 2000 passengers. Most modern Ferris wheels are only 45 feet tall.

The **circumference (C)** of a circle is the distance around the circle. Suppose a bicycle wheel makes one complete turn. The distance it travels is the same as the circumference of the wheel.

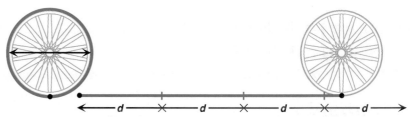

Notice in the drawing above that the circumference is a little more than three times the diameter of the wheel. In fact, the circumference of a circle is always 3.1415926 . . . times the diameter. The Greek letter π (pi) stands for this number. Although π is not a rational number, the rational numbers 3.14 and $\frac{22}{7}$ are two generally accepted approximations for π.

Circumference	In words: The circumference of a circle is equal to its diameter times π, or 2 times its radius times π.
	In symbols: $C = \pi d$ or $C = 2\pi r$

Examples

Find the circumference of each circle. Use the given approximation for π.

Why does C = πd but
C ≈ 3.14d?

1 $C = \pi d$ $\pi \approx \frac{22}{7}$

     $C \approx \frac{22}{7} \cdot 18$

     $C \approx \frac{396}{7}$ or $56\frac{4}{7}$

The circumference is about $56\frac{1}{2}$ inches.

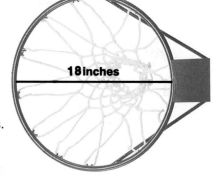

18 inches

Calculator Hint

π is such an important number in mathematics that it usually has its own key on a calculator. What is displayed on your calculator when you press $\boxed{\pi}$?

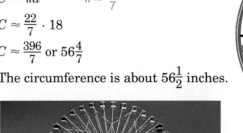

7.5 meters

2 $C = 2\pi r$ $\pi \approx 3.14$

     $C \approx 2 \cdot 3.14 \cdot 7.5$

     $2 \; \boxed{\times} \; 3.14 \; \boxed{\times} \; 7.5 \; \boxed{=}$

The circumference is about 47.1 meters.

Checking for Understanding

Communicating Algebra

1. In your own words, explain how to find the circumference of a circle.
2. Name two rational numbers that are approximations for π.
3. If the radius of a circle is 8 inches, find the diameter.
4. If you know the diameter of a circle, how can you estimate the circumference?

Guided Practice

State how to find the circumference of each circle described below.

5.
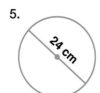
24 cm

6.
11 m

7.
$2\frac{1}{2}$ ft

8.

9 in.

9. $d = 18$ 10. $d = 15$ 11. $r = 1.3$ 12. $r = 21$

Exercises

Independent Practice

Find the circumference of each circle described below.

13.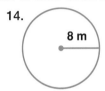
11 mm

14.
8 m

15.
63 ft

16.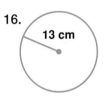
13 cm

17. The diameter is 16.4 km.

18. The radius is 0.5 ft.

19. The radius is 14 yd.

20. The diameter is 3.5 in.

Mixed Review

21. Express -7000 in scientific notation. (Lesson 6-11)

22. Which is greater, $-\frac{2}{3}$ or $-\frac{4}{5}$? (Lesson 6-12)

23. Which is greater, 45.4 or 4.54×10^{-1}? (Lesson 6-12)

Applications

24. **Recreation** A certain bicycle tire has a diameter of 26 inches. Find the distance the bicycle will travel in 10 rotations of the tire.

25. **Geography** The distance around the earth at the equator is about 25,000 miles. Find the approximate diameter of the earth at the equator.

Critical Thinking

26. **Make Up a Problem** Write a problem that can be solved using the information in Exercise 19.

Wrap-Up

27. Find a circular object and measure its diameter. Calculate the circumference and then measure to check your calculation.

History

π

One of the great challenges faced by mathematicians was trying to find an exact value for π. As early as 240 B.C., Archimedes calculated π to four decimal places using a very clever method. His method was to draw two polygons, one inside and one outside a circle. By increasing the number of sides of the polygon, the circumference of the circle was *"trapped"* between the perimeters of the two polygons. Using a 96-sided polygon, he found that π was between 3.140 and 3.142.

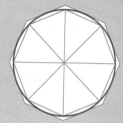

Chinese mathematicians soon surpassed Archimedes' accomplishment. Father and son mathematicians, Tsu Ch'ung-Chih and Tsu Keng-Chih, drew polygons in a circle ten feet across. They calculated π to ten places as 3.1415929203. In Europe, π wasn't calculated to even seven decimal places until the 1600s, twelve hundred years after the Chinese.

Review

Language and Concepts

Choose the correct term to complete each sentence.

1. The fraction $\frac{1}{3}$ written as a decimal is an example of a (*terminating, repeating*) decimal.

2. The decimal 0.125 is an example of a (*terminating, repeating*) decimal.

3. Numbers like 2 and $\frac{1}{2}$, whose product is 1, are called (*reciprocals, integers*).

4. To find $2\frac{3}{4} \div 1\frac{1}{6}$, multiply $\frac{11}{4}$ by $\left(\frac{6}{7}, \frac{7}{6}\right)$.

5. $\frac{3}{4} \times \frac{4}{3} = 1$ is an example of the (*identity, inverse*) property of multiplication.

Skills

Express each fraction as a decimal. Use a bar to show repeating decimals.
(Lesson 6-1)

6. $\frac{13}{25}$ 7. $4\frac{3}{4}$ 8. $\frac{5}{8}$ 9. $\frac{2}{9}$

Estimate each product or quotient. (Lesson 6-2)

10. $\frac{5}{9} \times 6\frac{1}{3}$ 11. $\frac{2}{3} \times 16$ 12. 9.6×2.02 13. $9\frac{5}{7} \times 12\frac{8}{9}$

Solve each equation. Write each solution in simplest form.
(Lesson 6-3, 6-4, 6-6, 6-7)

14. $\frac{37}{40} \cdot \frac{8}{12} = t$ 15. $\left(-\frac{7}{8}\right)\left(2\frac{4}{7}\right) = x$ 16. $(2.01)(0.04) = d$

17. $\frac{3}{5} \div \frac{13}{65} = y$ 18. $a = 2\frac{1}{9} \div 4$ 19. $5\frac{10}{12} \div \left(-1\frac{2}{3}\right) = c$

20. $2.13 \div (-0.3) = a$ 21. $-1.4 \div (-7) = b$ 22. $4.498 \div (-1.73) = y$

Compute using the distributive property. (Lesson 6-5)

23. $50 \times 3\frac{1}{2}$ 24. $8 \times 4\frac{3}{4}$ 25. $\frac{1}{2} \times 4\frac{1}{2}$ 26. $2\frac{1}{4} \times 12$

Solve each equality or inequality. (Lesson 6-8)

27. $-3x = 2.4$ 28. $\frac{a}{6} = 0.12$ 29. $\frac{1}{4}d = 1.7$ 30. $7a > \frac{14}{15}$

31. $-4.95a \geq 89.1$ 32. $\frac{r}{23} < 5.7$ 33. $\frac{2x}{5} = -10$ 34. $2z > \frac{9}{2}$

Determine whether each sequence is a geometric sequence. Then write the next three terms. (Lesson 6-10)

35. 6.6, 5.7, 4.8, . . .   36. $\frac{1}{2}, \frac{1}{3}, \frac{1}{4}, . . .$

37. 768, 192, 48, . . .   38. AN, OB, CP, QD, . . .

Express each number in scientific notation.   (Lesson 6-11)

39. 65,000   **40.** 198,000,000   **41.** 0.0021   **42.** 0.00000743

Replace each ● with <, >, or = to make a true sentence.   (Lesson 6-12)

43. $\frac{6}{2}$ ● $\frac{12}{4}$ **44.** 2.67 ● 2.76   **45.** $\frac{3}{4}$ ● $\frac{2}{3}$ **46.** 1.33 ● 1.25

Find the circumference of each circle.   (Lesson 6-13)

47.

6 cm

$\pi \approx 3.14$

48.

1.2 m

$\pi \approx 3.14$

49.

28 ft

$\pi \approx \frac{22}{7}$

Applications and Logic

50. A wagon wheel has spokes that are each 0.5 meters long. What is the circumference of the wheel?   (Lesson 6-13)

Determine whether each exercise is an example of inductive or deductive reasoning.   (Lesson 6-9)

51. Numbers whose last 3 digits are divisible by 8 are divisible by 8. So, 31,816 is divisible by 8.

52. Every student who came into the planetarium was wearing sunglasses. Professor Urban decided it was sunny outside.

ortfolio Suggestion

Place your favorite word problem from this chapter in your portfolio and attach a note explaining why it is your favorite.

Curriculum Connection

● **Chemistry**   Find the meaning of *Avogadro's number*. How does it relate to the concepts in this chapter?

● **Current Events**   Find examples of fractions in a newspaper or magazine. Find others that are decimals. Explain why each number was used.

Read More About It

Luce, Marnie.   *One is Unique.*

McHale, Thomas J.   *Introductory Algebra: Programmed.*

Temple, Robert.   *The Genius of China.*

CHAPTER 6

Test

Express each fraction as a decimal. Use a bar to show repeating decimals.

1. $\frac{15}{12}$
2. $\frac{17}{40}$
3. $\frac{4}{9}$
4. $\frac{17}{3}$

Estimate each product or quotient.

5. $\frac{9}{10} \times 21$
6. $16.542 \div 78$
7. 28.6×7.6
8. $27\frac{5}{6} \div 3\frac{4}{5}$

Solve each equation. Write each solution in simplest form.

9. $\frac{16}{3} \times \frac{18}{2} = n$
10. $\frac{7}{12} \div \frac{2}{3} = y$
11. $4 \cdot 7.07 = n$
12. $b = 62.9 \div 1000$

13. $f = 1.3 \div 1000$
14. $b = 13.98 \div 6$
15. $p = 4\frac{1}{8} \cdot 5\frac{1}{3}$
16. $\frac{14}{20} \div 2\frac{2}{5} = d$

Solve each inequality.

17. $2n \le 7$
18. $\frac{p}{12} \ge -6$
19. $-7s > 6.3$
20. $7 < \frac{x}{-8}$

Determine whether each sequence is a geometric sequence. Then write the next three terms.

21. $17, 13, 18, 14, 19, \ldots$
22. $3, 9, 27, \ldots$

Express each number in scientific notation.

23. $13{,}490{,}000$
24. 0.00000674
25. 0.00032
26. 5810

Replace each ● with <, >, or = to make a true statement.

27. $\frac{1}{4}$ ● $\frac{3}{8}$
28. 4.6 ● 4.75
29. 9.5 ● 9.50
30. $\frac{12}{7}$ ● $\frac{5}{3}$

Find the circumference of each circle.

31.

$\pi \approx \frac{22}{7}$

32.

$\pi \approx 3.14$

Determine whether the exercise is an example of inductive or deductive reasoning.

33. If you are a freshman, then you cannot attend the prom. Joni is a freshman. She is not allowed to attend the prom.

BONUS

How do you know that $10 \div \frac{1}{2} > 10$?

Academic Skills Test

Cumulative, Chapters 1-6

1. What is the value of $y + (-7)$ if $y = 8$?

   A -56
   B -15
   C -1
   D 1

2. If $14 = w - (-20)$, what is the value of w?

   A -6
   B 6
   C 34
   D Not Here

3. If $\frac{h}{0.7} = -2.8$, what is the value of h?

   A -4
   B -2.8
   C -1.96
   D 0.7

4. What is the value of $3t^5$ if $t = 2$?

   A 7776
   B 96
   C 32
   D 30

5. How is the product $2 \cdot 2 \cdot 2 \cdot 2$ expressed in exponential notation?

   A 2^2
   B $2 \cdot 4$
   C 4^2
   D 2^4

6. Which number should come next in this pattern?

   $$0.5, 2, 3.5, 5, \ldots$$

   A 8   C 7
   B 7.5   D 6.5

7. Jan plans to put a fence around a square garden that measures 12 feet by 12 feet. How many feet of fence will she need?

   A 144 ft   C 48 ft
   B 96 ft   D 24 ft

8. Carl bought a bike that cost $532.16. He paid for the bike in equal payments for 12 months. Estimate: the amount of each payment was between—

   A $10 and $20
   B $25 and $35
   C $40 and $50
   D $55 and $65

9. Lee is practicing for a 2500 meter run. He runs the distance twice a day for 5 days. How many kilometers has he run?

   A 25 km
   B 250 km
   C 25,000 km
   D Not Here

10. One red blood cell is about 7.5×10^{-4} centimeters long. What is another way to express this measure?

   A 0.000075 cm
   B 0.00075 cm
   C 0.0075 cm
   D 0.075 cm

CHAPTER OBJECTIVES

In this chapter you will learn to:

- write and solve two-step equations and problems
- solve multi-step equations and inequalities
- convert within the metric system
- solve problems by using the work backwards strategy

Solving Equations and Inequalities

Math at the mall!—It's more than paying for your purchases and receiving correct change. Every time you make a tough buying decision, you are using mathematics.

Imagine that you need to buy a new pair of shoes, and you have $90 to spend. However, you'd also like to buy two CDs. So you look at the shoes. The ones you really, really like cost $70. Do you have enough to buy two CDs and the shoes you like?

When you make your decision, you are mentally solving this inequality.

Let x = cost of CDs
$2x + 70 \leq 90$

What's *your* buying decision?

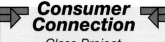
Consumer Connection

Class Project
Collect advertisements from newspapers or magazines. Make up a situation similar to the one described above. Make a collage showing the situation. Include the inequality that describes it.

7-1 Strategy: Work Backwards

Objective:
Solve problems by
working backwards.

Willis brought some of his baseball cards to Tom's house for the purpose of trading cards. Willis traded Tom half of the cards he brought in exchange for two Jose Canseco cards. Then Willis gave Tom's little sister Mary 8 cards as a gift. If Willis left Tom's house with 48 baseball cards, how many cards did he bring to Tom's house?

In most problems, a set of conditions or facts is given and an end result must be found. However, this problem tells the result and asks for something that happened earlier. The strategy of working backwards can be used to solve this problem and similar problems. To use this strategy, start with the end result and *undo* each step.

Explore Willis had some baseball cards. He traded away half the cards. Then he received 2 cards and gave away 8 cards. In the end, he had 48 cards. The problem asks how many cards he had at the beginning.

Plan Since this problem gives the end result and asks for something that happened earlier, start with the result and work backwards. *Undo* each step.

Why is 8 added to 48, but 2 is subtracted from the result?

Solve In the end Willis had 48 cards.

	48
Undo the 8 cards that he gave to Mary. ⟶	+ 8
	56
Undo the 2 cards that Tom gave to him. ⟶	− 2
	54
Undo the half that he gave to Tom. ⟶	× 2
	108

In the beginning Willis had 108 baseball cards.

Examine Assume that Willis started with 108 cards. After he gave Tom half his cards, he had 54 cards. Tom gave him 2 cards, so he had 54 + 2 or 56 cards. He gave Mary 8 cards, so he had 56 − 8 or 48 cards. In the end Willis had 48 cards, so the answer is correct.

Checking for Understanding _____

**Guided
Practice**

Solve. Work backwards.

1. A certain number is multiplied by 3 and then 5 is added to the result. The final answer is 41. Find the number.

2. A certain number is divided by 5 and then 1 is subtracted from the result. The final answer is 32. Find the number.

3. A certain number is added to 13 and the result is multiplied by 5. The final answer is 125. Find the number.

4. Ursula is conditioning for soccer. She does 7 minutes of stretching exercises followed by 2 minutes of push-ups and 3 minutes of jumping jacks. Finally, she runs for 18 minutes. If Ursula finishes her run at 2:07 P.M., when did she start the stretching exercises?

Exercises

Independent Practice

Solve. Use any strategy. Use mental math, estimation, or a calculator, as appropriate.

5. Zack was assigned some math exercises for homework. He did half the exercises in study hall. After school he did 5 more exercises. If he still has 13 exercises to do, how many exercises were assigned?

6. On Monday Buffy told a joke to 3 of her friends. On Tuesday each of those friends told 3 other friends the joke. On Wednesday each person who heard the joke on Tuesday told 3 other people. If this pattern continues, how many people will hear the joke on Saturday?

7. Samuel had some grapes. He gave half of his grapes to Chin. Chin then gave a third of his grapes to Rita. Rita gave a fourth of her grapes to Sue. If Sue has 9 grapes, how many grapes did Samuel have in the beginning?

8. Millie had some money. She baby-sat her nieces and earned $15. She took all her money on a shopping trip and spent half of it on some new jeans. She spent another $5 on some earrings. If she has $32 left, how much did Millie have in the beginning?

9. A certain bacteria doubles its population every 12 hours. After 3 full days, there are 1600 bacteria. How many bacteria were there at the beginning of the first day?

10. The members of the camera club need to raise $300 to buy some new equipment for a darkroom. They plan to sell 1000 school badges to encourage school spirit. If each badge costs the club 95¢, how much should the members charge for each badge in order to meet their goal?

Wrap-Up

11. **Make Up a Problem** Write a problem that can be solved by working backwards. Trade problems with a friend and solve your friend's problem.

Exploration

Two-Step Equations

Materials: cups, counters, mats

In this Exploration, you will use cups and counters as models for building and solving two-step equations.

▶ Consider the equation $2x + 2 = 6$. Let's use the model of the cups and counters to find all values of x for which $2x + 2 = 6$ is true. First, build the equation $2x + 2 = 6$.

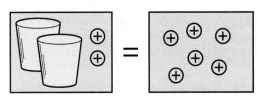

▶ How many counters are in each cup? To find this answer, first pair off and remove counters.

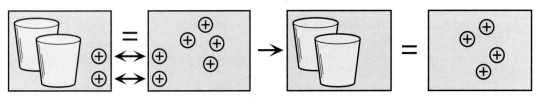

▶ Now your equation is $2x = 4$. Since there are two cups, match an equal number of counters with each cup. Therefore, each cup must contain two counters and $x = 2$.

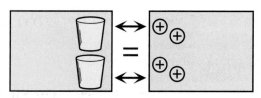

Your Turn: Model $2x + (-2) = 6$ and $2x + 2 = -4$. Add zero pairs as needed.

Model each equation and solve.

1. $2x + 3 = 13$ **2.** $2x + (-2) = -4$

3. $3x + 2 = -4$ **4.** $3x + -2 = 4$

Analysis

5. Explain how this model uses the work backwards strategy presented in Lesson 7-1.

6. Explain how you can use models to solve $-2x + 2 = 4$.

7-2   Solving Two-Step Equations

Objective:
Solve equations that involve more than one operation.

Judi mailed some photographs to her cousin Misae. She paid $1.44 to send them first-class. If the Post Office charges 29 cents for the first ounce and 23 cents for each additional ounce, how much did Judi's package weigh?

This problem can be solved by using the problem solving strategy from Lesson 7-1 and the equation $29 + 23w = 144$. This equation contains more than one operation and is solved by undoing these operations.

$$29 + 23w = 144$$
$$29 - 29 + 23w = 144 - 29 \qquad \text{First undo the addition.}$$
$$\text{Subtract 29 from each side.}$$
$$23w = 115$$
$$\frac{23w}{23} = \frac{115}{23} \qquad \text{Next undo the multiplication.}$$
$$\text{Divide each side by 23.}$$
$$w = 5$$

Check: $\quad 29 + 23w = 144$
$$29 + 23 \cdot 5 \overset{?}{=} 144 \qquad \text{Replace } w \text{ with 5.}$$
$$29 + 115 \overset{?}{=} 144 \qquad \text{Do the multiplication first.}$$
$$144 = 144 \; \checkmark$$

There were 5 additional ounces, so Judi's package weighed 6 ounces.

Why do you compute the multiplication first?

Example

1 **Solve $-2k - 3 = 16$. Check your solution.**
$$-2k - 3 + 3 = 16 + 3 \qquad \text{Add 3 to each side.}$$
$$-2k = 19$$
$$\frac{-2k}{-2} = \frac{19}{-2} \qquad \text{Divide each side by -2.}$$
$$k = -\frac{19}{2} \text{ or } -9\frac{1}{2}$$

Check: $-2k - 3 = 16$
$$-2\left(-\frac{19}{2}\right) - 3 \overset{?}{=} 16 \qquad \text{Replace } k \text{ with } -\frac{19}{2}.$$
$$19 - 3 \overset{?}{=} 16 \qquad \text{Do the multiplication first.}$$
$$16 = 16 \; \checkmark \quad \text{The solution is } -\frac{19}{2} \text{ or } -9\frac{1}{2}.$$

FYI

Benjamin Franklin was appointed the first American postmaster general in 1775. Nearly 80 years later, Congress issued the first postage stamps.

Examples

Solve each equation.

2 $\dfrac{t}{-2} + 4 = -10$ *Note that $\dfrac{t}{-2} = -\dfrac{1}{2}\,t$.*

$\dfrac{t}{-2} + 4 - 4 = -10 - 4$ Subtract 4 from each side.

$\dfrac{t}{-2} = -14$

$-2 \cdot \dfrac{t}{-2} = -2 \cdot (-14)$ Multiply each side by -2.

$t = 28$

Check: $\dfrac{t}{-2} + 4 = -10$

$\dfrac{28}{-2} + 4 \overset{?}{=} -10$ Replace t with 28.

$-14 + 4 \overset{?}{=} -10$ Do the division first.

$-10 = -10$ ✓   The solution is 28.

3 $\dfrac{b + 3}{10} = -4$

$10 \cdot \dfrac{(b + 3)}{10} = 10 \cdot (-4)$ Because $\dfrac{b+3}{10}$ means $(b + 3) \div 10$,

$b + 3 = -40$ undo the division first.

$b + 3 - 3 = -40 - 3$ Subtract 3 from each side.

$b = -43$ Check the solution.

Many times, real-life situations can be described by two-step equations. Suppose your family wants to buy 15 square yards of carpeting for your family room. In addition to the cost of the carpet, there is a $40 installation charge. The total bill is $234.25. You want to find the price of one square yard of carpet.

Let p = price of one square yard of carpet.
The equation $15p + 40 = 234.25$ can be used to solve this problem.
You will learn more about writing equations in the next lesson.

Estimation Hint

THINK:
$15p + 40 \approx 240$
$15p \approx 200$

Since $15(10) = 150$ and $15(15) = 225$, the solution should be between 10 and 15.

Example

4   **Solve $15p + 40 = 234.25$.**

$15p + 40 = 234.25$ Subtract 40. Then divide by 15.

$p = 12.95$ Compare with the estimate. The answer is reasonable.

The price of the carpet is **$12.95** per square yard.

Checking for Understanding

Communicating
Algebra **Describe the first step you should take to solve each equation.**

   1. $5r - 6 = 27$ **2.** $2 + 5m = -19$ **3.** $\frac{z - 5}{-3} = 9$

   4. Explain how you use the work backwards strategy to solve the equation $2x - 3 = 23$.

   Complete the directions for solving each equation. Then solve each equation.

   5. $2r - 7 = 1$ First add ■ to each side. Then divide each side by ■.

   6. $\frac{y}{5} + 8 = 7$ First ■ 8 from each side. Then multiply each side by ■.

   7. $4 - 2b = -8$ First subtract ■ from each side. Then ■ each side by -2.

Guided
Practice **Name the first step you would take to solve each equation. Then solve each equation.**

   8. $3m - 4 = 11$ **9.** $4z - 7 = -15$ **10.** $3 + 5m = -22$

   11. $\frac{a}{7} + 4 = 18$ **12.** $-\frac{x}{3} - 5 = -23$ **13.** $\frac{b}{4} - 19 = 17$

   14. $\frac{k + 5}{4} = 9$ **15.** $\frac{n - 6}{-5} = 3$ **16.** $\frac{2m + 7}{5} = 9$

Exercises

Independent
Practice **Solve each equation. Check your solution.**

   17. $-4y + 3 = 19$ **18.** $-8k - 21 = 75$ **19.** $9 - 4z = 57$

   20. $12 - z = 28$ **21.** $-8 - t = -25$ **22.** $-36 = 24 - w$

   23. $54 = 39 - x$ **24.** $-4 - y = 24$ **25.** $18 = 6 - y$

   26. $\frac{b}{8} - 17 = 13$ **27.** $\frac{g}{12} - 4 = 7$ **28.** $8 = \frac{h}{-3} + 19$

   29. $-12 + \frac{j}{4} = 9$ **30.** $13 + \frac{p}{-3} = -4$ **31.** $-3 = -31 + \frac{c}{6}$

   32. $-3b + 5 = 20$ **33.** $4m + 3 = -73$ **34.** $47 = 12 - 7m$

   35. $\frac{a}{7} + 21 = -12$ **36.** $\frac{d}{-6} - 17 = -8$ **37.** $\frac{x}{12} + 7 = 28$

   38. $\frac{a + 3}{5} = -11$ **39.** $\frac{6 + c}{-13} = -3$ **40.** $\frac{d - 5}{7} = 14$

Mental Math **Solve each equation mentally.**

   41. $2x - 8 = 12$ **42.** $-4a + 4 = 0$ **43.** $\frac{n}{2} + (-3) = 5$

Language
Skill **Translate each sentence into an equation.**

   44. Three times a number less sixteen equals twenty-five.

   45. Twenty more than twice c is negative thirty.

   46. The sum of fourteen and a number, divided by eight, equals five.

   47. The quotient of a number and four, decreased by five, is seven.

48. Name the operation that should be done first in the expression $15 \div 3 + 2 \cdot 9$.   (Lesson 1-1)

49. *True* or *false:* $-3.54 < -3.51$.   (Lesson 5-1)

50. Round 12.556 to the nearest tenth.   (Lesson 5-3)

51. Find the perimeter of a rectangle with a length of 4 ft 3 in. and a width of 2 ft.   (Lesson 5-10)

52. Write $\frac{8}{11}$ as a decimal.   (Lesson 6-1)

53. Find $\left(-\frac{2}{3}\right)^2$.   (Lesson 6-3)

54. Name the multiplicative inverse of $1\frac{5}{7}$.   (Lesson 6-5)

Application

55. Nutrition   Big Jake claims that his new potato chips have about 8 calories a piece. If you eat c chips, you will have consumed $8c$ calories. A Big Jake Sub has about 700 calories. Together, a Big Jake Sub and chips have about $700 + 8c$ calories. How many chips can you eat with your Big Jake Sub if you want to consume 1100 calories?

Connection

56. Geometry   The area of a trapezoid can be found by multiplying the height and one-half the sum of the lengths of the bases. The formula is: $A = h \cdot \frac{1}{2}(b_1 + b_2)$. The area of the trapezoid at the right is 64 square inches. The trapezoid is 8 inches high and the length of one base is 7 inches. What is the length of the other base?

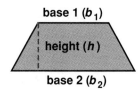

base 1 (b_1)

height (h)

base 2 (b_2)

Critical Thinking

57. Make Up a Problem   Write a two-step equation using the numbers 2, 3, and 6, in which the solution is $-\frac{1}{6}$.

Wrap-Up

Journal Entry

58. Briefly explain the general procedure for solving any two-step equation.

Challenge

Find the Equation

1. Find the missing values in the table below.

First Number	-2	-1	0	1	2	3	4	10	-10
Second Number	-7	-5	-3	-1	1	?	?	?	?

2. How are the second numbers related to the first numbers?

3. If y is the second number and x is the first number, write an equation that represents the relationship between x and y.

7-3　Writing Two-Step Equations

Objective:
Solve verbal problems by writing and solving equations.

Is 10 months a reasonable estimate? Why or why not?

Jason wants to buy a portable CD player that costs $127. He has saved $57 from money he made by recycling. If he can save $14 per month, how long will it take him to save enough money to buy the CD player?

This problem can be solved by writing and solving a two-step equation.

Explore You know that Jason has $57 saved and that he can save $14 a month. You need to find out how many months it will take him to save enough money to buy the CD player. Let m = the number of months.

Plan In one month, he will have $57 + 1(14)$ dollars.
In two months, he will have $57 + 2(14)$ dollars.
In three months, he will have $57 + 3(14)$ dollars.
⋮
In m months, he will have $57 + m(14)$ dollars.

Solve $57 plus $14 a month for m months is $127.

$$57 \quad + \quad 14m \quad = 127$$

$$57 + 14m = 127$$

$$57 - 57 + 14m = 127 - 57 \quad \text{Subtract 57 from each side.}$$

$$14m = 70$$

$$\frac{14m}{14} = \frac{70}{14} \quad \text{Divide each side by 14.}$$

$$m = 5$$

It will take Jason 5 months to save enough money to buy the CD player.

Examine Is 5 times $14 plus $57 equal to $127?　Yes

FYI

Aluminum recycling saves 95% of the energy needed to make aluminum from raw ore. In 1989 alone, the amount of energy saved was equal to 20 million barrels of oil.

Science Connection

The Celsius (C) temperature scale is part of the metric system. In the United States, temperature is often measured using the Fahrenheit (F) scale.

There is one temperature that reads the same in both systems. Substitute C for F in the equation to find it.

Example

1 **In a science lab, Seve needs to express 86° Fahrenheit in degrees Celsius. He remembers the formula for finding equivalent Fahrenheit temperatures when a Celsius temperature is known, $F = \frac{9}{5}C + 32$. What is the Celsius temperature?**

Explore You know the temperature in Fahrenheit and the formula for finding Fahrenheit. You need to find the Celsius temperature.

Plan Use the formula $F = \frac{9}{5}C + 32$ to write an equation.

$86 = \frac{9}{5}C + 32$ Replace F with 86.

Solve $86 = \frac{9}{5}C + 32$

$86 - 32 = \frac{9}{5}C + 32 - 32$ Subtract 32 from each side.

$54 = \frac{9}{5}C$

$\frac{5}{\cancel{9}} \cdot \cancel{54}^{6} = \frac{\cancel{5}}{\cancel{9}}^{1} \cdot \frac{\cancel{9}}{\cancel{5}}^{1}C$ Multiply each side by $\frac{5}{9}$.

$30 = C$

The equivalent Celsius temperature is 30°.

Examine Check to see if $86 = \left(\frac{9}{5} \cdot 30 \right) + 32$.

Checking for Understanding

Communicating Algebra

1. A TV repairman charges $20 to come to the house. He then charges $30 an hour for the time spent on the job.
   a. Choose a variable to represent the number of hours the repairman takes to repair a TV.
   b. Write an expression that represents the total cost to repair a TV.
   c. Write an equation to determine the number of hours worked if the repair bill is $65.
   d. What would be the first step in solving the equation you wrote in Exercise **1c**?

Guided Practice

Define a variable and write an equation for each situation. Then solve.

2. If twice a number is subtracted from 8, the difference is 12. What is the number?

3. The population of Lafayette is 4800. Each year the population decreases by 35. In how many years will the population be 4520?

Exercises

ESTIMATION
MENTAL MATH
CALCULATOR
PAPER/PENCIL

Define a variable. Then write and solve an equation.

4. One season Pablo scored 7 more runs than twice the number he batted in. He scored 133 runs last season. How many runs did he bat in?

5. A dump truck in a copper mine near Salt Lake City is capable of carrying 170 tons of ore. It has 50 tons already loaded. The power shovel loading the truck can lift 15 tons per scoop. How many scoops will it take to fill the truck?

6. Augie's bowling handicap is 7 less than half her average. Her handicap is 53. What is Augie's bowling average?

7. When the meter in a taxi is first turned on it reads $1.10. As the taxi travels, $1.40 is added for each mile driven. How far would you ride for $12.30?

8. Jeff bought a used 10-speed bike for $20 more than one-half its original price. If he paid $90 for the bike, what was the original price?

9. Angela scored 15 points less than twice the lowest score on a math test. If her score was 95, what was the lowest score on the test?

Critical Thinking

10. The telephone company charges $35 for the first 15 minutes on a repair job and $10 for each additional 15 minutes. If the repair bill totaled $85, how long did it take to make the repairs?

Mixed Review

11. Solve the equation $y = 25.5 \div 3.4$.   (Lesson 6-7)

12. Translate into an equation: *Three times the sum of a number and five equals twenty-seven.*   (Lesson 7-2)

13. Find the product of the first 99 terms of the sequence $\frac{1}{2}, \frac{2}{3}, \frac{3}{4}, \frac{4}{5}, \ldots$.

Wrap-Up

14. **Make Up a Problem**   Write a problem that could be solved using the equation $3x - 5 = 15$.

Team Problem Solving

Jack and Jill went up the hill to fetch a liter of water. Jack has a 2-liter pail and Jill has a 5-liter pail.

1. How can they measure out exactly 1 liter of water?

2. Which of the amounts below can Jack and Jill measure?

1 liter	2 liters	3 liters	4 liters
5 liters	6 liters	7 liters	8 liters

Exploration

More Equations

Materials: cups, counters, mats

In this Exploration, you will use cups and counters as models for building and solving equations in which variables appear on each side.

▶ Consider the equation $2x + 3 = x + 1$. To solve this equation, pair off and remove items from each side. This time you will remove both cups and counters. Let's remove the counters first, then the cups.

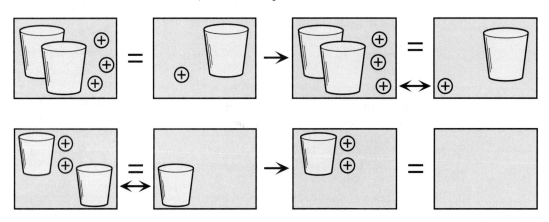

▶ Now your equation is $x + 2 = 0$. To find out how many counters are in the cup, pair off the positive counters on the left with counters on the right. Since there are no counters on the right, add zero pairs. Then pair off and remove the counters.

Therefore, $x = -2$.

Your Turn: Model $2x + 3 = x + 7$.

Model each equation and solve.

1. $3x + 2 = x + 6$ 2. $3x + 7 = x + 1$ 3. $3x + (-2) = x + 6$

4. $3x + (-7) = x + 1$ 5. $3x + 2 = -x + 6$ 6. $3x + 7 = -x + (-1)$

Analysis

7. Would it matter if you removed the cups first, then the counters?

8. Explain how you can use models to solve $-3x + 2 = -x + 6$.

7-4 Variables on Each Side

Objective:
Solve equations with variables on each side.

Key Terms:
null set
empty set

Every day in the United States, about 215,000,000 canned drinks are consumed. Think back to yesterday or last week. Did you throw away an aluminum can?

By itself, one can may not be a problem, but think what would happen if all 215,000,000 cans were thrown away. There may soon be a shortage of many natural resources because these materials are now being buried as solid waste. One way to make the resources go further is to recycle them. Recycling not only helps the environment; it is also a source of spending money.

Consider this situation:

- Cash-For-Trash Recycling Center pays 42¢ per pound for 10 or more pounds of aluminum cans.

- Cash-Can Recycling Center pays a flat fee of 75¢ if the cans are washed, plus 37¢ per pound for 10 or more pounds of cans.

You may reason that for some weights you earn more at Cash-For-Trash, and for other weights you earn more at Cash-Can. Is there any weight that would earn the same amount at either place?

You can solve this problem by solving the following equation.

Let p = number of pounds of aluminum cans.

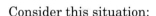

Cash-For-Trash   Cash-Can

$$42p \qquad = \qquad 37p + 75$$

This equation is different from the equations in previous lessons because there is a variable on each side of the equation. To solve this type of equation, use the properties of equality to eliminate the variable from one side. The equation is solved on the next page.

FYI

On average, each American consumes 320 canned drinks per year. A recycled can appears on the shelf as a new can in about 4 months.

$$42p = 75 + 37p$$

$$42p - 37p = 75 + 37p - 37p$$ To eliminate 37p from the right side, subtract 37p from each side.

$$5p = 75$$ 42p − 37p = 5p

$$\frac{5p}{5} = \frac{75}{5}$$ Divide each side by 5.

$$p = 15$$

Check: $42p = 75 + 37p$

$$42(15) \overset{?}{=} 75 + 37(15)$$ Replace p with 15.

$$630 \overset{?}{=} 75 + 555$$

$$630 = 630 \quad \checkmark$$

If you recycle 15 pounds of aluminum cans, you would earn the same amount, $6.30, at either Cash-For-Trash or Cash-Can.

Examples

1   **Solve 8x − 3 = 1 + 6x.**

$$8x - 3 = 1 + 6x$$

$$8x - 6x - 3 = 1 + 6x - 6x$$ Subtract 6x from each side.

$$2x - 3 = 1$$

$$2x - 3 + 3 = 1 + 3$$ Add 3 to each side.

$$2x = 4$$

$$\frac{2x}{2} = \frac{4}{2}$$ Divide each side by 2.

$$x = 2$$ Check this solution.

2   **Solve 3h + 9 = 4.5h + 5.**

$$3h + 9 = 4.5h + 5$$

$$3h - 3h + 9 = 4.5h - 3h + 5$$ Subtract 3h from each side.

$$9 = 1.5h + 5$$

$$9 - 5 = 1.5h + 5 - 5$$ Subtract 5 from each side.

$$4 = 1.5h$$

$$\frac{4}{1.5} = \frac{1.5h}{1.5}$$ Divide each side by 1.5.

$$4 \; \boxdot \; 1.5 \; \boxminus \; 2.6666667$$

$$2.6666667 = h$$

Calculator Hint

Use the Store and Recall keys to check solutions that involve decimals like 2.6666667.

Sometimes an equation has *no* solution. The solution set is the **null** or **empty set.** It is shown by the symbol { } or $\varnothing$.

Example

3 Solve $5t + 7 = 5t + 3$.

$$5t + 7 = 5t + 3$$

$$5t - 5t + 7 = 5t - 5t + 3 \qquad \text{Subtract } 5t \text{ from each side.}$$

$$7 = 3$$

This sentence is *never* true. The solution set is $\varnothing$.

Other times, the solution of an equation may be all numbers.

Example

4 Solve $2(x + 4) - 8 = 2x$.

$$2(x + 4) - 8 = 2x$$

$$2x + 8 - 8 = 2x \qquad \text{Distributive property}$$

$$2x = 2x \qquad 8 - 8 = 0$$

$$\frac{2x}{2} = \frac{2x}{2} \qquad \text{Divide each side by 2.}$$

$$x = x$$

This sentence is *always* true. The solution set is all numbers.

Checking for Understanding

Communicating Algebra

1. To solve $8a = 3a + 20$, __?__ $3a$ from each side.

2. What is the first step you would take to solve $5y - 2 = -2y + 8$?

3. Name the property of equality that allows you to subtract a number from each side of an equation.

Guided Practice

Simplify each expression.

4. $7 + 9y + y$

5. $6d - d + 15$

6. $6m - 4m + 3$

Name the first two steps you should take to solve each equation. Then solve.

7. $3k + 10 = 2k - 21$

8. $4m - 9 = 5m + 7$

9. $8y - 6 = 5y + 12$

10. $x - 4 = 6x - 19$

11. $-3p + 8 = 2p - 2$

12. $-t + 10 = t + 4$

Exercises

Solve each equation. Check your solution.

13. $5x - 17 = 4x + 36$ 14. $10g = 11g - 48$
15. $2f - 12 = 3f + 9$ 16. $y + 3 = 7y - 21$
17. $6m - 2 = m + 13$ 18. $20c + 5 = 5c + 65$
19. $18 + 4p = 6p + 11$ 20. $-3k + 10 = k + 3$
21. $13 - t = -t + 7$ 22. $3n + 7 = 7n - 13$
23. $-9r = 20 - r$ 24. $4a - 9 = 6a + 7$
25. $19b + 20 = 20 + 19b$ 26. $11h - 14 = 7 + 14h$
27. $-43 - 3z = 2 - 6z$ 28. $0.5x - 7 = 0.3x + 1$
29. $8z - 13 = 6z + 7$ 30. $7y + 7 = 3y - 5$
31. $31 - 4b = 10b + 10$ 32. $2a - 1 = 3.5a - 3$
33. $5x + 4 = 7x + 8$ 34. $6x + 17 = 4x - 31$
35. $12x - 24 = -14x + 28$ 36. $8x - 4 = -10x + 50$

Mixed Review

37. *True* or *false:*   The sum of a positive and a negative integer is always positive.   (Lesson 2-3)

38. Solve the equation $n = \frac{14}{11} - \frac{9}{11}$.   (Lesson 5-5)

39. Which type of logical reasoning makes a rule after seeing several examples?   (Lesson 6-9)

40. Solve the equation $-5t - 30 = 25$.   (Lesson 7-2)

Connection

41. **Age Problem**   In three years, Joe's age will be 9 less than twice his present age. Solve the equation $x + 3 = 2x - 9$ to find Joe's age now.

Critical Thinking

42. The greater of two numbers is 7 more than the lesser. Three times the greater number is 5 more than 4 times the lesser number. Find the numbers.

Wrap-Up

43. **Make Up a Problem**   Write an equation with variables on each side whose solution is 5.

Mid-Chapter Quiz

Solve each equation.   (Lessons 7-2 and 7-4)

1. $2y - 3 = 17$ 2. $-k + 4 = -9$ 3. $4x + 3 = 11$
4. $\frac{b}{4} + 7 = 2$ 5. $m + 3 = 7m - 21$ 6. $5x - 6 = 3x + 4$

Solve.   (Lesson 7-1)

7. Seth decides to give away his baseball cap collection. He gives one-fourth of his caps to Jovita. Then he gives one-half of the remaining caps to Karl. If he has 9 caps left, how many did he have in the beginning?

Define a variable and write an equation. Then solve.   (Lesson 7-3)

8. Three more than twice a number is equal to negative nine. Find the number.

7-5   Solving Multi-Step Equations

Objective:

Solve equations that involve more than one operation.

A team's score in cross-country is the sum of the place numbers of the first five finishers. If five runners for a team finished in 2nd, 5th, 6th, 9th, and 12th place, the team's score is

$$2 + 5 + 6 + 9 + 12 = 34$$

How can you estimate the solution?

Kristina and Ginnie are members of the Oswego cross-country team. In one meet Kristina finished three places ahead of Ginnie. Their combined score was 27. In what place did each girl finish?

You can solve this problem by using the techniques from the previous lessons.

Let k = Kristina's place.   *Kristina finished three*
$k + 3$ = Ginnie's place.   *places ahead of Ginnie.*

Solve the equation $k + (k + 3) = 27$.   *Their combined score was 27.*

$$k + (k + 3) = 27$$
$$2k + 3 = 27 \qquad \text{Combine like terms.}$$
$$2k = 24 \qquad \text{Subtract 3 from each side.}$$
$$k = 12 \qquad \text{Divide each side by 2.}$$

Kristina finished in 12th place. Ginnie finished in 15th place.

Example

1   **Solve $2k + 3(k + 1) = 6k - 3$.**

$$2k + 3k + 3 = 6k - 3 \qquad \text{Distributive property}$$
$$5k + 3 = 6k - 3$$
$$5k - 5k + 3 = 6k - 5k - 3 \qquad \text{Subtract } 5k \text{ from each side.}$$
$$3 = k - 3$$
$$3 + 3 = k - 3 + 3 \qquad \text{Add 3 to each side.}$$
$$6 = k$$

Check: $2k + 3(k + 1) = 6k - 3$
$$2 \cdot 6 + 3(6 + 1) = 6 \cdot 6 - 3 \qquad \text{Replace } k \text{ with 6.}$$
$$12 + 3(7) = 36 - 3$$
$$12 + 21 = 33$$
$$33 = 33 \quad \text{✓}$$

Checking for Understanding

Communicating Algebra
1. Explain your strategy for estimating the equation $x + (x + 1) + (x + 2) = 99$. What is your estimate?

2. Solve the equation in Exercise 1. How does your solution compare with the estimate?

Guided Practice

State the steps you would take to solve each equation. Then solve each equation.

3. $4x + 1 = 3x - 5$

4. $7a - 6 = 19 + 2a$

5. $4(s + 3) - 1 = 3(2s + 1) + 5$

6. $7(x - 2) = 5(3x + 1) + 5$

7. $5(b + 7) - 2 = 4(b + 3) + 5$

8. $2.3n + 1 = 1.3n + 7$

9. $\frac{5}{7}x - 4 = \frac{3}{7}x + 1$

10. $\frac{3}{4}m - 2 = \frac{1}{4}m + 7$

Exercises

Independent Practice

Solve each equation. Check your solution.

11. $4(x + 5) = 2x + 12$

12. $-5(y + 7) = 3y + 29$

13. $a + 3 = 5(2a - 3)$

14. $2m - 4 = 3(4m - 3)$

15. $7 + 3(x + 1) = 2x + 9$

16. $3(z + 5) = 5z + 7$

17. $2(x - 3) + 5 = 3(x - 1)$

18. $5b + 4 = 7(b + 1) - 3b$

19. $3.5x + 6 = 1.5x$

20. $\frac{5}{9}a + 7 = \frac{1}{9}a - 5$

21. $\frac{4}{7}y - 8 = \frac{2}{7}y + 10$

22. $\frac{7}{3}m + 3 = \frac{5}{3}m$

23. $-12 + 7(x + 3) = 4(2x - 1) + 3$

24. $-8 + 5(3a - 2) = 4(2a + 1) + 6$

Mixed Review

25. Use the formula $d = rt$ to find the time it takes a train to travel 434 miles at a rate of 62 mph.   (Lesson 3.5)

26. Write 0.000336 in scientific notation.   (Lesson 6-11)

27. Solve the equation $3r - 18 = -2r + 7$.   (Lesson 7-4)

Connections

Statistics   Use the information in the graph at the right to answer each question.

28. Let x represent the number of visitors to Harper's Ferry National Park. The expression $2x + 100,000$ is an estimate of the number of visitors at which park?

29. Let y represent the number of visitors to Manassas National Battlefield Park. Write an expression involving two operations that represents the number of visitors to Chickamauga and Chattanooga Military Park.

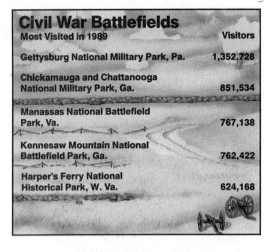

Civil War Battlefields
Most Visited in 1989   **Visitors**

Gettysburg National Military Park, Pa.   1,352,728

Chickamauga and Chattanooga
National Military Park, Ga.   851,534

Manassas National Battlefield
Park, Va.   767,138

Kennesaw Mountain National
Battlefield Park, Ga.   762,422

Harper's Ferry National
Historical Park, W. Va.   624,168

Application

30. **Sports**   The captain of a five-member cross-country team placed second in a meet. The other members placed in consecutive order but farther behind. The team score was 40. In what places did the other members finish?

Critical Thinking

31. Consecutive integers are integers in counting order, such as 2, 3, 4. The sum of three consecutive integers is -57. Find the integers.

Wrap-Up

32. **Make Up a Problem**   Write and solve a multi-step equation.

Algebra in Action-Cooking

Boiling Point

In home economics class you learn that to hard-boil an egg, you place it in boiling water for 5 minutes. But if you try this in a city with a high altitude, such as Denver, after 5 minutes the egg is still runny. Why?

Water boils at 212°F at sea level. However, as the altitude increases, the air pressure drops. The lower air pressure causes water to boil at a lower temperature. Mathematically, for every 550 feet above sea level, the boiling point is lowered 1°F.

Find the boiling point of water at each multiple of 550 feet.

Altitude, in feet	0	550	1100	1650	2200	2750
Boiling Point, in °F	212	211	210	?	?	?

You can use algebra to devise a formula for finding the boiling point.

boiling point $=$ 212°F lowered by 1°F for every multiple of 550 feet

$$\text{boiling point} = 212 - 1(\text{number of multiples of } 550)$$

$$= 212 - 1\left(\frac{\text{altitude}}{550}\right)$$

$$= 212 - \frac{A}{550} \qquad \text{Let } A \text{ represent the altitude.}$$

Therefore, the boiling point at Denver, where the altitude is 5280 feet above sea level, can be found as follows.

$$\text{boiling point} = 212 - \frac{A}{550}$$

$$= 212 - \frac{5280}{550} \qquad \text{Replace } A \text{ with 5280.}$$

$$212 \;\boxed{-}\; 5280 \;\boxed{\div}\; 550 \;\boxed{=}\; 202.4$$

The boiling point at Denver is 202.4°F.

Find the boiling point at each location.

1. Pike's Peak, 14,110 feet above sea level
2. Death Valley, 282 feet below sea level
3. your city or town
4. **Research** Look in a cookbook or on a package of cake mix to find how altitude affects baking.

7-6 Solving Multi-Step Inequalities

Objective:
Solve inequalities that involve more than one operation.

Why should you solve an inequality instead of an equation?

On page 251 you were presented with a situation that could happen to you while shopping at the mall.

Here is a summary of the problem.
• You have $90 to spend.
• You want to buy shoes that cost $70.
• You also want to buy two CDs.
Do you have enough money? You can find the answer by solving the inequality $2x + 70 \leq 90$, where x is the cost of one CD.

$2x + 70 \leq 90$

$\quad\quad 2x \leq 20$ Subtract 70 from each side.

$\quad\quad\quad x \leq 10$ Divide each side by 2.

If you can find CDs that cost less than or equal to $10, you can buy the shoes and two CDs.

Solving inequalities with more than one step or with variables on each side of the inequality requires you to apply the methods you have used to solve simple inequalities.

Examples

1   **Solve $13 - 6y > 49 - 2y$.**

$$13 - 6y > 49 - 2y$$

$$13 - 6y + 2y > 49 - 2y + 2y \quad\quad \text{Add } 2y \text{ to each side.}$$

$$13 - 4y > 49$$

$$13 + (-13) - 4y > 49 + (-13) \quad\quad \text{Add } -13 \text{ to each side.}$$

$$-4y > 36$$

$$\frac{-4y}{-4} < \frac{36}{-4} \quad\quad\quad\quad \text{Divide each side by } -4.$$

$$y < -9 \quad\quad \text{Any number less than } -9 \text{ is a solution.}$$

Why was the order symbol reversed? See page 114 for a review.

2   **Solve $605.7 + 2s \geq 754.41$.**

$$605.7 + 2s \geq 754.41 \quad\quad \text{Subtract 605.7 from each side.}$$
$$\text{Divide each side by 2.}$$

$$754.41 \; \boxed{-} \; 605.7 \; \boxed{=} \; \boxed{\div} \; 2 \; \boxed{=} \; 74.355$$

$$s \geq 74.355$$

Any number greater than or equal to 74.355 is a solution.

Checking for Understanding

Communicating Algebra

1. Explain how to check the solution for Example 2.
2. Explain why the solution of $-4x > 20$ is $x < -5$, not $x > -5$.

Guided Practice

State the steps that are necessary to solve each inequality. Then solve each inequality.

3. $9x + 2 > 20$
4. $4y - 7 < 21$
5. $-7y + 6 < 48$
6. $-5 < \frac{x}{4} - 7$
7. $10y + 3 < 7y$
8. $\frac{m}{3} - 7 > 11$

Exercises

Independent Practice

Solve each inequality.

9. $2a - 5 > 17$
10. $-5d + 7 > -38$
11. $7g - 3 \le 46$
12. $\frac{b}{-2} - 12 \le 11$
13. $\frac{z}{-3} - 4 > 27$
14. $\frac{k}{5} + 36 \ge 51$
15. $32 + 7t > 4$
16. $5 + 9a < -19$
17. $-12 + 11y < 54$
18. $\frac{c}{4} + 7 < -6$
19. $0.47 > \frac{t}{-9} + 0.6$
20. $-2.2 < \frac{b}{-10} - 2.4$
21. $-18 \ge 3(k + 5)$
22. $2(k + 4) \le 10$
23. $-3(m - 2) > 12$
24. $2k + 7 > k - 10$
25. $0.8x + 6 > 0.5x - 9$
26. $10y + 4 < 7y - 29$
27. $3x + 5 > 7x + 4$
28. $-5x + 3 < 3x + 23$
29. $11y + 9 > 5y - 21$

Language Skill

Translate each sentence into an inequality.

30. The sum of twice a number and 9 is less than 37.
31. The difference of three times a number and 5 is greater than 16.
32. Six more than the product of 3 and an integer is less than 39.
33. Five less than half an integer is greater than sixteen.

Mixed Review

34. Find the LCM of 18 and 4.   (Lesson 4-7)
35. Solve the inequality $w - 3.8 \le 9.4$.   (Lesson 5-7)
36. Estimate $\frac{1}{7} \times 29$.   (Lesson 6-2)
37. *True* or *false:* $1.87 \ge 18.7 \times 10^{-1}$.   (Lesson 6-13)

Solve each equation.   (Lessons 7-4 and 7-5)

38. $-3n - 9 = 15$
39. $a + 7 = 13$
40. $4(x - 1) + 7 = 5(-x + 6)$

Application

41. **Hobbies**   Glen has $10 to spend for two model airplanes and some supplies. The airplanes cost $3.49 each. Write an inequality that describes this situation. Then solve the inequality to find out how much Glen can spend on supplies.

Critical Thinking

42. The sum of the squares of two positive integers is 289. What are the integers?

Wrap-Up

43. **Make Up a Problem**   Write a problem that has $x \le 13$ as the solution.

7-7 Writing Inequalities

Objective:
Write and solve inequalities involving rational numbers.

Katie has scores of 7.5, 6.9, and 6.7 in the all-around competition at a gymnastics meet. She has one more event. The leading opponent has completed the competition with an average score of 7.12. What must Katie score on her final trial if she is to win the competition?

Why might an estimate not be appropriate?

Explore You must find Katie's fourth score so that her average score will be greater than 7.12.
Let s = Katie's fourth score.

Plan

Katie's average	is greater than	average of the leader
$\dfrac{7.5 + 6.9 + 6.7 + s}{4}$	$>$	7.12

Solve

$$\frac{21.1 + s}{4} > 7.12$$

$$4 \cdot \frac{21.1 + s}{4} > 4 \cdot 7.12 \qquad \text{Multiply each side by 4.}$$

$$21.1 + s > 28.48$$

$$21.1 - 21.1 + s > 28.48 - 21.1 \qquad \text{Subtract 21.1 from each side.}$$

$$s > 7.38$$

Katie's score in the last event must be greater than 7.38.

Examine Check by substituting a score greater than 7.38 for s in the original inequality.

FYI

The events in the women's all-around include the floor exercise, the vault, the uneven bars, and the balance beam.

Problems containing the phrases *greater than* or *less than* can often be solved using inequalities. Other phrases that suggest inequalities are *at least, at most,* and *between.* Study the inequalities that correspond to each statement below.

The number x is *at least* 8.	$x \geq 8$	*At least 8 means 8 or greater.*
The number y is *at most* 4.	$y \leq 4$	*At most 4 means 4 or less.*
A number z is *between* 7 and 9.	$7 < z < 9$	*Between 7 and 9 means greater than 7 and less than 9.*

Example

1 The owner of a lawn mower repair shop charges $5 to diagnose the problem. He then charges $30 per hour to fix the problem. If he estimates that the total cost will be no more than $50, how many hours can he work and still be within his estimate?

Explore The total bill will be the diagnosis fee plus labor. No more than $50 means the total will be less than or equal to $50.

Plan Let h represent the number of hours of labor. Then $30h$ is the total charged for the labor.

diagnosis fee	plus	labor	is less than or equal to	$50
5	+	$30h$	$\leq$	50

Solve
$$5 + 30h \leq 50$$
$$5 - 5 + 30h \leq 50 - 5 \quad \text{Subtract 5 from each side.}$$
$$30h \leq 45$$
$$\frac{30h}{30} \leq \frac{45}{30} \quad \text{Divide each side by 30.}$$
$$h \leq 1\tfrac{1}{2}$$

The owner can work no more than $1\tfrac{1}{2}$ hours to be within his estimate.

Examine Check by substituting times greater than, less than, and equal to $1\tfrac{1}{2}$ hours for h in the original inequality.

Checking for Understanding

Communicating Algebra **For each problem, answer the related questions.**

1. Jenny has scores of 87, 92, 89, and 97 on pre-algebra quizzes. There will be one more quiz this grading period. What must Jenny's fifth score be to have an average score of at least 90?
   a. What is asked?
   b. What does *at least 90* mean?
   c. What is the first operation that will be done to solve this problem?
   d. Write an inequality that represents this problem.

2. Nine less than six times a number is at most 33. What is the number?
   a. Write *nine less than six times a number* using symbols.
   b. What does *at most 33* mean?
   c. Write an inequality that represents this problem.
   d. How would you know if $7\tfrac{1}{2}$ is a member of the solution set?

Write an inequality for each situation.

3. You earn $8 per hour plus a bonus of $10. You wish to earn at least $60.

4. You can hike along the Appalachian Trail at 3 miles per hour. You will stop for an hour for a lunch break. You want to walk at least 18 miles.

5. You buy some candy bars at 55 cents each plus one newspaper for 35 cents. You can spend no more than $4.

6. Four times a number increased by 4 is at least 16. What is the number?

7. Twelve times a number decreased by $\frac{1}{20}$ of the number is less than 3250. What is the number?

Exercises

Independent
Practice

ESTIMATION
MENTAL MATH
CALCULATOR
PAPER/PENCIL

Write an inequality to solve each problem. Then solve each inequality.

8. Sue plans to spend no more than $100.00 on blouses and one sweater. She buys 3 blouses at $17 each. What is the most she can spend on the sweater?

9. Cyril is going to buy a coat and a hat. The coat costs 3 times as much as the hat. He cannot spend more than $120. What is the most he can spend for the coat?

10. A stove and a freezer weigh at least 260 kg. The stove weighs 115 kg. What is the weight of the freezer?

11. Tia is buying a new car. She can make a down payment of $1,200. She wants to pay off the car in two years. She cannot afford monthly payments over $250. What is the most she can pay for the car?

12. If 4.05 times an integer is increased by 3.116, the result is between 13 and 25. What is the number?

13. Estimate $(-2.03)(0.6)$.   (Lesson 6-2)

Solve each inequality.   (Lesson 7-6)

14. $-4 < -2(b + 7)$

15. $3x - 14 \geq 2(3 - x)$

16. Nancy has a stack of less than 50 pennies but more than one penny. When she divides the number of pennies by 2, there is 1 penny left over. There is also 1 penny left over when she divides by 3 or 7. How many pennies does she have?

  17. **Make Up a Problem**   Write a problem that would be solved using the inequality $215 - 3x \leq 175$.

7-8   Using the Metric System

Objective:
Convert measures within the metric system.

Key Terms:
meter
kilometer
gram
kilogram
liter

While on a family vacation in Canada, Chris noticed that the highway speed limit was 100! He reasoned that the speed limit couldn't be 100 miles per hour. Then he remembered that Canada uses the metric system of measurement. The metric system is also the system of measurement used by scientists in the United States.

The **meter** (m) is the basic unit of *length* in the metric system. All units of length in the metric system are defined in terms of the meter. The chart below shows how metric prefixes are related to decimal place-value positions.

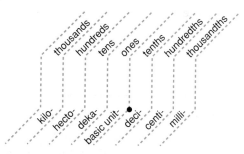

Note that each place value is 10 times the place value to its right.

Note that the value of each metric prefix is 10 times the value of the prefix to its right.

One **kilometer** (km) is 1000 meters. The speed limit that Chris noticed was 100 kilometers per hour, which is about 60 miles per hour.

Converting units within the metric system follows the same procedure as multiplying or dividing powers of ten. You may want to refer to Chapter 4 to review these procedures.

Larger units to smaller units: MULTIPLY →

Units of Length

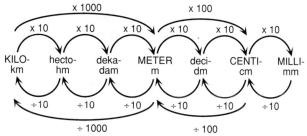

Smaller units to larger units: ← DIVIDE

Examples

Complete.

Mental Math Hint

Remember, you can multiply or divide by a power of ten by moving the decimal point.

1 7 mm = ■ cm

Smaller to larger means fewer units. Divide by 10.

$7 \div 10 = 0.7$
7 mm = 0.7 cm

2 5.3 km = ■ m

Larger to smaller means more units. Multiply by 1000.

$5.3 \times 1000 = 5300$
5.3 km = 5300 m

The *mass* of an object is the amount of matter that it contains. The basic unit of mass in the metric system is the **gram** (g). Kilogram, gram, and milligram are related in a manner similar to kilometer, meter, and millimeter.

$$1 \text{ kg} = 1000 \text{ g} \qquad 1 \text{ g} = 1000 \text{ mg}$$

A pair of shoes has a mass of about 1 **kilogram** (kg).

Capacity is the amount of liquid or dry substance a container can hold. The basic unit of capacity in the metric system is the **liter** (L). Four average size water glasses hold about 1 liter. A liter and milliliter are related in a manner similar to meter and millimeter.

$$1 \text{ L} = 1000 \text{ mL}$$

Converting metric units of mass and capacity is similar to converting metric units of length.

Examples

Complete.

3 **8 kg = ▮ g**

Larger to smaller
means more units.
Multiply by 1000.

$8 \cdot 1000 = 8000$
$8 \text{ kg} = 8000 \text{ g}$

4 **40 mL = ▮ L**

Smaller to larger
means fewer units.
Divide by 1000.

$40 \div 1000 = 0.040$
$40 \text{ mL} = 0.040 \text{ L}$

Checking for Understanding

Communicating Algebra

1. Explain how to convert 6 meters to centimeters.
2. Explain how to convert 850 milliliters to liters.
3. Explain how to convert 3.9 grams to milligrams.
4. Write a general rule for converting in the metric system.

Guided Practice

Complete.

5. 200 cm = ▮ m
6. 3.2 kg = ▮ g
7. 9.4 L = ▮ mL

Exercises

Independent Practice

Complete.

8. 4 m = ▮ cm
9. 30 mm = ▮ cm
10. 5 km = ▮ m
11. 500 cm = ▮ m
12. 1200 mm = ▮ m
13. 0.6 km = ▮ cm

14. 8 cm =  mm   **15.** 3 m = ▨ mm   **16.** 32,000 cm = ▨ km

17. 5 kg = ▨ g   **18.** 3000 mg = ▨ g   **19.** 2000 g = ▨ kg

20. 6.7 g = ▨ mg   **21.** 18,000 mg = ▨ kg   **22.** 1.9 kg = ▨ g

23. 7300 g = ▨ kg   **24.** 0.006 kg = ▨ mg   **25.** 53 g = ▨ kg

26. 5 L = ▨ mL   **27.** 2000 mL = ▨ L   **28.** 3.4 L = ▨ mL

29. 250 mL = ▨ L   **30.** 7 mL = ▨ L   **31.** 10.6 L = ▨ mL

32. 350 mL = ▨ L   **33.** 11 L = ▨ mL   **34.** 8.5 L = ▨ mL

Mixed Review

35. Find the next three terms in the arithmetic sequence 5, 2, -1, -4, . . .
(Lesson 5-9)

36. Solve the equation $q = 3\frac{5}{6} + 1\frac{1}{3}$.   (Lesson 6-6)

37. Solve the inequality $12 - 5y > 3y - 4$.   (Lesson 7-6)

38. Translate into an inequality. *The sum of a number and two more than the number is at least 12.*   (Lesson 7-6)

39. The maximum load of a cargo elevator is 1600 pounds. If the current load is 1362 pounds, at most how many pounds can be added?   (Lesson 7-7)

Application

40. Consumer Awareness   Which has a greater capacity, a 2-liter bottle of cola or a six-pack of cans each containing 354 mL of the same cola?

Connection

41. Geometry   The perimeter of a triangle is 1.6 meters. The length of the shortest side is 40 centimeters. Another side is 150 millimeters longer than the shortest side. Find the lengths of the other sides.

Critical Thinking

42. Four volumes of International Recipes are in order on a shelf. The total pages of each volume are 5 cm thick. Each cover is 5 mm thick. A bookworm started eating at page 1 of Volume I and ate through to the last page of Volume IV. What distance did the bookworm cover?

Wrap-Up

43. Explain how you use mental math skills to convert within the metric system.

Career

Laser Technician

Lasers have emerged as very valuable tools with applications ranging from very powerful cutting tools in industry to very delicate surgical instruments in medicine.

Laser technicians apply principles of laser engineering to the construction, testing, installation, and maintenance of laser devices. In the 1990s, the need for laser technicians is expected to grow 30%.

Review

Language and Concepts

Choose a word or symbol from the list at the right to correctly complete each statement.

1. To change larger units of measure to smaller units of measure, it is necessary to __?__ .

2. To change smaller units of measure to larger units of measure, it is necessary to __?__ .

3. __?__ is the mathematical symbol that means the solution set is empty or null.

4. The __?__ is based on powers of ten.

5. The symbol __?__ means *at least*.

6. The basic unit of capacity in the metric system is the __?__ .

customary system
divide
gram
liter
meter
metric system
multiply
$\varnothing$
$\neq$
$\leq$
$\geq$

Skills

Solve each equation. Check your solution.   (Lessons 7-2, 7-4, 7-5)

7. $2x + 4 = -32$

8. $3m + 8 = 71$

9. $-2r + 8 = 24$

10. $3y - 14 = 37$

11. $6z - 1.6 = -5.2$

12. $-2t - 1.5 = 1.3$

13. $\frac{y}{3} - 6 = -9$

14. $\frac{x}{5} + 3 = 6$

15. $\frac{c + 8}{4} = 16$

16. $5x - 4 = 4x + 3$

17. $2y - 6 = 3y - 5$

18. $7t + 6 = 3t - 14$

19. $3(x + 2) = 12$

20. $3(x - 5) = -6$

21. $6(x + 2) - 4 = -10$

22. $3x + 8 = 3x - 4$

23. $3(x + 2) - 6 = 3x$

24. $4y + 8 = 15$

Solve each inequality.   (Lesson 7-6)

25. $2m + 1 < 9$

26. $-3k - 4 \leq -22$

27. $-2 \geq 10 - 2x$

28. $1 - 3x > 7$

29. $-3.2 + 14z < 15z$

30. $3(y - 2) > 5(y - 7)$

Complete.   (Lesson 7-8)

31. $5m = $ ▮ cm

32. $3500 m = $ ▮ km

33. $2900 g = $ ▮ kg

34. $8.3 L = $ ▮ mL

35. $0.7 kg = $ ▮ g

36. $1860 mL = $ ▮ L

Applications and Problem Solving

Solve. Use the work backwards strategy. (Lesson 7-1)

37. Diana makes a salary of $250 a week. In addition to her weekly salary, she receives a bonus of $50 for every 25 boxes of books she sells. If she earned a total of $450 in salary and bonuses this week, how many boxes did she sell?

38. Frank walks 10 blocks to work every day. It takes him between 15 to 20 minutes to get to work. At this rate, what is the least amount of time it would take for Frank to walk 50 blocks?

Write and solve an equation. (Lesson 7-3)

39. Mr. Stinson bought a used boat for $30 more than half its original cost. He paid $150. What was the original cost?

40. Mrs. Hawn bought a washing machine on sale. She paid $25 more than half of the regular price. She paid $265. What is the regular price?

Use an inequality to solve each problem. (Lesson 7-7)

41. If 8 times a number is decreased by 2, the result is less than 15. What is the number?

42. Linda plans to spend at most $85 on jeans and shirts. She bought 2 shirts for $15.30 each. How much can she spend on jeans?

ortfolio Suggestion

Review the items in your portfolio. Make a list of the items, noting why each item was chosen. Replace any items that are no longer appropriate.

Curriculum Connection

• **Science** Do research on the relationship between balancing chemical equations and solving equations with a variable on both sides of the equal sign.

• **History** Do research to find out how Thomas Jefferson helped in the development of the metric system.

• **Language Arts** Write a composition titled *Math at the Mall*.

Read More About It

Barr, George. *Entertaining with Number Tricks.*

Ecker, Michal W. *Getting Started in Problem Solving and Math Concepts.*

Luce, Marnie. *Zero is Something.*

Test

Solve each equation.

1. $4 - b = -2$

2. $3(p - 5) = 9$

3. $\frac{h}{-2} + 5 = -12$

4. $\frac{y}{3} - 17 = 8$

5. $8x - 3 = 13$

6. $10k - 31 = 11k + 76$

7. $8p + 29 = 7p + 16$

8. $3x + 5 = -2x + 10$

9. $6 - x = -3x + 10$

10. $-4y + 2 = 32$

11. $\frac{m + 5}{6} = -19$

12. $\frac{5t - 9}{4} = 14$

13. $4(2y - 1) = -10(y - 5)$

14. $-2(3n - 5) + 3n = 2 - n$

15. $3(n + 5) - 6 = 3n + 9$

Solve each inequality.

16. $-6a + 2 < 14$

17. $2y + 1 \le -7$

18. $\frac{d}{-3} + 4 > -12$

Complete.

19. $8.4 \text{ km} = \blacksquare \text{ m}$

20. $250 \text{ mL} = \blacksquare \text{ L}$

21. $4 \text{ g} = \blacksquare \text{ mg}$

22. $18 \text{ cm} = \blacksquare \text{ m}$

Solve. Work backwards.

23. Duane passes out rulers for a class exercise. He gives one half of the rulers to Todd, who will help to pass them out. He then gives one third of the remainder to Pat, who will pass them down her row. Duane has 12 rulers left. How many rulers did he start with?

Write and solve an equation or inequality.

24. George plans to spend at most $40 for shirts and ties. He bought 2 shirts for $13.95 each. How much can he spend for ties?

25. Juanita lost her tennis racquet. She bought a new one for $30. This is $6 less than three times the price of her old one. How much did she pay for her old racquet?

BONUS

A plumber charges $36 for the first hour of work and $16 for every half hour or any part of a half hour thereafter. Find the longest amount of time this plumber can work for Tim Evans without having the bill exceed $150.

Academic Skills Test

Cumulative, Chapters 1-7

1. Which expression is equivalent to $-14ab - 27ab$?

   A $-41ab$ **C** $13ab$
   B $-13ab$ **D** $41ab$

2. Which is equivalent to $a^5 \cdot a^2$?

   A $2a^{10}$ **C** $2a^7$
   B a^{10} **D** a^7

3. Which fraction is less than $\frac{5}{12}$?

   A $\frac{1}{2}$ **C** $\frac{3}{4}$

   B $\frac{5}{8}$ **D** $\frac{2}{6}$

4. If $6.0 = b - 2.2$, what is the value of b?

   A 8.2   **C** 4.8
   B 6.0   **D** 3.8

5. If $3.2y = 80$, what is the value of y?

   A 256   **C** 25
   B 76.8   **D** 2.5

6. Which equation is equivalent to $2y + 6 = 12$?

   A $2y = 12$
   B $2y + 6 - 6 = 12 - 6$
   C $2y + 1 = 2$
   D $2y + 6 - 6 = 12 + 6$

7. If $5y - 4 = 3y + 12$, what is the value of y?

   A 1
   B 2
   C 4
   D 8

8. 3542 runners entered the Merrill Marathon. 1054 runners did not finish the race. The number of runners that finished the race is between—

   A 3500 and 3600
   B 2400 and 2600
   C 2000 and 2200
   D 1900 and 2100

9. The labor charge for repairing a car is $32.50 per hour. If it takes 2.5 hours to repair the car, what will be the charge for labor?

   A $812.50
   B $81.25
   C $13.00
   D Not Here

10. George bought a used stereo for $10 more than half its original cost. He paid $150. What was the original cost?

    A $70
    B $85
    C $280
    D $310

CHAPTER 8

CHAPTER OBJECTIVES

In this chapter you will learn to:

- graph the solution of equations and inequalities on a number line
- graph equations and inequalities in the coordinate plane
- use slope and intercepts to graph equations
- solve systems of linear equations by graphing
- solve problems by drawing a graph

Graphing Equations and Inequalities

Imagine that your math class decides to participate in a Bike-A-Thon to raise money for a local charity. You all seek pledges; that is, promises from community businesses to give the Bike-A-Thon donations. Three of your classmates each convince a business to sponsor them for different amounts.

Jamie:	$1.50 for each kilometer
Kenji:	$2.00 for each kilometer
Enrique:	a $15 contribution

On the day of the Bike-A-Thon, Jamie rode 15 kilometers, Kenji rode 14 kilometers and Enrique rode 10 kilometers.

Explain how you can use the graph to determine who raised the most money.

Language Connection

Class Project
Find examples of graphs in a newspaper or magazine. Write a paragraph that explains the purpose of the graph.

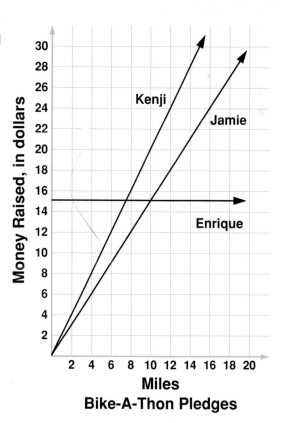

Money Raised, in dollars

Kenji

Jamie

Enrique

Miles
Bike-A-Thon Pledges

8-1   Equations and the Number Line

Objective:

Graph the solution of an equation on a number line.

Key Terms:

coordinate

graph

When you hear the word *graph,* what do you think of? You may think of a bar graph or a circle graph. The purpose of these graphs is to display data in a concise way. In algebra, we will study other kinds of graphs. These graphs will display data about equations and inequalities.

Let's start by assigning a number to each point on a line. The result is called a number line. Recall that the number is the coordinate of the point. The **coordinate** of a point tells its *distance* and *direction* from the 0-point of the line. The dot marking the point is called the **graph** of the number.

How would you use a number line to graph the number of United States Senators? Since there are 100 Senators, draw a number line and mark a dot on 100.

What do the arrowheads on each end of the number line indicate?

You can also show the solution of an equation on a number line.

Examples

1 Solve $x + 3 = -1$ and graph the solution.

$$x + 3 = -1$$
$$x + 3 - 3 = -1 - 3$$
$$x = -4$$

Check: $x + 3 = -1$
$$-4 + 3 \stackrel{?}{=} -1$$
$$-1 = -1 \;\checkmark$$

The solution is -4.

-7 -6 -5 -4 -3 -2 -1 0 1 2 3 4 5 6 7

2 Solve $2x + 1 = 4$ and graph the solution.

$$2x + 1 = 4$$
$$2x + 1 - 1 = 4 - 1$$
$$2x = 3$$
$$\frac{2x}{2} = \frac{3}{2}$$
$$x = \frac{3}{2}$$

Check: $2x + 1 = 4$
$$2(\tfrac{3}{2}) + 1 \stackrel{?}{=} 4$$
$$3 + 1 \stackrel{?}{=} 4$$
$$4 = 4 \;\checkmark$$

The solution is $\frac{3}{2}$.

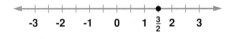

-3   -2   -1   0   1 $\frac{3}{2}$ 2   3

Checking for Understanding

Draw a graph for each situation.

1. There are 435 members in the House of Representatives.
2. The low temperature last night was -5°F.
3. You wrote a check for $10.50 yesterday.
4. You bought 6 tickets for the concert.
5. Gasoline costs $1.29 per gallon.

Solve each equation and graph the solution.

6. $x + 3 = 5$

7. $-5y = 10$

8. $a + 5\frac{2}{3} = 9\frac{1}{3}$

Exercises

Solve each equation and graph the solution.

9. $2x - 1 = 5$
10. $\frac{a}{3} = -1$
11. $3x - 1 = 4$
12. $n + 3.5 = 4$
13. $y = -5 + -3$
14. $2z - 6.5 = -1.5$
15. $x - 1.25 = 3.75$
16. $-7 + x = 9$
17. $2.5a = -15$
18. $5y + 2y - 3 = 18$
19. $2a + a + 5 + 1 = 18$
20. $4x - 7 = 2x + 15$
21. $1 + 10x = 8x - 3$

Choose the equation whose solution is graphed.

22.
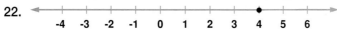

    a. $3x + 5 = 12$ **b.** $-2x = 8$ **c.** $x - (-2) = 6$

23.

    a. $2y + 8 = 3y + 9$ **b.** $5y = 5$ **c.** $4y - 3 = 1$

24.

    a. $7 + m = 4.5$ **b.** $2m - (-2) = 3$ **c.** $6m = 15$

25. Rewrite the expression $12 + (7 + m)$ using the associative property of addition. (Lesson 1-3)
26. Write the next term in the geometric sequence $1, -3, 9, -27. \ldots$ (Lesson 6-10)
27. Write an inequality for the sentence: *The product of 6 and a number increased by 11 is at most 35.* (Lesson 7-6)

28. **Health** A thermometer can be thought of as a physical representation of a number line. A healthy person has a body temperature of 98.6° on a Fahrenheit thermometer. Graph this number on a number line.

29. **Make Up a Problem** Write an equation whose solution is graphed on the number line below. Then write a different equation that has the same solution as the first equation.

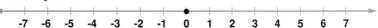

30. **Research** Find the coordinate for the normal body temperature on a Celsius thermometer. Then graph it on a number line.

8-2   Inequalities and the Number Line

Objective:
Graph the solutions of an inequality on a number line.

If a driver operates a car on an interstate highway at a speed that is greater than 65 mph, there is a risk of getting a speeding ticket. You can describe this situation algebraically by writing an inequality.

Suppose s represents the speed of the car. Then the inequality $s > 65$ is true for all values of s that are greater than 65. The solutions are graphed on the number line shown below.

Why is 65 not a solution of $s > 65$?

This point is not included in the solution.

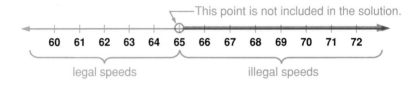

60 61 62 63 64 65 66 67 68 69 70 71 72

legal speeds   illegal speeds

Now let's consider another situation. Suppose the driver always operates the car at speeds that are less than or equal to 65 mph. The inequality $s \leq 65$ describes these speeds. The graph of $s \leq 65$ is shown below.

The dot shows that this point is included in the solution.

60 61 62 63 64 65 66 67 68 69 70 71 72

Just as you can solve and graph equations, you can solve and graph inequalities.

Examples

1   Solve $3x + 6 \leq 15$ and graph the solution.

$$3x + 6 \leq 15$$

$$3x + 6 - 6 \leq 15 - 6 \qquad \text{Subtract 6 from each side.}$$

$$3x \leq 9$$

$$\frac{3x}{3} \leq \frac{9}{3} \qquad \text{Divide each side by 3.}$$

$$x \leq 3 \qquad \text{The number 3 or any number less than 3 can be substituted for } x \text{ to make } 3x + 6 \leq 15 \text{ a true statement.}$$

How can you check your answer?

-8 -7 -6 -5 -4 -3 -2 -1 0 1 2 3 4 5 6 7 8

2 **Solve -3x − 2 > 10 and graph the solution.**

$$-3x - 2 > 10$$
$$-3x - 2 + 2 > 10 + 2 \quad \text{Add 2 to each side.}$$
$$-3x > 12$$
$$\frac{-3x}{-3} < \frac{12}{-3} \quad \begin{array}{l}\text{Divide each side by -3 and}\\ \text{reverse the order symbol.}\end{array}$$
$$x < -4 \quad \text{Any number less than -4 is a solution.}$$

Why do you need to reverse the order symbol when you divide by -3?

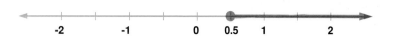

-10 -9 -8 -7 -6 -5 -4 -3 -2 -1 0 1 2 3 4 5 6 7 8 9 10

Notice that -4 is *not* a solution.

3 **Solve 3y − 4.7 ≥ -3.2 and graph the solution.**

$$3y - 4.7 \geq -3.2$$
$$3y - 4.7 + 4.7 \geq -3.2 + 4.7 \quad \text{Add 4.7 to each side.}$$
$$3y \geq 1.5$$
$$\frac{3y}{3} \geq \frac{1.5}{3} \quad \text{Divide each side by 3.}$$
$$y \geq 0.5 \quad \begin{array}{l}\text{Any number greater than or}\\ \text{equal to 0.5 is a solution.}\end{array}$$

Is 0.4999 a solution?

-2   -1   0   0.5   1   2

Checking for Understanding

Communicating Algebra

Draw a graph for each inequality.

1. He is more than 60 inches tall.
2. Juan spent less than $5.50.
3. The flight of the glider lasted more than 15 minutes.
4. She drove at least 100 miles today.
5. Karen earned $20 or more.

Guided Practice

Write an inequality that describes each graph.

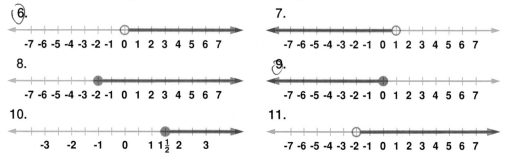

6.

-7 -6 -5 -4 -3 -2 -1 0 1 2 3 4 5 6 7

7.

-7 -6 -5 -4 -3 -2 -1 0 1 2 3 4 5 6 7

8.

-7 -6 -5 -4 -3 -2 -1 0 1 2 3 4 5 6 7

9.

-7 -6 -5 -4 -3 -2 -1 0 1 2 3 4 5 6 7

10.

-3   -2   -1   0   1 $1\frac{1}{2}$ 2   3

11.

-7 -6 -5 -4 -3 -2 -1 0 1 2 3 4 5 6 7

Solve each inequality and graph the solution.

12. $n + 3 \geq 8$

13. $a - 5 \leq -5$

14. $b + 8 > 9$

Exercises _____

Independent
Practice

Solve each inequality and graph the solution.

15. $x - 1.3 < 2.7$ **16.** $y + 2\frac{1}{2} \geq 4$ **17.** $2x - 1 \leq 3$

18. $3y - 5 < 4$ **19.** $3x + 3.2 > 4.7$ **20.** $-3a \geq 18$

21. $5c > -20$ **22.** $\frac{1}{2}a - 5 \geq -7$ **23.** $-4m - 3\frac{2}{3} > 8\frac{1}{3}$

24. $13a - 12 > 11a + 4$ **25.** $9r + 3 < 6r + 6$

26. $8n + 10 > 7n + 5$ **27.** $3(x + 6) > 2x + 12$

Logical
Reasoning

28. Graph the solutions for the compound inequality $y < -1$ or $y > 3$. Remember that an *or* sentence is true if either part is true.

29. Graph the solutions for the compound inequality $x > -5$ and $x \leq 4$. An *and* sentence is true only if both parts are true.

Mixed Review

30. Simplify $|14| - |-9|$.   (Lesson 2-1)

31. *True* or *false*: $-3\frac{1}{2} \geq -3.2$.   (Lesson 6-12)

32. **Business**   A plumber charges $35 to make a house call. For each hour of labor, he charges $25. If he works 3 hours, how much will he charge? (Lesson 7-3)

33. Solve the equation $-3m - 3.5 = 14.5$ and graph the solution.   (Lesson 8-1)

Applications

34. **Business**   A bookstore makes a profit of $5 on each two-volume set of books sold. The owner of the store wants to make a profit of at least $150 on these books. How many sets of books should the owner sell to meet her goal? Write your answer as an inequality.

35. **Sports**   Cecilia has a total score of 45.9 in five trials of a skating competition. The leading opponent has scores of 9.1, 8.7, 9.5, and 9.3 and has one more trial. What can you say about Cecilia's opponent's score in the last trial if Cecilia wins first place? Write your answer as an inequality.

Critical
Thinking

36. Suppose there are fewer than 35 students in your math class. Explain why this graph is not an appropriate graph.

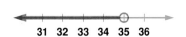

Wrap-Up

37. Devise a way to remember when to use a solid dot and when to use an open dot for the endpoint of the graph of an inequality.

Algebra in Action–Manufacturing

Tolerance

Parts used in automobiles must have very precise measurements or they will not work properly. However, it is impossible to produce parts with exact measurements. Thus, the dimensions of the parts must be between specified limits.

For example, a certain ball bearing that is specified to be 1 centimeter in diameter will work if it is only slightly larger or slightly smaller than 1 centimeter. The diameter may not differ from 1 centimeter by more than 0.001 centimeter. The 0.001 centimeter is called the **tolerance** of the ball bearing. The diameter of the ball bearing must be 1 ± 0.001 centimeter. That is, the diameter can vary between 1 + 0.001 centimeter and 1 − 0.001 centimeter. The acceptable diameter is shown by the following inequality.

$$1 - 0.001 \le x \le 1 + 0.001$$ The *tolerance interval* ranges
$$0.999 \le x \le 1.001$$ from 0.999 cm to 1.001 cm.

That is, the least possible diameter is 0.999 cm, and the greatest possible diameter is 1.001 cm. The solution is graphed on the number line shown below.

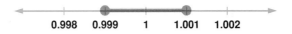

Write each expression as an inequality. Graph the solution.

1. $x = 3 \pm 0.01$
2. $x = 5 \pm 0.003$
3. $x = 7 \pm 0.0002$
4. $y = 6 \pm 0.0015$
5. $y = 1 \pm 0.15$
6. $y = 2 \pm 0.003$
7. $r = 0.5 \pm 0.0001$
8. $r = 1.5 \pm 0.001$
9. $d = \frac{1}{2} \pm 0.0035$

10. **Chemistry** A chemical supply company guarantees the precision weighing of its products. They advertise that a certain product weighs 8 oz ± 0.03 oz. What is the tolerance interval?

11. **Manufacturing** A pane of glass should be 26 inches wide by 32 inches long. The tolerance is $\frac{3}{16}$ inch. Find the tolerance interval for each dimension.

8-3 Ordered Pairs

Objective:
Identify the ordered pair
of numbers associated
with a point on a grid.

Key Term:
ordered pair

*Are the numbers in
an ordered pair
commutative?*

In 1791 President George Washington
selected the site for the new capital
that would be named in his honor. He
chose a Frenchman, Pierre L'Enfant, to
plan the city. The map at the right
shows L'Enfant's design. Notice how
the streets seem to form a grid pattern.

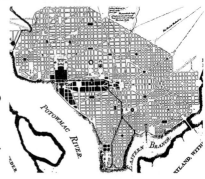

In mathematics, the location of any point on a grid can be indicated by an
ordered pair of numbers. The grid below is a simplified version of the
map of present-day Washington, D.C. To locate the Supreme Court on
this grid, begin at 0, and move horizontally (east) 4 units. Then move
vertically (north) 1 unit. The location of the Supreme Court is indicated
by the ordered pair (4, 1).

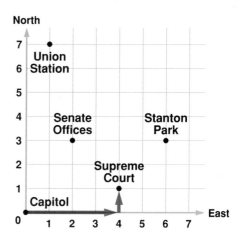

The location of any point on the grid can be indicated by the ordered pair
(x, y), where x represents the number of horizontal units from 0 and y
represents the number of vertical units from 0.

Examples

1 Use the grid at the right to name the point for the ordered pair (3, 7).

Move 3 units horizontally, then 7 units vertically.

The point for (3, 7) is *B*.

2 Use the grid at the right to find the ordered pair for point *E*.

Move 9 units horizontally, then 5 units vertically.

The ordered pair for point *E* is (9, 5).

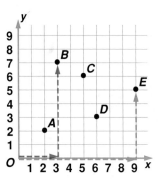

Checking for Understanding

Communicating Algebra

1. Locate Union Station on the grid on page 290. Name its ordered pair.

2. On this grid, what landmark is located at *O*?

3. If RFK Stadium were shown on the map, its ordered pair would be (24, 0). Describe its position with respect to the Capitol.

Guided Practice

Use the map at the right to answer each question.

4. Which ordered pair indicates the location of Harding Pass?

5. Is (2, 10) the correct ordered pair for Willard Junction?

6. Which ordered pair indicates the location of Red Clay Hill?

7. Is (3, 4) the correct ordered pair for Sun Valley Lookout Tower?

8. Which location does (8, 7) indicate?

9. Which ordered pair indicates the location of the forest service station?

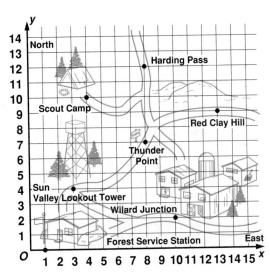

Exercises

Independent Practice

Use the grid at the right to name the point for each ordered pair.

10. (3, 4)   11. (5, 1)

12. (1, 7)   13. (8, 2)

14. (0, 2)   15. (5, 6)

16. (7, 5)   17. (9, 3)

18. (2, 0)   19. (5, 9)

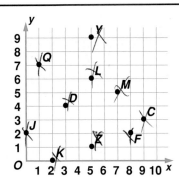

Use the grid to find the ordered pair for each labeled point.

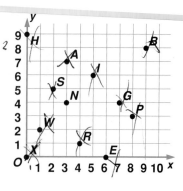

20. *W*   21. *R*   22. *A*

23. *P*   24. *S*   25. *I*

26. *G*   27. *B*   28. *N*

29. *E*   30. *X*   31. *H*

Mixed Review

32. Simplify the expression $-9ab - 13ab$.   (Lesson 2-5)

33. Solve the equation $9(y + 3) = 4y - 13$.   (Lesson 7-5)

34. Solve the equation $-4 + y = -3$ and graph the solution.   (Lesson 8-1)

35. Write an inequality for the graph: (Lesson 8-2)

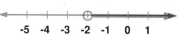

Connections

36. **Geometry** Graph (2, 1), (2, 4), and (5, 1) on a grid. Connect the points with line segments. What figure is formed?

37. **Geometry** Using the ordered pairs from Exercise 36, multiply each number by 2. Graph the new ordered pairs. What figure is formed?

Critical Thinking

38. Compare your figures from Exercises 36 and 37. Write a sentence that tells how the figures are the same. Write a sentence that tells how the figures are different.

Wrap-Up

39. An example of an ordered pair in your daily life might be (put on your socks, put on your shoes). Write another ordered pair from your life.

Biography

Benjamin Banneker

Benjamin Banneker, the grandson of a black slave, was born a free man on a farm outside Baltimore, Maryland, in 1731. He was a brilliant student in mathematics and quickly excelled beyond the ability of his teachers. At the age of twenty, he built the nation's first American made clock. It was keeping perfect time when he died in 1806.

At the suggestion of Thomas Jefferson, Banneker was appointed to a three-man team charged with designing and building Washington, D.C. When Pierre L'Enfant resigned, taking all of the plans and maps with him, Banneker was able to reproduce the entire set of plans completely from memory. Washington, D.C., stands today as a monument to the genius of a self-educated, African-American, mathematician, Benjamin Banneker.

8-4 The Coordinate System

Objective:
Graph points in all four quadrants.

Key Terms:
coordinate system
x-axis
y-axis
origin
quadrant

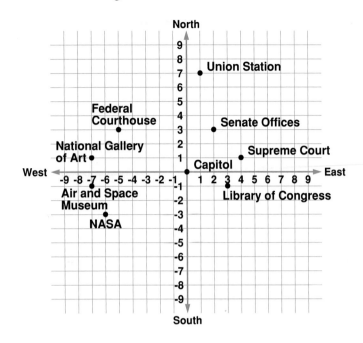

The map of Washington, D.C., can be extended to include the northwest, southwest, and southeast sections of the city. The location of the Library of Congress can be indicated by the ordered pair (3, -1). The location of the National Gallery of Art can be indicated by the ordered pair (-7, 1).

In mathematics, ordered pairs are used to locate points in a plane. A horizontal number line and a vertical number line intersect at their zero points to define a **coordinate system** for the plane. The horizontal number line is called the ***x*-axis;** the vertical number line is called the ***y*-axis.** The point where the two lines intersect is called the **origin.** The number lines separate the plane into four **quadrants.**

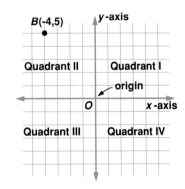

The numbers in an ordered pair are called coordinates. The *x-coordinate* of the ordered pair (-4, 5) is -4 and the *y-coordinate* is 5. The dot at (-4, 5) is the **graph** of point B.

The origin and the two axes do not lie in any quadrant. Axes is the plural of axis.

Examples

1 **Find the ordered pair for point _A_.**

Think of a vertical line and a horizontal line passing through point _A_. Since the vertical line intersects the _x_-axis at -2, the _x_-coordinate is -2. Since the horizontal line intersects the _y_-axis at 3, the _y_-coordinate is 3.

So, the ordered pair for point _A_ is (-2, 3).

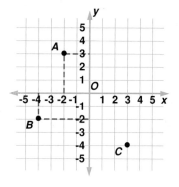

2 **Find the ordered pair for point _B_.**

The _x_-coordinate is -4. The _y_-coordinate is -2. The ordered pair for point _B_ is (-4, -2).

In which quadrant is point A? point B? point C?

3 **Find the ordered pair for point _C_.**

The ordered pair for point _C_ is (3, -4).

Remember that to graph a point means to place a dot at the point named by the ordered pair. This is sometimes called *plotting the point.*

Examples

4 **Graph point _D_(3, -1).**

Start at O. Move 3 units to the right. Then move 1 unit down to locate the point. Label the dot _D_(3, -1).

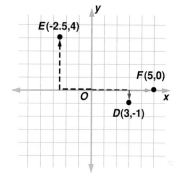

5 **Graph point _E_(-2.5, 4).**

In which quadrant is point F?

Start at O. Move 2.5 units to the left. Then move 4 units up to locate the point. Label the dot _E_(-2.5, 4).

6 **Graph point _F_(5, 0).**

Start at O. Move 5 units to the right. The graph is on the _x_-axis. Label the dot _F_(5, 0).

Checking for Understanding

Communicating
Algebra
1. Explain how the quadrants in a coordinate plane are numbered.

2. Suppose the x-coordinate of an ordered pair is negative. In which quadrants might the point be graphed? Give two examples.

3. Suppose the y-coordinate of an ordered pair is positive. In which quadrants might the point be graphed? Give two examples.

Guided
Practice
Name the x-coordinate and the y-coordinate of each ordered pair. Then tell which quadrant contains the graph of each ordered pair.

4. $(-3, 2)$ 5. $\left(\frac{1}{2}, -5\right)$ 6. $(3, 8)$ 7. $(-1, -4.7)$

8. $\left(-\frac{2}{3}, 5\right)$ 9. $(-2.25, 0)$ 10. $(0, -6)$ 11. $(0, 0)$

Exercises

Independent
Practice
On graph paper, draw coordinate axes. Then graph and label each point.

12. $K(0, 7)$ 13. $L(2, -8)$ 14. $M(-5, -5)$ 15. $N(-4, 7)$

16. $O(-1, 7)$ 17. $P(3, 4)$ 18. $Q(4, 3)$ 19. $R(-4, -3)$

20. $A(-2.5, 4)$ 21. $B\left(-1\frac{1}{2}, 6\right)$ 22. $C(-3, 0)$ 23. $D(5.5, 0.5)$

Logical
Reasoning
Name the quadrant for the graph of (x, y) given each condition.

24. $x > 0, y < 0$ 25. $x < 0, y < 0$ 26. $x < 0, y > 0$ 27. $x > 0, y > 0$

Mixed Review
28. Solve the equation $-7g = 84$.   (Lesson 3-3)

29. Factor $28a^2b^3$ completely.   (Lesson 4-4)

30. Write $5\frac{7}{8}$ as a decimal.   (Lesson 6-1)

31. Solve the inequality $-4m \leq 56$ and graph the solution.   (Lesson 8-2)

Application
32. **Physics**   Paulo and Elise conducted an experiment to see how the mass of an object affected the distance a spring stretched. The data are shown in the table at the right. Graph the ordered pairs (mass, distance) in the coordinate plane. Use the x-axis for mass and the y-axis for distance.

Stretching of a Spring	
Mass	**Distance**
100 g	3 cm
200 g	6 cm
300 g	9 cm
400 g	12 cm
500 g	15 cm

Connections
33. **Geometry**   Graph $(2, 1)$, $(2, 4)$, and $(5, 1)$ in the coordinate plane. Connect the points with line segments. What figure is formed?

34. **Geometry**   Using the ordered pairs from Exercise 33, add 5 to each x-coordinate. Graph the new ordered pairs. What figure is formed?

Critical
Thinking
35. Compare your figures from Exercises 33 and 34. Write a sentence that tells how the figures are the same and how the figures are different.

Wrap-Up   Journal Entry
36. Write two ordered pairs whose graphs are in Quadrant III. Write two ordered pairs whose graphs are *not* in Quadrant III.

8-5 Equations with Two Variables

Objective:
Find solutions for equations with two variables.

Forests serve as the only source of timber. Sometimes the production of timber conflicts with the need to conserve the environment and wildlife. In an effort to conserve both the forests and the wildlife, foresters replace the trees that have been cut down. About half of the seedlings planted survive until they are full grown.

Seedlings planted	One-half of seedlings planted	Full-grown trees
30 maples	$\frac{1}{2}(30)$	15
44 blue spruce	$\frac{1}{2}(44)$	22
46 aspen	$\frac{1}{2}(46)$	23

Suppose 50 pine seedlings were planted. How many would be expected to become full-grown trees?

The chart above could be written algebraically.

Let x represent the number of seedlings planted. Then $\frac{1}{2}x$ represents one-half the seedlings planted.

Let y represent the number of full-grown trees.

You know that one-half of the seedlings survive until full grown. This relationship can be shown with the equation $y = \frac{1}{2}x$. This equation has two variables, x and y.

x	$\frac{1}{2}x$	y
30	$\frac{1}{2}(30)$	15
44	$\frac{1}{2}(44)$	22
46	$\frac{1}{2}(46)$	23

Recall that solving an equation means to replace the variable so a true sentence results. The solution for an equation with two variables consists of two numbers, one for each variable. Usually, the solution is expressed as an ordered pair.

FYI

The United States Congress has set aside about 155 forest areas to be managed and protected by the federal government. These national forests cover a total area of about 183 million acres, larger than the state of Texas.

Example

1 **Which ordered pair, (10, 20) or (8, 4), is a solution of $y = \frac{1}{2}x$?**

Substitute the values for x and y into the equation.

$y = \frac{1}{2}x$ Replace x with 10 and y with 20.

$20 \overset{?}{=} \frac{1}{2} \cdot 10$

$20 \neq 5$

$y = \frac{1}{2}x$ Replace x with 8 and y with 4.

$4 \overset{?}{=} \frac{1}{2} \cdot 8$

$4 = 4$ ✓

(10, 20) is *not* a solution. (8, 4) is a solution.

An equation with two variables has an infinite number of solutions. To find a solution of such an equation, choose any value for x, substitute that value into the equation, and find the corresponding value for y.

It is often convenient to organize the solutions in a table.

Example

2 **Find four solutions of $y = 3x + 1$.**

x	$3x + 1$	y
-1	3(-1) + 1	-2
0	3(0) + 1	1
1	3(1) + 1	4
3	3(3) + 1	10

Select four convenient values for x.
Substitute that value for x in the expression $3x + 1$.
Do the computation to find y.

Four solutions are (-1, -2), (0, 1), (1, 4), and (3, 10).

Checking for Understanding

Communicating Algebra

1. Explain why (2, -3) is a solution of $y = 2x - 7$.

2. Name four solutions of $x + y = 5$.

3. If $x = 3$, what is the value of y in the equation $y = 0.5x + 1$?

4. Explain why an equation with two variables has an infinite number of solutions.

Guided Practice

Complete the table for each equation. Then use the results to write four solutions for each equation.

5. $y = x + 5$

x	$x + 5$	y
-3	-3 + 5	
0	0 + 5	
4	4 + 5	
9	9 + 5	

6. $y = 2x + 3$

x	$2x + 3$	y
-1	2(-1) + 3	
-3	2( ) + 3	
2	2( ) + 3	
4	2( ) + 3	

7. $y = 5x - 8$

x	$5x - 8$	y
2	5( ) - 8	
-1	5( ) - 8	
0	5( ) - 8	
4	5( ) - 8	

Exercises

Independent Practice

Find four solutions for each equation. Write your solutions as ordered pairs.

8. $y = x + 1$

9. $y = x + 3.2$

10. $y = x - 7$

11. $y = 2x$

12. $y = 3x + 0.5$

13. $y = 6x$

14. $y = 3x + 1$

15. $y = 2x - 3$

16. $y = -2x$

17. $y = -5x + 3$

18. $y = x$

19. $y = \frac{1}{2}x + 5$

20. $y = \frac{1}{3}x - 3$

21. $y = \frac{2}{3}x + 1$

22. $y = -\frac{1}{2}x$

23. $x + y = 6$

24. $x - y = 4$

25. $2x + y = 5$

Find four solutions for each sentence. Let x represent the first number and y represent the second number. Write your solutions as ordered pairs.

26. The sum of two numbers is 10.

27. The first number is 4 more than the second.

28. The first number is one-half the second.

29. The sum of two numbers is 0.

30. The first number minus the second number is 5.

Calculator

31. Using a calculator, find four solutions for the equation $y = 2.25x + 8.5$.

32. If $x = 8.95$, what is the value of y in the equation $y = 5x - 85$?

Mixed Review

Solve each inequality. (Lessons 3-7, 7-6)

33. $r - (-24) < -31$

34. $2k - 9 > 3k + 1$

35. Evaluate the expression $a - bc$ if $a = \frac{1}{2}$, $b = -2$, and $c = 1\frac{1}{4}$. (Lesson 6-3)

Graph and label each point. (Lesson 8-4)

36. $K(-3, -1)$

37. $L(5, -2)$

Computer

The following BASIC program uses a loop to generate solutions for $y = 2x - 1$.

```
10 FOR X = -2 TO 2
20 LET Y = 2 * X - 1
30 PRINT "(";X;" ,";Y;")"
40 NEXT X
50 END
```

The FOR statement starts the loop. It sets the initial value of x at -2 and the final value at 2.

The NEXT statement ends the loop. It causes x to be increased by 1 and the program to return to line 20 to find the corresponding value of y.

38. Run the program and describe the output.

39. Which ordered pairs would be printed if line 10 were changed to
    `FOR X = -2 TO 2 STEP 2`?

40. Modify the program to generate solutions for $y = -4x + 6$.

Connections

41. **Geometry** In any right triangle, the sum of the measures of the two acute angles is 90°. Find five ordered pairs of numbers that satisfy this condition. Graph your solutions on a coordinate plane.

42. **Geometry** The formula for the area of a rectangle is $A = \ell w$. Suppose the area is 20m². Find five ordered pairs of numbers that satisfy the formula. Graph your solutions on a coordinate plane.

Critical Thinking

43. Explain why negative numbers are not appropriate solutions in Exercises 41 and 42.

Wrap-Up

44. In your own words, explain how to obtain a solution for an equation with two variables.

8-6   Graphing Equations

Objective:
Graph linear equations.

Key Term:
linear equation

Have you ever wondered what the curved lines on a weather map mean? These lines are called isobars. They are drawn by connecting points that have the same air pressure. The isobar shows the relationship among all of the points on the curved line.

You have already learned that, in algebra, an equation with two variables describes the relationship between the value of x and the value of y. Another way to show the relationship is to graph the equation.

The graph of the equation $y = 2x - 1$ can be obtained by following the steps outlined below.

Step 1   Find at least four solutions for the equation $y = 2x - 1$. Record each result as an ordered pair (x, y).

x	$2x - 1$	y	(x, y)
-1	2(-1) – 1	-3	(-1, -3)
0	2(0) – 1	-1	(0, -1)
1	2(1) – 1	1	(1, 1)
2	2(2) – 1	3	(2, 3)

Step 2   Graph the points associated with the ordered pairs you obtained in Step 1.

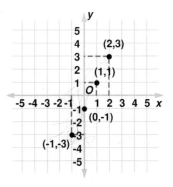

Step 3 Draw the line that contains these points.

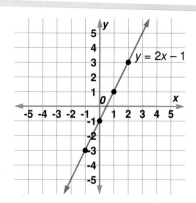

Every point on the line has coordinates that make the equation $y = 2x - 1$ true. It appears from the graph that $(3, 5)$ and $(1.5, 2)$ are also solutions of $y = 2x - 1$. You can check this by substitution.

The point (-2, 2) is not on the line. Is it a solution for y = 2x − 1?

Check $(3, 5)$.

$y = 2x - 1$

$5 \overset{?}{=} 2 \cdot 3 - 1$

$5 = 5 \ \checkmark$

Check $(1.5, 2)$.

$y = 2x - 1$

$2 \overset{?}{=} 2(1.5) - 1$

$2 = 2 \ \checkmark$

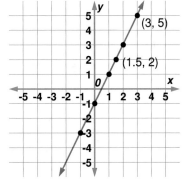

An equation like $y = 2x - 1$ is called a **linear equation** because its graph is a straight line.

Checking for Understanding

Communicating Algebra

1. List the steps you should use to graph a linear equation.

2. Which equation is graphed at the right?

a. $x - y = 3$

b. $x + y = 3$

c. $y = x + 3$

d. $y = x - 3$

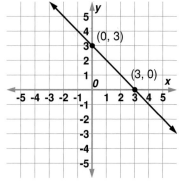

3. Define a **linear equation**.

Guided Practice

Graph each equation.

4. $y = x + 4$

5. $y = 2x + 1$

6. $y = 1.5x$

7. $y = -3x$

8. $y = -2x - 3$

9. $y = -x + 5$

Exercises

Independent Practice

Graph each equation.

10. $y = 4x + 3$

11. $y = \frac{1}{3}x$

12. $y = \frac{1}{2}x + 1$

13. $y = x + 1.5$

14. $y = -2x$

15. $y = x$

16. $x + y = 1$

17. $2x + y = 10$

18. $y - x = 1$

19. $x + y = 0$

20. $x = 5$

21. $y = -2$

Language Skill

Translate each sentence into an equation. Then graph the equation.

22. The sum of two numbers is 8.

23. One number is three more than a second number.

24. The length is twice the width.

25. The sum of two numbers is 0.

Mixed Review

26. Name the multiplicative inverse of $3\frac{2}{3}$. (Lesson 6-5)

27. Translate into an equation: *the sum of three and a number, divided by five is eleven.* (Lesson 7-2)

28. Name four solutions of $y = -3x + 1$. (Lesson 8-5)

29. If $n = 3$, what is the value of k in the equation $n = -\frac{1}{3}k + 5$? (Lesson 8-5)

Applications

30. **Business**   The ninth-grade student council sold candy as a fund-raising project. For each box sold, they earned $0.75. Let n represent the number of boxes sold and e represent the earnings. Write an equation that can be used to compute the earnings. Then graph the equation.

31. **Sports**   In a recent Indianapolis 500, the winner completed the race with an average speed of 167.6 mph. Graph the equation $d = 167.6t$.

Computer

32. Write a computer program similar to the one on page 298 to generate several solutions for $y = 12.5x - 20$. Then graph the equation.

Critical Thinking

33. Find an ordered pair that is a solution for both $x + y = 15$ and $x - y = (-1)$.

Wrap-Up

34. Explain why a computer is a good tool for generating ordered pairs.

Mid-Chapter Quiz

Graph the solution of each equation or inequality.   (Lessons 8-1, 8-2)

1. $x - 5 = -1$

2. $2x - 5 \leq 11$

3. $-5a > 15$

Graph and label each point.   (Lessons 8-3, 8-4)

4. $T(5, -3)$

5. $A(-1, 5)$

6. $S(0, 4)$

7. $N(-1.5, -2)$

Find four solutions for each equation. Then draw its graph.   (Lessons 8-5, 8-6)

8. $y = 4x - 2$

9. $y = \frac{1}{2}x + 5$

10. $x + y = 7$

8-7 Strategy: Draw a Graph

Objective:
Solve problems by using graphs.

Have you ever heard the expression *a picture is worth a thousand words?* What does this expression mean to you?

In mathematics, the *picture* can be a graph. It gives you the same information as an equation or table, but it is visual. It allows you to see patterns that may not be obvious from the equation. Drawing a graph is a powerful tool in problem solving.

Example

1 An area has been reforested. Certain trees are marked with special tags so they can be used as samples to measure growth. After five years, one of the sample trees is 2 meters tall. Six years later the same tree is 3.2 meters tall. The trees are harvested when the sample tree is about 7 meters tall. About how many years after planting can the trees be harvested? Assume the sample tree grows at a constant rate.

Explore > You need to find out how long it takes for a sample tree to reach a height of 7 meters. You know the height after five years and after six more years. You also assume that a tree continues to grow at a constant rate.

Plan > One way to solve the problem is to graph the given information. Then read the graph to find the number of years that corresponds to a height of 7 meters.

Solve > Let the horizontal axis of the graph represent time in years. Let the vertical axis represent the height of the sample tree in meters. Find and graph the ordered pairs (5, 2) and (11, 3.2). Then draw the line that contains these points.

Will this graph continue indefinitely?

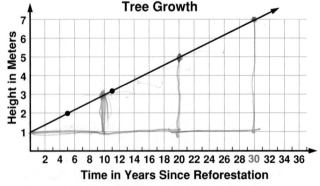

On the graph, 7 meters corresponds to 30 years. The trees can be harvested about 30 years after planting.

Examine One way to check the solution is by working backwards. Graph the ordered pairs (30, 7) and (11, 3.2). The line that contains the points also contains (5, 2).

Another way to check is to see if the solution is reasonable. The sample tree grew a little more than 1 meter in 6 years. It needs to grow about 4 more meters until harvest. Therefore, it will be about 6×4 or 24 more years. Since $11 + 24 = 35$, a solution of 30 seems reasonable.

Checking for Understanding

Communicating Algebra

1. Explain why it is important to take the time to plan before you solve a problem.

Use the graph in Example 1 to answer each question.

2. What was the height of the sample tree when it was planted?

3. What was the height of the tree 17 years after it was planted?

4. How many years did it take until the tree reached a height of 5 meters?

5. What height would you expect the tree to reach after 31 years?

Guided Practice

Name two ordered pairs (x, y) that can be used to graph the information in each problem.

6. David measures the heights of the steps going into his house. The 2nd step is 1 foot above ground. The 5th step is $2\frac{1}{2}$ feet above ground. What is the height of the 11th step?

7. Mrs. Madison drives at a constant rate for 6 hours. After $\frac{1}{2}$ hour she has driven 25 miles. After 2 hours she has driven 100 miles. How many miles does she drive in 6 hours?

Exercises

Independent Practice

Use a graph to solve each problem. Assume that the rate is constant in each problem.

8. After 2 hours, Kaiyo checks the odometer in her car. She has traveled 160 kilometers. After 3 more hours, she has traveled a total distance of 400 kilometers. How many more hours will it take for Kaiyo to travel a total distance of 560 kilometers?

9. Larry used a chart to see if he was close to the average weight for his height. The chart gave a weight of 130 pounds for a male 60 inches tall. The average weight for a male 66 inches tall was 143 pounds. Larry is 6 feet in height. What should Larry's weight be?

10. A temperature of 32° Fahrenheit corresponds to a temperature of 0° Celsius. A temperature of 100° Celsius corresponds to a temperature of 212° Fahrenheit. About what temperature in degrees Celsius corresponds to a temperature of 0° Fahrenheit?

11. Diego cuts lawns during the summer to earn extra money. After working for $2\frac{1}{2}$ hours, Diego has earned $10. After working for an additional $1\frac{1}{2}$ hours, Diego has earned a total of $16. How much is Diego paid per hour? How much will Diego earn if he works a total of 6 hours?

Decision Making

Solve using the graph and either mental math skills, estimation, or paper and pencil. Explain your method.

ESTIMATION
MENTAL MATH
CALCULATOR
PAPER/PENCIL

12. Kelly knows that her car can travel about 30 miles on one gallon of gasoline. The graph shows this relationship. Kelly has 6 gallons of gasoline in her car. She is starting on a 250-mile trip. Will she have to buy more gasoline sometime during her trip? If so, approximately when?

Wrap-Up

13. **Make Up a Problem** Write a problem that can be solved by using the graph for Exercise 12.

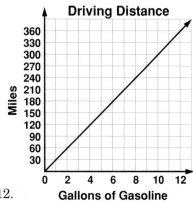

Driving Distance

Miles: 360, 330, 300, 270, 240, 210, 180, 150, 120, 90, 60, 30

Gallons of Gasoline: 0, 2, 4, 6, 8, 10, 12

Challenge

Parabolas

Not all equations are linear equations. Consider the equation $y = x^2$.

x	y
-3	9
-2	4
-1	1
0	0
1	1
2	4
3	9

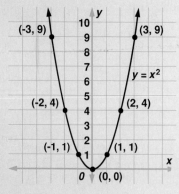

$y = x^2$

(-3, 9) (3, 9)
(-2, 4) (2, 4)
(-1, 1) (1, 1)
(0, 0)

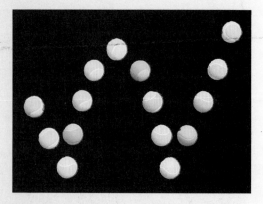

This curve is called a **parabola.** In real life, the path of a bouncing ball is a parabola. Research other areas of real-life where parabolas occur.

8-8 Slope

Objective:
Find the slope of a line.

Key Terms:
slope
change in *y*
change in *x*

Probably one of the most thrilling rides at an amusement park is the roller coaster. Imagine you are about to ride this one!

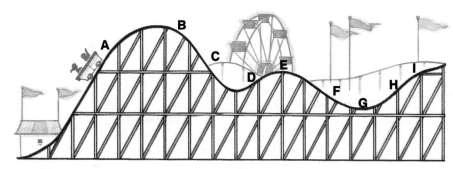

Let's compare position A with position D. In both cases, you are climbing. But at position A the climb is definitely steeper than at position D.

What about positions B and C? In both cases, you are dropping. At position C the drop is much steeper than at position B.

Think about what happens at position E. Here you are neither climbing nor dropping.

The steepness of the roller coaster depends on the vertical change and the horizontal change. It can be expressed as a ratio.

$$\text{steepness} = \frac{\text{vertical change}}{\text{horizontal change}}$$

Lines in a coordinate plane also have steepness. In mathematics, the steepness of a line is called its **slope.** The vertical change is called the **change in *y*,** and the horizontal change is called the **change in *x*.**

$$\text{slope} = \frac{\text{change in } y}{\text{change in } x}$$

The longest roller coaster in the world is The Beast at King's Island near Cincinnati, Ohio. It has reached a speed of 64.77 mph at the base of its 141-foot drop.

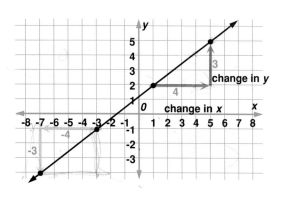

In Quadrant I, the change in *y* is 3 and the corresponding change in *x* is 4. Therefore, the slope of the line is $\frac{3}{4}$. Is the slope of the line the same in Quadrant III?

$$\frac{\text{change in } y}{\text{change in } x} = \frac{-3}{-4} \text{ or } \frac{3}{4}$$

The slopes are the same.

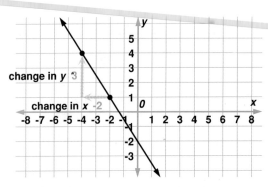

The slope of a line can also be negative. At the left the change in y is 3 and the corresponding change in x is -2. Therefore, the slope is $\frac{3}{-2}$ or $-\frac{3}{2}$.

Why is the slope of a horizontal line 0?

The slope of a line can be determined by using the coordinates of any two points on the line. The change in y can be found by subtracting the y-coordinates. Likewise, the change in x can be found by subtracting the corresponding x-coordinates.

Why does a vertical line have no slope?

$$\text{slope} = \frac{\text{difference of } y\text{-coordinates}}{\text{difference of } x\text{-coordinates}}$$

Example

1 **Find the slope of the line that contains $A(4, 3)$ and $B(-2, -5)$.**

$$\text{slope} = \frac{\text{difference of } y\text{-coordinates}}{\text{difference of } x\text{-coordinates}}$$

$$\text{slope of line } AB = \frac{3 - (-5)}{4 - (-2)}$$

$$= \frac{8}{6} \text{ or } \frac{4}{3}$$

Is the slope the same if you find $\frac{-5 - 3}{-2 - 4}$?

The slope is $\frac{4}{3}$.

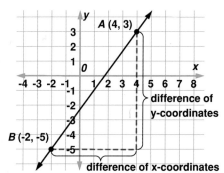

Checking for Understanding

Communicating Algebra

1. Using the drawing of the roller coaster, name all positions that represent a positive slope. Name all positions that represent a negative slope.

2. The slope of a line is 3. What does that mean?

3. On the drawing of the roller coaster, what is the slope at position G?

Guided Practice

Find the slope of each line.

4.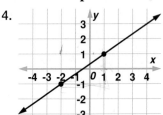

5.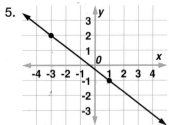

6.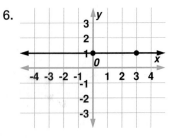

Exercises

Independent Practice

Find the slope of the line that contains each pair of points.

7. $A(5, 3), B(-1, 1)$ 8. $C(-3, 1), D(4, 5)$ 9. $F(2, 3), G(-1, 3)$

10. $L(-1, -2), M(2, -5)$ 11. $P(5, -2), Q(4, -3)$ 12. $R(7, 3), S(3, -9)$

13. $Q(5, 6), S(2, -2)$ 14. $V(1, 0), X(-2, -3)$ 15. $A(5, -1), C(-3, -4)$

16. $S(-7, -3), T(-4, -5)$ 17. $C\left(\frac{3}{2}, 1\right), D\left(\frac{5}{2}, \frac{1}{2}\right)$ 18. $G\left(3, \frac{7}{2}\right), H\left(3, \frac{11}{2}\right)$

Mixed Review

19. Find the LCD of $\frac{5}{9x}$ and $\frac{2}{3x^3}$.   (Lesson 4-8)

20. Simplify the expression $2(m + 3c) - 5c$.   (Lesson 1-4)

21. If $r = -2$, what is the value of s in the equation $2r + s = 6$?   (Lesson 8-5)

22. Graph the equation $y = \frac{1}{2}x - 1$.   (Lesson 8-6)

Challenge

Graph the line that contains the given point and has the given slope.

23. $M(0, 0); \frac{1}{2}$ 24. $R(-1, 2); -\frac{1}{4}$ 25. $T(3, 0); 0$

26. $D(-3, 1); 2$ 27. $K(2, 5);$ no slope   28. $E(3, -4); -1$

Applications

29. **Science** An inclined plane is a slanted surface, which may be used for raising objects to higher places. Find the slope of the inclined plane pictured at the right.

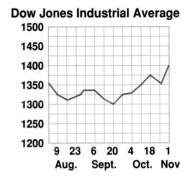
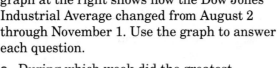

3 m

0.3 m

30. **Surveying** A surveyor talks about slope in terms of *rise* and *run*. Use a surveyor's terms to define the slope of a line.

Connection

31. **Statistics** Line graphs are used to show changes in data over a period of time. The graph at the right shows how the Dow Jones Industrial Average changed from August 2 through November 1. Use the graph to answer each question.

Dow Jones Industrial Average

a. During which week did the greatest increase occur?

b. How many weeks did the average decrease?

c. Name a week in which the average did not change.

Critical Thinking

32. Explain how you can tell from just looking at the graph of a linear equation whether its slope is positive or negative.

Wrap-Up

33. Find a coordinate graph in the newspaper. Explain how your knowledge of slope can help you analyze the graph.

8-9   Intercepts

Objective:
Graph a linear equation by using the *x*-intercept and the *y*-intercept.

Key Terms:
x-intercept
y-intercept

When a defensive back intercepts a pass in football, he generally crosses the path of the ball on the way to the receiver.

In mathematics, the graph of a linear equation may cross either the *x*-axis, the *y*-axis, or both axes. The **x-intercept** is the *x*-coordinate of the point where the graph crosses the *x*-axis. The **y-intercept** is the *y*-coordinate of the point where the graph crosses the *y*-axis.

The equation is $y = -2x + 4$.

The ordered pair for the point on the *y*-axis is (0, 4). So the *y*-intercept is 4.

The ordered pair for the point on the *x*-axis is (2, 0). So the *x*-intercept is 2.

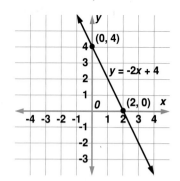

The *y*-intercept is the value of *y* when *x* is 0. The *x*-intercept is the value of *x* when *y* is 0. The *y*-intercept and *x*-intercept can be used to graph a linear equation.

Example

1   **Graph $y = \frac{1}{2}x - 1$ by using the *y*-intercept and the *x*-intercept.**

How many points determine a line?

Find the *y*-intercept.

Find the value of y when x = 0.

$$y = \frac{1}{2}x - 1$$
$$y = \frac{1}{2}(0) - 1$$
$$y = -1$$

The *y*-intercept is -1.
The ordered pair is (0, -1).

Find the *x*-intercept.

Find the value of x when y = 0.

$$y = \frac{1}{2}x - 1$$
$$(0) = \frac{1}{2}x - 1$$
$$0 + 1 = \frac{1}{2}x - 1 + 1$$
$$1 = \frac{1}{2}x$$
$$2 = x$$

The *x*-intercept is 2.
The ordered pair is (2, 0).

Graph the ordered pairs (0, -1) and (2, 0). Draw the line that contains the two points.

To test, you can graph a third point to see if it is on the line. For example, try $x = 4$. Then $y = \frac{1}{2}(4) - 1$ or 1. The ordered pair (4, 1) is on the line.

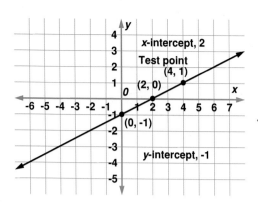

Checking for Understanding

Communicating Algebra

1. Graph the line that has an x-intercept of 3 and a y-intercept of -2.
2. Graph the line that has an x-intercept of -4 and a y-intercept of 1.
3. An x-intercept is 3. State the ordered pair describing its coordinates.
4. A y-intercept is -5. State the ordered pair describing its coordinates.

Guided Practice

Find the x-intercept and y-intercept for the graph of each equation.

5. $y = x - 3$
6. $y = x + 5$
7. $y = -x - 2$
8. $y = 2x + 1$
9. $y = 3x - 4$
10. $y = -5x + 6$
11. $y = 2x - \frac{2}{3}$
12. $y = 2 - 3x$
13. $y = 7x - 4.2$

Exercises

Independent Practice

Use the x-intercept and the y-intercept to graph each equation.

14. $y = x + 2$
15. $y = x - 1$
16. $y = x + 3$
17. $y = 2x + 3$
18. $y = 3x - 6$
19. $y = -5x + 10$
20. $y = \frac{1}{2}x - 5$
21. $y = 3 - 0.5x$
22. $y = \frac{1}{3}x - 2$

Mixed Review

23. Solve the equation $z = 37.2 \div 4$. (Lesson 6-7)
24. Graph the equation $2x + y = 1$. (Lesson 8-6)
25. Find the slope of the line that contains the points $P(7, -3)$ and $Q(-1, 4)$. (Lesson 8-8)

Application

26. **Business**   Suppose you have an after-school baby-sitting service. You charge a flat fee of $10 per week plus a fee of $3 per hour. The equation $y = 3x + 10$ represents your earnings per week for one child. Graph this equation. Describe the y-intercept.

Critical Thinking

27. Explain why you cannot graph the equation $y = 2x$ by using intercepts only.

Wrap-Up

28. Draw a graph of a line that has an x-intercept, but no y-intercept.

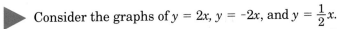

Exploration

Slope

Materials: graphing calculator

In this Exploration, you will investigate several families of graphs to determine their characteristics.

▶ Consider the graphs of $y = 2x$, $y = -2x$, and $y = \frac{1}{2}x$.

 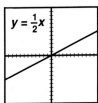

Alone the graphs may not appear to be related. But let's look at them together to discover some interesting patterns..

▶ Consider the graphs of $y = 2x$ and $y = -2x$.
They appear to have the same steepness but in different directions.

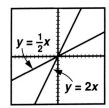

Your Turn: **Graph $y = 3x$ and $y = -3x$.**
Then graph $y = 1x$ and $y = -1x$.
Describe the graphs.

▶ Now consider the graphs of $y = 2x$ and $y = \frac{1}{2}x$.
They appear to have different steepness but in the same direction.

Your Turn: **Graph $y = 3x$ and $y = \frac{1}{3}x$.**
Then graph $y = 4x$ and $y = 2x$.
Describe the graphs.

Analysis

1. Write a sentence that explains the difference between the graphs of $y = 10x$ and $y = -10x$.
2. Write the equation of a line whose graph is between the graph of $y = 5x$ and $y = 1x$.

Extension

3. Graph $y = 2x$, $y = 2x + 1$, $y = 2x + 2$, and $y = 2x + 3$. How are the graphs the same? How do they differ?

8-10 Systems of Equations

Objective:
Solve systems of linear
equations by graphing.

Key Term:
system of equations

The owner of any business, big or small, knows that in order to stay in
business, the income from goods or services that are sold needs to
balance with the cost of producing the goods or services. This is
sometimes called the *break-even point*.

The break-even point can be found graphically. One equation is written
to represent the income. Another equation is written to represent the
cost. Then both equations are graphed. The point where the graphs
intersect is the *break-even point*.

In mathematics, two equations with the same two variables form a
system of equations. The solution of the system of equations is any
ordered pair that is a solution of both equations. One way to solve a
system of equations is by graphing.

Examples

1 **Use the graph at the right to
solve the system of equations
$y = x + 2$ and $y = 3x$.**

The graphs intersect at point A.
Because point A is on the graph
of each equation, the ordered
pair for A is a solution of each
equation. Thus, the solution of
the system is $(1, 3)$.

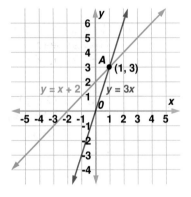

To check the solution, replace
y in both equations with 3
and replace x in both equations
with 1.

Check: $y = x + 2$ $y = 3x$

$3 \overset{?}{=} 1 + 2$ $3 \overset{?}{=} 3(1)$

$3 = 3$ ✓ $3 = 3$ ✓

2 **Solve $x + y = 3$ and $y = 2x$ by
graphing.**

Use intercepts to graph $x + y = 3$.

$x + 0 = 3$ Find the x-intercept.

$x = 3$

The ordered pair is $(3, 0)$.

$0 + y = 3$ Find the y-intercept.

$y = 3$

The ordered pair is $(0, 3)$.

Graph the points and draw the
line that contains them.

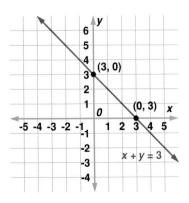

Use ordered pairs to graph $y = 2x$.

x	2x	y	(x, y)
-1	2(-1)	-2	(-1, -2)
0	2(0)	0	(0, 0)
1	2(1)	2	(1, 2)
2	2(2)	4	(2, 4)

Graph the points and draw the line that contains them.

The lines intersect at the point (1, 2). Therefore, the solution of the system of equations is (1, 2).

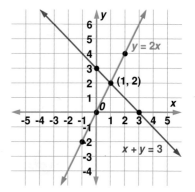

To check the solution, replace y in both equations with 2 and replace x in both equations with 1.

Check: $x + y = 3$ $\qquad$ $y = 2x$

$\qquad\qquad$ $1 + 2 \overset{?}{=} 3$ $\qquad$ $2 \overset{?}{=} 2(1)$

$\qquad\qquad\qquad$ $3 = 3$ ✓ $\qquad$ $2 = 2$ ✓

Checking for Understanding

1. In business, the ___?___ is where the costs of goods and services sold equals income.

2. Two equations with the same two variables form a ___?___ .

3. Explain why the ordered pair (2, -2) is not a solution of the system of equations $3x + 2y = 10$, $x + y = 0$.

4. Which ordered pairs are solutions of the equation $x + 2y = 7$?
   a. (3, 2) $\qquad$ **b.** (6, 1) $\qquad$ **c.** (9, -1) $\qquad$ **d.** (-1, 4)

Guided Practice

The graphs of several equations are shown at the right. State the solution of each system of equations.

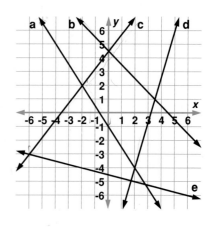

5. a and c

6. b and c

7. c and e

8. a and d

9. b and d

10. a and the y-axis

11. b and the x-axis

Exercises

Independent Practice

Use a graph to solve each system of equations.

12. $y = 3x + 5$
    $y = x + 5$

13. $y = x - 2$
    $y = 4x + 1$

14. $y = 4x$
    $y = -x$

15. $y = x - 5$
    $y = -2x + 4$

16. $x + y = 6$
    $y = 2x$

17. $y - 4x = 3$
    $y = x$

18. $x = y + 7$
    $2x + y = 2$

19. $y = \frac{1}{2}x$
    $x + 2y = 0$

20. $2x + y = 8$
    $x - y = 4$

21. $x + 3y = 5$
    $-x + y = 3$

22. $3x + y = 6$
    $4x + y = 7$

23. $5y + 4x = 12$
    $3y - 4x = 4$

Mixed Review

24. Solve the equation $7a + 2 = (-4)(a - 6)$.   (Lesson 7-6)

25. **Science**   Find the slope of the inclined plane pictured at right. (Lesson 8-8)

2 ft
4 ft

26. Graph $y = -3x + 2$ by using the x-intercept and the y-intercept.   (Lesson 8-9)

Application

27. **Business**   The Murphy Company, a leading gadget maker, has fixed costs of $900 per week. Each item produced by the company costs $2 to manufacture and can be sold for $5. If x is the number of gadgets produced each week, the cost of producing them can be represented by the equation $y = 900 + 2x$. The weekly income from selling the gadgets can be represented by $y = 5x$. Solve this system of equations by graphing. What is the company's *break-even point*?

Critical Thinking

28. In this lesson, all of the systems have one solution. Is it possible for a system of two linear equations to have no solution or more than one solution? If so, draw a graph to show each situation.

Wrap-Up

29. Write a system of equations whose solution is (5, 8).

Team Problem Solving

Sometimes a graph is used to describe a series of events. This graph shows the height of a football above the ground from the time the ball was snapped for a field goal attempt.

Over which time periods did these events occur?
(a) The holder received the snap.
(b) The ball was increasing in height.
(c) The ball was kicked.
(d) The ball lands in the stands behind the goal posts.

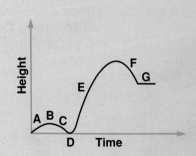

Draw a graph that shows how the speed of a skier changes as she waits for the ski lift, goes up the mountain, waits her turn, and skis down the mountain.

Graphing Inequalities

Materials: TI-81 graphing calculator

In this Exploration, you will use a graphing calculator to investigate the graphs of inequalities.

▶ Consider the inequality $y < x + 4$. The graph of the inequality must show all of the points whose ordered pairs satisfy the inequality $y < x + 4$. Type the following keystroke sequence into your calculator.

Notice that the graph is a shaded region. This indicates that all ordered pairs satisfy the inequality $y < x + 4$.

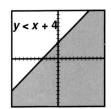

Your Turn: **How does this differ from the graph of $y = x + 4$?**

▶ Now consider the inequality $y > 2x - 1$. Type the following keystroke sequence into your calculator.

Once again, this graph is a shaded region. All ordered pairs satisfy the inequality $y > 2x - 1$.

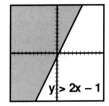

Your Turn: **How does the graph of $y > 2x - 1$ differ from the graph of $y < x + 4$?**

Application

1. The following keystroke sequence is used to graph the inequality $y \; \bullet \; x + 3$. Graph the inequality, then replace the ● with $<$ or $>$.

| 2nd | QUIT | 2nd | DRAW | 1 | ENTER | 2nd | DRAW | 7 |

| X|T | + | 3 | ALPHA | (,) | 1 | 0 |) | ENTER | .

8-11   Graphing Inequalities

Objective:
Graph linear inequalities.

Key Term:
boundary

The timberline separates a mountain into two regions. The region below the timberline can sustain tree growth. Above the timberline, the climate is too severe for tree growth.

The graph of a linear equation like $y = x + 1$ separates the coordinate plane into two regions, one above the line and one below the line. The line is called the **boundary** of the two regions. Point $A(2, 3)$ is on the line and $(2, 3)$ is a solution of $y = x + 1$.

The graph of the inequality $y > x + 1$ is the region *above* the boundary line. Point $B(2, 4)$ is in this region and $(2, 4)$ is a solution of $y > x + 1$.

The graph of the inequality $y < x + 1$ is the region *below* the boundary line. Point $C(4, 0)$ is in this region and $(4, 0)$ is a solution of $y < x + 1$.

How is graphing inequalities on the coordinate plane similar to the way in which you graph inequalities on a number line?

Since the graph of an inequality must show *all* of the points whose coordinates are solutions of the inequality, you shade a region of the coordinate plane to graph an inequality. A dashed line is used when the boundary line is *not* part of the graph. A solid line is used when the boundary line is part of the graph. Two examples are shown.

FYI

At the timberline, or tree line, forests are replaced by low shrubs or no greenery at all. Temperature has the greatest effect on the location of the timberline, but soil, moisture, drainage, and sunlight are also factors.

$y < -2x + 2$

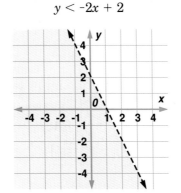

$y \geq x - 3$

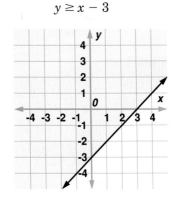

Example

1 Graph $y > 2x - 3$.

First graph $y = 2x - 3$. Draw a dashed line because the boundary line is not part of the graph of $y > 2x - 3$.

The graph is the region above the boundary. Shade this region.

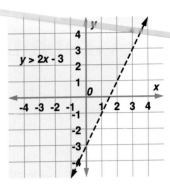

Check: Choose a point in the region. Substitute its coordinates for x and y in the inequality. The origin is usually chosen because the coordinates of the ordered pair $(0, 0)$ are easy to work with.

$$y > 2x - 3$$

$$0 \overset{?}{>} 2(0) - 3 \quad \text{Replace } y \text{ with 0 and } x \text{ with 0.}$$

$$0 > \text{-}3 \quad \text{✓} \quad \text{The inequality is true.}$$

Since the resulting inequality is true, the shading is correct.

Checking for Understanding

Communicating Algebra

1. A line that separates the coordinate plane into two regions is called a ___?___ line.

2. Explain when a dashed line is used for the boundary line of the graph of an inequality.

If each inequality were graphed, would the boundary line be dashed or solid?

3. $y > x + 3$ 4. $y \leq 3$ 5. $y < 5x$

Guided Practice

Copy each graph. Shade the region that represents the solution of the inequality.

6.
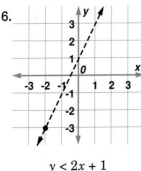

$$y < 2x + 1$$

7.
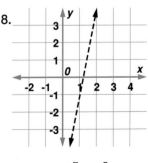

$$y \geq -\tfrac{3}{4}x - \tfrac{1}{2}$$

8.

$$y < 5x - 6$$

Exercises

Independent Practice

Graph each inequality.

  9. $y > x + 2$ **10.** $y \le 2x$ **11.** $y > -3x - 3$

12. $y > -4x$ **13.** $y \le -1$ **14.** $y \ge -\frac{1}{2}x + 3$

15. $y < \frac{1}{4}x - 2$ **16.** $y < -x - 1.5$ **17.** $y > -2x + 1$

18. $y < \frac{1}{3}x$ **19.** $y \ge x + 0.5$ **20.** $y \le 1.5x - 0.5$

Mixed Review

21. Name the additive inverse of $-\frac{2}{3}$.   (Lesson 5-1)

22. Find the x-intercept and y-intercept for the graph of the equation $y = -x - 3$. (Lesson 8-9)

23. Graph $y = -3x$ and $y = x - 8$. Then find the solution of the system of equations.   (Lesson 8-10)

Application

24. **Consumer Awareness**   Suppose you are shopping for cassettes and CDs at Music City. Cassettes cost $7, CDs cost $14, and you have $28 to spend. First list all the combinations of cassettes and CDs you can purchase. Then graph the inequality $7c + 14d \le 28$ where c is the number of cassettes and d is the number of CDs.

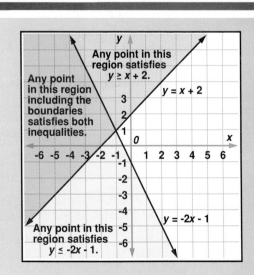

Critical Thinking

25. **Make Up a Problem**   Write an inequality whose graph has a solid boundary line.

Wrap-Up

26. Explain how you can check to see if you have shaded the correct region in the graph of an inequality.

Challenge

Systems of Inequalities

Consider the following system of inequalities.

$$y \ge x + 2$$
$$y \le -2x - 1$$

To solve this system, you find the ordered pairs that satisfy *both* inequalities. One way is to graph each inequality and find the overlap of the two graphs.

Solve each system of inequalities by graphing.

1. $y > x - 3$ **2.** $y > x + 2$
    $y \le -1$ $y < x - 3$

Review

Language and Concepts

Choose the letter of the correct word or words to complete each statement.

1. __?__ are graphed on a number line.

2. __?__ are graphed on a coordinate plane.

3. In the coordinate plane, the axes intersect at the __?__ .

4. When you graph an equation, all of the ordered pairs for the points on the line are __?__ of the equation.

5. The coordinate of a point tells its distance and __?__ from the zero point of the line.

6. The steepness of a line is called its __?__ .

7. Two equations with the same two variables form a __?__ of equations.

8. The coordinate axes separate a plane into four __?__ .

9. In a coordinate system, the __?__ is a horizontal line.

10. An equation whose graph is a straight line is called a __?__ equation.

a. direction
b. linear
c. numbers
d. ordered pairs
e. origin
f. quadrants
g. slope
h. solutions
i. system
j. x-axis
k. y-axis

Skills

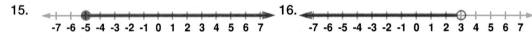

Solve each equation and graph the solution. (Lesson 8-1)

11. $3y = -12$

12. $x + 6 = 11$

13. $h - 3 = -12$

14. $3a - 7 = -10$

Write an inequality that describes each graph. (Lesson 8-2)

15.

16.

17.

18.

Solve each inequality and graph the solution. (Lesson 8-2)

19. $2a \geq -4$

20. $4b - 5 < 7$

21. $2x + 8 \geq -4$

22. $a - 4 < -8$

On graph paper, draw coordinate axes. Then graph and label each point.
(Lessons 8-3, 8-4)

23. $A(6, 0)$

24. $P(1, -6)$

25. $C(-2, 3)$

26. $D(-3, -2)$

Find four solutions for each equation. (Lesson 8-5)

27. $y = \frac{3}{2}x$

28. $y = -4x - 3$

29. $y = 5x - 3$

30. $y = 6x - 5$

Graph each equation. (Lesson 8-6)

31. $y = 3x - 1$ **32.** $y = -5x - 2$ **33.** $y = -3x - 4$ **34.** $y = -4x + 18$

Find the slope of the line that contains each pair of points. (Lesson 8-8)

35. $A(-3, 4)$, $B(-2, -2)$ **36.** $C(-1, 0)$, $D(5, -1)$ **37.** $F(6, 7)$, $G(-4, 3)$

Use the x-intercept and y-intercept to graph each equation. (Lesson 8-9)

38. $y = x - 6$ **39.** $y = -5 + x$ **40.** $y = -3 + x$ **41.** $y = -2x - 9$

Use a graph to solve each system of equations. (Lesson 8-10)

42. $y = x$
    $y = 2 - x$

43. $x + y = 6$
    $x - y = 2$

44. $y = x - 1$
    $x + y = 11$

Graph each inequality. (Lesson 8-11)

45. $y < x - 2$ **46.** $y \geq -2x + 1$ **47.** $y > \frac{1}{2}x$ **48.** $y \leq -x + 8$

Applications and Problem Solving

Solve by using a graph. Assume the rate is constant. (Lesson 8-7)

49. A cricket begins chirping when the temperature is about 37° Fahrenheit. When the temperature is 70° Fahrenheit, the cricket chirps about 132 times per minute. What is the approximate temperature when the cricket chirps 185 times per minute?

ortfolio Suggestion

Select some of your work from this chapter that shows how you used a calculator or computer. Place it in your portfolio.

Curriculum Connection

- **Geography**   Find the latitude and longitude of your town. Write your answer as an ordered pair.

- **Meteorology**   An isobar is a curved line on a weather map that connects points with the same air pressure. What does an isotherm show?

Read More About It

Arnold, Caroline. *Charts and Graphs: Fun, Facts, and Activities.*

Catherall, Ed. *Investigating Mathematics: Graphs.*

Haber, Louis. *Black Pioneers of Science and Invention.*

Test

Solve each equation and graph the solution.

1. $y - 5 = -7$ 2. $a + 3 = 2$ 3. $-13 = h - 11$ 4. $3b - 3 = 6$

Solve each inequality and graph the solution.

5. $x + 8 \geq -3$ 6. $y - 1 \leq -6$ 7. $-2y \geq -3$ 8. $\frac{y}{5} \geq -2$

Use the grid at the right to find the ordered pair for each labeled point.

9. A 10. C

11. D 12. G

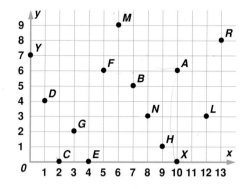

On graph paper, draw coordinate axes. Then graph and label each point.

13. $A(8, 9)$ 14. $B(-8, 9)$

15. $C(8, -9)$ 16. $D(-8, -9)$

17. $E(7, -6.5)$ 18. $F(5, -2.5)$

19. $G(0, -7)$ 20. $H(-5, 0)$

Find four solutions for each equation.

21. $y = x - 6$ 22. $y = x + 7$ 23. $y = 3x$ 24. $y = -4x$

Graph each equation.

25. $y = -3x + 1$ 26. $y = 3 - 4x$ 27. $y = -7x - 8$

28. $y = 9 - 8x$ 29. $y = 12 - 3x$ 30. $y = 5x$

Find the slope of the line that contains each pair of points.

31. $A(2, -3), B(-1, 0)$ 32. $C(-5, 6), D(-7, 8)$ 33. $E(4, 3), F(-5, 4)$

Use the x-intercept and y-intercept to graph each equation.

34. $y = -x + 1$ 35. $y = -x - 2$ 36. $y = 2 - \frac{1}{2}x$

Use a graph to solve each system of equations.

37. $x + y = 8$ 38. $x + 2y = 6$
    $x - y = 8$ $-x + 3y = 4$

Graph each inequality.

39. $y \leq \frac{1}{2}x$ 40. $y < 3 - x$

BONUS

Write equations, list solutions in a table, and draw a graph to illustrate the following relationship: The sum of two numbers is 42. Their difference is 18.

Academic Skills Test
Cumulative, Chapters 1-8

1. $5(6 + 8) =$

   A $56 + 58$
   B $5 \cdot 6 + 5 \cdot 8$
   C $5 \cdot 6 \cdot 5 \cdot 8$
   D $5 \cdot 6 + 8$

2. Which equation is equivalent to $y - 24 = -12$?

   A $y = -12$
   B $y - 2 = -1$
   C $y - 24 + 24 = -12 - 24$
   D $y - 24 + 24 = -12 + 24$

3. Which is equivalent to $4^3 \cdot 3^2$?

   A $12 \cdot 6$
   B $4 \cdot 4 \cdot 4 \cdot 3 \cdot 3$
   C $4 \cdot 4 \cdot 3 \cdot 3$
   D $3 \cdot 3 \cdot 3 \cdot 3 \cdot 2 \cdot 2 \cdot 2$

4. The mileage reading on the Garcia's car was 256.8 before they left for vacation. When the family returned, the reading was 739.4. How many miles did the Garcia family travel on their vacation?

   A 482.6 mi   C 583.5 mi
   B 483.6 mi   D 996.2 mi

5. Alejandra's car averages 30.5 miles per gallon. The gas tank holds 13.2 gallons of fuel. The best estimate of the distance she can drive on one tankful is—

   A 200 mi   C 400 mi
   B 300 mi   D 500 mi

6. The diameter of a grain of sand is about 0.00135 meters. How is this measure expressed in scientific notation?

   A 135×10^{-5} C 1.35×10^{-3}
   B 13.5×10^{-4} D 1.35×10^{-2}

7. If $4x - 8 = 28$, what is the value of x?

   A 5   C 20
   B 9   D 36

8. The sum of two consecutive even integers is less than 20. Which inequality can be used to find the integers?

   A $x + x + 2 < 20$
   B $x + 2 < 20$
   C $x + (x + 2) > 20$
   D $2x > 20$

9. Which number line shows the solution of the inequality $-2a + 3 \le -3$?

   A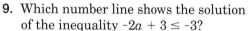

   B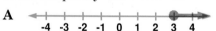

   C

   D

10. Which point is the y-intercept for the graph of $y = x + 2$?

    A $(-2, 0)$ C $(2, 0)$
    B $(0, -2)$ D $(0, 2)$

CHAPTER 9

CHAPTER OBJECTIVES

In this chapter you will learn to:

- [ ] write equivalent expressions for ratios, decimals, and percents
- [ ] use ratios and proportions to solve problems
- [ ] solve equations involving percents
- [ ] solve problems involving percents
- [ ] solve problems by making a table

Proportion and Percent

The chambered nautilus is one of the clearest examples of a spiral found in nature. In mathematics it is related to a very special rectangle called a *golden rectangle.*

A golden rectangle is visually pleasing, being neither too fat nor too skinny. It is somewhere between a square and two squares side by side. In fact, the ratio of the width to the length is 1 to about 1.618.

A golden rectangle has a unique characteristic. If a square is cut away from one end of a golden rectangle, the remaining rectangle is also a golden rectangle. This process can be repeated endlessly.

Can you find the spiral within the golden rectangle?

Computer Connection

Class Project

The numbers of the Fibonacci sequence are also related to the golden rectangle.

Write a computer program that calculates the ratio of each Fibonacci number to the next number in the sequence.

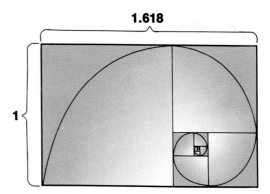

9-1 Ratios and Rates

Objective:
Write ratios as fractions in simplest form and determine unit rates.

Key Terms:
ratio
rate
unit rate

Is 24 to 32 the same as 32 to 24? Why or why not?

During a recent year it rained on 24 of the first 32 weekends in Columbus, Ohio. You can compare these two numbers using a ratio. A **ratio** is a comparison of two numbers by division. The ratio that compares 24 to 32 can be written as follows.

24 to 32   24:32   24 out of 32 $\frac{24}{32}$

A common way to express a ratio is as a fraction in simplest form or as a decimal.

$$\frac{24}{32} = \frac{3}{4}$$
÷ 8

The GCF of 24 and 32 is 8.

or

24 ÷ 32 = 0.75

It rained 3 out of every 4 weekends. Another way to state this fact is that it rained three fourths or 0.75 of the weekends.

In the situation described above, you compared 24 weekends to 32 weekends. Notice that the units are the same. Often it is necessary to compare two quantities with different units. For example, $\frac{187 \text{ miles}}{5.5 \text{ gallons}}$ compares the number of miles a vehicle is driven to the number of gallons of gasoline used. A ratio of two measurements having different units is called a **rate.**

To find the number of miles the vehicle travels on 1 gallon of gasoline, simplify the rate so that the denominator is 1 unit. A rate with a denominator of 1 is called a **unit rate.**

Example

1 **Express 187 miles on 5.5 gallons of gas as a unit rate.**

$$\frac{187 \text{ miles}}{5.5 \text{ gallons}} = \frac{\blacksquare}{1 \text{ gallon}}$$
÷ 5.5

Divide the numerator and denominator by 5.5.

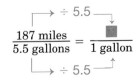

187 ÷ 5.5 = 34

The unit rate is 34 miles per gallon.

Checking for Understanding

Communicating
Algebra
1. In your own words, explain the difference between a ratio and a rate.
2. Write a ratio about the students in your class.

Guided
Practice
Express each ratio as a fraction in simplest form.
3. 5 out of 7 people
4. 20 out of 25 free throws
5. 196:7
6. 2 cups to 12 cups
7. 12 out of 20 people
8. 13 to 91

Express each ratio as a rate.
9. 100 miles in 4 hours
10. 24 pounds lost in 8 weeks

Exercises

Independent
Practice

?
ESTIMATION
MENTAL MATH
CALCULATOR
PAPER/PENCIL

Express each ratio as a fraction in simplest form.
11. 11 out of 121
12. 49:77
13. 72 to 24
14. 18 out of 27
15. 84:156
16. 65 to 105
17. 165:200
18. 117 to 351
19. 96:24
20. 118 out of 354
21. 27 to 15
22. 21 out of 28

Express each ratio as a unit rate.
23. 299.2 miles on 8.5 gallons
24. 378.4 miles in 8 hours
25. $105 for 15 tickets
26. $7.70 for 11 pounds
27. 12 inches of rain in 5 hours
28. 8 meters in 10 seconds
29. $25,000 per 100 employees
30. 96¢ per dozen

Mixed Review
31. Write an open sentence that has 7 as its solution.   (Lesson 1-5)
32. A hamburger in the cafeteria costs 35 cents less than twice the price of a slice of pizza. If a hamburger is $1.25, how much is a slice of pizza?   (Lesson 7-3)
33. Use a graph to solve the system of equations: $y = -3x, y = -2x + 1$.   (Lesson 8-10)
34. Graph the inequality $y \le 2.5x - 1.5$.   (Lesson 8-11)

Applications
35. **Consumer Awareness**   Unit prices are shown for products at grocery stores. A 14-ounce box of cereal costs $2.54. What is the unit price, that is, the cost per ounce?

36. **Aviation**   The first non-stop around-the-world flight started and ended at Carswell Air Force Base, Texas. The flight of 23,452 miles was made in 94 hours. What was the average rate in miles per hour?

Critical
Thinking
37. The Metro-Liner run from Baltimore, Maryland to Wilmington, Delaware, is regularly scheduled to travel the distance of 68 miles in 40 minutes. What is the unit rate in miles per hour?

Wrap-Up
38. **Research**   Find two examples of ratios and two examples of rates in a newspaper.

Algebra in Action-Art

The Golden Ratio

We see a good estimate of the golden ratio every time we use a 3×5 or a 5×8 index card.

$$\frac{3}{5} = 0.6 \qquad\qquad \frac{5}{8} = 0.625$$

The golden ratio lies somewhere between 0.6 and 0.625, at about 0.618.

The golden ratio has influenced mathematicians, artists, and architects for over 4000 years. One of the earliest examples of the golden ratio in architecture is the Great Pyramid of Gizeh in Egypt, built about 2600 B.C. The ratio of the altitude to the length of one side of the base is about 5 to 8 or 0.625.

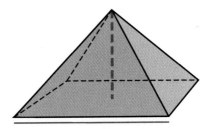

The great painter, Leonardo da Vinci, called the golden ratio the *divine proportion*. It is visible in the outside dimensions of the *Mona Lisa* as well as in the composition of the painting.

1. **Research** Find out how the golden ratio was used in the Parthenon in Athens, Greece, and in the United Nations Headquarters in New York, New York.

2. **Research** Write a report about Leonardo da Vinci.

9-2   Proportions

Objectives:
Determine if a pair of ratios form a proportion. Solve proportions.

Key Terms:
proportion
cross products

Have you ever heard a statement like this on TV?

> 2 out of 3 Americans approve of the job the President is doing.

The ratio *2 out of 3* can be expressed as the fraction $\frac{2}{3}$. The statement means that $\frac{2}{3}$ of the people surveyed approve of the job the President is doing.

Hopefully, more than 3 people were surveyed. Suppose 300 people were surveyed. You can estimate that $\frac{2}{3}$ of 300, or 200 people, approve. Notice that the two fractions shown below are equivalent.

$$\frac{2}{3} = \frac{200}{300}$$

An equation stating that two ratios are equivalent is called a **proportion.** One way to determine if two ratios form a proportion is to check their **cross products.** In the proportion shown at the right, $2 \cdot 300$ and $3 \cdot 200$ are cross products.

$$\frac{2}{3} = \frac{200}{300}$$

$$2 \cdot 300 = 3 \cdot 200$$

$$600 = 600$$

Property of Proportions	In words: The cross products of a proportion are equal.
	In symbols: If $\frac{a}{b} = \frac{c}{d}$, then $ad = bc$. If $ad = bc$, then $\frac{a}{b} = \frac{c}{d}$.

Examples

Use cross products to determine whether each pair of ratios forms a proportion.

1 $\frac{2}{3}, \frac{8}{12}$

$$2 \cdot 12 \stackrel{?}{=} 3 \cdot 8 \qquad \text{Property of Proportions}$$

$$24 = 24$$

So, $\frac{2}{3} = \frac{8}{12}$.

2 $\frac{3.5}{8}, \frac{2.5}{4.6}$

8 ⊗ 2.5 ⊜ 20

3.5 ⊗ 4.6 ⊜ 16.1

$20 \neq 16.1$ So, $\frac{3.5}{8} \neq \frac{2.5}{4.6}$.

Cross products can also be used to solve proportions.

Examples

Solve each proportion.

3 $\dfrac{d}{24} = \dfrac{15}{60}$

$\qquad d \cdot 60 = 24 \cdot 15$ Write the cross products.

$\qquad\quad 60d = 360$ $24 \cdot 15 = 360$

$\qquad\quad \dfrac{60d}{60} = \dfrac{360}{60}$ Divide each side by 60.

$\qquad\qquad d = 6$

The solution is 6.

4 $\dfrac{12}{1.3} = \dfrac{5}{c}$

$\qquad 12c = 1.3 \cdot 5$

$$\boxed{1.3}\;\boxed{\times}\;\boxed{5}\;\boxed{\div}\;\boxed{12}\;\boxed{=}\;\boxed{0.541666666}$$

Therefore, c is about 0.54.

Mental Math Hint

Sometimes you can solve a proportion mentally by using equivalent fractions.

$$\dfrac{2}{3} = \dfrac{x}{27}$$

THINK: $3 \times 9 = 27$
$\qquad\quad\; 2 \times 9 = 18$

So, $x = 18$.

How is this similar to the paper-and-pencil method?

Checking for Understanding

Communicating Algebra

1. Name the property that enables you to determine whether two ratios are equivalent.

2. Explain how to determine whether two ratios are equivalent.

3. Write the cross products for $\dfrac{2}{5} = \dfrac{0.02}{0.09}$.

4. Explain how to solve $\dfrac{x}{2.5} = \dfrac{3}{4}$ with a calculator.

5. Explain how to solve $\dfrac{2}{5} = \dfrac{m}{35}$ mentally.

Guided Practice

Replace each ▮ with = or ≠ to make a true statement.

6. $\dfrac{2}{3}$ ▮ $\dfrac{8}{12}$ 7. $\dfrac{6}{15}$ ▮ $\dfrac{3}{7}$ 8. $\dfrac{8}{9}$ ▮ $\dfrac{16}{17}$ 9. $\dfrac{4}{5}$ ▮ $\dfrac{12}{15}$

Solve each proportion.

10. $\dfrac{1}{5} = \dfrac{x}{35}$ 11. $\dfrac{1}{3} = \dfrac{6}{s}$ 12. $\dfrac{3}{a} = \dfrac{18}{24}$ 13. $\dfrac{m}{3} = \dfrac{14}{21}$

Exercises

Independent Practice

ESTIMATION
MENTAL MATH
CALCULATOR
PAPER/PENCIL

Replace each ▮ with = or ≠ to make a true statement.

14. $\dfrac{6}{4}$ ▮ $\dfrac{21}{14}$ 15. $\dfrac{8}{6}$ ▮ $\dfrac{28}{22}$ 16. $\dfrac{12}{8}$ ▮ $\dfrac{40}{28}$ 17. $\dfrac{7}{14}$ ▮ $\dfrac{20}{45}$

18. $\dfrac{1.3}{2}$ ▮ $\dfrac{16}{25}$ 19. $\dfrac{2.1}{3.5}$ ▮ $\dfrac{5}{7}$ 20. $\dfrac{1.5}{2}$ ▮ $\dfrac{1.8}{2.4}$ 21. $\dfrac{0.5}{1.5}$ ▮ $\dfrac{0.03}{0.09}$

22. $\dfrac{0.8}{1.6}$ ▮ $\dfrac{4.3}{9}$ 23. $\dfrac{1.3}{2}$ ▮ $\dfrac{3.055}{4.7}$ 24. $\dfrac{1.3}{2}$ ▮ $\dfrac{16.25}{25}$ 25. $\dfrac{8.6}{25.3}$ ▮ $\dfrac{1}{3}$

Solve each proportion.

26. $\frac{3}{7} = \frac{m}{49}$

27. $\frac{8}{6} = \frac{x}{27}$

28. $\frac{2}{5} = \frac{k}{35}$

29. $\frac{3}{39} = \frac{y}{13}$

30. $\frac{r}{3} = \frac{8}{15}$

31. $\frac{1}{2} = \frac{s}{7}$

32. $\frac{8}{6} = \frac{z}{14}$

33. $\frac{18}{12} = \frac{24}{k}$

34. $\frac{s}{9.6} = \frac{7}{1.6}$

35. $\frac{7}{16} = \frac{x}{4.8}$

36. $\frac{5.1}{1.7} = \frac{7.5}{a}$

37. $\frac{1.6}{2.4} = \frac{3.4}{p}$

Mixed Review

38. Solve the equation $d = (\text{-}7)(3)(\text{-}1)$.   (Lesson 2-7)

39. Solve the equation $7x + 4(x - 3) = 9x + 2$.   (Lesson 7-5)

40. What do x and y represent in the ordered pair (x, y)?   (Lesson 8-3)

41. Express the ratio 36:27 as a fraction in simplest form.   (Lesson 9-1)

42. Express the ratio $\frac{185 \text{ calories}}{5 \text{ grams}}$ as a unit rate.   (Lesson 9-1)

Challenge

Solve each proportion.

43. $\frac{2a}{5} = \frac{12}{15}$

44. $\frac{0.28}{4n} = \frac{1.4}{4}$

45. $\frac{3}{2} = \frac{x + 1}{8}$

46. $\frac{1.5}{h - 3} = \frac{3}{14}$

Applications

47. **Cooking**   A recipe calls for 3 cups of flour for 48 cookies. How much flour is needed for 72 cookies? Use the proportion $\frac{3}{48} = \frac{c}{72}$.

48. **Farming**   A 5-acre field has a yield of 140 bushels of wheat. What yield can be expected for a 42-acre field? Use the proportion $\frac{140}{5} = \frac{b}{42}$.

49. **Consumer Awareness**   Lauren's car averages 33 miles for each gallon of gasoline. How many gallons are needed for a trip of 313.5 miles? Use the proportion $\frac{33}{1} = \frac{313.5}{g}$.

Critical Thinking

50. If three bakers can prepare 15 cakes in 1 hour, how long will it take six bakers working at the same rate to prepare 20 cakes?

Wrap-Up

51. In your own words, write two examples in everyday life where proportions could be used.

Career

Actuary

Most states have laws requiring all drivers to have automobile insurance. Did you ever wonder who determines the cost of the insurance? An actuary does. Actuaries design insurance and pension plans for companies. They must make sure that the price charged for insurance will enable the company to pay all claims and expenses as they occur in the future. Employment for actuaries is expected to rise 35% or more through the 1990s as insurance sales increase and companies introduce new forms of insurance.

Capture-Recapture

Materials: small bowls, dried lima beans

Did you know that there is a way to estimate how many salmon are in Lake Ontario? Often naturalists want to know such a population but it would be impossible or impractical to make an actual count.

One method of estimating a population is the **capture-recapture** technique. In this exploration, you will model this technique using lima beans as "fish" and a bowl as "Lake Ontario."

Your Turn: Work in small groups to complete this activity.

▶ CAPTURE   Fill a small bowl with dried lima beans. Grab a small handful of beans. Mark each bean with an X on both sides. Count the "tagged" beans and record this number. This number is the number *captured*.

Return the "tagged" beans to the bowl and mix well.

▶ RECAPTURE   Grab another small handful of beans. Count the total number of beans. This number is the number *recaptured*. Count the number of "tagged" beans. Record these numbers. This is sample A.

Return all the beans to the bowl and mix.

▶ Repeat RECAPTURE nine more times, samples B through J. Find the total tagged and the total recaptured.

▶ Use the proportion shown below to estimate the number of lima beans in your bowl.

$$\frac{\text{original number captured}}{\text{number in bowl}} = \frac{\text{total tagged in samples}}{\text{total recaptured}}$$

Analysis

1. Why is it a good idea to base your prediction on several samples instead of just one sample?

2. What would happen to your estimate if some of your tags fell off or wore off?

3. Count the number of beans in your bowl. How does your estimate compare to the actual number?

9-3 Using Proportions

Objective:
Use proportions to solve verbal problems.

Ralph Ramos is building a base for a television satellite dish. He needs 1.5 cubic yards of concrete. It takes 200 pounds of sand to make 4 cubic yards of concrete. How much sand does he need?

This problem can be solved using a proportion. In the proportion below, s represents the amount of sand needed to make 1.5 cubic yards of concrete.

$$\text{sand} \rightarrow \frac{200}{4} = \frac{s}{1.5} \leftarrow \text{sand}$$
$$\text{concrete} \rightarrow \qquad\quad \leftarrow \text{concrete}$$

Using a calculator, $200 \; \boxed{\times} \; 1.5 \; \boxed{\div} \; 4 \; \boxed{=} \; 75$

Mr. Ramos needs 75 pounds of sand.

Examples

1 **Bill Lee took a 96-mile trip to visit the Johnson Space Center. His car used 6 gallons of gasoline for the trip. How many gallons of gasoline would the car use for a 152-mile trip?**

$$\frac{96 \text{ miles}}{6 \text{ gallons}} = \frac{152 \text{ miles}}{x \text{ gallons}} \qquad \text{Notice that both rates compare miles to gallons.}$$

$$\frac{96}{6} = \frac{152}{x} \qquad \text{Solve for } x.$$

$$152 \; \boxed{\times} \; 6 \; \boxed{\div} \; 96 \; \boxed{=} \; 9.5$$

$$x = 9.5$$

A trip of 152 miles would require 9.5 gallons of gasoline.

Could you use the proportion $\frac{20}{16} = \frac{15}{w}$?

FYI

98 out of 100 homes in the United States own at least one television set.

2 **A diagram measuring 20 cm long is reduced on a copying machine to 15 cm long. If the width of the original diagram is 16 cm, what is the width of the reduced copy?**

$$\text{original length} \rightarrow \frac{20}{15} = \frac{16}{w} \leftarrow \text{original width}$$
$$\text{reduced length} \rightarrow \qquad\quad \leftarrow \text{reduced width}$$

$$20 \cdot w = 16 \cdot 15 \qquad \text{Cross products are equal.}$$

$$20w = 240$$

$$\frac{20w}{20} = \frac{240}{20} \qquad \text{Divide each side by 20.}$$

$$w = 12 \qquad \text{The width is reduced to 12 cm.}$$

Chapter 9 331

Express $\frac{1}{4}$ as a decimal.
Do you get the
same result?

3 **Marcia draws a floor plan for the house she and her father
will build. On her plans, $\frac{1}{4}$-inch represents 1 foot of the real
house. If the house is to be 54 feet long, what is the length on
her plans?**

$$\frac{\frac{1}{4}\text{ inch}}{x\text{ inches}} = \frac{1\text{ foot}}{54\text{ feet}} \quad \rightarrow \quad \frac{\frac{1}{4}}{x} = \frac{1}{54}$$

$$\frac{1}{4} \cdot 54 = 1x \qquad \text{Find the cross products.}$$

$$13\frac{1}{2} = x$$

The length 54 feet is represented by $13\frac{1}{2}$ inches on the plans.

Checking for Understanding

**Communicating
Algebra**

1. In your own words, explain how to solve a proportion using a calculator.

Choose the proportion that could be used to solve each problem below.

2. If 2 liters of fruit juice costs \$3.98, how much does 5 liters cost?

  a. $\frac{2}{\$3.98} = \frac{5}{d}$ **b.** $\frac{2}{\$3.98} = \frac{d}{5}$ **c.** $\frac{2}{5} = \frac{d}{\$3.98}$ **d.** $\frac{5}{d} = \frac{\$3.98}{2}$

3. If 64 feet of rope weighs 20 pounds, how much will 80 feet of the same kind of
rope weigh?

  a. $\frac{p}{20} = \frac{64}{80}$ **b.** $\frac{80}{20} = \frac{64}{p}$ **c.** $\frac{64}{20} = \frac{80}{p}$ **d.** $\frac{64}{80} = \frac{p}{20}$

4. If a 10-pound turkey takes 4 hours to cook, how long will it take a 14-pound
turkey to cook?

  a. $\frac{4}{10} = \frac{14}{h}$ **b.** $\frac{4}{10} = \frac{h}{14}$ **c.** $\frac{10}{20} = \frac{h}{4}$ **d.** $\frac{h}{10} = \frac{4}{14}$

**Guided
Practice**

Write a proportion that could be used to solve for each variable. Then solve.

5. 40 ounces at \$3.00
   25 ounces at x dollars

6. 3 liters at \$7.00
   y liters at \$6.20

7. 20 by 30 cm enlarged
   to 25 by x cm

8. 16 by 24 cm reduced
   to 14 by x cm

9. $\frac{1}{4}$ in. represents 1 ft
   3 in. represents x ft

10. 1 cm represents 2.5 m
    2.5 cm represents x m

Exercises

**Independent
Practice**

**Write a proportion that could be used to solve for each variable.
Then solve.**

11. 250 bushels for 2 acres
    x bushels for 5 acres

12. 3 pounds for \$15
    x pounds for \$45

13. 15 by 21 cm enlarged
    to 20 by x cm

14. 5 liters at \$6.15
    x liters at \$8.00

15. Determine whether the following conjunction is true or false. *The product of 3x and 2x is 6x and the quotient of 8n and 2 is 4n.* (Lesson 3-9)

16. Find four solutions of the equation $y = 2x - 3$. (Lesson 8-5)

17. Express as a unit rate: *6 inches of rain in 4 hours.* (Lesson 9-1)

18. Solve the proportion $\frac{r}{4} = \frac{6}{16}$.
(Lesson 9-2)

Applications

19. **Science** The ratio of weight on Earth to weight on the moon is 6:1. If you weigh 135 pounds on Earth, how much would you weigh on the moon?

20. **Geography** The scale on a map is $\frac{1}{4}$ inch = 10 miles. On the map the distance from Pittsburgh to Philadelphia is about $6\frac{3}{4}$ inches. Estimate the actual distance between the two cities.

21. **Photography** In simple cameras, like the one at the right, light from an object passes through a lens and makes an image on film. The object and its image are always in proportion. This relationship between the object and its image can be expressed by the proportion

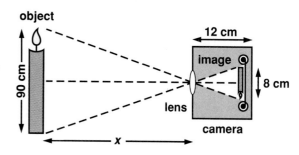

$$\frac{\text{image size}}{\text{object size}} = \frac{\text{image distance from lens}}{\text{object distance from lens}}.$$

Find the distance represented by x in the diagram.

22. **Consumer Awareness** Tai works 5 hours as a painter and earns $53.25. If he works 8 hours, how much does he earn?

23. **Cooking** A stew recipe uses 2 pounds of meat for 12 servings. How much is needed to make 9 servings?

24. **Hobbies** The mast on a model ship is $6\frac{7}{8}$ inches tall. The scale of the model is $\frac{1}{8}$ inch = 1 foot. How tall is the mast on the real ship?

Critical Thinking

25. A 3-inch by 5-inch photograph is enlarged so that the area of the enlargement is 9 times the area of the photograph. Find the width and length of the enlargement.

Wrap-Up

26. **Make Up a Problem** Write a problem that can be solved using a proportion.

9-4 Using the Percent Proportion

Objectives:
Use the percent proportion to write fractions as percents. Solve problems using the percent proportion.

Key Terms:
percent
percentage
base
rate
percent proportion

Advertising is a $61 billion industry in the United States. It encourages people to eat certain foods, drive certain cars, buy certain products, and so on. In order to gather information about the customers, advertisers often take surveys.

The results of surveys are usually reported as fractions or percents. A **percent** is a ratio that compares a number to 100. Percent also means *hundredths,* or *per hundred.* The symbol for percent is %.

Suppose the SuperCola Company conducted a taste test of its product. Five hundred people were asked their preference for Brand A or Brand B. Of these people, 230 preferred Brand A. What percent is this?

Is $\frac{230}{500}$ greater than $\frac{1}{2}$ or less than $\frac{1}{2}$?

$$\frac{230}{500} = \frac{x}{100}$$ 230 out of 500 preferred Brand A.

$230 \cdot 100 = 500x$ Write the cross products.

 230 ☒ 100 ⊟ 500 ☐ 46

$x = 46$

So, $\frac{46}{100}$ or 46% preferred Brand A.

In the proportion shown above, 230 is called the **percentage** (P). The number 500 is called the **base** (B). The ratio $\frac{46}{100}$ is called the **rate.**

$$\frac{230}{500} = \frac{46}{100} \rightarrow \frac{\text{Percentage}}{\text{Base}} = \text{Rate}$$

Note that the percentage, P, is a number that is compared to another number called the base, B.

If r represents the number per hundred, this equation can be rewritten as $\frac{P}{B} = \frac{r}{100}$. This proportion is called the **percent proportion.**

You can use the percent proportion to express fractions as percents and to solve percent problems.

Examples

1 **Express $\frac{5}{8}$ as a percent.**

$$\frac{P}{B} = \frac{r}{100} \rightarrow \frac{5}{8} = \frac{r}{100}$$ Replace P with 5 and B with 8.

$$5 \cdot 100 = 8 \cdot r$$ Find the cross products.

$$\frac{500}{8} = \frac{8r}{8}$$ Divide each side by 8.

$$62\frac{1}{2} = r$$

$\frac{5}{8}$ is equivalent to $62\frac{1}{2}\%$.

2 **What number is 35% of 263?**

Why should you replace r with 35?

$$\frac{P}{B} = \frac{r}{100} \rightarrow \frac{P}{263} = \frac{35}{100}$$ Replace B with 263 and r with 35.

$$P \cdot 100 = 263 \cdot 35$$ Find the cross products.

$$P = 92.05$$

92.05 is 35% of 263.

3 **The purchase price of a camera is $84. The state tax rate is 5.5% of the purchase price. Find the total cost.**

Is the total cost greater than $84 or less than $84?

$$\frac{P}{B} = \frac{r}{100} \rightarrow \frac{P}{84} = \frac{5.5}{100}$$ Replace B with 84 and r with 5.5.

$$P \cdot 100 = 84 \cdot 5.5$$ Find the cross products.

84 ⊠ 5.5 ⊡ 100 ⊟ 4.62

$$P = 4.62$$

The tax is $4.62.

The total cost is $84 + $4.62 or $88.62.

The percent proportion can also be used to find the base when the percentage and rate are given.

Example

Is the regular price greater than $48 or less than $48?

4 **School jackets went on sale for $48. This is 60% of the regular price. What is the regular price?**

$$\frac{P}{B} = \frac{r}{100} \rightarrow \frac{48}{B} = \frac{60}{100}$$ Replace P with 48 and r with 60.

$$48 \cdot 100 = B \cdot 60$$ Find the cross products.

$$\frac{4800}{60} = B$$ Divide each side by 60.

$$80 = B$$

The regular price is $80.

___

In summary, there are three basic types of percent problems. Using the proportion $\frac{3}{4} = \frac{75}{100}$, you can see that the types are related as shown below.

Proportion		Word Form	
$\frac{3}{4} = \frac{75}{100}$	⬌	3 is 75% of 4.	

$\frac{\blacksquare}{4} = \frac{75}{100}$	What number is 75% of 4? Percentage	Find the percentage.
$\frac{3}{4} = \frac{\blacksquare}{100}$	3 is what % of 4? Rate	Find the rate.
$\frac{3}{\blacksquare} = \frac{75}{100}$	3 is 75% of what number? Base	Find the base.

Checking for Understanding

Communicating Algebra

1. A __?__ is a ratio that compares a number to 100.

2. Explain how to express a fraction as a percent.

3. Set up a proportion to find 16% of 90.

Guided Practice

Express each fraction as a percent.

4. $\frac{31}{100}$ 5. $\frac{1}{25}$ 6. $\frac{3}{5}$ 7. $\frac{7}{20}$ 8. $\frac{3}{8}$

Match each question with its corresponding proportion.

9. 48 is 60% of what number?   a. $\frac{48}{60} = \frac{r}{100}$

10. 48 is what percent of 60?   b. $\frac{P}{48} = \frac{60}{100}$

11. What number is 60% of 48?   c. $\frac{48}{B} = \frac{60}{100}$

Identify each percentage, base, and rate. Then write a proportion and solve.

12. Find 7.5% of 405.

13. Find 81% of 32.

14. Twenty-eight is 20% of what number?

15. Sixteen is 40% of what number?

16. 19 is what percent of 76?

17. 37 is what percent of 296?

Exercises

Independent Practice

ESTIMATION
MENTAL MATH
CALCULATOR
PAPER/PENCIL

Express each fraction as a percent.

18. $\frac{2}{5}$

19. $\frac{7}{8}$

20. $\frac{3}{2}$

21. $\frac{15}{4}$

22. $\frac{9}{4}$

Use a proportion to solve each problem.

23. What is 40% of 60?

24. Find 37.5% of 80.

25. Twenty-one is 35% of what number?

26. Seventy-five is what percent of 250?

27. Fifty-two is what percent of 80?

28. Thirty-six is 45% of what number?

Mixed Review

29. **Personal Finance** Alison has developed a monthly budget for herself. She has allotted twice as much money for clothing as for eating out. If the total for both is $72, what is the most Alison can spend on eating out?   (Lesson 7-7)

30. Find the x-intercept and y-intercept for the graph of the equation $y = x - 4$. (Lesson 8-9)

31. *True* or *false*: $\frac{7}{21} = \frac{0.5}{1.5}$.   (Lesson 9-2)

32. Write a proportion that could be used to solve for the variable x: *8 by 12 cm enlarged to 10 by x cm.*   (Lesson 9-3)

Logical Reasoning

33. If $\frac{1}{3}$ is equivalent to $33\frac{1}{3}$%, what percent is equivalent to $\frac{2}{3}$?

34. If 22.2% is equivalent to $\frac{2}{9}$, what number is equivalent to 44.4%?

Applications

35. **Banking** Mika's savings account earned $9.48 in interest in one year. This is equal to $6\frac{1}{4}$% of his savings. What are his savings?

36. **Sports** Ten out of sixteen members of the softball team take the field. What percent of the team members are playing?

37. **Consumer Awareness** The regular price of a CD is $12.98. Find the discount if the CD is marked 25% off.

Critical Thinking

38. Three 24-hour clocks show the correct time, 12 noon. One of the clocks is always correct, one loses a minute every 24 hours, and one gains a minute every 24 hours. How many hours will pass before all three clocks again show the correct time?

Wrap-Up

39. **Research** Use newspaper articles or ads to help you create three percent problems. One of the problems should require finding the base, one should require finding the percentage, and one should require finding the rate.

9-10   Percent of Change

Objective:
Find the percent of increase or decrease.

Last year, a new bicycle cost $120. This year, the same model costs $135. You can express this increase in price using percents.

To find the percent of increase, you can follow these steps.

Mental Math Hint

THINK: $\frac{15}{120} = \frac{1}{8}$

$\frac{1}{8} = 12.5\%$

Step 1   Subtract to find the amount of change.

$135 - 120 = 15$ The cost of the bicycle increased $15.

Step 2   Solve the percent proportion. Compare the amount of increase to the original amount.

$$\text{percentage} \rightarrow \frac{15}{120} = \frac{r}{100} \Big\} \text{rate}$$
$$\text{base} \rightarrow$$

$15 \cdot 100 = 120r$ Find the cross products.

$1500 = 120r$ Divide each side by 120.

$$1500 \;\boxdot\; 120 \;\boxminus\; 12.5$$

$r = 12.5$

The cost of the bicycle increased 12.5%.

The percent of decrease can be found in a similar way.

Example

1 **When Andrea started bicycling last year, she completed the 5-mile course in 45 minutes. Now, she can complete the course in 40 minutes. Find the percent of decrease.**

$45 - 40 = 5$ Find the amount of decrease.

$\frac{5}{45} = \frac{r}{100}$ Write the percent proportion. The original time is 45 minutes.

$5 \cdot 100 = 45r$ Find the cross products.

$500 = 45r$ Divide each side by 45.

$$500 \;\boxdot\; 45 \;\boxminus\; 11.1111111$$

$r \approx 11.1$

Estimation Hint

THINK: $\frac{5}{45} = \frac{1}{9}$

Since $\frac{1}{9} > \frac{1}{10}$, the percent of decrease is greater than 10%.

The percent of decrease is about 11%.

Checking for Understanding

Communicating Algebra

1. What is the first step in finding the percent of change?

2. What part of the percent proportion do you solve for when finding a percent of change?

3. What is the name of the amount used as the base in the percent proportion when finding a percent of change?

Guided Practice

Estimate the percent of increase or decrease in the prices below.

4. old: $5
    new: $6

5. old: $12
    new: $15

6. old: $20
    new: $18

7. old: $14
    new: $13

8. old: $26
    new: $20

9. old: $0.49
    new: $0.56

Exercises

Independent Practice

Find the percent of change in the prices below. Round to the nearest whole percent.

10. old: $65.00
    new: $68.25

11. old: $79.18
    new: $74.00

12. old: $62.00
    new: $65.72

13. old: $156.90
    new: $194.55

14. old: $139.40
    new: $164.49

15. old: $65.48
    new: $60.24

16. old: $475
    new: $450

17. old: $800
    new: $835

18. old: $615
    new: $650

Mixed Review

19. Find the slope of the line that contains the points $A(1,2)$ and $B(-3,0)$. (Lesson 8-8)

20. Twenty-six is what percent of 130?   (Lesson 9-7)

Find the discount rate.   (Lesson 9-8)

21. a $10 CD on sale for $8

22. a $3.50 notebook on sale for $2.25

Applications

23. **Economics**   The Consumer Price Index (CPI) shows the relative costs of goods and services. If the CPI is 233.2 in April and 236.4 in May, what is the percent of increase?

24. **Sports**   At the start of wrestling season, Jorge weighed 150 pounds. At the end of the season, he weighed 144 pounds. What was the percent of change in his weight?

25. **Consumer Awareness**   Last year the value of Patty's car was $6150. This year the value is $4920. What is the percent of change in the car's value?

Critical Thinking

26. A pair of hiking boots originally cost $42. The price was increased by 25%. The boots were sold later with the price reduced $\frac{1}{4}$. What was the final sales price? Explain why the final sales price is not $42.

Wrap-Up

27. In your own words, explain how to find the percent of change.

Review

Language and Concepts

Choose the correct term or number to complete each sentence.

1. A __?__ compares two numbers and can be written as follows: 2 out of 3, 2 to 3, 2:3, or $\frac{2}{3}$.

2. In the equation 15% · $22 = $3.30, the rate is __?__ .

3. A __?__ is an equation in the form $\frac{a}{b} = \frac{c}{d}$, which states that two ratios are equivalent.

4. A ratio that has two measurements with different units of measure is called a __?__ .

5. In the proportion $\frac{441}{882} = \frac{50}{100}$, the base is __?__ .

6. To find the percent of increase or decrease, you compare the amount of change to the __?__ .

7. __?__ is equivalent to $66\frac{2}{3}\%$.

> base
> rate
> proportion
> original
> ratio
> 441
> 882
> $\frac{2}{3}$
>
> unit rate
> discount
> 15%

Skills

Express each ratio as a fraction in simplest form. (Lesson 9-1)

8. 5 out of 6
9. 3 to 6
10. 15:10

Express each ratio as a unit rate. (Lesson 9-1)

11. 450 miles in 9 hours
12. $12 in 1.5 hours

Solve each proportion. (Lesson 9-2)

13. $\frac{1}{3} = \frac{n}{12}$
14. $\frac{18}{48} = \frac{6}{y}$
15. $\frac{2}{3} = \frac{x}{0.6}$
16. $\frac{8}{2.5} = \frac{10.4}{c}$

Write a proportion that could be used to solve for each variable. Then solve. (Lesson 9-3)

17. 4 pounds costs $89.
    9 pounds costs n dollars.

18. Al earns $160.95 in 37 hours.
    Al earns $26.10 in n hours.

Express each fraction as a percent. (Lesson 9-4)

19. $\frac{7}{8}$
20. $\frac{9}{5}$
21. $\frac{2}{5}$
22. $\frac{3}{8}$
23. $1\frac{3}{50}$

Express each decimal as a percent. (Lesson 9-5)

24. 0.012
25. 1.05
26. 0.086
27. 1.21

Express each percent as a decimal and as a fraction in simplest form. (Lesson 9-5)

28. 25%
29. $33\frac{1}{3}\%$
30. 8%
31. 23.5%
32. 38%

Estimate the percent. (Lesson 9-6)

33. 12 out of 25   **34.** 9 out of 26   **35.** 16 out of 50   **36.** 28 out of 295

Solve. (Lesson 9-7)

37. 120 is 60% of what number?   **38.** 50% of what number is 46?

39. $83\frac{1}{3}\%$ of 48 is what number?   **40.** What is 0.6% of 59?

41. 250 is what percent of 100?   **42.** What percent of 400 is 2?

43. 1.2% of what number is 0.06?   **44.** 450% of what number is 49.5?

Applications and Problem Solving

Find the amount of discount and the sale price of each item. (Lesson 9-8)

45. $50 jacket, 15% off   **46.** $119.50 lamp, $\frac{1}{4}$ off   **47.** $3.59 tennis balls, $\frac{1}{3}$ off

Find the interest to the nearest cent. (Lesson 9-8)

48. $650 at 12% for 2 years   **49.** $1250 at 10% for 3 months

Find the percent of increase or decrease. (Lesson 9-10)

50. A $50 vase is now worth $87.50.   **51.** Half-time intermission went from 15 to 12 minutes.

Solve. (Lesson 9-9)

52. Anne Zody buys a car for $100, sells it for $110, buys it back for $120, and sells it again for $130. How much does Ms. Zody make or lose?

Portfolio Suggestion

Select an item from this chapter that you feel shows your best work and place it in your portfolio.

Curriculum Connection

- **Science**   Find three examples of spirals in nature.
- **Geology**   Two of the saltiest bodies of water are the Dead Sea and the Great Salt Lake. Determine how percents are used in describing how much salt is in salt water.

Read More About It

Charuhas, Mary S., et al.   *Essential Mathematics for Life: Percents, Graphs, and Measurement.*

Gardner, Martin.   *Perplexing Puzzles and Tantalizing Teasers.*

Schmitt, Lois.   *Smart Spending: A Young Consumer's Guide.*

Test

Express each ratio as a fraction in simplest form.

1. 14 to 28
2. 21:9
3. 8 out of 18

Solve each proportion.

4. $\frac{1}{a} = \frac{11}{22}$
5. $\frac{3}{8} = \frac{y}{12}$
6. $\frac{4}{3} = \frac{10}{w}$
7. $\frac{6}{5} = \frac{f}{17.5}$

Write a proportion that could be used to solve for each variable. Then solve.

8. 2 gallons costs $2.38
   n gallons costs $10.71

9. 6 packages costs 57¢
   4 packages costs n cents

Express each percent as a decimal and as a fraction in simplest form.

10. 1.2%
11. 105%
12. 7%

Express each decimal or fraction as a percent.

13. 0.37
14. 0.061
15. $\frac{31}{100}$
16. $\frac{19}{25}$

Estimate.

17. 9% of 11
18. 35% of 150
19. 19% of 250
20. 250% of 48

Solve.

21. What number is 35% of 120?
22. 28% of 70 is what number?
23. 30 is what percent of 250?
24. What percent of 32 is 8?
25. 2.25 is 50% of what number?
26. 40% of what number is 35.8?

Find the sale price of each item.

27. $29.95 jeans, 25% off
28. $11.50 album, 15% off

Find the principal plus interest to the nearest cent.

29. $560 at 8% for 1 year
30. $4000 at 9.25% for 6 months

Solve. Make a table.

31. How many 4-digit numbers can be made using the digits 1, 2, 3, and 4? Use each digit only once.

Find the percent of increase or decrease.

32. 40 tickets to 60 tickets
33. a $440 canoe on sale for $396

BONUS

How do you know that 1% does not always mean $1?

Academic Skills Test

Cumulative, Chapters 1-9

1. A group of divers needs to descend to a depth 3 times their present depth of -25 m. At what depth do they need to be?

   A 75 m
   B 28 m
   C -25 m
   D -75 m

2. Which is equivalent to $n^3 \cdot n^4$?

   A n^{12}
   B n^7
   C n^1
   D n^{-1}

3. Connie ordered computer equipment from a mail-order catalog. The components and prices were: a printer for $300, a modem for $150, and a disc drive for $75. She added $71.50 for tax, shipping, and handling. What was the total cost of Connie's order?

   A $525.00 **C** $767.50
   B $596.50 **D** Not Here

4. Joe Smith decided to make storm windows instead of buying them. The materials for twelve storm windows cost $72.48. He could buy storm windows for $21.50 each. How much will he save per window by making them?

   A $15.46 **C** $93.98
   B $50.98 **D** $185.52

5. If $3x + 4(x + 1) = 5x - 8$, what is the value of x?

   A -6 **C** -2
   B -4.5 **D** 6

6. Sami plans to spend no more than $50 on shirts and pants. He buys 2 shirts at $15 each. Which inequality shows the most he can spend on pants?

   A $2 \cdot 15 - x \le 50$
   B $2 \cdot 15 + x \ge 50$
   C $2 \cdot 15 + x \le 50$
   D Not Here

7. Which is the ordered pair for point N?

   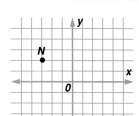

   A (2, -3)
   B (-2, 3)
   C (-3, 2)
   D (3, 2)

8. Which is a description of the graph of the equation $y = -3$?

   A The point (0, -3)
   B All points 3 units below the x-axis
   C All points 3 units left of the y-axis
   D The point (-3, 0)

9. A sweatshirt is on sale for $15. This is 80% of the regular price. Which equation could be used to find the regular price?

   A $x = \frac{80}{100} \times 15$ **C** $15 = \frac{80}{100} \times x$
   B $x = \frac{20}{100} \times 15$ **D** $15 = \frac{20}{100} \times x$

10. A sweater that normally sells for $35 is on sale at 25% off. The best estimate of the sale price is—

    A $9 **C** $32
    B $26 **D** $43

CHAPTER 10

CHAPTER OBJECTIVES

In this chapter you will learn to:

- gather and record data
- describe data using measures of central tendency and measures of variation
- display data using stem-and-leaf, box-and-whisker, and scatter plots
- use statistics to predict and recognize misleading statistics
- solve problems by first solving a simpler problem

Statistics and Graphs

Do you know someone who owns a CD player? Today's compact disc player is certainly a long way from Alexander Graham Bell's 1876 graphophone or Thomas Alva Edison's 1877 phonograph. The laser beam and microcomputer of the CD player are virtually a quantum leap in technology from the stylus and diaphragm of Edison's original phonograph.

If you spend money like the "average" teenager, you probably spend about $1.70 per week on records, tapes, or CDs. How do you measure up against this statistic?

Consumer Connection

Class Project

Design a survey in which you ask classmates about their spending habits. Display the data in several kinds of graphs.

For each graph, write a paragraph in which you summarize the results of the survey.

"HIS MASTER'S VOICE"

REG. U.S. PAT. OFF.

Gathering Data

Major advertisers, political parties, and media such as television and radio stations are concerned with **demographics,** the numbers that describe the characteristics of a population. Advertisers, for example, often find out about their public through surveys.

In this Exploration, you will find out about the students in your school by having some of them fill out a survey form. The pieces of information you will gather are called **data.** You will then analyze the data.

▶ Work in small groups. Pretend your group is an advertising company that has been hired to convince the students in your school to buy a new product.

▶ Decide what product you want to advertise.

▶ Decide how many people you want to survey. You probably cannot ask everyone's opinion, so use a smaller sample.

▶ Decide what questions to ask.

▶ Decide on the most effective way to gather the data.

Your Turn: **Write your survey and gather the data. Record the data from the surveys using a chart for each question.**

Analysis

1. Write a few sentences analyzing what you found out about the students in your school.
2. Explain how your advertising company can use the data to help convince students to buy your product.
3. Prepare a report, complete with charts and graphs, that summarizes your findings.

Extension

4. Survey the cars in the teachers' parking lot. Determine what characteristics you will study, collect the data, and analyze it.

10-1   Gathering and Recording Data

Objective:
Gather and record data using a frequency table or histogram.

Key Terms:
statistics
frequency table
sample
histogram

About how much television do students watch each week? **Statistics** involves collecting, analyzing, and presenting data such as this. To answer the question about television viewing, Andrea surveyed some students at her school. She recorded their responses, rounded to the nearest hour, on a **frequency table** as shown below.

Number of Hours Watched	Tally	Frequency
0-2	IIII	4
3-5	ЖІ III	8
6-8	ЖІ ЖІ ЖІ ЖІ II	22
9-11	ЖІ ЖІ ЖІ ЖІ ЖІ ЖІ II	32
12-14	ЖІ ЖІ ЖІ ЖІ ЖІ ЖІ	30
15-17	IIII	4

Andrea did not survey every student; she surveyed a smaller group, or **sample.** She chose the students in the sample group randomly, so the sample is assumed to be representative of the larger group.

What does "randomly" mean?

You can answer questions about the data by studying the frequency table.

Examples

1 **How many hours of TV do the greatest number of students watch?**

The greatest number of students (32) watch 9–11 hours per week. Almost as many (30) watch 12–14 hours.

2 **How many people were surveyed for this sample?**

$4 + 8 + 22 + 32 + 30 + 4 = 100$ The sum of the frequencies is the sample size.

One hundred people were surveyed.

3 **What percent of the sample watch fewer than 12 hours?**

$4 + 8 + 22 + 32$ or 66 people watch fewer than 12 hours. Compare 66 to the total (100). 66% watch fewer than 12 hours of TV per week.

4 **Would Andrea have a representative sample if all 100 people she surveyed were involved in evening sports practices?**

No, because the students sampled would not have as many hours available to watch TV.

FYI

The 1990 census was the 21st U.S. census. It gathered information from the estimated 250 million people living in the U.S. Census data is used to determine, among other things, amounts of federal funding and the number of representatives to Congress.

Another way to display the data from Andrea's frequency table is shown below. The graph is called a **histogram,** and it shows how the data are distributed.

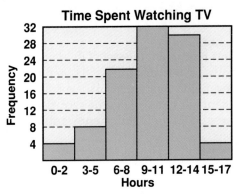

Time Spent Watching TV

In what way is a histogram a more effective display than a frequency table?

The width of the rectangular regions represents the time interval of the data. The height of the regions represents how many students are in the interval.

Checking for Understanding

Communicating Algebra

1. In your own words, describe a frequency table.

2. How are a frequency table and a histogram alike? How are they different?

3. When a sample is used, what does the surveyor need to consider?

Guided Practice

The frequency table below contains data from a favorite colors survey.

4. Copy the table and complete the frequency column.

5. What color was chosen most often?

6. What color was chosen least often?

7. How many people selected blue as their favorite color?

8. How many people gave information for this survey?

Favorite Color	Tally	Frequency
Yellow	JHÍ JHÍ IIII	14
Orange	IIII	
Blue	JHÍ JHÍ JHÍ JHÍ JHÍ JHÍ JHÍ JHÍ JHÍ	
Purple	II	
Green	JHÍ III	
Red	JHÍ JHÍ JHÍ JHÍ JHÍ II	

9. You would *not* make a histogram of the kind of data in this frequency table. Look at the information about histograms and tell what makes this data different from the kind of data in a histogram.

The Super Cola Company conducts a taste test. 800 people compare Super Cola to Brand X. 437 people prefer Super Cola. Which statements are true? Write *true* or *false*.

10. "Consumers prefer Super Cola 2 to 1."

11. "Over 50% of the people surveyed prefer Super Cola."

12. "More people always choose Super Cola over Brand X."

Exercises

Independent Practice

Refer to your completed frequency table of favorite colors on page 364.

13. What percent of people surveyed preferred red?

14. Would it be correct to say that 50% of the people surveyed preferred blue?

15. Would it be correct to say that 50% of the people surveyed preferred either blue or orange?

The scores on a 40-point test are given below.

16. Make a frequency table for the set of data.

17. What is the lowest score?

18. What is the highest score?

19. What is the frequency of the score that occurred most often?

20. What is the frequency of the score that occurred least often?

38	35	40	36	29
27	35	31	33	35
35	38	40	38	29
27	35	34	31	34
35	27	31	30	31

21. If 29 is the lowest passing score, how many scores are passing scores?

22. Make a histogram for the set of data.

23. Write a sentence that describes the test-score data.

Mixed Review

24. An elevator goes up 9 floors, then down 14 floors. If it started on the sixth floor, what floor is it now on?   (Lesson 2-9)

25. Use a graph to solve the following system of equations: $x + 2y = 5, x - y = 2$. (Lesson 8-10)

26. Express as a rate: *115 revolutions in 10 seconds.*   (Lesson 9-1)

27. Find the percent of change in taxes if the old taxes are $54.00 and the new taxes are $57.78.   (Lesson 9-10)

Application

28. **Retail Sales**   A women's clothing store conducted a survey that resulted in the frequency chart shown at the right.
    a. If the owner can only buy three colors of a certain item, what should they be?
    b. If the store usually serves 300 customers a day, what percent sample did the owner use?
    c. What are some ways the owner could be certain the sample was representative of the customers?

Favorite Clothing Color

Color	Tally	No.
Brown	II	2
Black	₩ll II	7
Teal	₩ll III	8
Plum	₩ll I	6
Orange	III	3
Green	IIII	4

Critical Thinking

29. Quality control engineers survey the performance of products. A light bulb designed to give 1000 hours of light is lighted at midnight on July 1. It goes out at 9 A.M. on August 15. Did it last at least 1000 hours? How many hours did it last?

Wrap-Up

30. Explain in a few sentences how to take a survey and make a frequency chart and histogram of your data.

Algebra in Action-Marketing

Market Data Surveys

Mary Felton is a market data collector. She collects information about the interests and tastes of different types of consumers. She enjoys surveying teenagers to find their tastes in music, clothing, colors, soft drinks, makeup, and sporting goods. From the information she collects, companies often make decisions about what products to manufacture, how to package them, and how to market them.

Mary works with samples of the population. At a basketball game, she tallied the soft drink preference of the fans who purchased soft drinks.

1. Copy and complete the table for her.

2. Would this be a good sample of the total high school population? Why or why not?

3. To find the favorite actor of a group of students, Mary surveyed every fifth person in a theater line. Why was this a poor sample?

4. Suppose your school has 2000 students in four grades. What advice could you give Mary to help her select a good sample of students for a general survey?

Soft Drink Purchases

Drink	Tally	No.
Cola	⅂⅂⅂⅂⅂⅂ IIII	
Diet Cola		47
Root Beer		29
Lemon-Lime	⅂⅂⅂⅂⅂⅂ ⅂⅂⅂⅂⅂⅂ II	
Orange		38
Grapefruit	⅂⅂⅂⅂⅂⅂ ⅂⅂⅂ II	

5. If Mary wanted to find people's choice for President before the election, should she survey people at a political party headquarters? Why or why not?

Would the following locations be good for a survey? Write *yes* or *no* and tell why.

	Survey	Location
6.	favorite detergent	laundromat
7.	number of dogs	apartment building
8.	favorite carpet color	carpet store
9.	favorite singer	homes during a weekday
10.	favorite lunch	school cafeteria
11.	favorite lunch	pizza parlor
12.	automobile owned	bus stop

13. Write some survey items that might interest a compact disk manufacturer. Suggest locations where the survey could be taken.

10-2   Measures of Central Tendency

Objective:
Use mean, median, and mode as measures of central tendency.

Key Terms:
central tendency
mean
mode
median

Various brands of a radio with similar features have the following prices: $49, $49, $50, $50, $52, $52, $52, $55, $57, $60, and $62. What is the "average" price for such a radio?

In analyzing sets of data, researchers often try to find a number or other datum that can represent the whole set. These numbers or pieces of data are **measures of central tendency**. Three that we will study are the mean, the mode, and the median.

The **mean** is what people usually are talking about when they say "average." It is the arithmetic average of the data. For the prices above, the mean is

$$\frac{49 + 49 + 50 + 50 + 52 + 52 + 52 + 55 + 57 + 60 + 62}{11} = \frac{588}{11} \approx 53.454545$$

The mean price is about $53.45.

Notice that the mean does not have to be a member of the set of data.

Definition of Mean	The mean of a set of data is the sum of the data divided by the number of pieces of data.

Is it possible for the mode to be greater than the mean?

In the prices above, the price $52 appears the most number of times (three), so $52 is the **mode**. If there was another radio priced at $50, then $50 would also be a mode and the data would have two modes.
The mode is always a member of the set of data.

Definition of Mode	The mode of a set of data is the number or item that appears most often.

The **median** is simply the middle number when the data are in order.

49	49	50	50	52	52	52	55	57	60	62

The number in the middle is 52. The median price is $52.

If the number of data is even, then the set has two middle numbers. In that case the median is the mean of the two numbers. Consider the set of data 5, 7, 9, 11, 12, 15, 23, 26, 28, and 30. The middle numbers are 12 and 15. Their mean is found as follows:

$$(12 + 15) \div 2 = 27 \div 2 \text{ or } 13.5$$

The median is 13.5.

The median does not have to be a member of the set of data.

Definition of Median	The median is the number in the middle when the data are arranged in order. When there are two middle numbers, the median is their mean.

Examples

Organize the data. Then find the median, mode, and mean.

1 Scores on a ten-question science quiz are given below.

3, 5, 7, 6, 8, 2, 9, 3, 7, 6, 5, 7, 7, 7, 8, 3, 2, 9, 9, 5, 8, 7, 2, 3, 5

One way to organize the data is to make a frequency table.

The median is the middle number. There are 25 scores, so the 13th score is the median. The median score is 6.

The mode score is 7. Why?

Score	Tally	Number
2	III	3
3	IIII	4
4		
5	IIII	4
6	II	2
7	IIII I	6
8	III	3
9	III	3

 To find the mean, add the scores and divide by 25. Use the frequency chart.

$$\frac{3(2) + 4(3) + 0(4) + 4(5) + 2(6) + 6(7) + 3(8) + 3(9)}{25} = 5.72$$

The mean score is 5.72.

2 In the ten games of basketball she played, Tammy scored 17, 3, 15, 8, 5, 13, 7, 9, 12 and 10 points.

Organize the data by putting them in order.

3, 5, 7, 8, 9, 10, 12, 13, 15, 17

Because there are ten pieces of data, the median is the mean of the 5th and 6th numbers.

$(9 + 10) \div 2 = 9.5$ The median is 9.5.

This data set has no mode because each piece of data occurs just once.

The mean is the sum of the data divided by 10.

$99 \div 10 = 9.9$ The mean is 9.9.

Which "average" do you think best represents Tammy's scores?

Computer Connection

You can use a data base program to organize data. Data base allows you to quickly sort data by categories and arrange the data in alphabetical or numerical order.

Checking for Understanding

Communicating Algebra

1. In your own words, tell how to find the median of a set of data.
2. In your own words, tell how to find the mode of a set of data.
3. Describe the mean. How is it different from the median?

Guided Practice

List the data in each set from least to greatest. Then find the median, mode, and mean.

4. 5, 6, 9, 11, 2, 11, 40
5. 3.2, 1.5, 2.6, 1.5, 2.6, 4.8
6. 0.2, 0.4, 0.1, 0.6, 1.2, 1.1
7. 4.5, 4.9, 4.7, 5.0, 4.7
8. 25.98, 30.00, 45.36, 25.00, 45.36
9. 105, 116, 125, 78, 78

Exercises

Independent Practice

Find the mean, median, and mode for each set of data. Round to the nearest tenth.

10. 36, 37, 41, 43, 43
11. 44, 48, 55, 56, 55, 68, 70
12. 2, 16, 21, 3, 8, 9, 8, 6, 7
13. 121, 130, 128, 126, 130, 121
14. 2.3, 3.6, 4.1, 3.6, 2.9, 3.0
15. 0.4, 1.6, 0.8, 0.9, 0.7, 1.1

Use the data at the right to answer questions 16-25.

16. What is the mode?
17. What is the mean?
18. What is the median?

Heights of Students in Class

Name	Height (cm)
Martha	130
Mary	155
Tom	148
Gene	184
Chip	172
Meg	155
Dottie	162
Sean	155
Kim	165
Ali	173
Jesse	155

Suppose Jean enrolls in the class and her height is 135 cm. Without computing:

19. How will Jean affect the mean?
20. How will Jean affect the mode?
21. How will Jean affect the median?

Suppose Tim now joins the class. His height is 155 cm. Without computing:

22. How will Tim affect the mode?
23. How will Tim affect the median?
24. How will Tim affect the mean?
25. When both Jean and Tim join the class, what are the new mean, mode, and median?

Logical Reasoning

26. Think about all the problems you have worked in this lesson. Which is affected more by very large or very small numbers in a set of data, the mean, the median, or the mode?

Mixed Review

27. Use the formula $F = \frac{9}{5}C + 32$ to find the equivalent Fahrenheit temperature for -25°C. (Lesson 7-8)

28. Solve the equation $n - 3.5 = -1$. Graph the solution. (Lesson 8-1)

29. **Cooking** A recipe that makes 8 muffins calls for 2 cups of flour. How much flour is needed to make 24 muffins? (Lesson 9-3)

30. Refer to the frequency table on page 363. What percent of the sample watch fewer than 9 hours of television? (Lesson 10-1)

Applications

31. **Sports** Julio bowled four games. His scores were 123, 108, 127, and 118. What is his mean score?

32. **School** Gary's mean score for three French tests is 77. His first two scores were 75 and 79. What was his other score?

33. **Meteorology** The high temperatures, in degrees Fahrenheit, for a week in May were 68, 70, 68, 66, 70, 74, and 72. Describe the temperatures using the mean, median, and mode.

34. **Business** At Wilcox Products, 7 employees earn $20,000, 3 earn $32,000, and 2 earn $60,000. List the salaries of the employees on a frequency chart. Find the mean, median, and mode. Which measure of central tendency would you use to describe the "average" salary and why?

35. Use the column from a phone book shown at the right.

    a. Make a tally of the last digits of the numbers. What is the mode? What is the median?

    b. Make a tally of the first digits of the numbers. What is the mode? What is the median?

    c. Is there a difference in parts **a** and **b**? Why or why not?

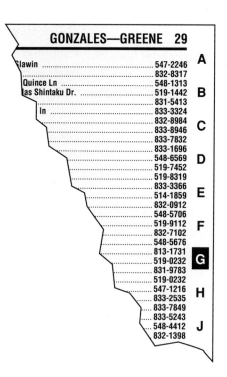

36. Here are Marilyn's test scores in mathematics for the semester: 78, 74, 85, 88, 93, 96, 98, and 100. If she needs a mean score of 90 to receive an A, what must she score on the final test? Write an inequality to solve this problem.

Critical Thinking

37. Construct a set of data with at least 7 items in which the mode, median, and mean are the same number.

Wrap-Up

38. **Collect Data** For one hour while you are watching television or listening to the radio, record the length of each commercial break. Find the mean, median, and mode. Compare your results with three other people. Find the mean, median, and mode for the four sets of data. Write about your findings.

10-3   Stem-and-Leaf Plots

Objective:
Display and interpret data on stem-and-leaf plots.

Key Term:
stem-and-leaf plot

To improve its advertising, a local theatre conducts a survey to see which age group of people is most likely to attend a Saturday matinee. The ages of the 35 people surveyed are listed below.

12, 11, 22, 67, 56, 32, 35, 45, 46, 14, 14, 16, 33, 30, 41, 7, 9, 25, 8, 51, 43, 55, 42, 17, 67, 58, 30, 18, 14, 24, 25, 26, 29, 12, 13

One way to condense the data and make it more usable is to make a **stem-and-leaf plot**. The greatest place value of the data can be used for the *stem*. The next greatest place value forms the *leaves*. Follow these steps to construct a stem-and-leaf plot from the data above.

Step 1 Find the least and the greatest item of data.

The least is 7 and the greatest is 67.

Step 2 Find the stems.

The least number, 7, has a 0 in the tens place. The greatest number, 67, has a 6 in the tens place. Therefore, the stems are digits from 0 to 6.

```
0|
1|
2|
3|
4|
5|
6|
```

Step 3 Put the leaves on the plot.

Record each of the data on the graph by pairing the units digit, or leaf, with the correct stem. For example, 25 is plotted by placing the units digit, 5, to the right of the stem 2.

```
0|798
1|2144678423
2|254569
3|25300
4|56132
5|6158
6|77
```

Step 4 Arrange the leaves so they are ordered from least to greatest.

```
0|789
1|1223444678
2|245569
3|00235
4|12356
5|1568
6|77
```

What is the mode of the data to the right?

Step 5 Include an explanation of the data.

2|5 means age 25.

By just observing the stem-and-leaf plot above, you can tell that the largest age group attending the Saturday matinee is the 10–19 age group, because this row has the most leaves in it.

Examples

1 Mrs. Martin's ninth grade homeroom had the following test scores.

Mathematics   75, 93, 87, 56, 60, 73, 78, 69,
                        83, 89, 94, 97, 65, 73, 87

History   68, 73, 98, 87, 65, 64, 70, 73,
                        72, 78, 81, 83, 68, 57, 63

She made stem-and-leaf plots for the scores, and used the same stem for both sets of data.

Mathematics		History
6	5	7
9 5 0	6	3 4 5 8 8
8 5 3 3	7	0 2 3 3 8
9 7 7 3	8	1 3 7
7 4 3	9	8

6|5 means 56.   5|7 means 57.

The stem-and-leaf plot shows that, overall, the class scored better in mathematics.

2 Here are the number of points scored by Shaquille O'Neal in ten games: 8, 30, 41, 27, 18, 22, 19, 37, 22, 26.
About how many points might you expect him to score in a game? Make a stem-and-leaf plot of the data.

The stems are
0, 1, 2, 3, and 4.

Shaquille O'Neal's Points

0	8
1	8 9
2	2 2 6 7
3	0 7
4	1

4|1 means 41 points.

From the leaves, you could expect Shaquille to score from 20 to 29 points.

Checking for Understanding

Communicating Algebra

1. When you make a stem-and-leaf plot, how do you know what to use for the stems and for the leaves?

2. Explain how a stem-and-leaf plot is similar to a horizontal bar graph or histogram. How is it different?

3. What are the stems for the following temperatures: 76, 83, 59, 71, 78?

Answer these questions about the stem-and-leaf plot in Example 1.

4. What were the highest and lowest history scores?

5. What is the median of the mathematics scores?

6. What information can you gain from the plot?

The heights (in inches) of members of a volleyball team are 64, 60, 72, 61, 73, 80, 68, 70, 65, 67, 70, and 80.

7. To make a stem-and-leaf plot of this data, what numbers will you use as stems?

8. Make a stem-and-leaf plot of the data.

9. What numbers are the leaves on the 8 stem?

10. What percent of the heights are between 60 and 70 inches? Why do you think that is true?

Exercises

The ages of the first twenty people into the museum on Saturday were 17, 9, 12, 25, 8, 39, 27, 14, 29, 40, 36, 8, 15, 41, 28, 29, 30, 31, 29, and 11.

11. Construct a stem-and-leaf plot for the data.

12. How old was the oldest person?

13. How young was the youngest?

14. What age group seemed most represented?

15. What might account for the limited span of years?

Work with a partner to complete the following activity.

16. Choose a topic and a group of people to gather numerical data about. Some examples are: basketball team—height, or class members—time it takes to get to school, or class members—number of hours of TV watched in a week.

17. Construct a stem-and-leaf plot for the data.

18. Make two or three statements about the data. (For example, most people in our class take between 10 and 20 minutes to get to school.)

19. Find four solutions for the equation $y = -\frac{1}{2}x + 1$.   (Lesson 8-5)

20. Write $\frac{7}{8}$ as a percent.   (Lesson 9-5)

21. Find the mean, median, and mode for the data set: 7, 2, 3, 9, 11, 8, 6, 3, 14. (Lesson 10-2)

22. **Teaching**   Mrs. Hunt taught English and Social Studies to the same class. The most recent test scores are plotted at the right. What was the highest grade in each subject? the lowest? In which subject did the class do better? Explain your answer.

English		Social Studies
8 7	5	1 8
9 2 0	6	0 3 5 7
8 7 5 3 1	7	5 6 7 8 8 8 9
9 6 6 5 4 1	8	3 4 5 8 9
8 8 4 3	9	2 5

7|5 means 57.   5|1 means 51.

23. **Research**   Find out the age at inauguration of all the presidents of the United States. Make a stem-and-leaf plot to display the data. Are the ages evenly distributed or clustered in the middle?

24. Suppose you have seven coins in your pocket totaling $1. What is one possible combination of coins? (Hint: How can a stem-and-leaf plot or frequency chart help?)

25. Give an example of data that might be represented on a stem-and-leaf plot. How would you use the plot to better understand the data?

10-4 Measures of Variation

Objective:
Use measures of variation to compare data.

Key Terms:
variation
range
interquartile range
upper quartile
lower quartile

Suppose you and a friend are in different Spanish classes, but have the same teacher. On a test, both classes have the same median score, the same mode score, and the same mean score. Does that mean that all the scores are the same? Not necessarily!

Here are the actual scores of the two classes:

First Period: 20, 28, 35, 38, 45, 50, 60, 60, 72, 75, 88, 89, 91, 96, 98

Second Period: 43, 45, 52, 54, 55, 56, 60, 60, 60, 68, 72, 75, 78, 81, 85

Using stem-and-leaf plots, we can see the differences quite easily.

First Period		Second Period
2	0 8 2|0 means 20.	2	
3	5 8	3	
4	5	4	3 5
5	0	5	2 4 5 6
6	0 0	6	0 0 0 8
7	2 5	7	2 5 8
8	8 9	8	1 5
9	1 6 8	9	

For both classes the mode and median are both 60, and the mean is 63. But as you can see, the classes had quite different scores. What differs is the **variation** of the scores.

One measure of variation is called the **range**.

Definition of Range	The range of a set of numbers is the difference between the least and the greatest number in the set.

The ranges of the Spanish test scores are as follows.

First Period: 98 − 20 or 78   Second Period: 85 − 43 or 42

The grades of the first period class are more spread out, or dispersed, than those of the second period.

Another measure of variation is the **interquartile range.**

Definition of Interquartile Range	The interquartile range is the range of the middle half of the data.

To find the interquartile range, you must first find the middle half of the data. Here's how to find the interquartile range of the First Period scores. (See Example 2 for the Second Period scores.)

Step 1 Find the median of the data since the median separates the data into two halves.

20 28 35 38 45 50 60 | 60 | 72 75 88 89 91 96 98
                                Median

How could you use the stem-and-leaf plot to find the median and upper and lower quartiles?

Step 2 Find the median of the upper half. This number is called the **upper quartile**, indicated by UQ.

20 28 35 38 45 50 60 | 60 | 72 75 88 | 89 | 91 96 98
                                Median   UQ

Step 3 Find the median of the lower half. This number is called the **lower quartile**, indicated by LQ.

20 28 35 | 38 | 45 50 60 | 60 | 72 75 88 | 89 | 91 96 98
              LQ   Median   UQ
                    └──── middle half of data ────┘

Step 4 The middle half of the data goes from 38 to 89. Subtract the lower quartile from the upper quartile.

89 − 38 = 51   The interquartile range of the scores is 51.

Examples

1 These are Kevin's scores on eight 25-point tests:

12, 15, 17, 20, 14, 18, 11, 21.

Find the range and the interquartile range.

First order the scores.   11   12   14   15   17   18   20   21

The range is 21 − 11 = 10.

The median is (15 + 17) ÷ 2 = 16.   *Recall that when there are two middle numbers, the median is their mean.*
The upper quartile is (18 + 20) ÷ 2 = 19.
The lower quartile is (12 + 14) ÷ 2 = 13.

Notice how the median and quartiles are shown below.

           13   16   19
11   12 ┊ 14   15 ┊ 17   18 ┊ 20   21

The interquartile range is 19 − 13 or 6. The middle half of Kevin's scores varied by 6 points.

Mental Math Hint

You can find the mean of pairs like 15 and 17 without dividing. Just think "what number is halfway between?" What is the mean of 14 and 18? of 5 and 10?

2 Find the interquartile range for the second period scores shown on page 374. Compare to the first period interquartile range.

43 45 52 | 54 | 55 56 60 | 60 | 60 68 72 | 75 | 78 81 85
              LQ   Median   UQ

The interquartile range is 75 − 54 or 21. Since the interquartile range is less for the second period class, its scores are more closely grouped around the median.

Checking for Understanding

Communicating Algebra

1. Name and describe the two measures of variation used in this lesson.

2. If a group of scores has a small interquartile range, what can you say about the scores?

Guided Practice

Given the set of data 5, 6, 9, 11, 24, 27, 29, find the following.

3. the range   4. the median   5. the upper and lower quartiles

Given the set of data 45, 47, 50, 51, 62, 68, 69, 72, 75, find the following.

6. the range   7. the median

8. the upper and lower quartiles   9. the interquartile range

Use the data in the stem-and-leaf plot shown at the right.

10. What is the lowest rate?

11. What is the highest rate?

12. What is the range?

13. What is the median?

14. What are the upper and lower quartiles?

15. What is the interquartile range?

Words Typed Per Minute

4|0 means 40.

4	02
5	159
6	3578
7	2378
8	01

Exercises

Independent Practice

Here are the final grades of the students in a French II class. 68, 70, 72, 75, 78, 80, 80, 82, 85, 89

16. What is the range of grades?   17. What is the median grade?

18. Did anyone receive the median grade?   19. What are the upper and lower quartile grades?

20. What is the interquartile range?   21. Write a paragraph describing the final grades.

The stem-and-leaf plots show the bowling averages of Mary and Tanya for the first ten weeks of the league.

22. How do their medians compare?

23. How do their ranges compare?

24. How do their interquartile ranges compare?

25. Which player is more consistent? Explain your answer.

Mary		Tanya
5	9	
3 2 1 1	10	3 5 6 8
9 8 8 7	11	0 0 3 5 7
9	12	2

5|9 means 95.   10|3 means 103.

26. Solve the equation $k - (-14) = 6$.   (Lesson 3-2)

27. Identify the percentage, base, and rate: *Fourteen is 40% of what number?* (Lesson 9-4)

28. Make a stem-and-leaf plot of the following test scores: 75, 78, 92, 68, 81, 83, 95, 77, 61, 90, 88, 78. In what interval do most of the scores lie?   (Lesson 10-3)

Applications

29. Meteorology   The table at the right lists the daily mean temperatures for San Francisco and Wichita. Construct stem-and-leaf plots for the temperatures of the two cities, using the same stem.

    a. How do the medians compare?

    b. How do the interquartile ranges compare?

    c. What conclusions can you draw from your answers to **a** and **b**?

Mean Daily Temperature (°F)		
Month	**San Francisco**	**Wichita**
January	49	30
February	52	35
March	53	44
April	55	56
May	58	66
June	61	76
July	62	81
August	63	80
September	64	71
October	61	59
November	55	44
December	49	34

30. Research   Find the winning speeds at the Indianapolis 500 for the past 20 years. Find the median, range, and interquartile range. As a reporter, how would you write about the 1990 winning speed of 164 mph compared to the previous 20 years?

Critical Thinking

31. Produce a set of at least 10 pieces of data that has an interquartile range of zero.

Wrap-Up

32. Explain how comparing the median, range, and interquartile range can help determine how consistent two sets of test scores are.

Challenge

Mean Variation

The mean variation of a set of data is the average amount each number differs from the mean. Suppose the high temperatures (°F) in Indianapolis on April 20 for the last 10 years have been:

    70 65 45 80 55 62 81 50 74 48

Find the mean variation.

Step 1   Find the mean of the temperatures.

Step 2   Find the difference between each of the 10 temperatures and the mean.

Step 3   Find the mean of the differences.

What is the likelihood that this April 20 the temperature will be about 65°?
What is the likelihood that this April 20 the temperature will be within 11° of 65°?

10-5   Box-and-Whisker Plots

Objective:
Construct box-and-whisker plots.

Key Terms:
box-and-whisker plot
outliers

How many girls are at least 6 feet tall?

So far, you have made frequency tables and stem-and-leaf plots to *display* sets of data, and you have used measures of central tendency and measures of variation to *summarize* data. Now we will combine the ideas of summarizing and displaying data. A **box-and-whisker plot** summarizes data using the median, the upper and lower quartiles, and the *extreme* (highest and lowest) *values.*

The heights of the eleven girls on the Towne High girls' volleyball team are displayed in the stem-and-leaf plot at the right. The median and quartiles are marked on the plot. The lower extreme is 59 inches, and the upper extreme is 74 inches.

Players' Heights

5	9	5	9 means 59 inches.
6	3 5̲ 7 8 8̲ 8 9	
7	1̲ 2 4	

Here's how to construct a box-and-whisker plot to display the data given above.

Step 1   Above a number line, mark points for the extreme, median, and quartile values.

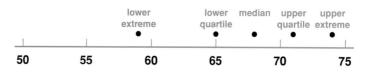

From observing the box-and-whisker plot, would you expect the mean of these data to be greater or less than the median?

Step 2   Next, draw a *box* around the quartile values. Draw a vertical line through the median value. Finally, extend *whiskers* from each quartile to the extreme data points.

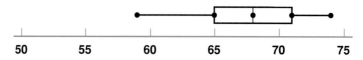

Thus, the box-and-whisker plot gives you five pieces of information about the data: lower extreme, lower quartile, median, upper quartile, and upper extreme.

Sometimes the data will have such great variation that one or both of the extremes will be far beyond the other data. Data that are more than 1.5 times the interquartile range from the quartiles are called **outliers.**

Examples

1 **Given the box-and-whisker plot below, answer these questions.**

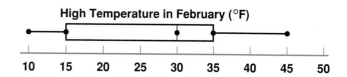

High Temperature in February (°F)

What was the highest temperature in February?   45°F

Exactly half of the days had highs over what temperature?   30°F

What fraction of the month had high temperatures less than 15°F? $\frac{1}{4}$

2 **Draw a box-and-whisker plot for the following test score data.**

20   60 | 66   70 |70| 70   76 | 80   100

Step 1   The median is 70; the upper quartile is (76 + 80) ÷ 2 or 78, and the lower quartile is (60 + 66) ÷ 2 or 63. Draw a box to show the median and the quartiles.

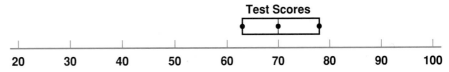

Test Scores

Step 2   The interquartile range is 78 − 63 or 15. So, data more than 1.5 times 15 from the quartiles are outliers.

$$1.5(15) = 22.5$$

Find the limits for the outlier.

Subtract 22.5 from the lower quartile.   63 − 22.5 = 40.5
Add 22.5 to the upper quartile.   78 + 22.5 = 100.5

So, 40.5 and 100.5 are the limits for outliers. There is one outlier in the data, 20. Plot the outlier with an asterisk. Draw the lower whisker to the last value that is not an outlier, 60, and the upper whisker to the upper extreme, 100.

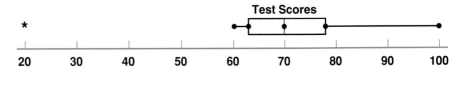

Test Scores

Checking for Understanding

Communicating Algebra

1. What five pieces of information can you learn from a box-and-whisker plot?
2. In your own words, describe an outlier.
3. How is the data shown on a box-and-whisker plot different from a stem-and-leaf plot?

Guided Practice

Use the box-and-whisker plot shown below to answer each question.

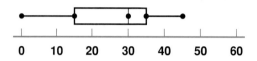

4. What is the median?
5. What is the upper quartile?
6. What is the lower quartile?
7. What is the interquartile range?
8. What are the extremes?
9. Are there outliers?
10. What are the limits on the outliers?

Exercises

Independent Practice

Use the box-and-whisker plot shown below to answer each question.

11. What is the median?
12. What is the range?
13. What is the upper quartile?
14. What is the lower quartile?
15. What is the interquartile range?
16. What are the extremes?
17. What are the limits of the outliers? Are there any outliers?

The box-and-whisker plot below represents last month's spending money for 31 students.

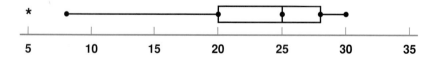

18. How many students spend less than $25 a month?
19. How many students spend between $20 and $28?
20. What is the range of spending money for the middle half of the students? (What is the interquartile range?)
21. What are the limits for the outliers?
22. Would you expect the mean spending money to be greater or less than the median? Why?

23. If you used computer software to construct a box-and-whisker plot of the data on players' heights, describe the values that you would expect to enter. Then describe the values you would enter to construct a stem-and-leaf plot.

Logical Reasoning

24. Is it possible to have a box-and-whisker plot with only one whisker? Explain your answer.

Mixed Review

25. Find the x-intercept and the y-intercept to graph the equation $y = 3 - 2x$. (Lesson 8-9)

26. Find 85% of $165. (Lesson 9-7)

27. Find the range and median of the data set: 3, 4, 8, 8, 12, 14, 15, 15. (Lesson 10-4)

Applications

28. Teaching Mr. Stone has two classes of World History. He made a stem-and-leaf plot of the scores for the last test. Use the data to make a box-and-whisker plot. (Use the same number line for both; make one plot above the other.) What do the plots tell you about the classes?

29. Research Find the average temperatures each month for two cities of your choice. Make a box-and-whisker plot for each using the same number line. Write a paragraph comparing the year-round temperatures of the cities.

Mr. Stone's History Classes		
Third Period		Fifth Period
8 0	2	
8 5	3	
5	4	3 5
0	5	2 4 5 6
0 0	6	0 0 0 8
5 2	7	2 5 8
9 8	8	1 5
8 6 1	9	

0|2 means 20.   4|3 means 43.

Critical Thinking

30. In making a box-and-whisker plot of 23 scores with no outliers, Sarah found the median to be 70, the upper quartile to be 78, and the lower quartile to be 68. She then discovered four more scores: 100, 75, 69, and 67. When she added these to the data, what happened to the box-and-whisker plot? Hint: You may want to sketch the plots.

Wrap-Up

31. How does a box-and-whisker plot separate the data into fourths?

Mid-Chapter Quiz

Use the data at the right to complete the following.

1. Make a frequency table for the data.   (Lesson 10-1)

2. Find the mean, median, and mode.   (Lesson 10-2)

3. Construct a stem-and-leaf plot.   (Lesson 10-3)

4. Find the range and the interquartile range.   (Lesson 10-4)

5. Construct a box-and-whisker plot.   (Lesson 10-5)

Scores on a Mathematics Quiz

21	30	17	19	21
18	27	30	21	21
22	30	28	29	21
29	25	28	29	22
27	17	30	17	21

Scatter Plots

Materials: tape measures, graph paper

In this Exploration, you will try to determine whether there is any relationship between two given variables. You will use a **scatter plot** to plot the data.

▶ Work with a partner. Measure the circumference of your partner's head, wrist, and neck in inches. Measure your partner's height in inches. Record these numbers.

▶ You will complete this part with the whole class. You will try to determine whether there is a relationship between the circumference of your head and your height. Write an ordered pair (circumference, height).

▶ Your teacher will instruct you how to display the data in a scatter plot. A sample is shown at the right. In this sample, the point shown represents a person whose head circumference is 21 inches and whose height is 60 inches.

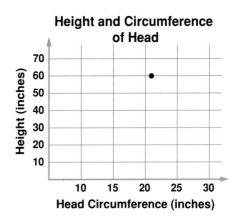

Height and Circumference of Head

Your Turn: Complete the scatter plot.

Analysis

1. Describe the scatter plot.

2. Is there a relationship between the head circumference and height? If so, write a sentence that describes the relationship.

3. In Chapter 8, you graphed linear equations on a coordinate plane. How is the scatter plot like those graphs? How does the scatter plot differ?

4. Why do you suppose this kind of graph is called a scatter plot?

Extension

5. Using similar methods, determine if there is a relationship among the circumference of your head, wrist, neck, and your height.

10-6 Scatter Plots

Objective:
Interpret data displayed in a scatter plot.

Key Term:
scatter plot

Louise wants to know if there is a relationship between the grades earned on a test and the amount of time spent studying for the test. She collected the data shown in the chart.

Student	Study Time	Test Score
Doug	10 min	65
Rebecca	15 min	68
Bradley	70 min	87
Justine	60 min	92
Allison	45 min	73
Tami	90 min	95
Mick	60 min	83
Montega	30 min	77
Christy	120 min	98

A **scatter plot** is a graph that shows the general relationship between two sets of data. To make a scatter plot of the study time data, draw a graph with Study Time along the horizontal axis and Test Scores along the vertical axis. Then plot points for each time and the corresponding score.

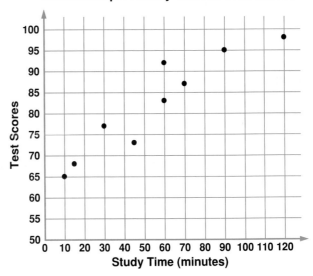

Relationship of Study Time to Test Scores

Why is this called a "positive" relationship?

From the scatter plot, you can see that there appears to be a relationship between test scores and time spent studying. In general, the more time spent studying, the higher the test score. This scatter plot shows a *positive* relationship.

Example

1 The scatter plot shows the cost of cars on a used car lot based on the age of the cars.
What happens to the cost as the age increases? The cost goes down.
What can you say about the data that do not seem to fit? These could be cars that had some major problems or were in better-than-average condition.

Why is this called a "negative" relationship?

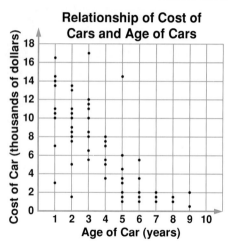

Relationship of Cost of Cars and Age of Cars

This scatter plot shows a *negative* relationship.

Checking for Understanding

Communicating Algebra

1. What is the major use of a scatter plot?

2. What do you think a scatter plot would look like if no relationship existed between the two sets of data?

Guided Practice

What type of relationship, *positive*, *negative*, or *none*, is shown by each scatter plot?

3.

4.

5.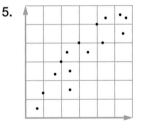

What relationship do you think a scatter plot of these data would show? Write *positive*, *negative*, or *none*.

6. outside temperature, sunburn

7. salary, years experience

8. outside temperature, frostbite

9. miles on odometer, cost of used car

10. hair color, weight

11. month of birth, years of college

Exercises

Independent Practice

A scatter plot of height and weight for a class is shown at the right.

12. Does this data show a relationship between height and weight?

13. Where on the plot are the points for people who grew taller but not heavier?

14. Where are the points for people who grew heavier but not taller?

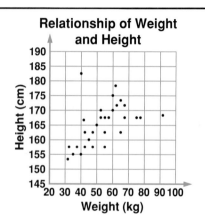

Relationship of Weight and Height

Juan wanted to find out whether there was any relationship between students' favorite classes and their best grades. He surveyed his class and used the data to construct the scatter plot at the right.

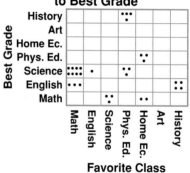

Relationship of Favorite Class to Best Grade

15. Do the data appear to be related?

16. Which class was the favorite of the most students?

17. In which class did the most students receive their best grades?

18. How many students are in the class?

Mixed Review

19. Write $\dfrac{6x^2y}{27xy}$ in simplest form.   (Lesson 4-6)

20. Find the upper and lower quartiles and the interquartile range of the data set: 38, 40, 41, 48, 52, 55, 56, 57, 61.   (Lesson 10-4)

21. Draw a box-and-whisker plot for the following bowling scores: 122, 185, 136, 139, 111, 142, 149, 133, 139, 157, 166, 153, 146. What are the extremes? (Lesson 10-5)

Applications

22. **Survey**   As a class, record the number of boys and girls in each family. Use the data to make a scatter plot. Plot the number of boys on the horizontal axis and the number of girls on the vertical axis. Use your scatter plot to answer these questions.

   a. Does there appear to be any relationship between the data in the scatter plot?
   b. How many families have three or more boys?
   c. How many families have three or fewer girls?
   d. How large is the largest family?
   e. What seems to be the mode?

23. **Medical Research**   A research doctor used a scatter plot to compare the number of doses of a new antibiotic to severity of a disease. How could a scatter plot help the doctor reach a conclusion on how effective the antibiotic is?

Critical Thinking

24. A scatter plot of monthly ice cream sales and monthly water sports accidents in Wisconsin shows a positive relationship. Why would this be true? Does a positive relationship necessarily mean that one factor causes the other? Why or why not?

Wrap-Up

25. List two sets of data that could give a positive relationship on a scatter plot. List two sets that could give a negative relationship. List two sets that would show no relationship.

10-7 Strategy: Simplify the Problem

Objective:
Solve problems by first solving simpler problems.

When you make graphs of information, you are taking large amounts of data and presenting them in a simplified manner. This helps you and others who view the graph to understand the data.

Solving a simpler problem is a strategy that can be applied to other kinds of problems as well.

Finding the sum of the whole numbers from 1 to 200 could be a tedious task, even on a calculator. Here's how to simplify the problem and look for a pattern that will help you find the answer.

Explore The problem asks for the sum of the whole numbers from 1 to 200.

Plan First find the partial sums.

$$1, 2, 3, 4, 5, \ldots, 196, 197, 198, 199, 200$$

$$2 + 199$$
$$1 + 200$$

Do this until you find a pattern that will allow you to solve the more difficult problem.

Solve

$$
\left.
\begin{array}{l}
1 + 200 = 201 \\
2 + 199 = 201 \\
3 + 198 = 201 \\
\quad \vdots \\
99 + 102 = 201 \\
100 + 101 = 201
\end{array}
\right\}
$$

There are 100 partial sums of 201.

$$100 \times 201 = 20{,}100$$

Examine Check to see if the pattern works for a small number of whole numbers. For example, try the numbers from 1 to 10.

$$1 + 2 + 3 + 4 + 5 + 6 + 7 + 8 + 9 + 10 = 55$$

Partial sums: $1 + 10, 2 + 9, 3 + 8, 4 + 7, 5 + 6$

There are 5 partial sums of 11: $5 \times 11 = 55$.

The pattern holds. So the answer 20,100 seems reasonable.

Checking for Understanding

Guided
Practice

Solve by first solving a simpler problem.

1. Find the sum of the first 20 odd numbers.

2. What simpler problem(s) did you use to solve exercise 1?

3. How many cuts are needed to divide a long rod into 25 smaller pieces?

4. What simpler problem(s) did you use to solve exercise 3?

Exercises

Independent
Practice

?
ESTIMATION
MENTAL MATH
CALCULATOR
PAPER/PENCIL

Solve. Use any strategy. Use mental math skills, estimation, or a calculator as appropriate.

5. Unit 2 of a science book starts on page 126 and ends on page 241. How many pages are in the unit?

6. The perimeter of a rectangle is 32 meters. Its area is 48 square meters. What are the dimensions of the rectangle?

7. Chef Martino made a huge pan of lasagna for a banquet. He makes 6 cuts along the length of the rectangular pan and 10 cuts along the width. How many pieces does he have?

8. Find the sum of the first 100 *even* positive numbers.

9. Iona started with a certain amount of money. She spent half of her money. She earned $6 the next day. Later, she lost $\frac{1}{6}$ of her money, but she still had $15 left. How much money did she have in the beginning?

10. Fifteen points are marked on a circle. Line segments are drawn connecting every pair of points. How many segments are drawn?

11. The original price of a pair of shoes was $56. The shoes were marked down 15% but didn't sell. Now, they are on final clearance at 30% off the sale price. If sales tax is 6%, what is the total cost of the shoes?

12. Elroy is on a diet. After 2 weeks on the diet, he weighs 173 pounds. After 4 weeks, he weighs 168 pounds. If he continues to lose weight at the same rate, what will he weigh after 10 weeks on the diet?

13. What is the total number of squares of any size shown in the checkerboard? (Hint: There are more than 64 squares.)

Journal Entry

Wrap-Up

14. Explain how working a simpler problem can sometimes help to solve a more difficult problem. Give some examples.

Team Problem Solving

In a town called Theirtown, there are 100,000 citizens. During the census, it was found that 45% of the males and 30% of the females were married to people from Theirtown. How many males and how many females live in Theirtown?

10-8   Using Statistics to Predict

Objective:

Use a sample to predict actions of a larger group.

Notice that this is an example of the problem-solving strategy, Simplify the Problem.

The 4-H Club will operate the soft drink booth at a big Fourth of July fireworks show at the stadium. Terry is in charge of ordering the soft drinks and needs to decide what kinds to order. She decided to *simplify the problem* and sampled a group of people at the school picnic. Because similar people would attend the show, she felt her sample would be representative. The bar graph shows the results of the survey.

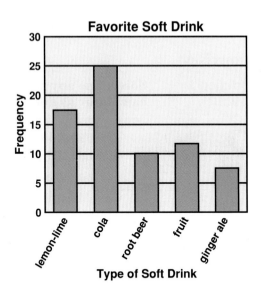

Terry estimates that every person at the stadium will buy one soft drink and she knows that the stadium holds about 7200 people. Using the results of the survey, she will order the drinks. Of the 72 people she surveyed, 12 chose a fruit drink. So she reasoned that $\frac{12}{72}$ or $\frac{1}{6}$ of her order should be fruit drink.

$$\frac{1}{6} \text{ of } 7200 = 1200$$

She could also have used a proportion to solve the problem.

$$\frac{12}{72} = \frac{x}{7200}$$

$$(12)(7200) = 72x$$

$$12 \; \boxed{\times} \; 7200 \; \boxed{\div} \; 72 \; \boxed{=} \; 1200 \qquad x = 1200$$

Terry should order 1200 cans of fruit drink.

FYI

The four Hs in 4-H stand for head, heart, hands, and health. Members of 4-H do projects and activities to help the community. More than 80 countries have 4-H or similar programs.

Example

1 **How many bottles of cola should Terry order? 25 of 72 preferred cola.**

Method 1

$$\frac{25}{72} \cdot \overset{100}{\cancel{7200}} = 2500$$

Method 2

$$\frac{25}{72} = \frac{x}{7200}$$

$$72x = (25)(7200)$$

$$25 \; \boxed{\times} \; 7200 \; \boxed{\div} \; 72 \; \boxed{=} \; 2500$$

$$x = 2500$$

She should order 2500 cans of cola.

Checking for Understanding

Communicating Algebra

1. Explain how a sample group survey can be used to predict the actions of a whole population.

2. What is the most important thing to consider in sampling?

Guided Practice

Use the survey on favorite color to answer each question.

3. What is the size of the sample?

4. What is the mode?

5. What fraction chose blue?

6. For 800 students, how many blue covers should the bookstore stock?

Favorite Color	
Red	13
Green	17
Blue	15
Yellow	5

Use the sample data on yearbook price ranges to answer each question.

7. What is the size of the sample?

8. If the price were less than $19, about how many would buy the yearbook?

9. If there are 600 in the senior class, about how many would buy a yearbook that cost $18.50?

10. About how many would buy at $16.95?

Amount Willing to Pay for Yearbook	
no more than $15	8
no more than $17	30
no more than $19	32
no more than $21	10

Exercises

Independent Practice

The bookstore sells physical education uniforms. All incoming ninth grade students will need uniforms. There are 825 students starting ninth grade. Of these, 425 are girls. The uniforms come in four sizes: small, medium, large, and extra large. The students who run the store decide to survey about 50 girls to find out what size they wear. Using this information, they will order uniforms to sell.

ESTIMATION
MENTAL MATH
CALCULATOR
PAPER/PENCIL

11. From an alphabetical list, the students survey every eighth girl. How many girls are surveyed? Is this a good sample? Why or why not?

12. Of the girls surveyed, 27 wear medium, 10 wear large, and 2 wear extra large. How many wear small?

13. For the 425 girls, how many uniforms of each size should be ordered?

14. The bookstore personnel estimated how much of each supply item would be needed during the year. They used the information in the chart below. They estimated based on 2450 students in the school and 10 months.
Find the number of each item that should be ordered.
   a. paper   b. pencils   c. pens   d. folders   e. covers

Supply Item	Amount Each Student Uses
Notebook paper	60 sheets per month
Pencils	1 per month
Pens	4 per school year
Pocket folders	2 per school year
Theme covers	3 per school year

15. The notebook paper comes in fillers of 50 sheets. How many fillers should be ordered?

16. There are 10 fillers in a package and 10 packages in a case. How many cases need to be ordered?

17. **Collect Data** Make a frequency table showing hair color for students in your class. Use the data to predict the number of students in your school with each hair color. Choose a time and place to observe at least 50 students in your school and record their hair color. Compare that data to your predictions. Write about your findings.

Mixed Review

18. Solve the equation $d = \frac{15}{8} - \frac{3}{8}$ and write the solution in simplest form. (Lesson 5-5)

19. Solve the proportion $\frac{3}{5} = \frac{t}{40}$. (Lesson 9-2)

20. What relationship (*positive, negative,* or *none*) do you think a scatter plot of the following data would show: calorie intake, weight gained? (Lesson 10-6)

Critical Thinking

21. Refer to the situation at the beginning of the lesson. Terry buys her soft drinks at a cost of 6 for $1. If she sells all 7200 at 75¢ each, how much profit will the 4-H club make?
If Terry sells only 90 percent of the drinks, what is the profit?

Wrap-Up

22. Explain how sampling plays an important part in using statistics to predict.

Career

Weather Forecasting

It would be difficult to imagine watching the evening news without the statistics, graphs, and maps of the local meteorologist. The meteorologist you see forecasting the weather on TV is most likely an *operational* meteorologist. *Physical* meteorologists collect data on the chemical and electrical properties of the atmosphere. *Climatologists* concentrate to a greater degree on past and present climate data. They use the statistical data collected over time to plot trends in our climate.

Meteorologists receive their weather data from satellites, radar, high altitude balloons, and rockets. They use computers to analyze weather data and to generate graphs that not only assist in forecasting weather, but in preparing warnings and advisories. In addition to working for radio and television stations, meteorologists are employed by government agencies, the military, private industry, educational institutions, and airlines. Employment opportunities for meteorologists are expected to grow at least 25% by the year 2000.

10-9　Misleading Statistics

Objective:
Recognize when statistics are misleading.

Tom wants to convince his parents that his math grades are improving. He decides to make bar graphs that show his math grades for each grading period. Graphs A and B show Tom's grades. Although they display the same data, each graph gives a different impression.

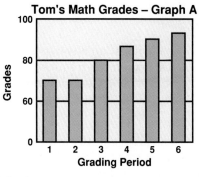 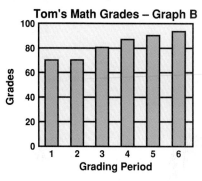

Which graph shows the greatest improvement in Tom's grades? Graph A does. However, it is misleading. Notice the vertical axis. The distance from 0 to 60 is the same as the distance from 60 to 80. This is an incorrect representation of the data and can lead to wrong conclusions.

To correct this graph, you can draw a broken line at the bottom of the vertical axis as shown at the right.

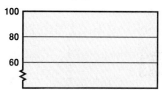

Example

1 The two graphs display the same ticket sales information for the Drexel Theater.

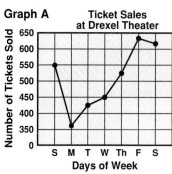

Is Graph A misleading? If so, why? Graph A is misleading because the distance between 0 and 350 is the same as the distance between 350 and 400 on the vertical axis.

Is Graph B misleading? If so, why? Graph B is misleading because there is no title and there are no labels on either scale.

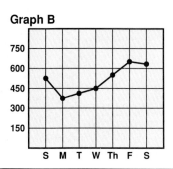

Another way statistics can be misleading is by using the wrong average. Following are some guidelines for best use of the measures of central tendency.

- The mode can be used with either items or numbers. Use the mode when the most frequent item or number best represents the data.
- Very large or very small numbers affect the mean. Therefore, use the mean when the range is not great.
- Very large or very small numbers have little effect on the median. Therefore, use the median when the range is great.

Example

2 Employee salaries at an advertising agency are shown in the frequency table at the right. The mode and the median wage is $24,000. The mean wage is $30,815.

Salary	Number
$100,000	1
$75,000	2
$50,000	3
$24,000	15
$12,000	6

Which "average" best represents the data? Why?

Which "average" would the union want to use in its wage talks? Why? They might use the mode or median because it makes the current salaries look lower.

Which "average" would the factory owner use? Why? They might use the mean wage because it makes current salaries look higher.

Checking for Understanding

Communicating Algebra

1. What are some ways in which graphs can mislead?

2. When should the mean be used instead of the median?

Which average, *mode*, *median*, or *mean*, would best describe the following? Explain each answer.

3. eye color

4. height of the players on a basketball team

5. ages of persons in a theater

6. height of the fans at a game

Guided Practice

Linda Kline made two different line graphs showing the monthly sales for her bakery.

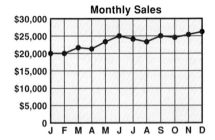

Answer each question about the graphs shown above.

7. Explain why these graphs made from the same data look different.

8. Which graph would she show to someone who wanted to buy her bakery? Explain.

9. Which graph would she show an employee who asked for a big raise? Explain.

Exercises

Independent Practice

The personnel manager of a local law firm made the following chart of its employees.

	Ana Sarafin	Doris Nethers	Jerry Anderson	Kimiko Chung	Brian Sommers
Age	29	32	35	38	31
Salary	$47,000	$13,000	$25,500	$66,000	$34,700
Height (cm)	154	160	165	157	149
Favorite Entertainment	movies	theater	books	sports	sports

Use the chart to find the following. State the type of average you used.

10. average age

11. average salary

12. average height

13. average favorite entertainment

14. How can you describe the "average" employee in this group?

Mixed Review

15. Write 93,200 in scientific notation.   (Lesson 6-11)

16. Find the mean, median, and mode for the data set: 2.4, 7.0, 3.8, 5.6, 4.1, 4.5. Round to the nearest tenth.   (Lesson 10-2)

17. Refer to the bar graph on page 388. How many cans of ginger ale should Terry order?   (Lesson 10-8)

Applications

Solve. State the type of average you used and why.

18. **Consumer Surveys**   A survey at a grocery store showed that 5 people preferred Brand A, 15 people preferred Brand B, 27 people preferred Brand C, and 10 people preferred Brand D. Which brand did the "average" customer prefer?

19. **Real Estate**   A subdivision has houses that are priced $70,000, $75,000, $82,000, $88,000, and $150,000. Which price gives a better description of the houses in that subdivision? Why?

Critical Thinking

20. **Research**   Find a graph in a magazine or newspaper. Redo the graph so that the data will appear to show different results.

Wrap-Up

21. Describe two uses of statistics that can mislead the consumer.

Review

Language and Concepts

Choose the correct term to complete each sentence.

1. The (median, mode) is the number or item that appears most often in a data set.

2. The (stem, leaf) of a stem-and-leaf plot is the greatest place value of the data.

3. Two *different* sets of data (can, cannot) have the same mean, median, and mode.

4. An (extreme value, outlier) in a box-and-whisker plot is any piece of data that is more than 1.5 times the interquartile range from the quartiles.

Skills

The table shows test scores for a biology class. Use the table to answer each question. (Lesson 10-1)

5. How many students are in the sample?

6. What scores occurred most often?

7. What scores occurred least often?

8. Would it be correct to say that 50% of the students scored at least 17?

9. Make a histogram of the data.

Score	Test Scores Tally	Frequency
19-20	ⵜ	5
17-18	ⵜ IIII	9
15-16	ⵜ	5
13-14	III	3
11-12	I	1
9-10	II	2

Find the mean, median, and mode for each set of data. (Lesson 10-2)

10. 156, 175, 212, 167, 628, 156

11. 2.6, 3.1, 6.8, 4.9, 5.7, 3.4, 4.3

Use the stem-and-leaf plot to answer each question. (Lessons 10-3, 10-4)

12. What were the least and greatest movie attendance?

13. What is the median of the data?

14. What are the lower and upper quartiles?

15. What is the interquartile range?

16. What is the range of the data?

Movie Attendance

```
0 | 7
1 | 7 9           2|1 means 21.
2 | 1 5 5 8
3 | 0 0 2 4 5 9
4 |
5 |
6 |
7 | 8
```

Use the box-and-whisker plot to answer each question.   (Lesson 10-5)

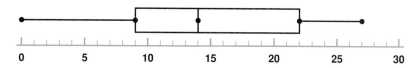

17. What is the median?

18. What is the upper quartile?

19. What is the lower quartile?

20. What is the interquartile range?

21. What are the extremes?

22. Are there any outliers?

Use the scatter plot to answer each question.   (Lesson 10-6)

23. What kind of relationship, if any, is shown by the scatter plot?

24. What does the point shown in red represent?

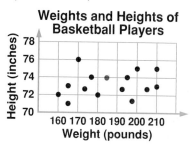

Applications and Problem Solving

25. The lockers in the hall outside Joanna's homeroom start with number 256 and end with number 375. How many lockers are in the hall?   (Lesson 10-7)

26. In a sample, 25 of 75 students surveyed would choose a blue notebook cover. If 500 students buy notebook covers, how many blue covers should be ordered? (Lesson 10-8)

ortfolio Suggestion

Select one of the assignments from this chapter that you found especially challenging and place it in your portfolio.

Curriculum Connection

- **Consumer Awareness**   Find at least two current ads for different brands of CD players. Determine which brand offers the better buy, technologically.

Read More About It

James, Elizabeth.   *What Do You Mean by "Average"?*

Robertson, Bruce.   *How to Draw Charts and Diagrams.*

Test

Use the data in the temperature chart to complete the following.

1. Make a frequency table of the data.

2. Find the range of the data.

3. Find the mean of the data.

4. Find the mode of the data.

5. Find the median of the data.

6. Make a stem-and-leaf plot of the data. Describe the distribution of the data.

High Temperature (°C) for Houston for a 20-day Period				
21	25	16	18	18
26	21	17	19	21
22	24	26	21	22
17	16	17	20	25

Use the data in the television-viewing chart to complete the following.

7. Make a scatter plot of the data.

8. Does there appear to be a positive, negative, or no relationship between age and number of hours of TV watched weekly?

9. Make a box-and-whisker plot of the hours.

10. What is the range of the hours?

11. What is the interquartile range?

12. What is the median number of hours of TV watched weekly?

13. What percent of the people surveyed watch more than 36 hours of TV per week?

Television Viewing Per Week

Age	Hours	Age	Hours
12	20	48	40
16	19	21	19
10	19	67	41
24	34	15	25
25	21	34	25
50	46	17	28
58	32	6	32
71	49	22	33
62	38	18	22
8	25	4	36

Solve.

14. Find the sum of the first 50 odd numbers. How does this compare to the sum of the first 50 even numbers?

15. Six out of 70 students surveyed at Midview School said they would like to start a Spanish club. The school's policy says there must be 30 interested students to start a new club. If there are 560 students in the school, predict how many students would be interested in a Spanish club. Is this enough to start a new club?

Use the graph to the right to complete the following.

16. What is wrong with this graph?

17. If the graph were corrected, what could you say about the profits from 1987 to the present?

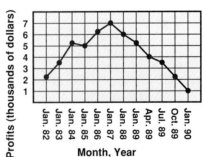

Academic Skills Test

Cumulative, Chapters 1-10

1. Which equation is equivalent to $1.8a = 36$?

   A $a = 36 \div 1.8$
   B $a = 36 - 1.8$
   C $1.8a - 1.8 = 36$
   D $1.8a = 36 \div 1.8$

2. If $\frac{n}{11} = 33$, what is the value of n?

   A 3   **C** 44
   B 22   **D** 363

3. Cindy needs $\frac{3}{4}$ cup of sliced almonds for a recipe. One bag contains about $1\frac{1}{2}$ cups. How much will be left?

   A $\frac{3}{4}$ cup

   B 1 cup

   C $1\frac{1}{4}$ cups

   D $2\frac{1}{4}$ cups

4. If $\frac{b}{456} = 1.2$, what is the value of b?

   A 38
   B 380
   C 547.2
   D 5472

5. A mail-order card company charges 50¢ for each greeting card plus a handling charge of $1.50. Which sentence could be used to find n, the number of cards ordered, if the total charge was $9?

   A $0.5n + 1.5 = 9$
   B $0.50n + 1.50n = 9$
   C $9 = 50n + 1.50$
   D $9 = (0.5 + 1.50)n$

6. Scott generally runs a mile in anywhere from 7.5 to 8.5 minutes. Which number line shows this range of times?

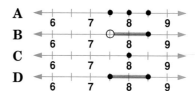

7. In a survey, 75 percent of the people surveyed like pizza. If 36 people were surveyed, which proportion will find n, the number who like pizza?

   A $\frac{75}{36} = \frac{n}{100}$ **C** $\frac{75}{100} = \frac{36}{n}$

   B $\frac{75}{n} = \frac{36}{100}$ **D** Not Here

8. The number of students in Sam's aerobics class increased from 15 to 24. What was the percent of increase?

   A 37.5%   **C** 62.5%
   B 60%   **D** 135%

9. The high temperatures for a week in April were 56°F, 58°F, 60°F, 63°F, 58°F, 62°F, and 70°F. What was the median temperature?

   A 58°F   **C** 61°F
   B 60°F   **D** 63°F

10. Chapter 10 of the book Milagros is reading starts on page 286 and ends on page 319. How long is the chapter?

    A 32 pages   **C** 34 pages
    B 33 pages   **D** Not Here

CHAPTER OBJECTIVES

In this chapter you will learn to:

- ■ count outcomes
- ■ find permutations and combinations
- ■ find probabilities of simple events and compound events
- ■ solve problems by making a table
- ■ use simulations to investigate problems

Probability

What are the chances that two Presidents of the United States share the same birthday? Considering that there are only 40 men in that group, you might think that the chances are pretty slim.

You may be surprised to know that in a group of only 24 people, the likelihood that two of them have birthdays on the same day is just about 50%, or one out of two. With 40, the likelihood is 85%!

So what about the group of presidents? Do any of them share a birthday? Two birthdays are shared!

- James Polk and Warren Harding were born on November 2.
- Andrew Johnson and Woodrow Wilson were born on December 29.

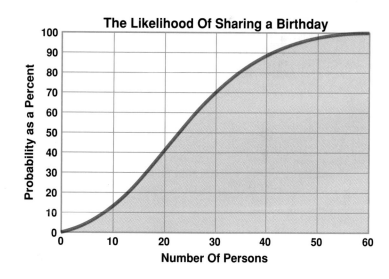

The Likelihood Of Sharing a Birthday

Statistics Connection
Class Project
Survey groups of 24 people in your school. Tally their birthdays.
Do 50% of your groups contain a shared birthday? Display your findings in a chart or graph.

11-1 Counting Using Tree Diagrams

Objective:
Use a tree diagram to count outcomes.

Key Terms:
outcome
tree diagram
event

Beth, Juan, and Tim are running for Student Council president at Monroe High School. Kelly and Hiroshi are running for vice-president. How many possible combinations are there for president and vice-president?

You can draw a diagram to find the number of possible combinations or **outcomes**.

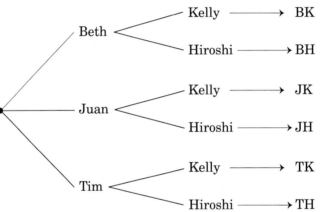

President	Vice-President	Outcomes
Beth	Kelly	BK
	Hiroshi	BH
Juan	Kelly	JK
	Hiroshi	JH
Tim	Kelly	TK
	Hiroshi	TH

There are six possible outcomes.

The diagram above is called a **tree diagram**. You can use a tree diagram to list all the possible outcomes of other **events** like tossing a coin or rolling a die. For example, if you toss a penny and a nickel at the same time, a tree diagram can show the possible outcomes.

How does TH differ from HT?

The possible outcomes are HH, HT, TH, and TT.
HH means heads on both coins.

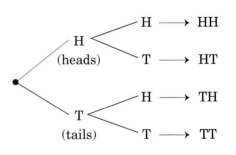

Penny	Nickel	Outcomes
H (heads)	H	HH
	T	HT
T (tails)	H	TH
	T	TT

If you toss a penny, a nickel, and a dime, how many possible outcomes are there?

There are four possible outcomes.

Example

1 Use a tree diagram to find all the possible outcomes in the situation described below.

The Corner Sub Shop serves two types of sub sandwiches, a ham sub and a turkey sub. They also serve three beverages, milk, coffee, and soda. Suppose you choose one sub sandwich and one beverage. How many possible choices are there?

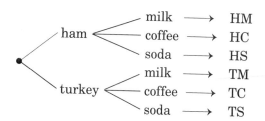

There are six possible choices, or outcomes.

Checking for Understanding

Communicating Algebra

1. Explain how the tree diagram at the right can be used to count the different sweatshirts available at the school store.

2. List the outcomes illustrated in the tree diagram at the right.

3. Suppose you toss a penny and get heads. What is the event and what is the outcome?

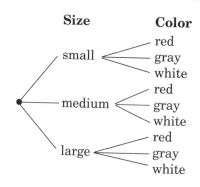

Guided Practice

State the number of possible outcomes for each event.

4. tossing a quarter

5. spinning the spinner

Refer to Example 1 above.

6. How many outcomes include a turkey sub?

7. How many outcomes include milk?

8. Include a pastrami sub with the sandwich choices and juice with the beverage choices. Draw the new tree diagram. What are the new outcomes?

2 Rosa's Deli makes sandwiches for their customers. Customers have a choice of four types of meat (ham, turkey, pastrami, and corned beef), four types of cheese (cheddar, Swiss, mozzarella, and American), and four types of bread (white, wheat, rye, and sour dough). How many different sandwiches can be made with one type of meat, one type of cheese, and one type of bread?

$$\begin{pmatrix}\text{number of}\\\text{choices for}\\\text{meat}\end{pmatrix} \times \begin{pmatrix}\text{number of}\\\text{choices for}\\\text{cheese}\end{pmatrix} \times \begin{pmatrix}\text{number of}\\\text{choices for}\\\text{bread}\end{pmatrix} = \begin{pmatrix}\text{number of}\\\text{possible}\\\text{outcomes}\end{pmatrix}$$

$$4 \quad \times \quad 4 \quad \times \quad 4 \quad = \quad 64$$

There are 64 possible sandwiches.

Checking for Understanding

Communicating Algebra

1. Write a problem that corresponds to the tree diagram at the right.

2. Use the Fundamental Counting Principle to write a mathematical equation that corresponds to the tree diagram.

3. What advantage does the Fundamental Counting Principle have over a tree diagram?

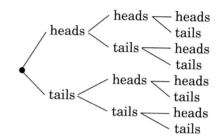

4. In literature class, each student must choose one short story and one poem to read for homework. The students must choose from a list of g short stories and h poems. Write an algebraic expression for the number of possible outcomes.

Guided Practice

Use multiplication to answer each question.

5. Suppose there are only two types of cheese in Example 2. How many sandwiches are possible?

6. Three dice are rolled. How many outcomes are possible?

7. Four coins are tossed. How many outcomes are possible?

8. A quiz has five true and false questions. How many outcomes for giving answers to the five questions are possible?

9. Judy has 5 blouses, 6 skirts, and 4 scarves. How many three-piece outfits are possible?

Exercises

Independent Practice

10. The spinner at the right is spun three times. How many outcomes are possible?

11. A die is rolled. Then a coin is tossed. How many outcomes are possible?

12. Draw a tree diagram that represents the situation described in Exercise 11. How many outcomes show heads? How many outcomes show 6? How many outcomes show both 3 and tails?

13. At Ben's Cafeteria, dinner consists of one main dish, one vegetable, and one beverage. The main dish choices are chicken, steak, or fish. The vegetable choices are corn, green beans, or broccoli. The beverage choices are juice or milk. How many dinners are possible?

14. Draw a tree diagram to represent the situation described in Exercise 13. How many outcomes show chicken? How many outcomes show steak and corn?

15. If you include iced tea as a beverage choice in Exercise 13, how many dinners would be possible?

16. A quiz has ten true-false questions. How many outcomes for giving answers to the ten questions are possible?

17. A multiple-choice quiz has five questions. Each question is answered with a, b, c, or d. How many outcomes are possible?

18. Suppose e is included as a possible answer for the questions in Exercise 17. How many outcomes are possible?

19. A quiz has five true-false questions and five multiple-choice questions. If each multiple-choice question has four choices, how many outcomes are possible?

Mixed Review 20. Express $3\frac{9}{20}$ as a decimal.   (Lesson 6-1)

21. Refer to the Favorite Color table on page 364. What percent of the people chose red as their favorite color?   (Lesson 10-1)

22. Find the number of possible outcomes: *red, green, pink, or blue wrapping paper with a yellow or white ribbon.*   (Lesson 11-1)

Applications 23. **Clothing**   Jeff is buying clothes to take to college. He buys three pairs of slacks, three shirts, and two sweaters. How many days could he wear a different three-piece outfit before repeating any outfits?

24. **Business**   In a cafeteria there are three different main courses, two vegetables, three kinds of potatoes, and three beverages. Each meal has a separate price programmed into the cash register. If each meal consists of a main course, a vegetable, a potato, and a beverage, how many different meals had to be programmed?

Critical Thinking 25. A test has all true-false questions. If there are x questions on the test, write an algebraic expression for the number of possible outcomes.

Wrap-Up 26. In your own words, explain the Fundamental Counting Principle.

Algebra in Action-Communication

Telephone Area Codes

Telephone numbers such as 213-555-1212 are made up of a three-digit area code (213), a three-digit exchange (555), and a four-digit extension (1212). You can use the Fundamental Counting Principle to find how many telephone numbers are possible in an area code.

How many extensions are possible in one exchange? Because any digit 0-9 can be used in each of the four digit places, there are $10 \cdot 10 \cdot 10 \cdot 10$ or 10,000 numbers in an exchange.

How many exchanges would you expect? Since 0 is used to dial the operator and 1 is used to start dialing a long distance number, the first digit of an exchange cannot be 0 or 1. It appears that there are $8 \cdot 10 \cdot 10$ or 800 exchanges in an area code. However, exchanges such as 911, 976, 800, and 555 cannot be used because they have other uses. There are actually only 774 exchanges available in each area code.

So, in each area code, there are $774 \cdot 10,000$ or 7,740,000 telephone numbers available. When the number of telephone numbers increases to a certain point, a new area code is formed.

1. How many exchanges are not available because of special numbers such as 911?
2. If the three-digit area code must contain a 0 or a 1 as the second digit, how many area codes appear to be available? Remember that 0 and 1 cannot be used for the first digit.

11-3   Permutations and Combinations

Objective:
Find permutations and combinations.

Key Terms:
permutation
factorial
combination

The faculty at Jackson Junior High School is sponsoring a career day. Amy wants to hear about the following careers.

> lab technician
> athletic director
> news reporter
> pilot
> dentist

Each student can only attend three sessions. Amy must choose three of the five sessions and then sign up in the guidance department. She will choose a first session, second session, and third session. As she is signing up, she wonders how many possible arrangements there are.

Amy decides that making a list is too time consuming. She reasons that there are 5 possible choices for the first session. After that session is chosen, there are 4 possible choices for the second session. Finally, there are 3 possible choices for the third session. Using the Fundamental Counting Principle, she decides that there are $5 \cdot 4 \cdot 3$ or 60 possible arrangements.

A **permutation** is an arrangement or listing in which order is important. In the above example, the symbol $P(5, 3)$ represents the number of permutations of 5 things taken 3 at a time.

$$P(5, 3) = 5 \cdot 4 \cdot 3$$

In general, $P(n, r)$ means the number of permutations of n things taken r at a time. Another way to write this is $_nP_r$.

Notice that $P(5, 5)$ means the number of permutations of 5 things taken 5 at a time.

$$P(5, 5) = 5 \cdot 4 \cdot 3 \cdot 2 \cdot 1$$

The mathematical notation 5! means $5 \cdot 4 \cdot 3 \cdot 2 \cdot 1$. The symbol 5! is read five **factorial.** In general, $n!$ means the product of all counting numbers beginning with n and counting backwards to 1. We define 0! as 1.

Calculator Hint

The factorial key $\boxed{n!}$ on your calculator provides a fast way to compute factorials. Find 6!

6 $\boxed{n!}$ 720

Check 0! on your calculator.

Suppose Amy decides that the order in which she hears the speakers is not important. In other words, {lab technician, pilot, dentist} is the same as {pilot, dentist, lab technician}.

Arrangements or listings where order is not important are called **combinations.** To find the number of combinations without duplication, divide the number of permutations by the number of orders of the speakers. There are 3! or $3 \cdot 2 \cdot 1$ ways to order three speakers. The symbol $C(5, 3)$ means the number of combinations of 5 things taken 3 at a time.

$$C(5, 3) = \frac{P(5, 3)}{3!}$$

$$= \frac{5 \cdot 4 \cdot 3}{3 \cdot 2 \cdot 1} \text{ or } 10$$

In general, $C(n, r)$ means the number of combinations of n things taken r at a time.

$$C(n, r) = \frac{P(n, r)}{r!}$$

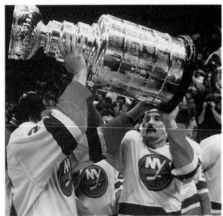

Examples

1 Six hockey teams are in the league. How many ways can the teams win the first and second place trophies? Assume there are no ties for the top places.

You must find the number of permutations of 6 teams taken 2 at a time. Of course, order of finish is important.

$$P(6, 2) = 6 \cdot 5 = 30$$

There are 30 different ways the teams can win the trophies.

2 Pizza Pizza Pizza has a choice of eight different toppings for their pizzas. They are having a special on pizzas with three toppings. How many different pizzas with three toppings can they serve?

Order does not matter, so you need to find the number of combinations of eight things taken three at a time.

$$C(8, 3) = \frac{P(8, 3)}{3!}$$

$$= \frac{8 \cdot 7 \cdot 6}{3 \cdot 2 \cdot 1} \text{ or } 56$$

There are 56 different pizzas with three toppings.

Why should you find the number of combinations rather than permutations?

Checking for Understanding

Communicating Algebra

1. Explain what 5! means.
2. What is the difference between $P(6, 3)$ and $C(6, 3)$?
3. Without computing, tell which is larger, $P(6, 3)$ or $C(6, 3)$. Why?

Guided Practice

Tell whether each exercise represents a *permutation* or *combination*.

4. first, second, and third chairs for five trumpets in a band
5. a five-person committee from your class
6. four new CDs from a group of eight
7. five outfits from 15 outfits to be used for a window display
8. five-letter "word" from 20 letters
9. arranging five out of ten dogs in a line

Find each value.

10. 4!
11. 6!
12. 3!
13. $P(5, 2)$
14. $P(8, 3)$
15. $C(5, 2)$

Exercises

Independent Practice

Find each value.

16. 8!
17. 10!
18. $\dfrac{7!3!}{5!0!}$
19. $\dfrac{9!5!0!}{10!}$
20. $P(6, 3)$
21. $P(8, 4)$
22. $C(8, 4)$
23. $C(10, 5)$

24. How many four-letter "words" can be made from C, A, R, and D?
25. How many of the "words" from Exercise 24 will be CARD?

Mixed Review

26. Evaluate $\dfrac{2(3a - b)}{a - 4}$ for $a = 6$ and $b = 11$.   (Lesson 1-2)

27. **Consumer Awareness**   How much would the discount be on a $54 pair of running shoes at 25% off?   (Lesson 9-8)

28. Two dice are rolled and a coin is tossed. How many outcomes are possible? (Lesson 11-2)

Applications

29. **Sports**   There are four runners in a race. In how many orders can the runners finish the race? Assume there are no ties.

30. **Business**   A candy salesperson has 10 different kinds of boxed candy. Only four kinds can be displayed in the counter window at any one time. How many ways can the products be displayed?

Connection

31. **Geometry**   Ten points are marked on a circle. How many line segments can be drawn between any two of the points?

Critical Thinking

32. Ten people are running for chair and vice-chair of the Social Studies Club. After they are selected, three directors will be elected from the remaining candidates. How many different ways can the offices be filled?

Wrap-Up

33. Explain how the permutations of a set of items help in finding the number of combinations of the same set.

Games

Materials: dice or number cubes

In this Exploration you will investigate fair and unfair games. A fair game is defined as one in which each side has an even chance of winning. If the game is weighted in some way so that one side has a better than even chance of winning, the game is unfair.

Suppose a dime is tossed 50 times. Each time the coin shows a head, a win is recorded. Each time the coin shows tails, a loss is recorded.

Wins	Losses
ⵏⵏⵏ ⵏⵏⵏ ⵏⵏⵏ ⵏⵏⵏ IIII	ⵏⵏⵏ ⵏⵏⵏ ⵏⵏⵏ ⵏⵏⵏ ⵏⵏⵏ I

Since there are approximately the same number of wins as losses, the game of tossing a dime seems to be fair.

Your Turn: Work with a partner and play each game.

▶ Roll two dice 50 times and find the sum of the numbers on the dice. Record an even sum as a win; record an odd sum as a loss.

▶ Roll two dice 50 times and find the product of the numbers on the dice. Record an even product as a win; record an odd product as a loss.

Analysis

1. Which game is fair? Which is unfair? Explain why.

2. Modify the rules of the unfair game in order to make it fair.

3. Jake is blindfolded. He picks one of the marbles at the right. If the marble is red, he wins. If the marble is blue, he loses. Do you think this is a fair game?

4. Two dice are rolled. If both dice show the same number, you win. If the dice show different numbers, you lose. Do you think this is a fair game?   Why or why not?

11-4 Probability

Objective:
Find the probability of a simple event.

Key Terms:
probability
sample space

A weather forecaster says that there is a 70% chance of rain. What does this mean? It means that the probability of rain is 70%. **Probability** is the chance that some event will happen.

The symbol $P(\text{rain})$ is used to mean the probability that it will rain.

$$P(\text{rain}) = 70\% \text{ or } 0.7$$

On any given day, the probability of its raining or not raining is 100%. That means that $P(\text{rain}) + P(\text{no rain})$ equals 1. So, on a day when there is a 70% chance of rain, we know that the chance of no rain is 30%.

$$P(\text{no rain}) = 0.3$$

If you look out the window and see that it is already raining, the probability of rain is 100% (or 1). In that case, the probability of a completely sunny day is 0% (or 0).

Probability is also defined as the ratio of the number of ways that a certain outcome can occur to the number of possible outcomes.

Business Connection

Insurance companies use probability theory when they issue life insurance. The companies need to estimate how long the policy holder will live. This estimate of probability is called life expectancy.

Definition of Probability

$$\text{Probability} = \frac{\text{number of ways that a certain outcome can occur}}{\text{number of possible outcomes}}$$

The set of all possible outcomes is called the **sample space.** You can find the sample space in many ways. You already know how to make a list, construct a tree diagram, use the Fundamental Counting Principle, and use permutations and combinations to find a sample space.

Examples

There are 2 red pens, 3 blue pens, 1 black pen, and 4 green pens in a desk drawer. All the pens are the same size and shape. Suppose you open the drawer and grab a pen without looking. In other words, you are choosing the pen at random and each outcome is equally likely.

Estimation Hint

Use estimation to determine if P(blue) is less than or greater than $\frac{1}{2}$.

1 **What is the probability that the pen chosen is blue?**

The number of blue outcomes is 3. The number of possible outcomes is $2 + 3 + 1 + 4$ or 10.

$$\text{Probability} = \frac{\text{number of ways that a certain outcome can occur}}{\text{number of possible outcomes}}$$

$$P(\text{blue}) = \frac{3}{10}$$

Can a probability be greater than 1? Why?

2 **What is the probability that the pen chosen is either red or green?**

$$P(\text{red or green}) = \frac{6}{10} \qquad \begin{array}{l}\text{number of red or green outcomes} \\ \text{number of possible outcomes}\end{array}$$

Can a probability be less than 0? Why?

3 **What is the probability that the pen is purple?**

The probability is 0.   Why?

Checking for Understanding

Communicating Algebra

1. What is a sample space?

2. In your own words explain what is meant by probability. Explain how to find the probability of an outcome.

3. Give an example of an outcome with a probability of 0.5.

Guided Practice

State the probability of each outcome.

4. The sun will rise tomorrow.

5. A coin is tossed and shows tails.

6. Your friend will live to be 300 years old.

7. Today is Monday.

The spinner shown below is equally likely to stop in any one of the eight regions. It is spun once. Find the probability of each outcome.

8. a two

9. an odd number

10. not a seven

11. a nine

12. prime number

13. a number less than four

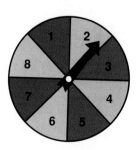

Exercises

Independent Practice

There are 3 blue marbles, 6 red marbles, 2 green marbles, and 1 black marble in a bag. Suppose you select one marble at random. Find each probability.

14. P(blue)
15. P(black)
16. P(green)
17. P(red)
18. P(not green)
19. P(blue or red)
20. P(green or black)
21. P(neither red nor green)
22. P(yellow)
23. P(not orange)

Suppose you roll a die and toss a coin.

24. List all the members of the sample space.
25. What is P(6, tails)?
26. What is P(5, heads)?
27. What is P(8, tails)?

Suppose you roll three dice.

28. How many outcomes are in the sample space?
29. List three members of the sample space.
30. What is P(6, 6, 6)?

Explain the meaning of each of the following.

31. P(heads) $= \frac{1}{2}$
32. P(red or green) $= \frac{1}{5}$

Mixed Review

33. Rename 0.88 as a fraction.   (Lesson 5-2)
34. Solve $18 - 7t \leq -3$.   (Lesson 7-6)
35. Find the percent of increase if the old price of a bus ticket was $0.60 and the new price is $0.75.   (Lesson 9-10)
36. Draw a stem-and-leaf plot of the following temperatures: 54, 71, 58, 66, 64, 67, 58, 69, 63, 51, 70.   (Lesson 10-3)
37. Find the value of P(7, 2).   (Lesson 11-3)

Applications

38. **Sports**   A batter hits 0.300. What is the probability that she will get a hit next time at bat? What is the probability that she will not get a hit?

39. **Sports**   Jane averages two strikes every ten frames of bowling. What is the probability that she will get a strike in the first frame of the next game?

40. **Space Exploration**   What is the probability that humans will explore the moon on foot? What is the probability that humans will explore the sun on foot?

Challenge

All of the possible outcomes that can occur when a die is rolled twice are listed in the chart below. Find the probability of each outcome.

41. 3, 5

42. not 2, 3

43. a sum of 5

44. a sum of 6 or 2

45. a sum of 12

46. a sum of 13

47. 2, 1 or 5, 3

48. both numbers odd

49. the same number on both dice

50. the first number is greater than the second number

51. a sum greater than 9

52. a sum less than 15

1, 6	2, 6	3, 6	4, 6	5, 6	6, 6
1, 5	2, 5	3, 5	4, 5	5, 5	6, 5
1, 4	2, 4	3, 4	4, 4	5, 4	6, 4
1, 3	2, 3	3, 3	4, 3	5, 3	6, 3
1, 2	2, 2	3, 2	4, 2	5, 2	6, 2
1, 1	2, 1	3, 1	4, 1	5, 1	6, 1

Critical Thinking

53. In a box you have cards with the letters W, E, A, T, H, E, and R. What is the probability of drawing the letters W E T in order? (Hint: Find the total number of outcomes and the number of outcomes in which the letters W E T are drawn in order.)

Wrap-Up

54. Write a problem in which the answer will be a probability of 0.75.

Biography

Blaise Pascal

Blaise Pascal was a French mathematician and physicist who lived from 1623 to 1662. Although he lived to be only 39 years old, he accomplished much in his short lifespan.

His father had wanted him to study ancient languages and denied him any books on mathematics. However, when he learned the first thirty-two theorems of Euclidean geometry at the age of 9, his father allowed him to study mathematics.

At 16 he published a book on the geometry of conic sections that far surpassed any work in this area since Apollonius in the third century B.C. When he was 19, he invented the ancestor of our pre-electronic calculating machines. However, it was too expensive to build at the time.

Pascal later corresponded with Pierre de Fermat, and together they became the founders of the modern theory of probability. Today, by applying the theory of probability, physicists can understand and develop theories about the atom.

11-5 Make a Table

Objective:
Solve problems by
making a table.

Parcheesi is a game that was first played in India. In this game 2 dice are rolled. A token can be put into play if a total of 5 is shown on the dice or if a 5 is shown on at least one of the dice. What is the probability that a token can be put into play in one roll of the dice (a favorable outcome)?

Explore There are 2 dice. You must find the probability of either rolling a total of 5 on both of the dice or rolling a 5 on at least one of the dice.

Is the probability of getting a token into play greater than or less than $\frac{1}{2}$?

Plan A table will help to find the number of favorable outcomes as well as the total number of outcomes. Using the information from the table, the probability can be determined.

Solve

First
Die

Second Die

	1	**2**	**3**	**4**	**5**	**6**
1	1 + 1 = 2	1 + 2 = 3	1 + 3 = 4	(1 + 4 = 5)	(1 + 5 = 6)	1 + 6 = 7
2	2 + 1 = 3	2 + 2 = 4	(2 + 3 = 5)	2 + 4 = 6	(2 + 5 = 7)	2 + 6 = 8
3	3 + 1 = 4	(3 + 2 = 5)	3 + 3 = 6	3 + 4 = 7	(3 + 5 = 8)	3 + 6 = 9
4	(4 + 1 = 5)	4 + 2 = 6	4 + 3 = 7	4 + 4 = 8	(4 + 5 = 9)	4 + 6 = 10
5	(5 + 1 = 6)	(5 + 2 = 7)	(5 + 3 = 8)	(5 + 4 = 9)	(5 + 5 = 10)	(5 + 6 = 11)
6	6 + 1 = 7	6 + 2 = 8	6 + 3 = 9	6 + 4 = 10	(6 + 5 = 11)	6 + 6 = 12

There are 4 ways of getting a total of 5 and 11 ways of getting a 5 on at least one die. This gives a total of 15 favorable outcomes. Since the sample space has 36 outcomes, the probability is $\frac{15}{36}$ or $\frac{5}{12}$.

Examine this solution.

Checking for Understanding

Communicating Algebra

1. How does the table above help you to solve the problem?

Guided Practice

Use the table above to answer Exercises 2 and 3.

2. In many games where 2 dice are thrown, doubles (the same number showing on both dice) are important. What is the probability of throwing doubles?

3. What is the probability that 2 dice will show a total less than 5?

4. Make a table that can be used to solve probability problems about tossing two coins.

Exercises

Independent
Practice

Solve. Use any strategy.

5. If 2 coins are tossed, what is the probability that both coins show heads?

ESTIMATION
MENTAL MATH
CALCULATOR
PAPER/PENCIL

6. In planning the big spring party, a chairperson is selected. The chairperson picks a person to head each of the 5 committees. Each person who heads a committee finds 6 people to work on his or her committee. How many people are helping to plan the spring party?

7. Pablo collects postage stamps from foreign countries. Half of his stamps are from France. Half of the remaining stamps are from Japan. After setting aside the stamps from France and Japan, one fifth of the remaining stamps are from Spain. If there are 5 stamps from Spain, how many stamps does Pablo have altogether?

8a. A tetrahedron is a 3-dimensional figure with 4 triangular faces the same size and shape. Assume there are 2 tetrahedron-shaped "dice" with a number from 1 to 4 marked on each face. The "dice" are rolled on the floor. What is the probability that the sum of the numbers facing the floor is 3?

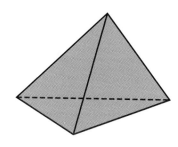

8b. What is the probability that the sum will be greater than 4?

9. The owner of Rosa's Deli charges $1.80 for 10 ounces of pasta salad. She charges $2.16 for 12 ounces of the same pasta salad. At this rate, what would a pound of pasta salad cost?

10a. A game has 2 spinners as shown at the right. After spinning both spinners, what is the probability that the sum of the numbers is 8?

10b. What is the probability that the sum is less than 8?

Mixed Review

11. Find the mean, median, and mode for the data 12, 17, 9, 8, 20, 10, 15, 10, 9, 7, 10, 14, 6, 7.   (Lesson 10-2)

12. There are 7 blue marbles, 5 red marbles, and 3 yellow marbles in a bag. If you select one marble at random, what is the probability it will be blue? (Lesson 11-4)

Critical
Thinking

13. The product of three consecutive numbers is 10,626. Find the numbers. Use a calculator.

Wrap-Up

Journal Entry

14. Look at Exercises 5-13. List the Exercises you solved by using a table. How did a table help you to solve each of these exercises?

Making Predictions

Materials: paper bags containing 10 colored golf tees

In this Exploration you will make predictions based on the results of an experiment.

Your Turn: Work with a partner to complete this experiment.

▶ Draw one golf tee from the bag, record its color, and replace it in the bag. Repeat this 10 times.

▶ The term **relative frequency** means what fraction of the time a certain color was drawn. Calculate the relative frequency for each color golf tee. Express the relative frequency as a decimal.

▶ Repeat both steps described above for twenty, thirty, forty, and fifty draws.

Analysis

1. Is it possible to have a certain color tee in the bag and never draw that color? Is this situation likely to happen if you make only two draws? ten draws? fifty draws?

2. Write a paragraph that describes how the relative frequencies changed as you increased the number of draws.

3. The **experimental probability** is an estimate based on the relative frequency you obtained in your experiment. Based on your experiment, what is the experimental probability of drawing each color tee?

4. Predict the colors of the tees in your bag. Open the bag and check your prediction against the tees in the bag.

11-6   Probability of Independent Events

Objective:
Find the probability of independent events.

Key Term:
independent event

Without looking, Poloma selects one ticket at random from each group of tickets. What is the probability that she will select a baseball ticket and a rock concert ticket?

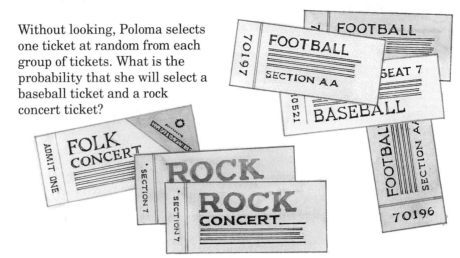

The ticket she chooses from the first group does not affect the ticket she chooses from the second group. We call these **independent events.** That is, the outcome of the one event does not affect the outcome of the other event.

Probability of Two Independent Events	The probability of two independent events can be found by multiplying the probability of the first event by the probability of the second event. $$P(A \text{ and } B) = P(A) \cdot P(B)$$

$P(\text{baseball}) = \dfrac{1}{4}$ $P(\text{rock concert}) = \dfrac{2}{3}$

$P(\text{baseball and rock concert}) = P(\text{baseball}) \cdot P(\text{rock concert})$

$$= \dfrac{1}{4} \cdot \dfrac{2}{3}$$
$$= \dfrac{2}{12} \text{ or } \dfrac{1}{6}$$

The probability that the two events will occur is $\dfrac{1}{6}$.

FYI

Are you afraid to fly? Don't worry too much. The probability of being killed on a commercial airliner is only one in almost two million. There is a greater chance of being injured in your own bathtub— 1 in 1028.

Examples

1 What is the probability that Poloma will select one football ticket and one rock concert ticket?

$P(\text{football}) = \dfrac{3}{4}$ $P(\text{rock concert}) = \dfrac{2}{3}$

$P(\text{football and rock concert}) = \dfrac{3}{4} \cdot \dfrac{2}{3}$

$$= \dfrac{6}{12} \text{ or } \dfrac{1}{2}$$

The probability that the two events will occur is $\dfrac{1}{2}$.

2 Two dice are rolled. Find the probability that an even number is rolled on one die and a prime number is rolled on the other.

P(even number) $= \frac{1}{2}$ P(prime number) $= \frac{3}{6}$ or $\frac{1}{2}$

P(even number and prime number) $= \frac{1}{2} \cdot \frac{1}{2}$

$$= \frac{1}{4}$$

How can the probability of three independent events be found?

The probability that the two events will occur is $\frac{1}{4}$.

Checking for Understanding

Communicating Algebra

1. What are independent events?

2. Two positive numbers, each less than one, are multiplied. What is the relationship between the product and each of the numbers?

Guided Practice

The chart below lists the number and type of chocolates found in two boxes of candy. A milk chocolate is chosen at random. Then a dark chocolate is chosen at random. Find the probability of each outcome.

3. a milk chocolate with nuts and a dark chocolate with nuts

4. a milk chocolate with nuts and a dark chocolate with fruit

5. a plain milk chocolate and a dark chocolate with nuts

6. a milk chocolate with fruit and a plain dark chocolate

Boxes	Nuts	Fruits	Plain
Milk chocolate	4	6	2
Dark chocolate	2	6	6

Exercises

Independent Practice

Each spinner is spun once. Find each probability.

7. P(1 and A)

8. P(2 and B)

9. P(3 and C)

10. P(2 and C)

11. P(1 and B)

12. P(an even number and a vowel)

13. P(an odd number and a consonant)

 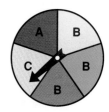

A coin is tossed, then a die is rolled. Find each probability.

14. P(heads and 6)

15. P(heads and number less than 5)

16. P(tails and 5)

17. P(tails and an odd number)

18. P(heads and 3)

19. P(heads and a number greater than 3)

20. P(tails and 2 or 3)

21. P(tails and a prime number)

One bag contains 3 red and 4 white balls. A second bag contains 6 yellow and 3 green balls. One ball is drawn from each bag. Find each probability.

22. P(red and yellow)

23. P(red and green)

24. P(white and green)

25. P(white and yellow)

26. *True* or *false:* $\frac{9}{11} < \frac{5}{7}$.   (Lesson 4-8)

27. Express as a unit rate: *$3.00 for 12 doughnuts.*   (Lesson 9-1)

28. Two dice are rolled. What is the probability that the sum of the numbers shown is equal to or less than 7?   (Lesson 11-5)

Applications   **29. Clothing**   On a rack in his closet, Amos has two green shirts, three red shirts, and four blue shirts. On another rack he has two blue slacks, two brown slacks, and two green slacks. Suppose Amos makes a selection from each rack without looking. What is the probability that he will have an all blue outfit?

30. Business   If 65% of a store's customers are men and 80% of the men have charge cards, what is the probability that a customer chosen at random is male and has a charge card?

Critical Thinking   **31.** The estimated probability that the Chargers will win a game is 0.7. The probability that the Colts will win is estimated as 0.6. If they do not play each other, what is the probability that they will both win? both lose? How do these two probabilities compare? Is this reasonable?

Wrap-Up   **32.** Explain how to find the probability of two independent events.

Mid-Chapter Quiz

1. A diner serves three breakfasts (scrambled eggs, cereal, and pancakes) and two types of juice (orange and apple). Draw a tree diagram that illustrates the outcomes. (Lesson 11-1)

2. For lunch Kelly can choose from six different sandwiches, three different beverages, and three different types of fruit. She picks one sandwich, one beverage, and one fruit. How many outcomes are possible?   (Lesson 11-2)

Find each value.   (Lesson 11-3)

3. 5!   **4.** $P(7, 3)$ **5.** $C(9, 4)$

The first bag has 4 red marbles and 3 green marbles. The second bag has 5 blue marbles and 2 white marbles.   (Lessons 11-4, 11-5)

6. A marble is picked at random from the first bag. Find P(green).

7. A marble is picked at random from the second bag. Find P(orange).

8. A marble is picked at random from each of the bags. Find P(red and white).

11-7 Probability of Dependent Events

Objective:
Find the probability of dependent events.

Key Term:
dependent event

Why is the probability on the second selection $\frac{1}{9}$ instead of $\frac{1}{10}$?

FYI

Imagine that you play baseball and are considering a career in professional baseball immediately after you graduate. You had better be very good. The probability that you'll make it is less than 1%.

The band members sold magazines to help pay for new uniforms. Prizes are awarded to eight students who sold the most magazines. The prizes are three concert tickets and five movie tickets. Starting with the top salesperson, each of the top eight salespersons draws one ticket from a bag. Tina and Jose finished first and second. What is the probability that both Tina and Jose will draw concert tickets?

This is an example of **dependent events** because what Tina draws affects what Jose draws.

For Tina, $P(\text{concert}) = \dfrac{\text{number of concert tickets}}{\text{total number of tickets}} = \dfrac{3}{8}$.

Assume that Tina drew a concert ticket.

For Jose, $P(\text{concert})\ \dfrac{\text{number of concert tickets left}}{\text{number of tickets left}} = \dfrac{2}{7}$.

Therefore, $P(\text{concert tickets for both}) = \dfrac{3}{8} \cdot \dfrac{2}{7}$

$$= \frac{6}{56} \text{ or } \frac{3}{28}.$$

Examples

1 From a deck of 10 cards (5 ten-point cards, 3 twenty-point cards, and 2 fifty-point cards), Nancy needs to pick the 2 fifty-point cards to win a game. What is the probability that she will win?

First selection: $P(\text{fifty-point card}) = \dfrac{2}{10}$

Second selection: $P(\text{fifty-point card}) = \dfrac{1}{9}$

$P(\text{2 fifty-point cards}) = \dfrac{2}{10} \cdot \dfrac{1}{9}$

$$= \frac{2}{90} \text{ or } \frac{1}{45}$$

The probability that Nancy will draw 2 fifty-point cards is $\frac{1}{45}$.

2 If there were 3 fifty-point cards, 4 ten-point cards, and 3 twenty-point cards, what is the probability that Nancy will win the game described in Example 1?

First selection: $P(\text{fifty-point card}) = \dfrac{3}{10}$

Second selection: $P(\text{fifty-point card}) = \dfrac{2}{9}$

$P(\text{2 fifty-point cards}) = \dfrac{3}{10} \cdot \dfrac{2}{9}$

$$= \frac{6}{90} \text{ or } \frac{1}{15}$$

The probability that Nancy will draw 2 fifty-point cards is $\frac{1}{15}$.

Checking for Understanding

Communicating Algebra
1. Give an example of dependent events.
2. Explain the difference between independent events and dependent events.

Guided Practice

Given the ten-card deck in Example 1, find the probability of each outcome.

3. 2 ten-point cards in a row
4. 2 twenty-point cards in a row
5. a ten-point card and then a twenty-point card
6. a twenty-point card and then a fifty-point card
7. a ten-point card and then a fifty-point card
8. 3 twenty-point cards in a row

Exercises

Independent Practice

In a bag there are 3 red marbles, 2 white marbles, and 4 blue marbles. Once a marble is selected, it is not replaced. Find the probability of each outcome.

9. a red marble and then a white marble
10. a white marble and then a blue marble
11. 2 white marbles in a row
12. 2 blue marbles in a row
13. a blue marble three times in a row
14. 2 red marbles in a row
15. a red marble, a white marble, and then a blue marble
16. a white marble three times in a row

Keith makes up a deck of forty cards. The cards are numbered from 1 to 10 and each number comes in four colors (orange, red, black, and blue). Keith selects cards from the deck. Once a card is selected, it is not replaced. Find the probability of each outcome.

17. a 5 and then a 9
18. a 10 and then a 7
19. two 7s in a row
20. three 5s in a row
21. an orange and then a blue
22. a red, a blue, and then an orange
23. a 4, a 5, and then a 6
24. a 10, a 9, and then an 8
25. the orange 10 and then the red 8
26. four 6s in a row

Mixed Review
27. Simplify $9x + (-14x) + 3x$.   (Lesson 2-4)
28. What type of relationship, *positive, negative,* or *none,* is shown by the scatter plot at the right? (Lesson 10-6)

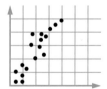

29. Find the value of $C(6, 4)$.   (Lesson 11-3)

30. A coin is tossed, then a die is rolled. Find P(heads and a number less than 3). (Lesson 11-6)

Applications

31. **Business**   A quality control inspector selects seventy samples from the conveyor. Thirty are defective. What is the probability that the next two selected are defective?

32. **Clothing**   Martin has five blue socks and eight black socks in his drawer. He picks one sock at random and then another. What is the probability that he will draw a pair of black socks?

Critical Thinking

33. **Decision Making**   Maria can't decide whether to go out or stay at home. If she goes out, she can go to dinner, to a school dance, to a concert, to a ball game, to a friend's house, or to a mall. She flips a coin to decide whether to go out or to stay home and then rolls a die to see which activity she will do if she goes out. What is the probability that she will go to the mall?

Wrap-Up

34. Give an example that finds the probability of independent events and another example that finds the probability of dependent events.

Team Problem Solving

Rhonda buys four stamps. The postal clerk tears four stamps from a large sheet of stamps as shown.

Rhonda wonders how many different ways the clerk can tear off four attached stamps. Draw all the different formations of stamps. How many formations are there?

11-8   Adding Probabilities

Objective:
Find the probability of compound events by using addition.

Key Term:
mutually exclusive event

For a school bond election, people in the community were surveyed to determine whether they would work for the campaign, donate money, or do both. Of the sample, 15% were willing to work, 27% were willing to give money, and 8% were willing to do both. If a member of the community were selected at random, what is the probability that he or she would be willing to work *or* give money?

A similar example arises when throwing a die. What is the probability of throwing a 5 or a 6? In this case both events cannot happen at the same time. That is, the events are **mutually exclusive.**

When events A and B are mutually exclusive,

$P(A \text{ or } B) = P(A) + P(B)$.

$$P(5 \text{ or } 6) = P(5) + P(6)$$
$$= \frac{1}{6} + \frac{1}{6}$$
$$= \frac{2}{6} \text{ or } \frac{1}{3}$$

Event A is throwing a 5.
Event B is throwing a 6.

In the opening example both events can happen at the same time. That is, a person can work on the campaign *and* give money. They are not mutually exclusive events. In this case,

$$P(A \text{ or } B) = P(A) + P(B) - P(A \text{ and } B).$$

So, $P(\text{work or give money}) =$

$$P(\text{work}) + P(\text{give money}) - P(\text{work and give money}).$$

$$P(\text{work or give money}) = 0.15 + 0.27 - 0.08 \text{ or } 0.34$$

There is a 34% chance that a member of the community chosen at random would be willing to work or give money.

Example

Why are the events in Example 1 not mutually exclusive?

1 A die is rolled. What is the probability of rolling a 2 or a prime number?

In this case the events are not mutually exclusive.

$$P(2 \text{ or prime}) = P(2) + P(\text{prime}) - P(2 \text{ and prime})$$
$$= \frac{1}{6} + \frac{3}{6} - \frac{1}{6}$$
$$= \frac{3}{6} \text{ or } \frac{1}{2}$$

The probability of rolling a 2 or a prime number is $\frac{1}{2}$.

Notice that $P(A \text{ or } B) = P(A) + P(B) - P(A \text{ and } B)$ can be used whenever an **or** situation occurs. However, when the events are mutually exclusive, $P(A \text{ and } B) = 0$.

Examples

Do you think that the probability of the Yankees or the Red Sox winning is greater than or less than $\frac{1}{2}$? Why?

2 The probability that the Yankees will win their next game is 0.4. The probability that the Red Sox will win their next game is 0.6. The probability that they will both win is 0.4×0.6 or 0.24. What is the probability that one or the other will win? (They don't play each other.)

These are not mutually exclusive events.

Therefore, $P(\text{Yankees or Red Sox}) =$

$$P(\text{Yankees}) + P(\text{Red Sox}) - P(\text{Yankees and Red Sox}).$$

$$P(\text{Yankees or Red Sox}) = 0.4 + 0.6 - 0.24 \text{ or } 0.76.$$

The probability that the Yankees or the Red Sox will win is 0.76.

3 A bag contains six blue marbles and three red marbles. A marble is drawn, it is replaced, and another marble is drawn. What is the probability of drawing a red marble and a blue marble in either order?

Consider two draws that are mutually exclusive.

$$P(\text{red and blue}) = P(\text{red}) \cdot P(\text{blue})$$

$$= \frac{3}{9} \cdot \frac{6}{9}$$

$$= \frac{18}{81} \text{ or } \frac{2}{9}$$

$$P(\text{blue and red}) = P(\text{blue}) \cdot P(\text{red})$$

$$= \frac{6}{9} \cdot \frac{3}{9}$$

$$= \frac{18}{81} \text{ or } \frac{2}{9}$$

$$P(\text{red and blue in either order})$$

$$= \frac{2}{9} + \frac{2}{9}$$

$$= \frac{4}{9}$$

The probability of drawing a red and blue is $\frac{4}{9}$.

Checking for Understanding

Communicating Algebra

1. Explain what is meant by mutually exclusive events. Give an example of mutually exclusive events in your own life. For example, walking the dog and practicing the piano are mutually exclusive events.

2. Give an example of two events in your life that are not mutually exclusive.

Guided Practice

A die is rolled. Which probabilities represent mutually exclusive events?

3. $P(3 \text{ or } 4)$

4. $P(4 \text{ or even})$

5. $P(3 \text{ or prime})$

A card is drawn from the cards below. Which events are not mutually exclusive?

6. 7 or odd

7. 6 or prime

8. 9 or greater than 5

$$\boxed{1} \quad \boxed{2} \quad \boxed{3} \quad \boxed{4} \quad \boxed{5}$$

$$\boxed{6} \quad \boxed{7} \quad \boxed{8} \quad \boxed{9} \quad \boxed{0}$$

Suppose $P(A) = \frac{1}{2}$, $P(B) = \frac{1}{3}$, and A and B are mutually exclusive.

9. What is $P(A \text{ and } B)$?

10. What is $P(A \text{ or } B)$?

Exercises

Independent Practice

11. A coin is tossed and a die is rolled. What is $P(\text{heads or } 3)$?

12. A die is rolled. What is $P(\text{even or less than } 5)$?

13. One bag contains three red balls and four white balls. A ball is selected at random, replaced, and another ball is drawn. What is $P(\text{red on either draw})$?

A card is drawn from the cards at the right.

14. What is $P(E \text{ or consonant})$?

15. What is $P(A \text{ or vowel})$?

$$\boxed{A} \quad \boxed{B} \quad \boxed{C} \quad \boxed{D} \quad \boxed{E} \quad \boxed{F} \quad \boxed{G}$$

The chart shows the ages and gender of the students in Melissa's homeroom. A student is to be selected for student council.

16. What is $P(14 \text{ or female})$?

17. What is $P(13 \text{ or male})$?

Age	Male	Female
13	5	8
14	8	9

The chart shows the number and types of chocolates in a box of candy. If a selection is made at random, find each of the following.

Chocolates	Nuts	Fruits	Plain
Light	5	3	7
Dark	7	2	6

18. P(light)

19. P(dark or nut)

20. P(light or fruit)

21. P(dark or plain)

22. P(fruit or plain)

23. P(light or dark)

Mixed Review 24. Graph the equation $y = \frac{1}{3}x + 1$.   (Lesson 8-6)

25. **Cooking**   If a 6-pound beef roast takes 2 hours to cook, at this rate how long will a 9-pound beef roast take to cook?   (Lesson 9-3)

26. A bag of candy contains 6 peppermints and 4 jaw breakers. Without looking, you select two pieces. What is the probability of selecting two peppermints in a row?   (Lesson 11-7)

Application 27. **Business**   An auto dealer finds that of the new cars returning for service, 70% need an oil change, 50% need lubrication, and 35% need both. What is the probability that a new car brought in for service needs either an oil change or a lubrication?

Critical Thinking 28. The Centerville High School volleyball team needs to beat both the Bulldogs and the Hawks to advance to the semifinals of the tournament. The coach believes that the probability of beating the Bulldogs is 0.6 and the probability of beating the Hawks is 0.7. She believes that the probability of beating at least one of them is 0.85. If these probabilities are valid, what is the probability of winning both games?

Wrap-Up 29. Make up a problem where the probability of one event is 0.45, the probability of another event is 0.70, and the probability of both events is 0.20.

Challenge

The referee tosses a coin. The probability of getting heads is $\frac{1}{2}$. Theoretically, this means that if the coin is tossed twice, heads will show once and if the coin is tossed four times, heads will show two times. This information can be graphed and the graph can be used to predict the number of heads for a given number of tosses.

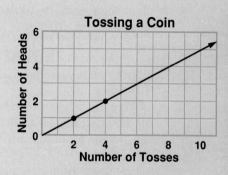

A die is rolled. Make a graph that shows how many times the die, in theory, should show 5 or 6. Use the graph to decide how many times a 5 or a 6, in theory, should show if the die is rolled 45 times.

Computer Simulation

Materials: computer

In this Exploration, you will use a computer to simulate the results of rolling a die and spinning a spinner.

▶ The following BASIC program simulates one hundred rolls of a die.

```
10   FOR R = 1 TO 100
20   LET D = INT(RND*6)+1
30   PRINT D
40   NEXT R
50   END
```

Your Turn: Run the program shown above and record your results.

▶ A game uses a spinner shown at the right. The following BASIC program simulates 100 spins.

```
10   FOR S = 1 TO 100
20   LET D = INT(RND*8)+1
30   PRINT D
40   NEXT S
50   END
```

Your Turn: Run this program and record your results.

Analysis

1. Look at the results from the simulation of rolling a die. Are the results similar to what you expected? Why or why not?

2. Look at the results from the simulation of spinning the spinner. Are the results similar to what you expected? Why or why not?

3. A local sports celebrity is planning a visit to Western Middle School. After she talks to all the students, 20 students will be selected at random to eat lunch with the guest. Design a computer simulation to choose the students.

4. Describe a situation in which a computer simulation is preferred to actually rolling a die or spinning a spinner.

11-9 Strategy: Use a Simulation

Objective:
Investigate problems using simulation.

Key Term:
simulation

Why is it impractical to have a real player act it out?

How is the circle divided?

Have you ever been to a basketball game in which free throws decided the winner? Suppose your team is losing by one point when Sal steps to the foul line.

• If he misses the foul shot, your team loses.
• If he makes the foul shot, he gets one point and another try.
• If he makes the second shot, your team wins the game.

Suppose Sal has a record of making an average of three out of every four free throws. What is the probability that Sal scores 2 points and wins the game?

One way to solve this problem is to do a **simulation**, that is, act it out. You might find a basketball player whose record of making free throws is three out of four. Have the player act out the problem.

A more efficient way is to make a model of the situation. For example, construct a spinner similar to the one shown at the right. Spin the spinner and record the results of several trips to the foul line.

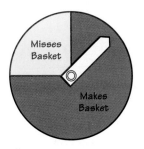

A sample simulation for 25 trips to the foul line is shown in the chart below.

Misses the first shot (0 points)	Makes the first shot, misses the second (1 point)	Makes both shots (2 points)
ⵘⵘ I	ⵘⵘ	ⵘⵘ ⵘⵘ IIII

Calculate the relative frequency for 0, 1, and 2 points.

0 points → $\frac{6}{25}$ = 24%   These are the relative
1 point → $\frac{5}{25}$ = 20%   frequencies for this
2 points → $\frac{14}{25}$ = 56%   simulation only.

Based on this simulation, the probability that Sal scores 2 points and wins the game is 56%.

Example

1   The history quiz has ten true and false questions. Sharon wants to
know if tossing a coin to decide the answers is a good strategy for
taking the quiz. She would write T (true) if tails show and F (false) if
heads show. Assume that the correct answers are F, T, T, T, T, T, F, T,
T, F and that 7 or more correct answers are needed to pass the quiz.

Explore   The quiz has ten questions. Tossing a coin is used to
simulate writing the answers. Would Sharon pass using
this strategy?

Plan   Toss a coin and record the answer for each question.
Check to see how many answers are correct. Repeat the
simulation three times.

Solve

Answers	F	T	T	T	T	T	T	F	T	T	F
Simulation 1	T	Ⓣ	Ⓣ	F	F	F	T	Ⓣ	Ⓣ	Ⓕ
Simulation 2	T	F	Ⓣ	Ⓣ	Ⓣ	Ⓣ	T	F	F	Ⓕ
Simulation 3	Ⓕ	Ⓣ	F	Ⓣ	Ⓣ	F	Ⓕ	F	F	T

*What other ways can you
simulate true and false
answers?*

Circle the correct answers. Since none of the simulations
results in a passing grade, this is not a good way to take
the quiz. Sharon decides to study for the quiz.

Examine   Try some more simulations to confirm the results.

Checking for Understanding

**Communicating
Algebra**

1. In your own words explain what a simulation is.

2. Will a simulation be exactly the same as the actual problem? Why?

3. Suppose Jake had gone to the foul line instead of Sal. Jake has a record of
making one basket out of every two tries. How would you modify the spinner to
simulate this situation?

**Guided
Practice**

4. Conduct three simulations for the basketball situation on page 429. Write a
sentence describing the results.

Exercises

**Independent
Practice**

5. Conduct three more simulations for Example 1. Do your simulations back up
Sharon's decision to study?

6. Colin kicks extra points for the football team. He makes 75% of his attempts for
extra points. How could you simulate the results of the next six attempts?

7. Laneeda is a forward on the freshman basketball team. She usually makes $\frac{2}{3}$ of her shots from the field and $\frac{5}{6}$ of her free throws. If she averages 10 shots a game and 5 free throws, describe a simulation that could give her probable number of points in the next game.

8. Conduct your simulation for the situation in Exercise 7. Record your total from the simulation. Compare your total with those of your classmates.

9. Mr. Namura runs a small gourmet restaurant. He has 12 tables. It is a popular restaurant and anyone wanting to eat at 6:00 on Saturday evening must have reservations. Mr. Namura knows that one out of six reservations usually does not show, so he takes reservations for 14 tables for 6:00. Describe a way to simulate the number of tables that will be filled at 6:00.

10. Conduct your simulation for Exercise 9 ten times. From your simulations, how many times is the restaurant overbooked? Are all 12 tables always filled?

11. At a certain restaurant, prizes are given with children's meals. During the spring promotional, three different prizes are given at random. Estimate how many children's meals must be purchased in order to get all three prizes.

12. Use simulation to test your estimate in Exercise 11. Conduct several simulations. Was your estimate big enough to ensure all three prizes at least 75% of the time? Do you think you could buy fewer meals and still be likely to get all three prizes?

13. Game wardens can estimate the number of fish in a lake by capturing, marking, and releasing fish and then catching fish again after a few days. The warden caught, marked and released 150 fish on Monday. On Thursday she caught 300 fish and 15 of them were marked. How many fish did she estimate were in the lake?

14. Opal and Kara play offense for the junior varsity soccer team. Opal normally scores on one out of every four shots on goal, and Kara normally scores on one out of every six shots on goal. In a typical game, Opal makes seven shots on goal and Kara makes eleven shots on goal. The varsity coach wants one of the girls to play offense for the varsity team. If the coach's goal is to choose the girl who will most likely score more points for the team, which girl should she choose? Why?

15. A football team plays 12 games during the season. A newspaper reporter predicts that the team has a 75% chance of winning each of its games. Devise and carry out a simulation that will determine the probability of the team winning three games in a row during the season.

16. Write a few sentences on why simulations are useful in probability.

Review

Language and Concepts

Choose the correct term to complete each sentence.

1. In probability, spinning a spinner is called an (event, outcome).

2. If one event affects the outcome of another event, the events are (independent, dependent).

3. The fraction, $\dfrac{\text{number of ways a certain outcome can occur}}{\text{number of possible outcomes}}$, defines (probability, the Fundamental Counting Principle).

4. The set of possible outcomes is called a (factorial, sample space).

5. Getting an A in English and getting an A in algebra (are, are not) mutually exclusive events.

Skills

Draw a tree diagram for each of the following. Find the number of possible outcomes. (Lesson 11-1)

6. Each spinner is spun once.

7. a choice of orange, tomato, or grapefruit juice with a choice of bacon or ham

Answer each question. (Lesson 11-2)

8. A die is rolled and a coin is tossed. How many outcomes are possible?

9. Four dice are rolled. How many outcomes are possible?

Find each value. (Lesson 11-3)

10. 9!

11. 0!

12. $P(8, 5)$

13. $C(8, 5)$

14. Andy needs to pack three shirts for his trip. How many ways can he choose the shirts from the eight shirts in his closet?

15. A baseball team has 13 players on the roster. How many ways can the coach pick the first three batters?

There are 4 black marbles, 2 yellow marbles, 3 green marbles, and 1 white marble in a bag. Find the probability of each outcome. (Lessons 11-4, 11-6, 11-7)

16. draw a yellow marble

17. draw a green or a white marble

18. draw a black, yellow, or green marble

19. draw a white marble

20. draw a yellow marble; replace it; draw a green marble

21. draw a black marble; replace it; draw a white marble

22. draw a green marble; do not replace it; draw a black marble

23. draw a white marble; do not replace it; draw a white marble

Solve. (Lesson 11-8)

24. The probability that Vera will get an A in algebra is 0.8. The probability that Ned will get an A in algebra is 0.6. If the probability that both will get an A is 0.5, what is the probability that one or the other will get an A?

Application and Problem Solving

25. Two dice are rolled. What is the probability that the product of the numbers shown on the dice is 12? (Lesson 11-5)

26. A certain restaurant gives out game cards. One out of six cards wins a small soft drink. Describe a simulation to predict how many soft drinks Ted will win with 14 game cards. (Lesson 11-9)

 ## ortfolio Suggestion

Review the items in your portfolio. Make a list of the items, noting why each item was chosen. Replace any items that are no longer appropriate.

Curriculum Connection

- **Meteorology** Check the weather forecast. Find out the probability of precipitation in your community for tomorrow.

- **History** Not only do some of the Presidents of the United States share a birthday, some also share a date of death. Report on this.

Read More About It

Anno, Mitsumasa. *Socrates and the Three Little Pigs.*

Razzell, Arthur G. *Probability: The Science of Chance.*

Riedel, Manfred G. *Odds and Chances for Kids: A Look at Probability.*

Test

Draw a tree diagram for each of the following. Find the number of possible outcomes.

1. a choice of floral or plaid wallpaper with a choice of beige or ivory paint

2. a choice of sweater in three colors (white, black, and navy) and in four sizes (small, medium, large, and extra large)

Answer each question.

3. A penny, a nickel, a dime, a quarter, and a half dollar are tossed. How many outcomes are possible?

4. Andy has 5 pairs of slacks, 6 shirts, and 3 sweaters. How many three-piece outfits are possible?

Find each value.

5. $6!$ 6. $\frac{6!2!}{4!0!}$ 7. $P(10, 3)$ 8. $C(12, 3)$ 9. $C(7, 4)$

The spinner at the right is equally likely to stop in any one of the six regions. It is spun once. Find the probability of each outcome.

10. a 2   11. a number less than 5

12. an odd number   13. a number greater than 6

A coin is tossed, then a die is rolled. Find the probability of each outcome.

14. tails and a 2   15. heads and a number less than 4

In a bag there are 4 red balls, 2 white balls, 3 black balls, and 1 green ball. Once a ball is selected, it is not replaced. Find the probability of each outcome.

16. a red ball and then a white ball   17. 2 black balls in a row

Answer each of the following.

18. The probability that the Bears will win their next game is 0.4. The probability that the Tigers will win their next game is 0.5. If the probability that both teams will win is 0.2, what is the probability that one or the other will win?

19. Two dice are tossed. What is the probability that the positive difference between the two numbers is 1?

20. A certain type of cereal advertises that there is a prize in each box. Actually there are six different prizes. Walt buys ten boxes hoping to get one of each of the prizes. Describe a way to simulate the situation.

BONUS

In a bag there are 3 red marbles, 2 white marbles, and 4 blue marbles. A marble is drawn and it is not replaced. A second marble is drawn. What is the probability that both marbles are the same color?

Academic Skills Test

Cumulative, Chapters 1-11

1. Which fractions are in order from least to greatest?

    A $\frac{1}{2}, \frac{1}{3}, \frac{1}{4}$

    B $\frac{3}{5}, \frac{7}{10}, \frac{9}{12}$

    C $\frac{3}{4}, \frac{5}{8}, \frac{9}{10}$

    D $\frac{5}{6}, \frac{5}{8}, \frac{5}{12}$

2. If your heart beats about once every second, it will beat about 442,000,000 times in 15 years. How is this number expressed in scientific terms?

    A 4.42×10^{-8}
    B 44.2×10^{-7}
    C 442×10^{6}
    D 4.42×10^{8}

3. What is the solution for the inequality $3y - 5 > 2y + 8$?

    A $y < 3$ C $y < 13$
    B $y > 3$ D $y > 13$

4. Which is an equation of line ℓ?

    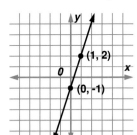

    A $y = -1$
    B $x - 1 = y$
    C $y = 3x$
    D $3x - 1 = y$

5. Which is the slope of the line in question 4?

    A 3   C $\frac{1}{2}$

    B 2   D $\frac{1}{3}$

6. The value of a certain car decreases by 20% after one year. To find the amount of decrease on a car originally costing $12,500, multiply 12,500 by—

    A $\frac{1}{20}$ C $\frac{1}{5}$

    B $\frac{1}{4}$ D $\frac{4}{5}$

7. The frequency table below contains data about students' test scores on a 25-point test.

Score	18	19	20	21	22	23	24	25
Number of Students	2	1	5	4	5	6	4	1

    How many students had a score greater than 20?

    A 4   C 20
    B 5   D 25

8. Using the data in problem 7, what was the range of the test scores?

    A 7 points   C 18 points
    B 8 points   D 25 points

9. In how many different ways can all four students be arranged in a row?

        Sue   Janet   Marty   Bill

    A 4 ways   C 24 ways
    B 12 ways   D 36 ways

10. A coin is tossed and a die is rolled. What is the probability of tossing tails and rolling a 5?

    A $\frac{1}{12}$ C $\frac{1}{4}$

    B $\frac{1}{10}$ D $\frac{1}{3}$

CHAPTER 12

CHAPTER OBJECTIVES

In this chapter you will learn to:

- use some basic terms of geometry
- construct a circle graph
- identify the relationships of intersecting and parallel lines
- identify properties of congruent and similar triangles
- classify and draw polygons
- identify conditional statements and write their converses

Applying Algebra to Geometry

The new cars of the 1990s owe a lot to computers. After automotive designers have sketched several new car designs, promising designs are modeled in clay at a scale of 1 to 5. A three-dimensional scanner is run over the surface of each model, measuring and storing lengths, widths, and heights automatically. A computer accepts the measurements and projects three-dimensional images of the car on its screen.

Using the computer, the designer can investigate the car's structural strength. First it divides the car's entire surface into triangles. By shading each triangle, the computer shows the load on the car's surface at that location. The darker the shading, the bigger the load. That means today's designers can plan lighter, stronger cars than even the most experienced automotive designers of the past who worked without computer assistance.

Language Connection

Class Project

Write letters to several automobile companies, asking for information about computer-aided design.

Write a report that summarizes the information.

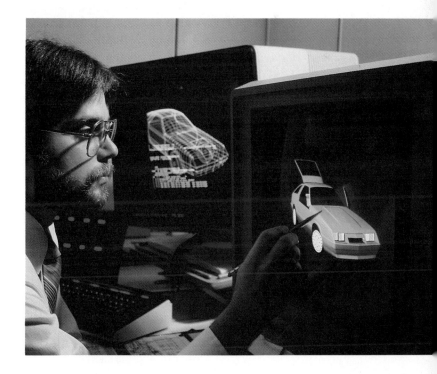

12-1 The Language of Geometry

Objectives:

Identify points, lines, planes, rays, segments, and angles.

Classify angles as acute, right, or obtuse.

Key Terms:

point
line
line segment
plane
ray
angle

Mr. Thompson uses a laser measuring device to detect bumps and hollows in his fields. This device sends signals to a land leveling machine which smooths the field to ensure even irrigation after the crop is planted.

Explain why the three points used to name a plane cannot lie on the same line.

The light from a laser is seen as a beam because it strikes and reflects off particles in the air. Each dust particle suggests a **point.** A point is a specific location in space, but the point itself has no size or shape. A point can be represented by a dot and named with a capital letter.

A laser beam travels in a straight line as can be seen in a collection of dust particles. In geometry, a **line** is a collection of points that extends indefinitely in two directions. Arrowheads are used to show that a line has no endpoints. A line can be named by a single lowercase letter or by using two points of the line.

The section of a laser beam that goes from the measuring device to the land leveler is a model of a **line segment.** A line segment is a part of a line containing two endpoints and all points between the endpoints. A line segment is named by its endpoints.

Mr. Thompson's field, after leveling, is a model of a **plane.** A plane is a flat surface that has no boundaries. A plane can be named by any three points of the plane. (The three points must *not* lie on the same line.)

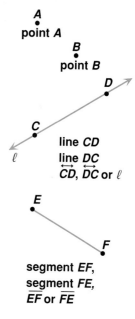

point *A*

point *B*

line *CD*
line *DC*
$\overleftrightarrow{CD}$, $\overleftrightarrow{DC}$ or ℓ

segment *EF*,
segment *FE*,
$\overline{EF}$ or $\overline{FE}$

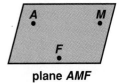

plane *AMF*

The path of the laser beam is a model of a **ray.** A ray extends from one point indefinitely in one direction. A ray is named by using its endpoint first and then any other point on the ray.

ray *DF* or $\overrightarrow{DF}$

Examples

1 **Give three names for the line shown below.**

What is another name for the line that was not given?

There are three points labeled on the line. Any two can be used to name it. So we can name it $\overleftrightarrow{PQ}$, $\overleftrightarrow{QR}$, or $\overleftrightarrow{RP}$.

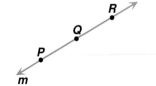

2 **Name two rays and three segments in the figure below.**

What is another name for $\overrightarrow{AC}$?
What segment was not named?

Two of the rays are $\overrightarrow{AC}$ and $\overrightarrow{DB}$. Three of the segments are $\overline{AB}$, $\overline{BD}$, and $\overline{BC}$.

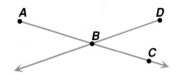

3 **Give three names for the plane shown below.**

How many other ways are there to name the plane?

There are four points labeled on the plane. Any three can be used to name it. So we can name it plane *EFG*, plane *FGH*, or plane *GHE*.

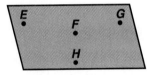

An **angle** is formed by two rays with a common endpoint called the **vertex.** The rays are called the **sides** of the angle. In the angle at the right, $\overrightarrow{QP}$ and $\overrightarrow{QR}$ form the sides of angle *PQR*. The vertex is point *Q*.

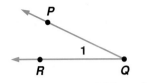

angle *PQR*, angle *RQP*, angle *Q*,
∠*PQR*, ∠*RQP*, ∠*Q*, or ∠1

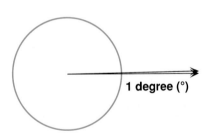

1 degree (°)

The most common unit of measure for angles is the **degree.** A circle can be separated into 360 arcs of the same length. An angle has a measurement of one degree if its vertex is at the center of the circle and its sides contain the endpoints of one of the 360 equal arcs.

You can use a **protractor** to measure angles.
- Place the center of the protractor on the vertex of the angle (B) with the straightedge along one ray ($\overrightarrow{BA}$).
- Use the scale that begins with 0 at $\overrightarrow{BA}$. Read where the ray ($\overrightarrow{BC}$) crosses this scale.

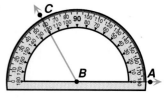

Say: The measure of angle ABC is 120 degrees.

Write: $m \angle ABC = 120°$

Example

Why is $\angle V$ not a good way to name an angle?

4 **Find the measure of $\angle RVZ$ and $\angle XVW$.**

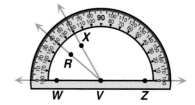

$m \angle RVZ = 140°$ Use the scale that begins with 0 at $\overrightarrow{VZ}$.

$m \angle XVW = 60°$ Use the scale that begins with 0 at $\overrightarrow{VW}$.

Angles are classified by degree measure.

Acute angles have measures greater than $0°$ and less than $90°$.

Right angles have measures of $90°$.

The small square indicates that $\angle JKL$ is a right angle.

Obtuse angles have measures greater than $90°$ but less than $180°$.

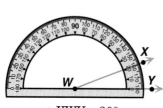

$m \angle XWY = 20°$

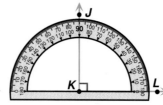

$m \angle JKL = 90°$

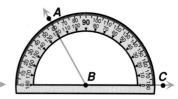

$m \angle ABC = 120°$

Checking for Understanding

Communicating Algebra

1. The line through points P and Q is named ___?___ .

2. Draw and label an obtuse angle with sides $\overrightarrow{GQ}$ and $\overrightarrow{GC}$.

3. Draw and label a $30°$ angle with vertex K.

4. Explain the difference between an obtuse angle and an acute angle.

Guided Practice

Use the figure at the right to name an example of each term.

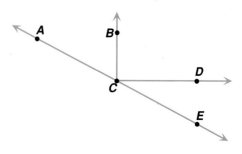

5. line

6. obtuse angle

7. ray

8. line segment

9. acute angle

10. point

11. Use a protractor to find the degree measure of $\angle ECD$.

Determine whether each model suggests a point, line, or plane.

12. corner of a box

13. guitar string

14. meeting of two walls

15. ice on an ice rink

16. cover of a book

17. sharp end of a thorn

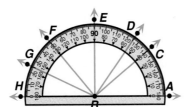

Use the figure at the left to find each measure. Classify each angle as *acute*, *right*, or *obtuse*.

18. $m \angle ABC$

19. $m \angle CBH$

20. $m \angle EBH$

Exercises

Independent Practice

Use the figure above to find each measure. Classify each angle as *acute*, *right*, or *obtuse*.

21. $m \angle GBH$

22. $m \angle ABD$

23. $m \angle FBH$

24. $m \angle GBA$

25. $m \angle ABE$

26. $m \angle DBH$

Draw a picture to represent each of the following.

27. line CD

28. point F

29. ray ST

30. plane WXY

31. ray EF

32. plane ABC

33. line GH

34. point M

Use a protractor to draw angles having the following measurements. Classify each angle as *acute*, *right*, or *obtuse*.

35. $30°$

36. $55°$

37. $90°$

38. $110°$

39. $145°$

Mixed Review

***True* or *false*.** (Lessons 1-10, 6-12)

40. $8(2) < 15$

41. $-38.12 > -38.22$

42. **Statistics** What fraction of people chose red as their favorite color? (Lesson 10-8)

43. **Probability** Draw a tree diagram to determine the outcomes for tossing a coin three times. (Lesson 11-1)

44. **Probability** Suppose $P(A) = \frac{1}{3}$, $P(B) = \frac{1}{4}$, and A and B are mutually exclusive. What is $P(A \text{ or } B)$? (Lesson 11-8)

Favorite Color	
Blue	11
Red	12
Green	7

Application

45. **Photography** Cameras are often mounted on tripods to give stability. Why do tripods give stability?

Critical Thinking

46. ***True* or *false*.**

   a. The intersection of any two rays is always a point.

   b. A line is part of a line segment.

Wrap-Up

47. Use two pencils and your desk to model the concepts of point, line, plane, ray, and angle.

12-2 Making Circle Graphs

Objective:
Make a circle graph from data.

Key Term:
circle graph

Thirty percent of Earth's surface is land. The other 70% is water. This information can be displayed in a circle graph. A **circle graph** is used to compare parts of a whole. Since 30% + 70% is 100%, the circle represents the whole surface of Earth.

Earth's Surface

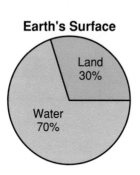

Check the sum of 108° and 252°. What should it be?

In Lesson 12-1, you learned that there are 360° in a circle.

30% of 360° is 108°.   70% of 360° is 252°.

Use your protractor to measure the angles in the circle graph. You should find that they measure 108° and 252°.

Sometimes you are given data that have not been expressed as a percent. Study the following example to find out how to make such a circle graph.

Example

1 Make a circle graph to display the following data.

FYI

About 4 billion years ago, water from deep within Earth was released at Earth's surface through volcanic activity. This water began to accumulate to form the oceans. Today, the mass of the oceans is about 1.4×10^{24} g.

Homework for April (in hours)	
Math	20
English	15
History	8
Science	5
Other	2

Step 1 Find the total number of hours.

20 + 15 + 8 + 5 + 2 = 50

Step 2 Find the ratio that compares the time spent on each subject to the total time.

Math: $\dfrac{20}{50} = 0.4$ Science: $\dfrac{5}{50} = 0.1$

English: $\dfrac{15}{50} = 0.3$ Other: $\dfrac{2}{50} = 0.04$

History: $\dfrac{8}{50} = 0.16$

Check the sum of the ratios. What should it be?

Step 3 Find the number of degrees for each section of the graph.

Math: $0.4 \times 360° = 144°$ Science: $0.1 \times 360° = 36°$

English: $0.3 \times 360° = 108°$ Other: $0.04 \times 360° = 14.4°$

History: $0.16 \times 360° = 57.6°$

Check the sum of the degrees. What should it be?

Step 4 Draw a circle graph.

Use a compass to draw a circle and a radius as shown.

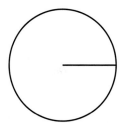

Start with the least number of degrees, in this case, 14.4°. Use your protractor to draw an angle of 14.4°.

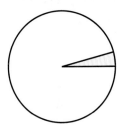

Repeat for the remaining sections. Label each section of the graph and give the graph a title.

0.4 = 40%

0.3 = 30%

0.16 = 16%

0.1 = 10%

0.04 = 4%

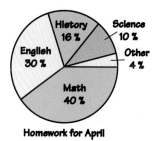

Checking for Understanding _____

Communicating Algebra

1. When is it appropriate to graph data in a circle graph?

2. Explain why it would not be appropriate to make a circle graph called "Population of the United States, 1900-2000."

Guided Practice

3. **Collect Data** Make a circle graph to represent your classmates' hair colors.

Exercises

Independent Practice

4. **Statistics** The chart shows the average number of births per day in the United States. Make a circle graph to display the data.

Average U.S. Births Per Day	
Sunday	8,532
Monday	10,243
Tuesday	10,730
Wednesday	10,515
Thursday	10,476
Friday	10,514
Saturday	8,799

Mixed Review

5. **Statistics** What percent of the sample practice the piano more than 4 hours a week?   (Lesson 10-1)

Piano Practice (hours)	Tally	Frequency
0-2	𝍇 𝍇	10
2-4	𝍇 𝍇 𝍇 𝍇 II	22
4-6	𝍇 𝍇 𝍇 I	16

6. Use a protractor to draw a 40° angle. Classify this angle as *acute, right,* or *obtuse.*   (Lesson 12-1)

Applications

Make a circle graph to display each set of data.

7. **Geography**

8. **Biology**

Areas (in square miles) of the Continents of the World	
Asia	17,012,000
Africa	11,785,000
North America	9,400,000
South America	6,883,000
Antarctica	5,100,000
Europe	4,071,000
Australia	2,966,000

Chemical Composition of the Human Body	
Oxygen	65%
Carbon	18%
Hydrogen	10%
Nitrogen	3%
Other	4%

Critical Thinking

9. Marcus had a pizza party. He had a large round pizza delivered for the occasion. He made four straight cuts from edge-to-edge to form 7 pieces. Make a drawing to show how he cut the pizza.

Wrap-Up

10. **Collect Data** Make a circle graph to represent how you spend an average 24-hour day. Include the following categories: school, sleep, eating, leisure, and miscellaneous.

Algebra in Action-Physics

Ray Optics

Light travels from the sun to Earth through a series of light waves. These light waves travel together in a straight line which, on a foggy day, can be seen in the form of a beam.

Light waves can be represented by rays. The rays show the direction the light is traveling. The study of light using ray diagrams is called **ray optics**.

When a ray of light strikes a flat reflecting object, such as a mirror, it bounces off the mirror in the same way a tennis ball bounces off the floor.

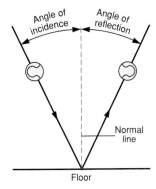

The angle a descending tennis ball makes with a line perpendicular to the floor (the *normal line*) is called the *angle of incidence*. The angle between the rebounding ball and the normal line is the *angle of reflection.*

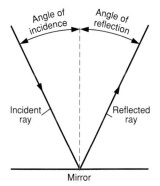

When a ray of light is reflected off a mirror, it also forms an angle of incidence and an angle of reflection. These two angles have the same measure.

A ray of light strikes a mirror at a 53° angle to the normal line.

1. What is the measure of the angle of reflection?

2. What is the measure of the angle between the incident ray and the reflected ray?

12-3 Angle Relationships

Objective:
Identify angle relationships.

Key Terms:
vertical angles
congruent
perpendicular
adjacent angles
complementary
supplementary

Did you know that there is a place in the United States in which you can stand in four different states at once? The map shows that the boundaries of Arizona, New Mexico, Colorado, and Utah intersect in exactly one point.

When two lines intersect, they form four angles with one point in common.

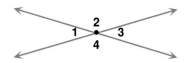

Name another pair of vertical angles shown in the figure.

Angles that are opposite to one another are called **vertical angles.** Angles 1 and 3, for example, are vertical angles. Vertical angles have the same measure. In other words, they are **congruent.**

If the vertical angles formed by two intersecting lines are right angles, the lines are said to be **perpendicular.**

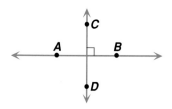

Line AB is perpendicular to line CD. This can be written $\overleftrightarrow{AB} \perp \overleftrightarrow{CD}$. The symbol $\perp$ means *is perpendicular to.*

FYI

Cliff dwellers were native Americans who lived in the southwestern United States between 1000 and 1300. The most famous were the Anasazi, who built their homes in canyon walls, using sandstone blocks and mud mortar. As many as 1500 people could live in some of these dwellings.

Two angles are called **adjacent angles** if they have a common side, the same vertex, and do not overlap. In the diagram at the right, $\angle 1$ and $\angle 2$ are adjacent angles.

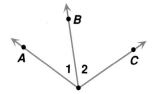

Two angles are **complementary** if the sum of their measures is 90°.

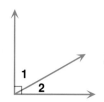

Two angles are **supplementary** if the sum of their measures is 180°.

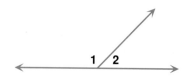

In geometry, we use the symbol $m\angle A$ to mean *the degree measure of angle A*. If $\angle A$ measures $50°$, we write $m\angle A = 50°$.

Examples

1 Angles A and B are complementary. If $m\angle A = 35°$, find $m\angle B$.

Since $\angle A$ and $\angle B$ are complementary,
$m\angle A + m\angle B = 90°$.

$$m\angle A + m\angle B = 90$$
$$35 + m\angle B = 90 \qquad \textit{Replace } m\angle A \textit{ with 35.}$$
$$m\angle B = 55$$

The measure of $\angle B$ is $55°$.

2 The angles shown below are supplementary. Find the measure of each angle.

$$m\angle PQR + m\angle RQT = 180°$$
$$(x + 5) + (2x - 11) = 180$$
$$(x + 2x) + (5 - 11) = 180$$
$$3x - 6 = 180$$
$$3x = 186$$
$$x = 62$$

Why is 62° not the measure of one angle?

$$
\begin{array}{cc}
\angle PQR & \angle RQT \\
x + 5 = 62 + 5 & 2x - 11 = 2(62) - 11 \\
= 67 & = 124 - 11 \\
& = 113 \\
m\angle PQR = 67° & m\angle RQT = 113°.
\end{array}
$$

Check: $67° + 113° = 180°$ The angles are supplementary.

Checking for Understanding

Communicating Algebra **Draw and label a diagram to show each of the following.**

1. Angles ABC and CBD are adjacent angles.
2. Angles XYZ and AYC are vertical angles.
3. Angles XYZ and ZYW are adjacent, complementary angles.
4. Angles RST and TSW are adjacent, supplementary angles.

Guided Practice **Angles M and P are supplementary. Find $m\angle M$ when $\angle P$ has each measure.**

5. $30°$ 6. $120°$ 7. $80°$ 8. $175°$

Angles A and B are complementary. Find $m\angle A$ when $\angle B$ has each measure.

9. $30°$ 10. $45°$ 11. $50°$ 12. $15°$

Exercises

Independent Practice

Find the value of x in each figure.

13.

14.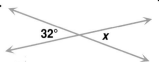

15.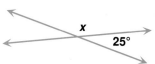

Each pair of angles is either complementary or supplementary. Find the degree measure of each angle.

16.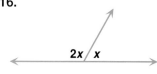

17.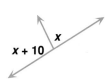

18.

19.

20.

21.

Mixed Review

22. Translate into an equation: *Seven times some number equals -28.* (Lesson 3-10)

23. Jennifer scored 5 points less than twice the number of points she scored in the last basketball game. If she scored 12 points in the last game, how many points did she score in this game? (Lesson 7-3)

Find each value. (Lesson 11-3)

24. 6!

25. C(8, 3)

26. Suppose your daily food intake includes 25% fruits and vegetables, 40% breads and cereals, 20% meat, and 15% dairy products. Make a circle graph of this data. (Lesson 12-2)

Challenge

27. The measure of an angle is 26° more than three times the measure of the complement. Find the measure of the angle and the measure of the complement.

Application

28. **City Planning** The Traverse City Street Commission is planning to have the curbs in the business district replaced. The construction company needs to know the angles of the street intersections ahead of time so they can build the forms that hold the cement. Two of the streets intersect at 42°. Draw a diagram of this intersection and label the angle measure of all four corners.

Critical Thinking

29. Find the measures of $\angle PQR$, $\angle RQS$, $\angle TQS$, and $\angle PQT$.

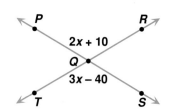

Wrap-Up

30. Use a protractor to draw adjacent, supplementary angles. One angle should measure 55°.

12-4 Parallel Lines

Objective:
Identify the relationships of angles formed by two parallel lines and a transversal.

Key Terms:
parallel
transversal
interior angles
exterior angles
corresponding angles

The Bigtown Transit Authority is expanding its bus service. They need to check the angles of the corners at various intersections to make sure the turns will not be too sharp for the buses to make. The map at the right shows two parallel streets intersected by another street, forming eight angles. One angle measure is already known. How can the transit authority find the measure of the remaining seven angles without having to measure each one?

In geometry, we think of **parallel** lines as two or more lines in a plane that do not intersect. Parallel lines are *always* the same distance apart. In the figure below, line ℓ is parallel to line m, or $\ell \parallel m$. The symbol $\parallel$ means *is parallel to*.

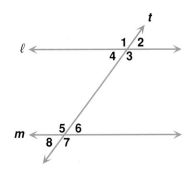

When a line, called a **transversal,** intersects two parallel lines, eight angles are formed. One way to classify these angles is as **interior angles** or **exterior angles.**

Interior angles: $\angle 3, \angle 4, \angle 5, \angle 6$
Exterior angles: $\angle 1, \angle 2, \angle 7, \angle 8$

When we study the relationship between the different angles, we can come up with further classifications.

- **alternate interior angles:** $\angle 4$ and $\angle 6$, $\angle 3$ and $\angle 5$
  Alternate interior angles are interior angles found on opposite sides of the transversal.

- **alternate exterior angles:** $\angle 1$ and $\angle 7$, $\angle 2$ and $\angle 8$
  Alternate exterior angles are exterior angles found on opposite sides of the transversal.

- **corresponding angles:** $\angle 1$ and $\angle 5$, $\angle 2$ and $\angle 6$, $\angle 3$ and $\angle 7$, and $\angle 4$ and $\angle 8$
  Corresponding angles are angles that hold the same position on two different parallel lines cut by a transversal.

Examples

Refer to the diagram at the right. Classify each pair of angles.

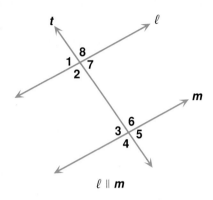

ℓ ∥ m

1 ∠1 and ∠5

Since ∠1 and ∠5 are exterior angles on opposite sides of the transversal, they are alternate exterior angles.

2 ∠6 and ∠8

∠6 and ∠8 are corresponding angles.

Use a protractor to measure the angles formed by the parallel lines shown above. You will find that certain angles have the same measure. These angle relationships are summarized below.

Parallel Lines Cut by Transversal	• Corresponding angles are congruent. • Alternate interior angles are congruent. • Alternate exterior angles are congruent.

Examples

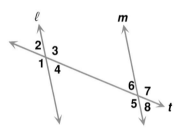

In the diagram at the right, ℓ is parallel to m. The measure of ∠5 is 120°.

3 Find $m\angle 1$.

∠1 and ∠5 are corresponding angles.

$m\angle 1 = m\angle 5$ Corresponding angles are congruent.

$m\angle 1 = 120$ Replace $m\angle 5$ with 120.

The measure of ∠1 is 120°.

4 Find $m\angle 4$.

∠1 and ∠4 are supplementary.

$m\angle 1 + m\angle 4 = 180$ The sum of the measures of two supplementary angles is 180°.

$120 + m\angle 4 = 180$ Replace $m\angle 1$ with 120.

$m\angle 4 = 60$

The measure of ∠4 is 60°.

Checking for Understanding

Communicating Algebra

1. In your own words, define parallel lines.

2. What is the relationship between any two corresponding angles?

3. Draw parallel lines a and b intersected by transversal c. Label the interior angles formed 9, 10, 11, and 12.

Guided Practice

Refer to the diagram at the right to complete Exercises 4-9.

4. Name the transversal.

5. Name the interior angles.

6. Name the exterior angles.

7. Name the pairs of alternate interior angles.

8. Name the pairs of alternate exterior angles.

9. Name the pairs of corresponding angles.

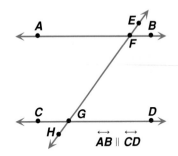

Exercises

Independent Practice

In the figure at the right, ℓ is parallel to m. If the measure of $\angle 1$ is 58°, find the measure of each angle.

10. $\angle 2$ 11. $\angle 3$

12. $\angle 5$ 13. $\angle 4$

14. $\angle 7$ 15. $\angle 8$

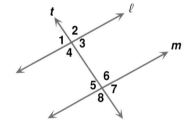

Mixed Review

16. **Probability**   If two dice are rolled, what is the probability that the sum is 7? (Lesson 11-5)

17. Find the measure of the complement and supplement of a 16° angle. (Lesson 12-3)

Application

18. **Public Service**   The Bigtown Transit Authority is studying the street intersection on page 449. If a bus cannot make a turn at an angle that is less than 70°, can the Transit Authority provide service on the proposed route? Why or why not?

Critical Thinking

19. In the figure at the right, $\ell \parallel m$. Find the measures of each of the eight angles formed when ℓ and m are cut by t.

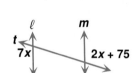

Wrap-Up

20. Explain the meaning of the statement "when parallel lines are cut by a transversal, angles that look equal are equal."

12-5 Constructing Segments and Angles

Objectives:
Construct congruent segments and angles using a compass and a straightedge.
Bisect segments and angles.

Key Terms:
straightedge
compass
bisect
midpoint
angle bisector

Segments and angles can be constructed using a straightedge and compass. A **straightedge** is any object that can be used to draw a straight line, such as an ID card, a piece of cardboard, or a ruler. A **compass** is used to draw a circle or part of a circle.

You can construct a line segment that has the same exact length as a given line segment by using a straightedge and a compass. Two line segments that have the same measure are said to be *congruent*.

Given $\overline{AB}$:

Step 1: Use a straightedge to draw $\overrightarrow{PS}$.

Step 2: Place the steel tip of the compass at A and the writing tip at B.

Step 3: Keep the same setting on the compass and place the steel tip at P. Draw an arc that intersects $\overrightarrow{PS}$ at Q.
$\overline{PQ}$ is congruent to $\overline{AB}$.

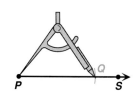

We can also use a compass and straightedge to **bisect** a line segment. Bisecting a segment means we draw a line segment through the **midpoint** of the given segment, forming two congruent segments.

Given $\overline{PQ}$:

Step 1: Place the steel tip of the compass at P. Set the compass so the writing tip is more than halfway to Q and draw two arcs as shown.

Step 2: With the same setting on the compass, place the steel tip at Q and draw two arcs as shown. There must be two intersection points.

Step 3: With a straightedge, draw the line determined by the intersection points. This line bisects $\overline{PQ}$ at M, the midpoint of $\overline{PQ}$.
$\overline{PM}$ is congruent to $\overline{MQ}$.

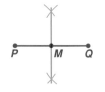

FYI
For over 200 years mathematicians tried to trisect an angle (divide an angle into three equal parts), using only a straightedge and a compass. In 1837 it was finally proved to be an impossible construction.

We can also construct congruent angles by using a compass and straightedge. Congruent angles have the same angle measure. Follow these steps for constructing $\angle TJK$ congruent to $\angle ABC$.

Given $\angle ABC$:

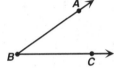

Step 1: Place the steel tip of the compass at B and draw an arc as shown. Label points R and S.

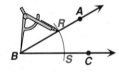

Step 2: Use a straightedge to draw $\overrightarrow{JK}$. Use the compass setting from Step 1 and construct an arc as shown. Label the intersection point M.

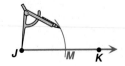

Step 3: Set the compass at points R and S as shown. Keep that setting and place the steel tip at M. Draw an arc as shown. Label the intersection point T.

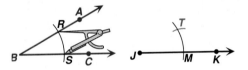

Step 4: With a straightedge, draw $\overrightarrow{JT}$. $\angle TJK$ is congruent to $\angle ABC$.

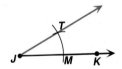

To bisect an angle, we construct a ray through the interior of the angle that forms two congruent angles. This ray is called the **angle bisector.**

Given $\angle RST$:

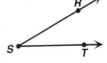

Step 1: Place the steel tip of the compass at S and draw an arc that intersects both sides of the angle. Use X and Y to name the points of intersection.

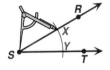

Step 2: Place the steel tip of the compass at X and draw an arc in the interior of $\angle RST$.

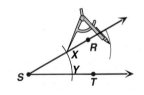

Step 3: With the same setting on the
compass, place the steel tip at Y
and draw an arc that intersects the
arc you drew in Step 2. Use W to
name the point of intersection of
the arcs.

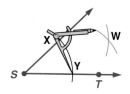

Step 4: With a straightedge, draw $\overrightarrow{SW}$.
$\overrightarrow{SW}$ is the bisector of $\angle RST$.
$\angle XSW$ is congruent to $\angle WSY$.

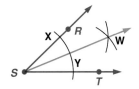

Checking for Understanding

Communicating 1. What tools, other than a sharp pencil and a piece of paper, are needed to do a
Algebra   geometric construction?

2. In your own words, explain how you would construct a line segment congruent
to a given line segment.

3. What point is always on the bisector of a line segment?

Guided   **Trace each segment. Then construct a segment congruent to it.**
Practice

4. ──── 5. ──────── 6. ────────────

Trace each angle. Then construct an angle congruent to it.

7.    8.    9.

Exercises

Independent   **Trace the drawing at the right. Then**
Practice   **construct a line segment or angle congruent**
                  to each segment or angle named.

10. $\overline{LM}$ 11. $\overline{QR}$ 12. $\overline{SM}$

13. $\angle QML$ 14. $\angle NRQ$ 15. $\angle QSN$

16. $\overline{LR}$ 17. $\angle PMQ$ 18. $\overline{QS}$

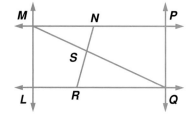

Use a ruler to draw a segment with the given measurement. Construct a
segment congruent to it. Then construct the line segment that bisects
that segment.

19. 2 inches   20. 7.5 cm   21. 43 mm

22. $1\frac{5}{8}$ inches   23. $2\frac{3}{4}$ inches   24. 67 mm

Use a protractor to draw an angle with the given measurement. Construct an angle congruent to it. Then construct the ray that bisects that angle.

25. 60°   **26.** 140°   **27.** 47°

28. 163°   **29.** 25°   **30.** 104°

Mixed Review   **31.** Convert 32 inches into feet.   (Lesson 5-10)

32. Graph $y \leq -2x + 1$.   (Lesson 8-11)

33. Statistics   Construct a stem-and-leaf plot of the heights (in inches) of members of a basketball team: 72, 70, 74, 66, 67, 71, 72, 66, 70, 69, 71, 66. (Lesson 10-3)

34. Two parallel lines are cut by a transversal. If an interior angle measures 64°, what is the measure of the alternate interior angle?   (Lesson 12-4)

Application   **35. Physics**   The *center of mass* of an object is the point that acts as though the entire mass of the object were concentrated there. If you have a flat triangular object like the one shown to the right, with mass equally distributed, the center of mass is the intersection of three segments, each drawn from the vertex of one angle to the midpoint of the opposite side. Trace each triangle below and find its center of mass.

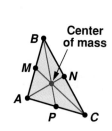

Center of mass

a.   b.   c.

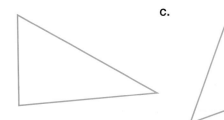

Critical Thinking

Using the figure below, construct each of the following.

36. a line segment that is twice as long as $\overline{MN}$

37. a line segment that is as long as the measure of $\overline{MN}$ plus the measure of $\overline{NQ}$

38. an angle whose measure is twice the measure of $\angle MNQ$

39. an angle whose measure is twice the measure of $\angle MNQ$ plus the measure of $\angle NQM$

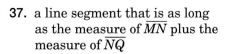

Wrap-Up   **40.** Draw an obtuse angle. Then bisect the angle. What can you say about the two angles created?

12-6 Triangles

Objectives:
Find the missing angle measure of a triangle. Classify triangles.

Key Terms:
triangle
vertices
acute
obtuse
right

In a plane, figures formed by three line segments, as shown, are called **triangles.** Triangles are named by their **vertices,** the endpoints of the line segments. The figure shown at the right is triangle *ABC*. Triangle *ABC* is written △*ABC*.

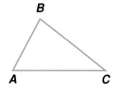

There are three angles in every triangle. There is also a unique relationship among the measures of these three angles. Try the following activity to try to discover the relationship.

1. Cut a large triangle out of a piece of paper. On one side label vertices *P, Q,* and *R.*
2. Fold the triangle so that point *P* lies on $\overline{QR}$ and the fold is parallel to the side. Label ∠*P* as ∠2.
3. Fold the triangle again so that points *Q* and *R* meet at the vertex of ∠2. Label these angles ∠1 and ∠3.

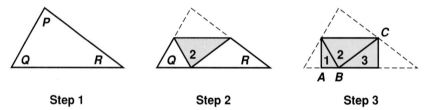

| Step 1 | Step 2 | Step 3 |

What do you notice about the relationship of ∠1, ∠2, and ∠3?

The activity above will work with any triangle. You can also use a protractor to measure each angle. This activity suggests the following relationship.

Angles of a Triangle ▶ The sum of the measures of the angles of a triangle is 180°.

Example

1 **In △*ABC,* the measure of ∠*B* is 38° and the measure of ∠*C* is 47°. Find the measure of ∠*A.***

$$m\angle A + m\angle B + m\angle C = 180$$ The sum of the measures of the angles of a triangle is 180°.

$$m\angle A + 38 + 47 = 180$$ Replace *m*∠*B* with 38 and *m*∠*C* with 47.

$$m\angle A + 85 = 180$$

$$m\angle A = 95$$ The measure of ∠*A* is 95°.

A triangle can be classified according to its angles.

What is the meaning of the red angle in the right triangle?

Acute	Obtuse	Right
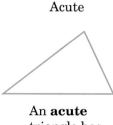	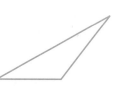	
An **acute** triangle has three acute angles.	An **obtuse** triangle has one obtuse angle.	A **right** triangle has one right angle.

Checking for Understanding

Communicating Algebra

1. Draw a sketch of an acute triangle, an obtuse triangle, and a right triangle.
2. Can a triangle have two right angles? Explain your answer.

Guided Practice

Find the value of x. Then classify each triangle as *acute, right,* or *obtuse*.

3.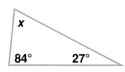

4.

5.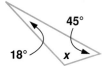

Exercises

Independent Practice

Find the value of x. Then classify each triangle as *acute, obtuse,* or *right*.

6.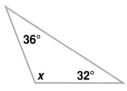

7.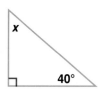

8.

Use the figure at the right to solve each of the following.

9. Find $m\angle 1$ if $m\angle 2 = 40$ and $m\angle 3 = 55$.
10. Find $m\angle 1$ if $m\angle 2 = 60$ and $m\angle 3 = 60$.
11. Find $m\angle 1$ if $m\angle 2 = 81$ and $m\angle 3 = 74$.
12. Find $m\angle 2$ if $m\angle 1 = 45$ and $m\angle 3 = 75$.
13. Find $m\angle 2$ if $m\angle 1 = 47$ and $m\angle 3 = 48$.

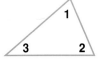

Mixed Review

14. **Probability** In a bag there are 4 red marbles and 8 white marbles. Once a marble is selected, it is not replaced. Find the probability of selecting two white marbles in a row.   (Lesson 11-7)

15. Use a ruler to draw a line segment 2.5 cm long. Construct a line segment that bisects that segment.   (Lesson 12-5)

Challenge   **Find the measures of the angles in each triangle.**

16.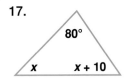
3x
2x   x

17.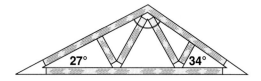
80°
x   x + 10

18.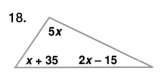
5x
x + 35   2x − 15

Applications   19. **Construction**   Find the missing angle measure for the roof support shown at the right.

27°   34°

20. **Public Utilities**   Brenda Sullivan works for the telephone company. One of her jobs is to attach a support cable to an upright telephone pole so that the cable makes a 20° angle with the pole. What angle does the cable have to make with the ground? Assume that the ground is level.

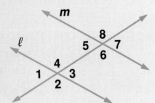
20°

Critical Thinking   21. Are the acute angles of a right triangle complementary? Explain.

Wrap-Up   22. If one angle of a triangle measures 34° and another angle measures 56°, what is the measure of the third angle? Classify this triangle by its angles. Draw the triangle.

Mid-Chapter Quiz

Use a protractor to draw an angle with the given measurements. Classify each angle as *acute*, *right*, or *obtuse*. Then construct an angle congruent to one of your angles.   (Lessons 12-1, 12-5)

1. 60°   2. 115°   3. 85°   4. 150°   5. 20°

6. 17% of Foodtown's sales comes from the produce department. Find the measure of the produce department angle for a circle graph showing Foodtown's sales.
(Lesson 12-2)

Find the measure of each angle using the following information.   (Lesson 12-3)

7. $\angle A$ and $\angle B$ are complementary.
$m\angle A = x + 25; m\angle B = 2x - 10$

8. $\angle F$ and $\angle G$ are supplementary.
$m\angle F = 3x - 50; m\angle G = 2x - 20$

In the figure at the right, ℓ is parallel to m. If the measure of $\angle 3$ is 34°, find the measure of each angle.   (Lesson 12-4)

9. $\angle 1$ 10. $\angle 4$ 11. $\angle 5$ 12. $\angle 6$ 13. $\angle 8$

12-7 Conditional Statements

Objectives:
Identify a conditional and its parts.
Write the converse of the conditional.

Key Terms:
conditional
converse

Miriam Frank writes copy for an advertising firm. She wrote the following slogan for Bob Dunn's campaign for city council. "If you want a booming city economy, vote for Dunn."

You can rewrite the slogan using the words *if* and *then*. "If you want a booming city economy, then you should vote for Dunn." A statement that can be written in *if-then* form is a **conditional**.

Conditionals have two parts.
- The part after *if* is the *hypothesis*.
- The part after *then* is the *conclusion*.

Examples

1 **Identify the hypothesis and the conclusion in the slogan given above.**

If you want a booming city economy, then you should vote for Dunn.
 hypothesis   conclusion

2 **Write a statement "All residents of Denver live in Colorado." in the form of a conditional.**

If you are a resident of Denver, then you live in Colorado.
 hypothesis   conclusion

You can interchange the hypothesis and the conclusion of a conditional. The new statement is called the **converse** of the conditional.

Examples

3 **Write the converse of the following conditional.**
Conditional: If a triangle is a right triangle, then the triangle has two acute angles.
Converse: If a triangle has two acute angles, then the triangle is a right triangle.

4 **In Example 3, determine whether each statement is true or false.**
Conditional: A right triangle always has two acute angles. The conditional is true.
Converse: As shown in the drawing, an obtuse triangle also must have two acute angles. So, the converse is false.

acute
angles

obtuse
angle

Checking for Understanding

Communicating Algebra

1. What are the two parts of a conditional?
2. Give an example of a conditional. Then write its converse.

Guided Practice

Identify the hypothesis and the conclusion in each conditional.

3. If it is warm, then it is summer.
4. If a polygon has three sides, then it is a triangle.
5. If a month has 31 days, then it is January.

Exercises

Independent Practice

Identify the hypothesis and the conclusion in each statement. Write the statement in the form of a conditional, if you need to.

6. If a triangle is a right triangle, then none of its angles are congruent.
7. It is morning if the sun is shining.
8. An angle that measures 90° is a right angle.

Write the converse of each conditional. Then determine whether each statement (conditional and converse) is true or false.

9. If an animal has four legs, then it is a cow.
10. If it is raining, the streets are wet.
11. When two lines intersect to form right angles, the lines are perpendicular.

Mixed Review

12. **Statistics** What is the median and range of the data in the box-and-whisker plot at the right? (Lesson 10-5)

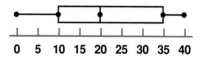

13. In △ABC, the measure of ∠B is 26° and the measure of ∠C is 55°. Find the measure of ∠A. (Lesson 12-6)

Computer

In BASIC, an if-then statement is used to compare two numbers. If the condition is true, the computer follows the instructions given after the word *then*. If the condition is false, the computer goes to the next line.

Example: `10 IF 7 < 9 THEN GOTO 40`   Since 7 is less than 9, the computer will go to line 40.

Determine whether the computer will go to line 40. Use the information in the chart at the right.

14. `10 IF 15.8 < 16 THEN GOTO 40`
15. `10 IF 23 <> 23 THEN GOTO 40`
16. `10 IF 3.2 <= 3.2 THEN GOTO 40`
17. `10 IF 9.16 >= 9.2 THEN GOTO 40`

Inequality Symbols	
Algebra	**BASIC**
$<$	$<$
$>$	$>$
$\leq$	$<=$
$\geq$	$>=$
$\neq$	$<>$

Critical Thinking

18. Shawn told his parents that if he earned an A in pre-algebra, he would be on the honor roll. Suppose Shawn is not on the honor roll. Does that mean that he did not earn an A in pre-algebra? Why or why not?

Wrap-Up

19. Give an example of a conditional and its converse where the conditional is false and the converse is true.

12-8 Congruent Triangles

Objective:

Identify congruent triangles and corresponding parts of congruent triangles.

Key Terms:

congruent
corresponding parts

The roof supports of the Spaceship Earth geosphere at Epcot Center have the same size and shape. Figures that have the same size and shape are **congruent.**

Try the following activity to learn more about congruent figures.

1. On a piece of graph paper, draw two triangles like the ones below. Label the vertices as shown. Cut the triangles out.

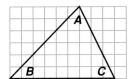

2. Put one triangle over the other so that parts with the same measures match up. Write down the pairs of angles and sides that match or correspond. Use the symbol ⟷ which means *corresponds to*.

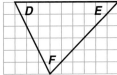

$$\angle A \longleftrightarrow \angle F \qquad \overline{AB} \longleftrightarrow \overline{FE}$$
$$\angle B \longleftrightarrow \angle E \qquad \overline{BC} \longleftrightarrow \overline{ED}$$
$$\angle C \longleftrightarrow \angle D \qquad \overline{AC} \longleftrightarrow \overline{FD}$$

3. Place the triangles beside each other so that the corresponding parts are in the same positions.

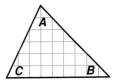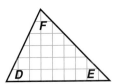

From the activity, you know that the **corresponding parts** have the same measure. Therefore, the corresponding parts are congruent. The symbol ≅ means *is congruent to*.

$$\angle A \cong \angle F \qquad \overline{AB} \cong \overline{FE}$$
$$\angle B \cong \angle E \qquad \overline{BC} \cong \overline{ED}$$
$$\angle C \cong \angle D \qquad \overline{AC} \cong \overline{FD}$$

These relationships help to define congruent triangles.

If two triangles are congruent, their corresponding sides are congruent and their corresponding angles are congruent.

Note that when writing $\triangle ABC \cong \triangle FED$, the corresponding vertices are written in the same order. For example, A is the first vertex listed in the first triangle. Since F corresponds to A, F is the first vertex listed in the second triangle. Likewise, B corresponds to E and C corresponds to F.

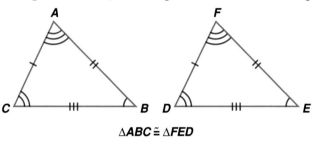

$\triangle ABC \cong \triangle FED$

Another way to show the corresponding parts is to use matching marks like those shown above in red.

Examples

1 **If $\triangle PQR \cong \triangle MLN$, name the congruent angles and sides.**

First, name the three pairs of congruent angles by looking at the order of the vertices.

$$\angle P \cong \angle M$$
$$\angle Q \cong \angle L$$
$$\angle R \cong \angle N$$

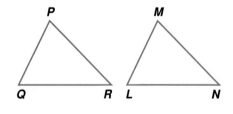

Now, draw the two triangles in the same position to confirm the corresponding sides. Name the three pairs of congruent sides.

$\overline{PQ} \cong \overline{ML}, \overline{QR} \cong \overline{LN}, \overline{PR} \cong \overline{MN}$

2 **The corresponding parts of two congruent triangles are given. Write a congruence statement for the triangles.**

$$\angle I \cong \angle K \qquad \overline{IH} \cong \overline{KH}$$
$$\angle G \cong \angle J \qquad \overline{GH} \cong \overline{JH}$$
$$\angle GHI \cong \angle JHK \qquad \overline{GI} \cong \overline{JK}$$

In this case, the congruence can be written by matching the vertices of the congruent angles. $\triangle IGH \cong \triangle KJH$

Checking for Understanding

Communicating Algebra

1. Explain what it means when two triangles are congruent.

2. In your own words, describe the corresponding parts of congruent triangles.

3. If $\triangle QRS \cong \triangle TUV$, name the angle that corresponds to $\angle R$.

Guided Practice

Use the congruent triangles at the right to complete Exercises 4-9.

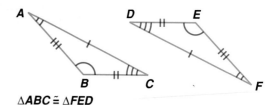

4. $\overline{AB} \cong$ ■

5. $\angle E \cong$ ■

6. $\angle C \cong$ ■

7. $\overline{BC} \cong$ ■

8. $\overline{DF} \cong$ ■

9. $\angle D \cong$ ■

$\triangle ABC \cong \triangle FED$

Exercises

Independent Practice

Complete each congruence statement. Then name the corresponding parts.

10.

$\triangle ABC \cong \triangle$ ___?___

11.

$\triangle CBA \cong \triangle$ ___?___

12.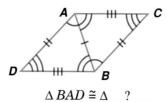

$\triangle BAD \cong \triangle$ ___?___

13.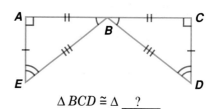

$\triangle BCD \cong \triangle$ ___?___

If $\triangle BCA \cong \triangle GFH$, name the part congruent to each angle or segment given.

14. $\angle F$

15. $\overline{BA}$

16. $\overline{FG}$

17. $\angle A$

18. $\overline{AC}$

Mixed Review

19. Find the discount rate for a $40 sweater on sale for $34. (Lesson 9-10)

20. **Probability** Three dice are rolled. How many outcomes are possible? (Lesson 11-2)

21. **Logic** Write as a conditional statement: *An angle that measures less than 90° is an acute angle.* (Lesson 12-7)

Connection

22. **Geometry** Two triangles are congruent and the perimeter of one triangle is 5 ft. What is the perimeter of the second triangle?

Critical Thinking

23. If two triangles have congruent, corresponding angles, are the triangles congruent? Explain your answer by making a drawing.

Wrap-Up

24. Make a sketch of two congruent triangles. Label the triangles and write a congruence statement.

Transformations

Materials: notebook paper, scissors, cardboard

Most of the technical art in this textbook was generated using a computer. The artist drew a figure and used a computer mouse to translate, reflect, or rotate the figure to the desired location.

In this Exploration, you will investigate a transformation called a translation.

Your Turn: Work with a partner as you complete this activity.

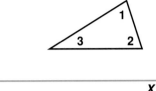

▶ Cut a triangle out of a thin piece of cardboard. Label its angles 1, 2, and 3. Place the triangle on a piece of notebook paper as shown. Trace the triangle on the paper. Label the vertices so that the vertex of ∠1 is X, the vertex of ∠2 is Y, and the vertex of ∠3 is Z.

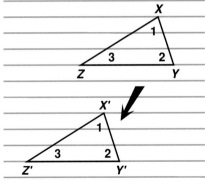

▶ Slide the cutout triangle to another place on the paper, making sure the common side of ∠2 and ∠3 is still aligned with a horizontal rule as shown. Trace the triangle. Label the vertices X′, Y′, and Z′, so they correspond to the first triangle you drew.

▶ Draw $\overline{XX'}$, $\overline{YY'}$, and $\overline{ZZ'}$.

Analysis

1. Measure $\overline{XX'}$, $\overline{YY'}$, and $\overline{ZZ'}$. What seems to be true about these segments?

2. What kind of figure does quadrilateral XX′Y′Y appear to be? Name two other figures of this type from your drawing.

3. Write a congruence statement for the triangles you traced.

4. Create a design by translating a figure.

12-9 Similar Triangles

Objectives:
Identify corresponding parts of similar triangles. Find missing measures by using lengths of corresponding sides.

Key Term:
similar

Rachel is on the yearbook staff at Wyandot Middle School. She must reduce the size of photographs to fit on the pages. The two photographs shown at the right have the same shape, but one is smaller than the other. Figures that have the same shape, but may differ in size are called **similar** figures.

Triangle *ABC* is similar to triangle *DEF*.

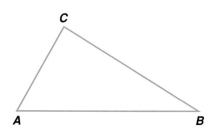 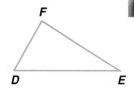

Say: Triangle *ABC* is similar to triangle *DEF*.

Write: $\triangle ABC \sim \triangle DEF$ The symbol $\sim$ means *is similar to*.

Use a protractor to measure each angle. Compare the measures of $\angle A$ and $\angle D$, $\angle C$ and $\angle F$, and $\angle B$ and $\angle E$. How do the measures of these angles compare?

> **Corresponding Angles of Similar Triangles**
>
> If two triangles are similar, then the angles of one triangle are congruent to the corresponding angles of the other triangle.

Remember that the converse of a conditional is formed by switching the hypothesis and the conclusion of the conditional.

The converse of the statement or conditional above is also true.

If the angles of one triangle are congruent to the corresponding angles of another triangle, then the triangles are similar.

Compare the measures of the corresponding sides of the similar triangles below by writing their ratios. The corresponding sides are $\overline{LN}$ and $\overline{RT}$, $\overline{MN}$ and $\overline{ST}$, and $\overline{LM}$ and $\overline{RS}$.

$$\triangle LMN \sim \triangle RST$$

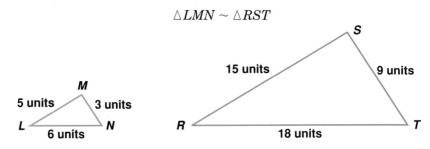

The notation $m\,\overline{LN}$ means the measure of $\overline{LN}$.

$$\frac{m\,\overline{LN}}{m\,\overline{RT}} = \frac{6\text{ units}}{18\text{ units}} \text{ or } \frac{1}{3} \qquad \frac{m\,\overline{MN}}{m\,\overline{ST}} = \frac{3\text{ units}}{9\text{ units}} \text{ or } \frac{1}{3} \qquad \frac{m\,\overline{LM}}{m\,\overline{RS}} = \frac{5\text{ units}}{15\text{ units}} \text{ or } \frac{1}{3}$$

Notice that the ratios are equivalent. So, the corresponding sides are proportional to each other. This example suggests the following property.

> **Corresponding Sides of Similar Triangles**
>
> If two triangles are similar, then their corresponding sides are proportional.

Is the converse of the conditional above true also?

Proportions can be used to find the measures of the sides of similar triangles when some measures are known.

Example

1 **Find the value of x.**

$\angle A \cong \angle D$, $\angle B \cong \angle E$, and $\angle ACB \cong \angle DCE$.

So, $\triangle ABC \sim \triangle DEC$ because their corresponding angles are congruent.

Which angles are vertical angles?
How are vertical angles related?

$$\frac{m\,\overline{BC}}{m\,\overline{EC}} = \frac{m\,\overline{AB}}{m\,\overline{DE}} \qquad \text{Corresponding sides of similar triangles are proportional.}$$

$$\frac{7.5}{12} = \frac{15}{x}$$

$$7.5 \cdot x = 12 \cdot 15 \quad \text{Cross multiply.}$$

$$7.5x = 180$$

$$\frac{7.5x}{7.5} = \frac{180}{7.5}$$

$$\boxed{180} \; \boxed{\div} \; \boxed{7.5} \; \boxed{=} \; \boxed{24}$$

$$x = 24$$

Checking for Understanding

Communicating Algebra

1. How do you know if two triangles are similar to each other?
2. Explain how to determine the corresponding parts of similar triangles.
3. Are congruent triangles also similar triangles? Explain.

Guided Practice

Use the similar triangles at the right to answer Exercises 4-6.

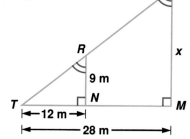

4. List three proportions for $\triangle NRT$ and $\triangle MST$.
5. What side corresponds to $\overline{SM}$?
6. Find the value of x.

Exercises

Independent Practice

Use the similar triangles at the right to answer Exercises 7-9.

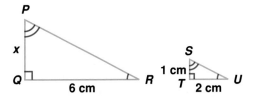

7. List three proportions for $\triangle PQR$ and $\triangle STU$.
8. What side corresponds to $\overline{PQ}$?
9. Find the value of x.

Find the value of x in each pair of similar triangles.

10.

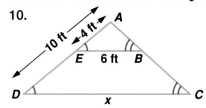

11.

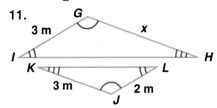

Mixed Review

12. What is the probability that it is 98°F at the North Pole?   (Lesson 11-4)
13. Are the two triangles in Exercise 11 congruent? Why or why not? (Lesson 12-8)

Application

14. **Civil Engineering**   The city of Marion plans to build a bridge across Pine Lake. Use the information in the diagram at the right to find the distance across Pine Lake.

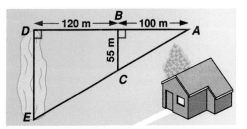

Critical Thinking

15. A photo negative is 1.5 cm wide by 2.2 cm long. The print is made 6 cm wide. How many times greater is the area of the print than the area of the negative?

Wrap-Up

16. **Make Up a Problem**   Write a problem involving the use of similar triangles.

12-10 Indirect Measurement

Objective:

Use proportions to solve problems involving similar triangles.

In ancient Egypt, mathematicians used a technique called *shadow reckoning* to determine the heights of tall objects, such as the pyramids. Shadow reckoning is based on the principle of similar triangles. The height of a staff and the length of its shadow are proportional to the height of an object and the length of its shadow.

How do you know the two triangles are similar?

You can use the same technique today by using a meterstick as your "staff". If the shadow of your meterstick is 0.7 meters and the shadow of a tree is 3.5 meters, you have enough information to find the height of the tree.

1 m

0.7 m

3.5 m

What is another proportion that could be used?

shadow of stick — height of stick

$$\frac{0.7}{3.5} = \frac{1}{h}$$

shadow of tree — height of tree

$0.7 \times h = 3.5 \times 1$ — Cross multiply. Cross products are equal.

$$\frac{0.7h}{0.7} = \frac{3.5}{0.7}$$ — Divide each side by 0.7.

$h = 5$ — The tree is 5 meters tall.

Example

1 Use similar triangles to find the distance across the river.

$\triangle TSR \sim \triangle QVR$ — Corresponding angles are congruent.

$$\frac{m\,\overline{RT}}{m\,\overline{RQ}} = \frac{m\,\overline{ST}}{m\,\overline{VQ}}$$

$$\frac{10}{16} = \frac{20}{x}$$

$10 \cdot x = 20 \cdot 16$

$10x = 320$

$$\frac{10x}{10} = \frac{320}{10}$$

$x = 32$

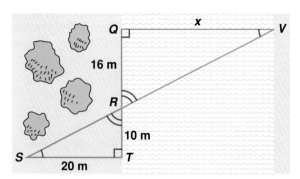

Checking for Understanding

Communicating Algebra

1. Liz wants to enlarge an embroidery pattern. She connects the centers of three flowers in the pattern to form a triangle. The sides of the triangle measure 3, 4, and 5 centimeters. Liz wants the longest distance between two flowers in the enlargement to be 12 centimeters. Where should Liz place the third flower in the enlargement?

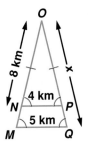

   a. What side corresponds to the longest side of the enlargement?

   b. What question is implied ?

   c. What assumption is made?

   d. What can be done to solve this problem?

   e. Draw and label a sketch to illustrate the enlargement.

   f. State a proportion you would use to find the length of the shortest side of the enlargement.

Guided Practice

Write a proportion to find each missing measure *x*. Then find *x*.

2.

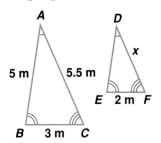

3.

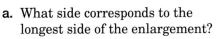

Exercises

Independent Practice

Write a proportion to find each missing measure *x*. Then find *x*.

4.

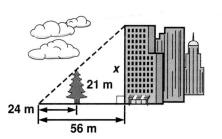

5.

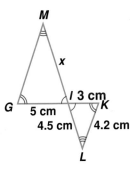

Mixed Review

6. Name two ways a graph can be misleading. (Lesson 10-9)

7. Find *x*. (Lesson 12-9)

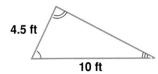

8. **Crafts** Mark wants to cut a triangular patch to make an emblem. The pattern for the emblem is a triangle with sides 8, 8, and 10 centimeters. If Mark wants to make the longest side of the emblem 25 centimeters, how long should the other sides be?

9. **Geography** On a map, the length from Cleveland to New York is 7 cm, from Cleveland to Atlanta is 10 cm, and from New York to Atlanta is 13 cm. If on a larger map the length from Cleveland to New York is 17.5 cm, what are the other two lengths?

10. Find the length of Kingly Lake.

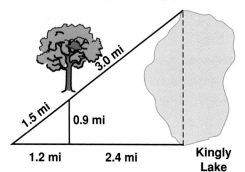

11. Find the length of the brace.

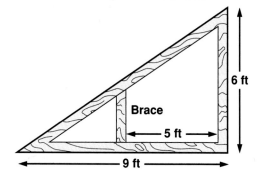

12. Find the length of Cedar Lane.

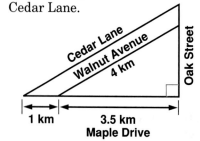

13. Find the distance across the pond from point A to point B.

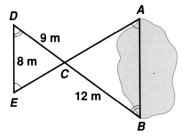

14. Written Exercises 8-13 above are examples of indirect measurement. Explain why it is called indirect measurement.

15. Explain how you could use indirect measurement to determine the height of your school building.

History

African Observers of the Universe

In the Republic of Mali in West Africa, there is a people called the Dogons who have an extremely complex knowledge of astronomy.

They are especially knowledgeable in the Sirius star system. Sirius is the brightest star in the sky. Without the use of a telescope, they have observed and studied the elliptical orbit of Sirius' small companion star, Sirius B, and have plotted its course and trajectory from the thirteenth century to the present.

Modern astronomy, with its most advanced observatories, has confirmed that Sirius B is in fact in orbit around Sirius. The fact that the Dogons have known this for 700 years, just on the basis of the naked eye, is truly amazing.

12-11 Quadrilaterals

Objectives:
Find the missing angle measure of a quadrilateral.
Classify quadrilaterals.

Key Terms:
quadrilateral
trapezoid
parallelogram
rectangle
rhombus
square

Geometric figures are often used to make patterns for quilts. All of the figures used to make the design in the quilt shown at the right are four-sided figures called **quadrilaterals.**

Just as with triangles, there is a relationship among the measures of the angles of a quadrilateral. Try the following activity.

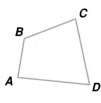

1. Draw a quadrilateral. Label its vertices *A, B, C,* and *D.*
2. Use a protractor to find the measure of each angle.
3. Find the sum of the angle measures.
4. Repeat steps 1-3 with other quadrilaterals.
5. Select one of the quadrilaterals and draw $\overline{AC}$. Note that two triangles are formed.

Since the sum of the measures of the angles of a triangle is 180°, what is the sum of the measures of the angles of the two triangles?

The activity above will work with any quadrilateral. This activity suggests the following relationship.

| **Angles of a Quadrilateral** | The sum of the measures of the angles of a quadrilateral is 360°. |

Example

1 **Find the measures of the angles in the quadrilateral.**

The sum of the measures of the angles is 360°.

$$75 + 40 + x + x = 360$$
$$115 + 2x = 360$$
$$2x = 245$$
$$x = 122\frac{1}{2}$$

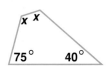

The measures of the angles are 75°, 40°, $122\frac{1}{2}$°, and $122\frac{1}{2}$°.

There are also special kinds of quadrilaterals. They can be classified by the relationship of their sides and angles. Some of the well-known members of the quadrilateral family and their definitions are shown in the chart below.

QUADRILATERALS

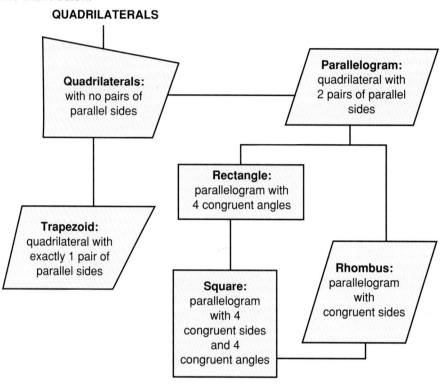

Checking for Understanding

Communicating Algebra

1. Which quadrilaterals have all four sides congruent?

2. Which quadrilaterals have two pairs of parallel sides?

3. Which quadrilaterals have all four angles congruent?

Guided Practice

Find the value of x.

4.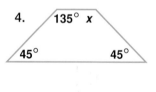

5.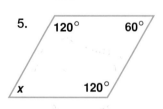

6.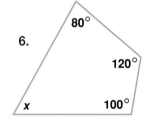

Classify each quadrilateral.

7.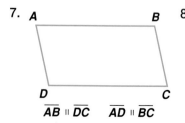

$\overline{AB} \parallel \overline{DC}$ $\overline{AD} \parallel \overline{BC}$

8.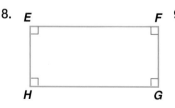

9.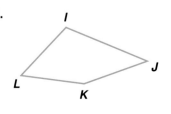

Exercises

Classify each quadrilateral.

10.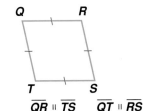

$\overline{QR} \parallel \overline{TS}$ $\overline{QT} \parallel \overline{RS}$

11.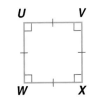

12.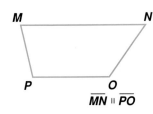

$\overline{MN} \parallel \overline{PO}$

Find the value of x.

13.

14.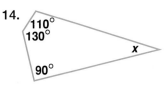

15.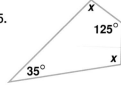

Logical Reasoning

Determine whether each statement is *always*, *sometimes*, or *never* true.

16. A rectangle is a parallelogram.

17. A rectangle is a square.

18. A rhombus is a square.

19. A trapezoid is a parallelogram.

20. A square is a parallelogram.

21. A square is a rhombus.

Mixed Review

22. Solve $a = \frac{-95}{5}$. (Lesson 2-8)

23. Factor 98 completely. (Lesson 4-4)

24. If $\angle A$ and $\angle B$ are complementary and $m\angle A = 70°$, find $m\angle B$. (Lesson 12-3)

25. Ed measures the shadow of an evergreen tree to be 8 feet and his shadow to be 3 feet. If Ed is 6 feet tall, how tall is the tree? (Lesson 12-10)

Challenge

26. The measures of the four angles of a quadrilateral are $x - 15$, $x + 35$, $x + 20$, and x. Find each measure.

Application

27. **Sports** The center portion of the main frame of the bicycle shown at the right is a trapezoid. Find x.

Critical Thinking

28. One angle of a parallelogram measures 40°. What are the measures of the other angles?

Wrap-Up

29. Fold a sheet of paper in half and draw a trapezoid. Keep the paper folded and cut out two congruent trapezoids. Use your trapezoids to form a parallelogram. Then form a different parallelogram.

12-12 Polygons

Objectives:
Classify polygons.
Determine the sum of the measures of the interior angles of a polygon.

Key Terms:
polygon
sides
vertices
regular polygon
equilateral
diagonal

The honeycomb shape of a beehive is a common example of a six-sided figure found in nature. The six-sided figure is a simple closed figure.

A simple closed figure can be traced in a continuous path without tracing any point other than the starting point more than once.

A **polygon** is a simple, closed figure formed by three or more line segments called **sides.** These sides are all in one plane and meet only at their endpoints. These points of intersection are called **vertices** (plural of **vertex**).

Polygons are classified by the number of sides. Some of the more common polygons are shown below.

What would you call a polygon with n sides?

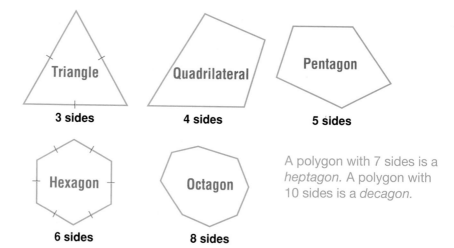

Triangle
3 sides

Quadrilateral
4 sides

Pentagon
5 sides

Hexagon
6 sides

Octagon
8 sides

A polygon with 7 sides is a heptagon. A polygon with 10 sides is a decagon.

A **regular polygon** is a polygon that has all sides congruent and all angles congruent. The triangle and hexagon above are examples of regular polygons. The regular triangle is called an **equilateral** triangle. A regular quadrilateral is a square.

A **diagonal** of a polygon is a line segment that joins two non-consecutive vertices. In each of the following polygons, all possible diagonals from one vertex are shown.

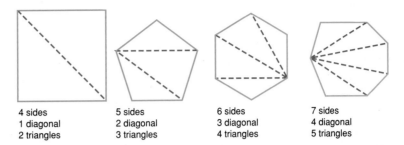

4 sides	5 sides	6 sides	7 sides
1 diagonal	2 diagonal	3 diagonal	4 diagonal
2 triangles	3 triangles	4 triangles	5 triangles

Notice that the number of triangles is equal to the number of sides minus two. You can use diagonals and the property of the sum of the angles of a triangle to determine the sum of the measures of the angles of any polygon without using a protractor.

If a polygon has n sides, then the sum of the degree measures of its angles is $(n - 2)180$.

Examples

1 Find the sum of the measures of the angles of a pentagon.

A pentagon has five sides. Therefore, $n = 5$.

$(n - 2)180 = (5 - 2)180$ Replace n with 5.
$$= 3 \cdot 180 \text{ or } 540$$

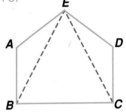

Check by drawing a pentagon and drawing the diagonals. There are three triangles. The solution checks.

The sum of the measures of the angles of a pentagon is $540°$.

2 Find the sum of the measures of the angles of an octagon.

An octagon has eight sides. Therefore, $n = 8$.

$(n - 2)180 = (8 - 2)180$ Replace n with 8.
$$= 6 \cdot 180 \text{ or } 1080$$ Check this solution.

The sum of the measures of the angles of an octagon is $1080°$.

Checking for Understanding

Communicating Algebra

1. In your own words, explain what a regular polygon is.

2. Draw a hexagon and its diagonals.

3. What is the relationship between the number of sides in a polygon and the number of triangles formed by the diagonals?

Guided Practice

Classify each polygon below and determine whether it appears to be *regular* or *not regular*.

4.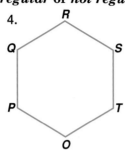

5.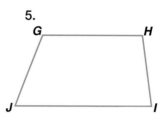

6.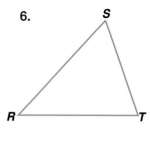

Exercises

Independent Practice

Find the sum of the measures of the angles of each polygon.

7. hexagon

8. heptagon

9. decagon

10. 25-gon

11. 50-gon

12. 100-gon

Find the perimeter of each regular polygon.

13. triangle with sides 30 feet long

14. pentagon with sides 25 inches long

15. quadrilateral with sides 16.7 meters long

16. hexagon with sides 34.5 meters long

Mixed Review

17. **Statistics**   Find the mean, median, and mode of the following data: 0.5, 0.7, 0.3, 0.4, 0.4, 0.6, 1.6, 0.4, 0.8, 0.3.   (Lesson 10-2)

18. **Probability**   A coin is tossed, then a die is rolled. Find P(tails and 6). (Lesson 11-6)

19. Using a protractor, draw a right triangle.   (Lesson 12-6)

20. *True* or *false*: A rectangle is a parallelogram with all angles congruent. (Lesson 12-11)

Application

21. **Manufacturing**   The S & D Company has an order from a flooring company for tiles to be cut in the shape of regular hexagons. At what angle should the cutting machine be set so that the measure of each angle of each tile is identical?

Critical Thinking

22. Trace the dot pattern shown at the right. Without lifting your pencil from the paper, draw four line segments that connect all the points.

```
• • •
• • •
• • •
```

Wrap-Up

23. Explain why a square is a regular polygon.

Team Problem Solving

Mrs. O'Grady buys a square building to use as a dog kennel. She thinks that she can house 9 dogs in this kennel in the formation shown at the right. However, she feels it would be best if there were a fence to separate the dogs. She calls in the fencing company and asks them to build two fences in the shape of squares so that no two dogs share the same area. The fencing company is confused. Can you help?

476   *Applying Algebra to Geometry*

Tessellations

Materials: tracing paper, cardboard, scissors

The repetitive pattern of regular polygons shown at the right is an example of a **tessellation.** In a tessellation, the polygons fit together with no holes or gaps.

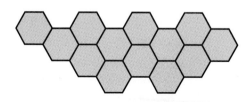

In this Exploration, you will investigate tessellations with regular polygons.

Your Turn: Work with a partner.

▶ Trace each regular polygon shown. Cut each shape and trace it onto a piece of cardboard. Cut each shape from the cardboard. Use the cardboard piece to try to draw a tessellation for each regular polygon.

Equilateral Triangle

Square

▶ For each polygon, calculate the sum of the angle measures. Then determine the measure of one angle in each polygon. Record your results in a table.

Regular Pentagon

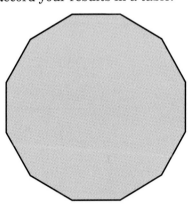

Regular Dodecagon

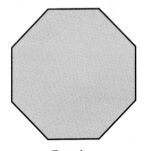

Regular Octagon

Analysis

1. How can you tell from the measure of one angle whether or not a polygon will tessellate?

2. Try to make tessellations from the combinations below.
   a. square and octagon
   b. square, triangle, and hexagon
   c. square, triangle, and dodecagon
   d. another combination you choose

3. Describe a pattern similar to the one in Exercise 1 that tells whether a combination of polygons will tessellate.

Review

Language and Concepts

Choose the correct letter to complete each sentence.

1. An angle that measures between 0° and 90° is __?__ .

2. Two angles are __?__ if the sum of their measures is 90°.

3. When a __?__ intersects two parallel lines, the corresponding angles are congruent.

4. When a line is perpendicular to another line, the angles formed are __?__ angles.

5. An angle that measures between 90° and 180° is __?__ .

6. A parallelogram with four congruent sides is a __?__ .

7. Two angles are __?__ if the sum of their measures is 180°.

a. obtuse
b. acute
c. vertical
d. trapezoid
e. rhombus
f. right
g. transversal
h. complementary
i. supplementary

Skills

Draw a picture to represent each of the following. (Lesson 12-1)

8. point C 9. $\overrightarrow{XY}$ 10. plane GHI 11. $\overleftrightarrow{RS}$

Refer to the diagram at the right to name each of the following.
(Lessons 12-1, 12-3, 12-4)

12. two obtuse angles

13. two acute angles

14. a right angle

15. two pairs of vertical angles

16. two pairs of complementary angles

17. two pairs of supplementary angles

18. two pairs of adjacent angles

19. a transversal

20. two pairs of alternate interior angles

21. two sets of corresponding angles

22. Trace $\angle IFC$ shown above. Construct an angle congruent to it. (Lesson 12-5)

23. Trace $\overline{IF}$ shown above. Construct a segment congruent to it. (Lesson 12-5)

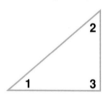

Use the figure at the right to solve. (Lesson 12-6)

24. Find $m\angle 1$ if $m\angle 2 = 50°$ and $m\angle 3 = 80°$.

25. Find $m\angle 2$ if $m\angle 1 = 72°$ and $m\angle 3 = 66°$.

If $\triangle XYZ \cong \triangle ABC$, name the part congruent to each angle or segment.

26. $\angle X$ 27. $\angle B$ 28. $\overline{YZ}$ 29. $\overline{AC}$

Find the value of x in each pair of similar triangles. (Lesson 12-9)

30.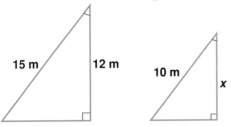

31.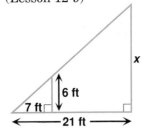

32. Find the value of x in the figure at the right.
    (Lesson 12-11)

33. Find the sum of the degree measures of the angles of
    a hexagon. (Lesson 12-12)

Applications and Problem Solving

34. Make a circle graph to display the
    data in the chart.
    (Lesson 12-2)

Water Use in the U.S.	
Agriculture	36%
Public Water	8%
Utilities	33%
Industry	23%

35. A 6-foot ladder touches the side of a
    building at a point 5 feet above the
    ground. At what height would a 15-
    foot ladder touch the building if it
    makes the same angle with the
    building as the shorter ladder?
    (Lesson 12-10)

**Write the converse of each conditional. Then determine whether each statement
(conditional and converse) is *true* or *false*.** (Lesson 12-7)

36. If an angle measures 90°, then it is a
    right angle.

37. If Mrs. Clark lives in Boston, then
    Mrs. Clark lives in a city.

Portfolio Suggestion

Select an item from your work in this chapter that
shows your creativity and place it in your portfolio.

Curriculum Connection

• **Biology** Research why honeycombs are made in
  the shape of regular hexagons.

Read More About It

Frieder, David. *Clear and Simple Geometry.*

Laithwaite, Eric. *Shape: The Purpose of Forms.*

Zaslavsky, Claudia. *Africa Counts.*

Test

Match the letter of each figure to its most exact description.

1. line segment
2. complementary angles
3. obtuse angle
4. vertical angles
5. supplementary angles
6. perpendicular lines
7. parallel lines
8. acute triangle
9. regular quadrilateral
10. rhombus
11. right triangle
12. ray

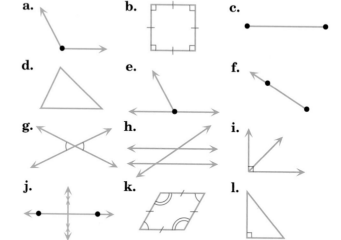

a.

b.

c.

d.

e.

f.

g.

h.

i.

j.

k.

l.

13. Find the value of x. Then classify the angle as *acute, right* or *obtuse*.

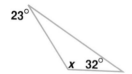

23°

x   32°

14. Find the value of x.

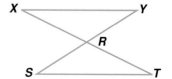

110°

40°

x

110°

15. In the figure at the right, $\triangle XYR \cong \triangle TSR$. Name the congruent angles and sides.

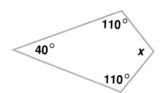

X   Y

R

S   T

16. An angle of a circle graph represents $30 out of a total of $120. What should the measure of the angle be?

17. Mara is 5 feet 6 inches tall. She notices that her shadow is 4 feet long and the shadow of a nearby tree is 24 feet long. How tall is the tree?

Write the converse of the following statement.

18. If the sum of the measures of two angles is 180°, then the angles are supplementary.

Find the sum of the measures of the angles of each polygon.

19. quadrilateral

20. decagon

BONUS

In your own words, explain why a triangle cannot have two right angles.

Academic Skills Test

Cumulative, Chapters 1-12

1. Lois wants to cut a piece of ribbon that is $5\frac{1}{2}$ feet long into strips that are 8 inches long for some awards. How many inches are equivalent to $5\frac{1}{2}$ feet?

   A $16\frac{1}{2}$ inches   **C** 55 inches

   B 44 inches   **D** 66 inches

2. Which equation is equivalent to $4n - 4 = 3n + 10$?

   A $4n - 4 - 3n = 10$
   B $4n + 3n = 10 - 4$
   C $4n = 10 - 4$
   D $7n = 10 + 4$

3. Which ordered pairs are solutions for $y = 2x - 3$?

   A (-3, -9), (2, 1), (4, 5)
   B (-3, 0), (-1, -5), (3, 3)
   C (-3, 3), (-2, 1), (-1, -1)
   D (-2, -7), (-1, 1), (0, -3)

4. As a team, the Wildcats made 24 shots in 40 times at the freethrow line. Which expression could be used to find the percent of shots the team made?

   A $\frac{16}{40} \times 100$ **C** $\frac{40}{24} \times 100$

   B $\frac{24}{40} \times 100$ **D** Not Here

5. The high temperatures (in °F) for January 15th during the past 10 years are: 15°, 18°, 25°, 38°, 6°, 25°, 10°, 30°, 18°, and 25°. What is the mode of these temperatures?

   A 18°   **C** 25°
   B 21.5°   **D** Not Here

6. What is the median of the test scores shown in the stem-and-leaf plot below?

   6|0 means 60

   ```
   6 | 0 5 6 8
   7 | 0 0 7 9 9
   8 | 3 4 7
   9 | 2 2 4 5 8
   ```

   A 38
   B 79
   C 79 and 92
   D 80

7. The fastest two of six runners will qualify for the final race. How many different combinations of two qualifiers are possible?

   A 2   **C** 15
   B 12   **D** 30

8. Mr. Garza has 4 black ties, 3 gray ties, 2 blue ties, and 1 brown tie in his closet. If he selects one tie without looking, what is the probability that it will be blue?

   A $\frac{1}{10}$ **C** $\frac{1}{4}$

   B $\frac{1}{5}$ **D** $\frac{1}{2}$

9. The shaded region represents part of—

   A an angle
   B a line
   C a ray
   D a plane

10. $\triangle QRS$ is similar to $\triangle XYZ$. What is the length of $\overline{RQ}$?

    A 16
    B 18
    C 20
    D 22

CHAPTER 13

CHAPTER OBJECTIVES

In this chapter you will learn to:

- find the area of polygons and circles
- find the surface area of prisms, cylinders, pyramids, and cones
- find the volume of prisms, cylinders, pyramids, and cones
- solve problems by making a model or drawing

Measuring Area and Volume

Engineers and builders in ancient Egypt, around 3000 B.C., pioneered many aspects of mathematics, and for good reasons. Their land experienced yearly floods, during which the Nile River's waters covered farmland for weeks. Every year after the waters drained away, farm boundaries had to be remeasured. It's no wonder that Egyptians were pioneers in many aspects of mathematics, particularly in measurement.

Khufu's Great Pyramid at Gizeh is evidence of the impressive mathematical skill of the Egyptians.

History Connection

Class Project

Choose one of these topics: history of mathematics, history of building, or archaeology. Research some of the important developments in your topic. Make a time line of those developments.

13-1 Area: Parallelograms and Triangles

Objective:
Find the area of parallelograms and triangles.

Key Terms:
area
base
height
altitude

If the area of each square on the map of Tennessee is about 1000 square miles, what is an estimate of the area of Tennessee?

If you were driving from Memphis, Tennessee, to the Great Smoky National Park in Gatlinburg, Tennessee, you might use a road map similar to the one shown below. The map has been separated into congruent, nonoverlapping squares to aid you in finding certain cities.

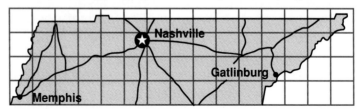

The surface covered by the squares is the **area** of the region. Area is measured by the number of units that it takes to cover the region exactly. Two common units of measure for area are the square centimeter (cm^2) and the square inch (in^2).

You can calculate the area of a rectangle by multiplying the measure of the length, or **base,** and the measure of the width, or **height.**

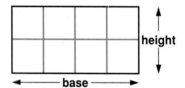

The rectangle shown at the left has a base of 4 units and a height of 2 units. The area is 8 square units.

History Connection

There is evidence that, as long ago as 2000 B.C., the Babylonians used some of the formulas we use today for measuring area, including formulas for the area of a rectangle and a right triangle.

A rectangle is a special kind of parallelogram. It should not be surprising to learn that finding the area of a parallelogram is closely related to finding the area of a rectangle.

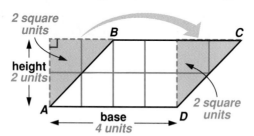

Parallelogram *ABCD* is made from the rectangle by moving the shaded triangle.

The areas of the parallelogram and the rectangle above are both 8 square units. This suggests that you can find the area of a parallelogram by multiplying the measures of the base and the height.

Area of a Parallelogram

If a parallelogram has a base of b units and a height of h units, then the area (A) is $b \cdot h$ square units.
$$A = bh$$

The base can be any side of the parallelogram. The height is the length of an **altitude,** a line segment perpendicular to the base with endpoints on the base and the side opposite the base.

Example

1 **Find the area of the parallelogram.**

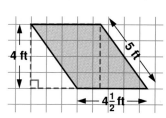

The base is $4\frac{1}{2}$ ft long. The height is 4 ft.

$A = bh$ Formula for area of a parallelogram

$A = 4\frac{1}{2} \cdot 4$ Replace b with $4\frac{1}{2}$ and h with 4.

$A = 18$

The area is 18 ft^2.

Notice that the diagonal and sides of a parallelogram form two triangles of the same size. You can use this fact to find the area of a triangle.

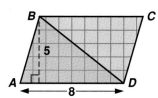

The area of the parallelogram at the left is 8 · 5 or 40 square units. The area of each triangle is one half the area of the parallelogram. So, each triangle has an area of 20 square units.

Area of a Triangle	If a triangle has a base of b units and a height of h units, then the area (A) is $\frac{1}{2} \cdot b \cdot h$ square units. $$A = \frac{1}{2}bh$$

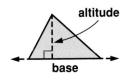

Any one of the sides of a triangle can be used as a base. The height is the length of the corresponding altitude, a line segment perpendicular to the base from the opposite vertex.

Example

2 **Draw a model of an equilateral triangle. Then identify the possible bases and draw the altitude for each base.**

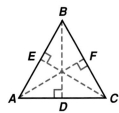

base		altitude
$\overline{AC}$	with	$\overline{BD}$
$\overline{AB}$	with	$\overline{CE}$
$\overline{BC}$	with	$\overline{AF}$

Examples

3 **Draw a model of an obtuse triangle. Then identify the possible bases and draw the altitude for each base.**

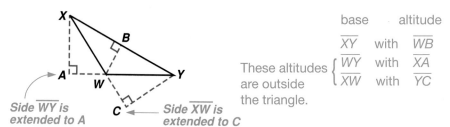

	base		altitude
These altitudes are outside the triangle.	$\overline{XY}$	with	$\overline{WB}$
	$\overline{WY}$	with	$\overline{XA}$
	$\overline{XW}$	with	$\overline{YC}$

Side $\overline{WY}$ is extended to A

Side $\overline{XW}$ is extended to C

Find the area of each triangle.

4

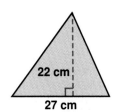

22 cm

27 cm

$A = \frac{1}{2}bh$

Replace b with 27 and h with 22.

$A = \frac{1}{2} \cdot 27 \cdot 22$

$$0.5 \;\boxtimes\; 27 \;\boxtimes\; 22 \;\boxminus\; 297$$

$A = 297$

The area is 297 cm².

5

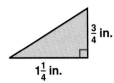

$\frac{3}{4}$ in.

$1\frac{1}{4}$ in.

In a right triangle, two of the altitudes are sides of the triangle.

$A = \frac{1}{2}bh$

$A = \frac{1}{2} \cdot 1\frac{1}{4} \cdot \frac{3}{4}$

$A = \frac{1}{2} \cdot \frac{5}{4} \cdot \frac{3}{4}$

$A = \frac{15}{32}$

The area is $\frac{15}{32}$ in².

Checking for Understanding

Communicating Algebra

1. In your own words, describe area.

2. Name two common units that are used to measure area.

3. Describe the altitude of a triangle.

State the formula used to find the area of each figure.

4. rectangle   5. parallelogram   6. triangle

Guided Practice

State the measures of the base and height, and how to find the area of each figure. Then find each area.

7.

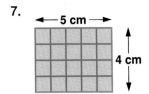

5 cm

4 cm

8.

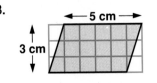

5 cm

3 cm

9.

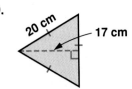

20 cm

17 cm

Exercises

Independent Practice

Draw a model of each triangle. Then identify the possible bases and draw a model of the altitude for each base.

10. acute

11. obtuse

12. right

Find the area of each figure.

13.

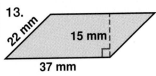

22 mm
15 mm
37 mm

14.

15 ft
12 ft
9 ft

15.

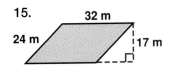

32 m
24 m
17 m

16.

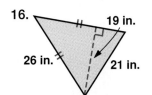

19 in.
26 in.
21 in.

17.

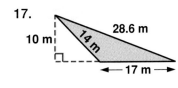

10 m
14 m
28.6 m
17 m

18.

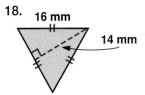

16 mm
14 mm

Find the area of each figure described below.

19. parallelogram; base, 12 cm; height, 18 cm

20. parallelogram; base $3\frac{1}{4}$ ft; height, 2 ft

21. triangle; base, 10 in.; height 7 in.

22. triangle; base, 0.8 km; height, 2 km

23. rectangle; length, $2\frac{1}{2}$ in.; width, $1\frac{1}{4}$ in.

24. rectangle; length, 11 yd; width, 25 yd

Mixed Review

25. Solve $g = -12 + 16$.   (Lesson 2-4)

26. *True* or *false:* $\frac{5}{9} > \frac{3}{5}$.   (Lesson 4-8)

27. **Probability**   A multiple choice quiz has 5 questions with 4 possible answers for each. How many outcomes are possible?   (Lesson 11-2)

28. **Geometry**   Use a protractor to draw an angle having a measure of 115°. Classify this angle as *acute, right,* or *obtuse.*   (Lesson 12-1)

Applications

29. **Gardening**   A triangular rose garden has a base of 5 meters and a height of 3 meters. If each rose plant needs at least 0.5 m^2 of space, how many roses can be planted in this garden?

30. **Geography**   The state of Tennessee as shown in the map on page 484 is shaped almost like a parallelogram. The distance along the north boundary is 447 miles and the north-south distance is 116 miles. Estimate the area of the state.

Critical Thinking

31. What is the effect on the area of a square if the length of the sides is doubled? What is the effect on the area of a triangle if both the base and height are doubled?

Wrap-Up

32. In your own words, explain how the areas of a rectangle, a parallelogram, and a triangle are related.

13-2 Area: Trapezoids

Objective:
Find the area
of trapezoids.

Key Term:
base

The design of the College Park
Pyramids in Indianapolis is based on
a quadrilateral called a trapezoid.

Remember, a trapezoid is a
quadrilateral with exactly two
parallel sides called the **bases.** The
height of a trapezoid is the distance
between the bases. Like a
parallelogram, an altitude is a
segment perpendicular to both
bases, with endpoints on the base
lines. The length of the altitude is
called the height.

Examples

Name the bases and an altitude of each trapezoid.

1

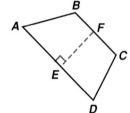

Sides $\overline{BC}$ and
$\overline{AD}$ are the bases.
$\overline{FE}$ is an altitude.

2

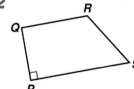

Sides $\overline{QR}$ and
$\overline{PS}$ are the bases.
$\overline{QP}$ is an altitude.

The area of a trapezoid can be found using the area of a triangle. The
diagonal, $\overline{BD}$, separates the trapezoid shown below into two triangles.
So, the area of the trapezoid is the sum of the areas of the triangles.

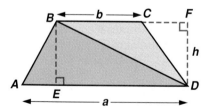

The two triangles are $\triangle ABD$ and
$\triangle BCD$. The altitudes of these
triangles, $\overline{BE}$ and $\overline{FD}$, are congruent.
Both are h units long. The base of
$\triangle ABD$ is $\overline{AD}$, a units long. The base
of $\triangle BCD$ is $\overline{BC}$, b units long.

Area of trapezoid $ABCD$ = area of $\triangle ABD$ + area of $\triangle BCD$

$$= \quad \frac{1}{2}ah \quad + \quad \frac{1}{2}bh$$

$$= \frac{1}{2}h(a + b) \qquad \text{Distributive property}$$

Area of a Trapezoid	If a trapezoid has bases of a units and b units and a height of h units, then the area (A) of the trapezoid is $\frac{1}{2} \cdot h \cdot (a + b)$ square units. $$A = \frac{1}{2}h(a + b)$$

Example

3 **Find the area of the trapezoid.**

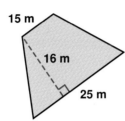

$A = \frac{1}{2}h(a + b)$ *a* is 15, *b* is 25, and *h* is 16.

$A = \frac{1}{2} \cdot 16(15 + 25)$

$A = \frac{1}{2} \cdot \overset{8}{\cancel{16}} \cdot 40$

$A = 320$

The area is 320 m^2.

Checking for Understanding

Communicating Algebra   **Use the figure at the right for Exercises 1 and 2.**

1. Name the bases of trapezoid $ABCD$.

2. Name an altitude of trapezoid $ABCD$.

3. In your own words, state how to find the area of a trapezoid.

4. Draw and label a trapezoid with bases $\overline{RS}$ and $\overline{TV}$.

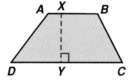

Guided Practice   **State the measures of the bases and altitude. Then find each area.**

5.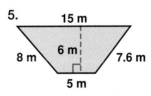

6.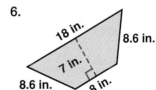

7.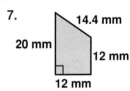

Exercises

Independent Practice

ESTIMATION MENTAL MATH CALCULATOR PAPER/PENCIL

Find the area of each trapezoid described below.

	base (*a*)	base (*b*)	height
8.	9 cm	24 cm	12 cm
9.	26 in.	14 in.	15 in.
10.	4.7 cm	5.9 cm	2 cm

	base (*a*)	base (*b*)	height
11.	12 ft	8 ft	7 ft
12.	0.3 km	0.5 km	0.2 km
13.	5.3 m	1.72 m	8.4 m

14. Express 44,010,000 in scientific notation. (Lesson 6-11)

15. If $m\angle T = 54°$, find the measure of its supplement. (Lesson 12-3)

16. Using the formula $A = \frac{1}{2}bh$, find the area of a triangle that has a base of 8 inches and a height of 7 inches. (Lesson 13-1)

Applications

17. **Architecture** The side of a building is in the shape of a trapezoid. The base of the building is 400 feet wide. The top is 150 feet wide. The building is 500 feet tall. If the side is completely glass-paned, how much glass is used?

18. **Agriculture** A parcel of farmland, in the shape of a trapezoid, has been divided into two fields as shown in Figure 1. Which field has the greater area? How much greater?

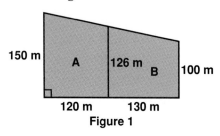

Figure 1

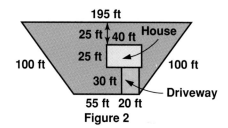

Figure 2

19. **Home Care** David Ling wants to re-seed his lawn. A diagram of his lot is shown in Figure 2. Before he buys the seed, he needs to know the area of the lawn. The lot is 85 feet deep. Calculate the number of square feet of lawn to be seeded.

20. **Geography** The state of Wyoming is shaped like a trapezoid. Use the information on the map to estimate the area of the state.

Critical Thinking

21. When a diagonal is drawn in a trapezoid, will one of the triangles formed always be an obtuse triangle? Will one always be an acute triangle? If not, draw a counterexample.

Wrap-Up

22. In your own words, tell how the formula for the area of a trapezoid is related to the formula for the area of a triangle.

Team Problem Solving

Someone has covered square *ABCD* with eight smaller squares, all the same size. The square marked 8 was the last square placed.

• Here is your challenge. Number the squares in the order they were placed.

• Make a puzzle similar to this one to share with your friends.

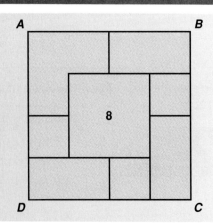

Algebra in Action-Design

Interior Design

The library at Sam Houston High School needs new carpeting. An interior designer has selected the carpet and measured the room. A drawing of the room is shown below. All the corners are right angles.

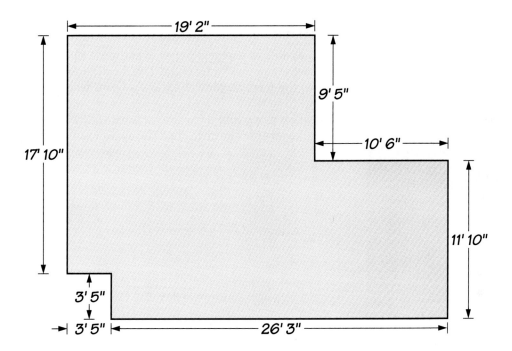

Work in a team to answer the following questions.

1. How could you find the area to be carpeted?

2. The carpeting can be purchased on a 12-foot wide roll or in square-foot pieces. How much carpeting will be needed if it is bought in the square-foot pieces?

3. If the carpeting is bought from a 12-foot wide roll, how much carpeting must be purchased?

4. The square-foot pieces cost $35 per dozen. The carpet on the roll costs $28 per square yard. Based on price, would you choose to buy the square pieces or the carpet from the roll? Why?

13-3 Area: Circles

Objective:
Find the area of circles.

Can you think of a way to estimate the area of a circle? Here's a hint. First use a compass to draw a circle on grid paper. Count the squares contained completely within the circle. Include in your estimate the squares that are partially contained within the circle.

It is not always possible to use grid paper to estimate the area of a circle. In these cases, you need a formula. The formula for the area of a circle is related to the formula for the area of a parallelogram.

Separate a circle into parts and make a figure that looks like a parallelogram. The circle has the same area as that figure.

$A = b \times h$ Formula for area of a parallelogram

$A = \left(\frac{1}{2} \times C\right) \times r$ The base of the parallelogram is one-half the circumference.

$A = \frac{1}{2} \times (2\pi r) \times r$ Remember, $C = 2\pi r$.

$A = \pi \times r \times r$ $\frac{1}{2} \cdot 2 = 1$

$A = \pi r^2$ The formula is $A = \pi r^2$.

Area of a Circle	If a circle has a radius of r units, then the area (A) is $\pi \cdot r \cdot r$ or $\pi \cdot r^2$ square units. $$A = \pi r^2$$

Example

1 **Find the area of the circle. Round your answer to the nearest whole number.**

$A = \pi r^2$ Formula for area of a circle

$A = \pi \cdot 4^2$ Since the radius is one-half the diameter, $r = \frac{1}{2}(8)$ or 4.

$A \approx 50$ Compare with the estimate.

The area is about 50 cm^2.

Checking for Understanding

Communicating Algebra

1. Estimate the area of the circle shown at the right by counting squares and by using the Estimation Hint on page 492. How do the estimates compare?

2. State the formula for finding the area of a circle.

Guided Practice

Tell how you should find the area of each circle. Then find each area.

3. radius, 3 cm   4. diameter, 10 ft   5. radius, 3.5 mi

Exercises

Independent Practice

Find the area of each circle shown or described below. Round answers to the nearest whole number.

6.
     8 cm

7.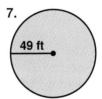
     49 ft

8.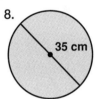
     35 cm

9. radius, 16 cm   10. diameter, 8 m   11. radius, 7 yd

12. diameter, 42 in.   13. radius, 5.2 km   14. radius, $3\frac{1}{2}$ in.

Mixed Review

15. **Sports** A 4-person relay team ran a race in 232 seconds. What was the average time for each person? (Lesson 1-9)

16. Using the formula $A = \frac{1}{2}h(a + b)$, find the area of a trapezoid with bases of 4 cm and 9 cm and a height of 6 cm. (Lesson 13-2)

Applications

17. **Horticulture** A gardener for a park district has to prepare three circular gardens for planting. One garden has a diameter of 12 feet, one has a radius of 8 feet, and the third has a diameter of 20 feet. If each bag of peat moss can cover 160 square feet of topsoil, how many bags of peat moss will he need?

18. **History** Stonehenge, an ancient monument in England, may have been used as a calendar. The stones are arranged in a circle 30 meters in diameter. Find the area of the circle.

Critical Thinking

19. The area and circumference of a circle have the same measure. Find the radius.

Wrap-Up

20. In your own words, explain how to find the area of a circle if you know its diameter.

Area and Probability

Materials: unlined paper, ruler, compass,
thumb tacks or straight pins

In this Exploration you will investigate the relationship between the area of irregular figures and probability.

Your Turn: Work in small groups to complete this activity.

▶ Draw a 6-inch square on your paper. Within the square, draw a figure with an irregular shape. Hold 20 thumb tacks about 3 inches above the paper and drop them onto the paper.

▶ Count the number of thumb tacks that landed within the square. (Include those thumb tacks that landed in the figure.) Then count the number that landed only in the figure. This is sample A.

▶ Repeat nine more times for samples B through J. Find the total within the square and the total within the figure.

Analysis

1. The experimental probability that a thumb tack landed only in the figure is given by the ratio $\frac{\text{total within figure}}{\text{total within square}}$. Based on your findings, calculate the probability.

2. The probability is also related to the area of the figure. Use the proportion below to estimate the area of the figure.

$$\frac{\text{total within figure}}{\text{total within square}} = \frac{\text{area of figure}}{\text{area of square}}$$

3. Devise a way to determine whether your answer is reasonable.

4. Suppose a skydiver parachutes onto a square field that contains a pond. Assume there is an equal chance of landing at any point within the field. Design a model to help estimate the probability that the diver has a dry landing.

13-4 Surface Area: Prisms

Objective:
Find the surface area of triangular and rectangular prisms.

Key Terms:
prism
base
rectangular prism
face
surface area
triangular prism

Rectangular regions ABCD and EFGH are bases. Could another pair of regions be the bases of this rectangular prism?

In Lessons 13-1, 13-2, and 13-3, you found the area of two-dimensional shapes. However, most of the shapes you use in everyday life have three dimensions, length, width, and height.

In geometry, solids such as the box in the photo are called **prisms.** A prism is a solid figure that has two parallel congruent sides, called **bases**. A prism is classified by the shape of its bases. The prism shown at the right is a **rectangular prism.**

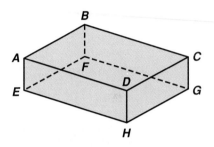

If you "open up" or "unfold" a rectangular prism, you will see the rectangular regions or **faces** that make up the surface of the prism. The **surface area** of a prism is the sum of the areas of its faces.

There are six faces in a rectangular prism.

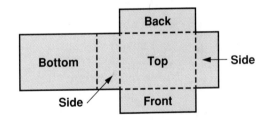

Example

1 **Find the surface area of a rectangular prism if the length is 0.8 m, the width is 0.5 m, and the height is 0.4 m.**

How can you use the properties of a rectangle to find the dimensions that are not marked?

Use the formula $A = \ell w$ to find the area of each face.

Front or Back $\quad$ 0.8 ⊠ 0.4 ⊟ 0.32

Top or Bottom $\quad$ 0.8 ⊠ 0.5 ⊟ 0.4

Sides $\quad$ 0.5 ⊠ 0.4 ⊟ 0.2

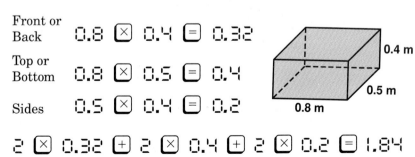

2 ⊠ 0.32 ⊞ 2 ⊠ 0.4 ⊞ 2 ⊠ 0.2 ⊟ 1.84

The surface area is 1.84 m^2.

In the prism at the right, △LMN is one base. What is the other?
Why are these the bases? Does a base always need to be on the "bottom?"

A prism is named by the shape of its bases. The parallel bases of the prism at the right are shaped like triangles, so it is called a **triangular prism.** Name something in everyday life that is the shape of a triangular prism.

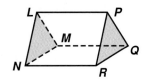

Example

2 **Find the surface area of the triangular prism shown below.**

There are five faces in a triangular prism.

Use the formula $A = \frac{1}{2}bh$.

Front Face $\frac{1}{2} \cdot 4 \cdot 3 = 6$

Back Face $\frac{1}{2} \cdot 4 \cdot 3 = 6$

Use the formula $A = \ell w$.

Side $8 \cdot 5 = 40$
Side $8 \cdot 4 = 32$
Side $8 \cdot 3 = 24$
Total   108

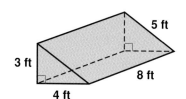

The surface area of the triangular prism is 108 ft².

Checking for Understanding

Communicating Algebra

1. A prism is named by its __?__.

2. A cube is a rectangular prism. Describe its faces.

3. Explain why surface area is important when you giftwrap a package.

Name each prism shown below. Justify your answer.

4.

5.

6.

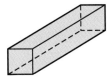

Guided Practice

Explain how to find the surface area of each prism. Then find the surface area.

7.

8.

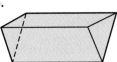

9.

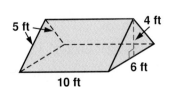

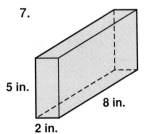

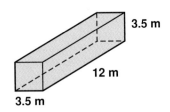

Exercises

Find the surface area of each prism. Round to the nearest whole number.

ESTIMATION
MENTAL MATH
CALCULATOR
PAPER/PENCIL

10.

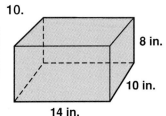

8 in.

10 in.

14 in.

11.

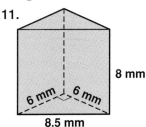

8 mm

6 mm   6 mm

8.5 mm

12.

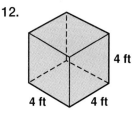

4 ft

4 ft   4 ft

Mixed Review

13. Solve $\frac{h}{-4} = 19$. (Lesson 3-4)

14. **Probability**   What is the probability of correctly guessing the answer on one true-false question on a quiz? (Lesson 11-4)

15. Using the formula $A = \pi r^2$, find the area of a circle that has a diameter of 14 cm. (Lesson 13-3)

Applications

16. **Woodworking**   Paula is building a storage chest as a project in woodworking class. To help her plan the project, Paula made the scale drawing shown at the right. When the chest is completed, she intends to use a stain for the outside finish. If a can of stain will cover 30 ft², how many cans of stain will Paula need?

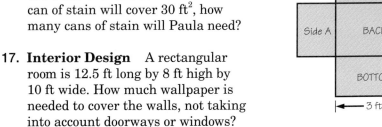

17. **Interior Design**   A rectangular room is 12.5 ft long by 8 ft high by 10 ft wide. How much wallpaper is needed to cover the walls, not taking into account doorways or windows?

18. **Make Up a Problem**   Write a situation in which you would need to know the surface area of a prism.

Challenge

19. Suppose a piece of paper $8\frac{1}{2}$ in. by 11 in. will be used to make a prism. Could the prism have a surface area of 95 in²? Explain your answer.

**Critical
Thinking**

20. Draw all the possible patterns that can be folded into a cube. (One is ⌐⌐⌐.)

Wrap-Up

*Journal
Entry*

21. Write an explanation of surface area that would help a younger student understand what surface area is.

13-5 Surface Area: Cylinders

Objective:
Find the surface area of circular cylinders.

Key Term:
circular cylinder

Many foods are packaged in recyclable cans. These cans are examples of cylinders. A cylindrical container is relatively easy to manufacture and, once filled, can be sealed so the contents are airtight. In addition to food, cylindrical containers are also used to store water, oil, and grain.

Most of the cylinders we see are called **circular cylinders.** The bases of a circular cylinder are two parallel, congruent circular regions. In this textbook, when we say *cylinder*, assume we are talking about a circular cylinder.

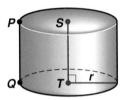

An altitude is any perpendicular line segment joining the two bases.
$\overline{ST}$ and $\overline{PQ}$ are altitudes.

If you open the top and bottom of a cylinder and then make a vertical cut in the curved surface, you could lay the cylinder flat in the same way you did with a rectangular prism. You would then have a pattern for making that cylinder.

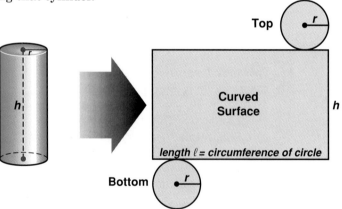

From the diagram above, you can see that:

Why is the length of the rectangle the same as the circumference of a base?

1. The bases of the cylinder are congruent circular regions.
2. The curved surface of the cylinder opens to form a rectangular region.
3. The width of the rectangle is the height of the cylinder.
4. The length of the rectangle is the circumference of a base.

You can use these observations to find the surface area of a circular cylinder.

Example

1 **Find the surface area of a cylinder if the radius of a base is 7 cm and the height is 30 cm.**

Top or Bottom Base

$A = \pi r^2$ Area of a circle

$A = \pi \cdot 7^2$

$\boxed{\pi}\,\boxed{\times}\ 7\ \boxed{x^2}\,\boxed{=}\ {153.93804}$

$A \approx 153.9$

The areas of the bases are both about 153.9 cm^2.

7 cm

30 cm

Curved Surface

$A = \ell w$ Area of a rectangle

$A = 2\pi r \cdot h$ Replace ℓ with the expression for the circumference of a circle.

$A = 2 \cdot \pi \cdot 7 \cdot 30$

$2\ \boxed{\times}\,\boxed{\pi}\,\boxed{\times}\ 7\ \boxed{\times}\ 30\ \boxed{=}\ {1319.4689}$

$A \approx 1319.5$

The area of the curved surface is about 1319.5 cm^2.

Add to find the total surface area.

$2\ \boxed{\times}\ 153.9\ \boxed{+}\ 1319.5\ \boxed{=}\ {1627.3}$

The surface area is about 1627.3 cm^2.

Checking for Understanding

Communicating Algebra

1. A circular cylinder has __?__ bases.

2. State the formula you should use to find the area of the bases of a cylinder.

Use the figure shown at the right.

3. Name the bases and two altitudes.

4. Explain how to find the curved surface area. Then explain how to find the total surface area.

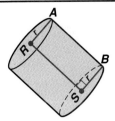

Guided Practice

Explain how to find the surface area of each cylinder. Then find the surface area.

5. 9 in.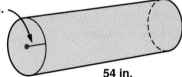

54 in.

6. 2 ft

$\frac{1}{2}$ ft

Exercises

Independent Practice

Find the surface area of each cylinder. Round each answer to the nearest whole number.

7.
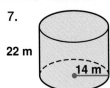

22 m

14 m

8.
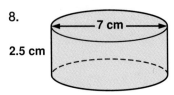

7 cm

2.5 cm

9.
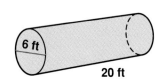

6 ft

20 ft

10.
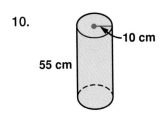

10 cm

55 cm

Mixed Review

11. Express as a unit rate: *$1.62 for 18 pounds.* (Lesson 9-1)

12. Find the surface area of the prism shown at the right. (Lesson 13-4)

10 cm

7 cm

5 cm

13. Identify the hypothesis and the conclusion in the statement: *If a polygon has six sides, then it is a hexagon.* (Lesson 12-7)

Applications

14. **Engineering** A water storage tank has a cylindrical shape. If the radius of the base is about 20 feet and the tank is 60 feet high, what is the area of the curved surface?

15. **Business** The ABC Company has been hired to paint the tank described in Exercise 14. How many gallons of paint would be needed to apply one coat of paint to the tank if one gallon of paint covers about 450 square feet?

16. How much would the paint cost if one coat is applied to the tank in Exercise 14 and one gallon of paint costs $10? How much would the paint cost if two coats are applied?

Challenge

17. Use algebra to show how steps 2, 3, and 4 on page 498 lead to the formula $2\pi rh$ for the curved surface area of a circular cylinder.

18. Show how to derive a formula for finding the total surface area of a circular cylinder.

Critical Thinking

19. Explain how you can find the surface area of a circular cylinder if you know its height and the circumference of a base.

Wrap-Up

20. In your own words, explain how to find the total surface area of a circular cylinder.

13-6 Surface Area: Pyramids and Cones

Objective:
Find the surface area of pyramids and cones.

Key Terms:
pyramid
vertex
square pyramid
lateral surface
slant height
circular cone

The pyramids of Egypt were built as burial tombs for the pharaohs and their relatives. The Great Pyramid at Gizeh, built about 2600 B.C. is one of the "Seven Wonders of the Ancient World." It measures 776 feet on each side of the square base and is about 450 feet high.

In geometry, **pyramids**, like prisms, are named by the shapes of their bases. A pyramid has only one base. All the other faces of a pyramid intersect at a point called the **vertex**. The pyramid shown below is a **square pyramid**.

A square pyramid has 5 faces.
The base is a square.
There are four triangular faces.

The altitude, $\overline{VP}$, is perpendicular to the base at its center.

If the base of a pyramid were shaped like a pentagon, how many faces would the pyramid have? Would the faces still be shaped like triangles?

The surface area of a pyramid can be found by adding the areas of the faces of the pyramid. In the diagram below, a square pyramid has been "unfolded" to help you identify the regions that make up the pyramid.

FYI

More than 2 million stone blocks, each weighing between 2 tons and 150 tons, were used in the construction of the Great Pyramid of Gizeh. To this day, historians have difficulty explaining how the blocks were transported from the quarries to the building site.

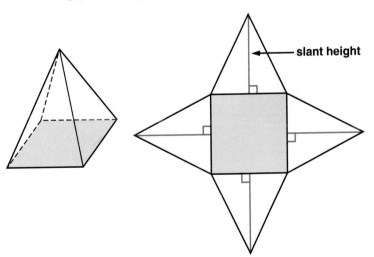

slant height

The unshaded regions of the square pyramid (the four faces) make up the **lateral surface** of the pyramid. As you can see, all of the faces are triangles. Because the base is a square region, the triangles are all congruent and their altitudes all have the same length. The length of any one of these altitudes is called the **slant height** of the pyramid.

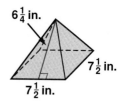
$6\frac{1}{4}$ in.

$7\frac{1}{2}$ in.

$7\frac{1}{2}$ in.

Example

1 **Find the surface area of the square pyramid shown at the right.**

Base	Each Triangular Side
$A = lw$	$A = \frac{1}{2}bh$
$A = 7\frac{1}{2} \cdot 7\frac{1}{2}$	$A = \frac{1}{2} \cdot 7\frac{1}{2} \cdot 6\frac{1}{4}$
$A = 56\frac{1}{4}$ or 56.25	$A = 23\frac{7}{16}$ or 23.4375

Add to find the total surface area.

area of base $+$ area of four triangular surfaces $=$ surface area

 56.25 ⊕ 4 ⊗ 23.4375 ⊜ 150

The surface area is 150 in². Compare with the estimate.

Estimation Hint

The area of the base is between 7×7 and 8×8 or about 56 in².

The area of each side is about one-half of 42 or 21 in². Since there are 4 sides, the area of the sides is about 84 in².

The total surface area is about $56 + 84$ or 140 in².

In this textbook, when we say *cone*, assume we are talking about a circular cone.

A cone is another three-dimensional shape that appears in everyday life as the cone for some ice cream cones, a support for cotton candy, or some paper drinking cups. Most of the cones that we see are called **circular cones** because the base is a circular region.

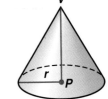

The base is a circular region with radius r. The altitude, $\overline{VP}$, is perpendicular to the base at its center.

In the following diagram, a cone has been opened and unrolled to help you identify the regions that make up the cone.

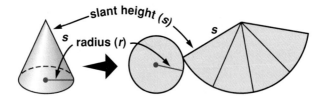

slant height (s)

s radius (r)

s

The curved edge is the length of the circumference of the base.

The base of the cone is a circular region, with area πr^2. The lateral surface can be cut into sections to form a parallelogram. The base of the parallelogram is half the circumference of the cone or $\frac{1}{2} \cdot 2\pi r$. Its height is the slant height of the cone, s. So the area of the lateral surface is $A = \pi rs$.

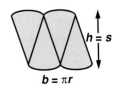

$h = s$

$b = \pi r$

Not all prisms are rectangular. Three other types are shown below.

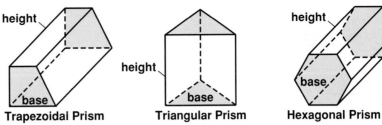

Trapezoidal Prism   **Triangular Prism**   **Hexagonal Prism**

In Example 1 you can see that the volume is equal to the area of the base times the height. In fact, the volume of any prism can be found by multiplying the area of the base times the height.

| **Volume of a Prism** | If a prism has a base area of B square units and a height of h units, then the volume (V) is $B \cdot h$ cubic units. $$V = Bh$$ |

Examples

How could you find the height of a prism if you know the area of the base and the volume?

2 **Find the volume of the rectangular prism.**

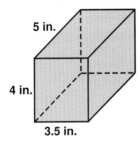

5 in.
4 in.
3.5 in.

$V = Bh$	Formula for volume of a prism
$V = \ell wh$	Since the base of the prism is a rectangle, $B = \ell w$.
$V = 5 \cdot 3.5 \cdot 4$	Replace ℓ with 5, w with 3.5 and h with 4.
$V = 70$	

The volume is 70 in³.

3 **Find the volume of the triangular prism.**

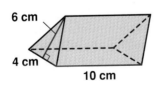

6 cm
4 cm
10 cm

$V = Bh$	Formula for volume of a prism
$V = \left(\frac{1}{2} \cdot 4 \cdot 6\right)10$	The base of the prism is a triangle. The area of the triangle is $\frac{1}{2} \cdot 4 \cdot 6$.
$V = 12 \cdot 10$	
$V = 120$	

The volume is 120 cm³.

Checking for Understanding ─────────────

Communicating Algebra

1. In the formula $V = Bh$, B represents the __?__ and h represents the __?__.

2. Volume is a __?__-dimensional measure.

3. Draw a model of a rectangular prism.

Name each prism and explain how to find the volume. Then find the volume.

4.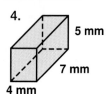
5 mm
7 mm
4 mm

5.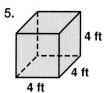
4 ft
4 ft
4 ft

6.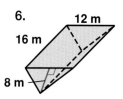
12 m
16 m
8 m

Exercises

Find the volume of each prism shown or described below.

7.
9 cm
8 cm
3 cm

8.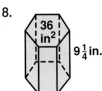
36 in²
$9\frac{1}{4}$ in.

9.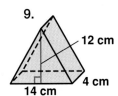
12 cm
4 cm
14 cm

10. rectangular prism: length, 4 m; width, 7 m; height 9 m

11. rectangular prism: length, 8 ft; width, 10 ft; height, 15 ft

12. rectangular prism: length, 16.2 cm; width, 9 cm; height, 1.24 cm

13. triangular prism: base of triangle, 5 cm; altitude, 5 cm; prism height, 5 cm

14. triangular prism: base of triangle, $2\frac{1}{2}$ ft; altitude, 1 ft; prism height, 2 ft

15. area of base, 17 cm²; height, 15 cm

16. area of base, 75 in²; height, 28 in.

17. **Statistics**   Find the mean, median, and mode for the data set: 3, 8, 11, 6, 9, 11, 12, 6, 7, 4, 11.   (Lesson 10-2)

18. Using the formulas $A = \pi r^2$ and $A = \pi rs$, find the surface area of a cone that has a radius of 3 m and a slant height of 5 m.   (Lesson 13-6)

19. **Manufacturing**   A leading manufacturer of sugar cubes packs a rectangular box so that there are six cubes along one edge, eleven cubes along a second edge, and three cubes along the third edge. How many sugar cubes are in the box?

20. **Pets**   A rectangular fish tank is 30 in. long by 16 in. wide by 12 in. high. If the tank is filled to a height of 10 in., what is the volume of water in the tank?

21. List all the possible combinations of whole number dimensions of a rectangular prism that result in a volume of 48 cm³.

22. **Make Up a Problem**   Write a problem that can be solved by finding volume.

13-8 Volume: Cylinders

Objective:
Find the volume of circular cylinders.

The stack of quarters shown at the right resembles a cylinder. To find the "value" of the quarters, multiply $0.25 by the number of quarters in the stack.

In a similar manner, you can find the volume of a cylinder. Think of a cylinder that has been "layered." The volume of the cylinder is the area of each layer times the height of the stack. Note that the base is a circular region and $A = \pi r^2$.

Volume of a Circular Cylinder	If a circular cylinder has a base with a radius of r units and a height of h units, then the volume (V) is $\pi r^2 h$ cubic units. $$V = \pi r^2 h$$

Examples

If you know the height of a cylinder and the circumference of one base, can you find the volume of the cylinder? Explain how.

1 **Find the amount of soil needed to fill a planter with a diameter of 3 feet and a height of 2 feet. Round to the nearest tenth.**

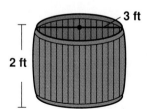

3 ft

2 ft

$V = \pi r^2 h$ Formula for volume
$V = \pi \cdot (1.5)^2 \cdot 2$ Substitute 1.5 for r and 2 for h.

$$\boxed{\pi}\boxed{\times} 1.5 \boxed{x^2}\boxed{\times} 2 \boxed{=} 14.137167$$

$V \approx 14.1$

Each planter will hold about 14.1 ft³ of soil.

2 **A smaller planter must hold 2.5 ft³ of soil and has a diameter of 2 ft. Use $V = \pi r^2 h$ to find the height the planter must be.**

$V = \pi r^2 h$ Write the formula.
$2.5 = \pi \cdot (1)^2 \cdot h$ $V = 2.5, r = 1$
$2.5 = \pi h$

$$2.5 \boxed{\div}\boxed{\pi}\boxed{=} 0.7957747$$

$0.8 \approx h$

The planter must be about 0.8 feet or about 9.5 inches tall.

Checking for Understanding

Communicating Algebra

1. What dimensions are involved in finding the volume of a circular cylinder?
2. How are the volume of a prism and the volume of cylinder similar?

Explain how to find the volume of each cylinder. Then find the volume.

3.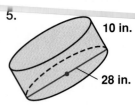
5 yd
3 yd

4.
10.1 mm
d = 1.4 mm

5.
10 in.
28 in.

Exercises

Find the volume of each cylinder. Round decimal answers to the nearest tenth.

ESTIMATION
MENTAL MATH
CALCULATOR
PAPER/PENCIL

6.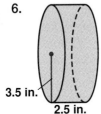
3.5 in.
2.5 in.

7.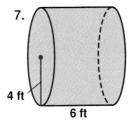
4 ft
6 ft

8.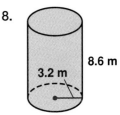
8.6 m
3.2 m

9. radius, 6 m; height, 9 m

10. radius, 7 yd; height, 11 yd

11. diameter, 1.4 m; height, 5 m

12. radius, $2\frac{1}{2}$ in.; height, 35 in.

Find the value of h.

13. $V = Bh;\ V = 156,\ B = 26$

14. $V = \pi r^2 h;\ V = 770,\ \pi = \frac{22}{7},\ r = 7$

15. Solve $\frac{m}{-8} + 7 < \text{-}4$.   (Lesson 7-6)

16. In the parallelogram shown at the right, which angle is congruent to $\angle C$?   (Lesson 12-11)

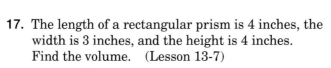

17. The length of a rectangular prism is 4 inches, the width is 3 inches, and the height is 4 inches. Find the volume.   (Lesson 13-7)

18. **Science**   Jack knows that 1 mL of water has a volume of 1 cm³. He needs 1125 mL of water for an experiment. He has an unmarked cylindrical container with a diameter of 12 cm. To what height should he fill the container to measure the water for the experiment?

19. **Agriculture**   A water trough is half of a circular cylinder with a radius of 6 ft and a height of 3 ft. If one gallon of water has a volume of about 230 in³, about how many gallons of water will the trough hold?

20. The formulas for finding the volume of a prism and a cylinder are similar. How are they different? Explain your answer.

21. **Research**   Make a chart of the height, diameter, and volume of the cans of food in your kitchen. Is one size more common than others?

13-9 Volume: Pyramids and Cones

Objective:
Find the volume of pyramids and cones.

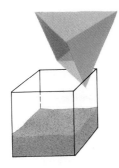

You probably have conducted experiments in science classes to find out how something works or to prove that a hypothesis is true. Gene and his friends conducted this experiment to find out if there was a relationship between the volume of a pyramid and a prism that it "fit" inside.

Here are the conditions.

1. The base of the pyramid and the base of the prism are congruent.

2. The altitude of the pyramid is congruent to the altitude of the prism.

You can see that the volume of the pyramid is less than the volume of the prism. The students used the models and colored sand to find out how the volumes compared.

As you can see from the drawings, three pyramids full of sand filled the prism. This suggests that the volume of the pyramid is one-third the volume of the prism.

Since the formula for the volume of the prism is $V = Bh$, the volume of the pyramid would be equal to $\frac{1}{3}Bh$.

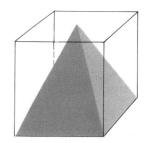

Volume of a Pyramid

If a pyramid has a base of B square units and a height of h units, then the volume (V) is $\frac{1}{3} \cdot B \cdot h$ cubic units.
$$V = \frac{1}{3}Bh$$

Calculator Hint

Recall that 3 and $\frac{1}{3}$ are multiplicative inverses. An easy way to enter $\frac{1}{3}$ into your calculator is to enter 3 and press the $\boxed{1/x}$ key.

Example

1 Find the volume of the pyramid.

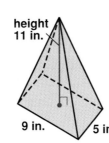

height 11 in.

9 in.   5 in.

$V = \frac{1}{3} Bh$ Formula for volume of a pyramid

$V = \frac{1}{3} \ell wh$ Replace B with ℓw.

$V = \frac{1}{3} \cdot 9 \cdot 5 \cdot 11$ Replace ℓ with 9, w with 5, and h with 11.

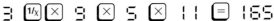 165

The volume is 165 in^3.

It makes sense to predict that the volume of a cone and a cylinder might be related in the same way as the volume of a pyramid and a prism. Here is an experiment to check that prediction.

The conditions are similar.

1. The base of the cone and the base of the cylinder are congruent.

2. The altitude of the cone is congruent to the altitude of the cylinder.

By using the models and colored sand, it is easy to perform an experiment like the one done earlier for prisms and pyramids.

You can see that the cylinder was filled by three cones of sand. So, the volume of the cone appears to be one-third of the volume of the cylinder.

Volume of a Cone	If a cone has a radius of r units and a height of h units, then the volume (V) is $\frac{1}{3} \cdot \pi \cdot r^2 \cdot h$. $$V = \frac{1}{3}\pi r^2 h$$

Example

2 Find the volume of the cone to the nearest tenth.

Estimation Hint

THINK: $\frac{1}{3} \times 3.14$ is a little greater than 1.

$3^2 \cdot 8 = 9 \cdot 8$ or 72

The volume should be a little greater than 72 m³.

8 m

3 m

$V = \frac{1}{3}\pi r^2 h$ Formula for volume of a cone

$V = \frac{1}{3} \cdot \pi \cdot 3^2 \cdot 8$ Replace r with 3 and h with 8.

3 [1/x] [×] [π] [×] 3 [x²] [×] 8
[=] 75.398224

$V \approx 75.4$

The volume is about 75.4 m³.

Checking for Understanding

Communicating Algebra

1. State the formula for finding the volume of a pyramid. Then tell what each variable represents.

2. State the formula for finding the volume of a cone. Then tell what each variable represents.

Guided Practice

Explain how to find the volume of each pyramid or cone shown. Then find the volume.

3. 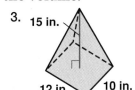 15 in. 12 in. 10 in.

4. 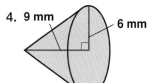 9 mm 6 mm

5. 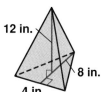 12 in. 8 in. 4 in.

Exercises

Independent Practice

Find the volume of each pyramid or cone. Round decimal answers to the nearest tenth.

? ESTIMATION
MENTAL MATH
CALCULATOR
PAPER/PENCIL

6. 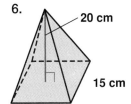 20 cm 15 cm 18 cm

7. 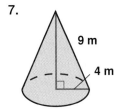 9 m 4 m

8. 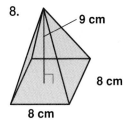 9 cm 8 cm 8 cm

9. 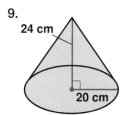 24 cm 20 cm

10. 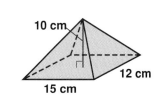 10 cm 15 cm 12 cm

11. 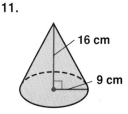 16 cm 9 cm

12. rectangular pyramid; length, 9 in.; width, 7 in.; height, 18 in.

13. square pyramid; length, 5 cm; height, 6 cm

14. circular cone; radius, 3 ft; height, 14 ft

15. circular cone; radius, 10 m; height, 18 m

Mixed Review

16. **Probability**   What is $P(A$ or $B)$ if $P(A) = \frac{1}{4}$ and $P(B) = \frac{2}{3}$ and A and B are mutually exclusive?   (Lesson 11-8)

17. Find the sum of the measures of the angles of a regular pentagon. (Lesson 12-12)

18. Using the formula $V = \pi r^2 h$, find the volume of a circular cylinder that has a radius of 8 inches and a height of 13 inches.   (Lesson 13-8)

Applications

19. **Food Service**   The diameter of the rim of an ice cream cone is 2 inches and the height of the cone is 6 inches. If the cone is filled even with the rim with soft ice cream, what is the volume of the ice cream?

20. If the cone in Exercise 19 is filled to a height of 3 inches above the top of the cone, in the shape of a cone, about how much ice cream is there above the cone? How much ice cream is there all together?

Challenge

Find the volume, to the nearest tenth, of each figure shown below.

21.
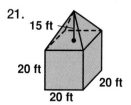

15 ft
20 ft
20 ft
20 ft

22.
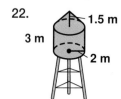

1.5 m
3 m
2 m

23.
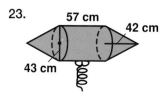

57 cm
42 cm
43 cm

Critical Thinking

24. Suppose you double the height of a cone. How does the volume change? Suppose you double the radius of the base. How does the volume change?

Wrap-Up

25. In your own words, explain the relationship between the volumes of pyramids and prisms and between the volumes of cones and cylinders.

Career

Computer-Aided Design Technician

Computer-Aided Design (CAD) technicians use computers plus graphics printers and plotters to produce blueprints, schematics, and drawings for a variety of fields. Many technicians work for architectural and engineering firms to produce construction blueprints of buildings. Others work for automotive designers, technical illustrators, and other companies that require product design and detailed technical drafting.

A CAD technician should have the ability to concentrate on details and be accurate. The outlook for opportunities in CAD design is excellent through the 1990s.

13-10 Precision and Significant Digits

Objective:
Use precision and significant digits to describe a measurement.

Key Terms:
precision
significant digits

Have you ever heard someone say that all measurements are approximate? What they mean is that no measurement can be any more exact than the scale the object is being measured with. The micrometer shown at the right is used by machinists to make measurements that need to be very precise. The micrometer can measure to the nearest 0.001 inch.

When you measure the length of a paper clip, you might use a metric ruler. Measuring to the nearest millimeter (0.1 of a centimeter) is all the precision you can expect.

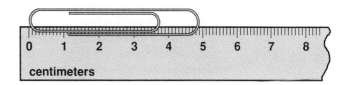

When you say that the length of the paper clip is 4.9 cm, you are really stating two things.

1. The unit being used (the precision unit) is 0.1 cm.
2. The number you selected for the measure is closer to 4.9 than it is to 4.8 or 5.0.

The **precision** of a measurement depends on the unit of measure. To improve the precision of a measurement, you must choose a measuring instrument that provides smaller units of measure. For example, a measurement obtained using a bathroom scale that reads to the nearest 0.5 kg would not be as precise as a measurement obtained using a balance scale that reads to the nearest 0.005 kg.

The digits you record when you measure are **significant.** These digits indicate the precision of the measurement.

Example

1 **The length of a piece of copper tubing is given as 43.58 cm. What can you tell from this information?**

   a. The unit of measure is 0.01 cm.
   b. The measurement is not exact. However, you can be reasonably certain that the actual length is between 43.57 cm and 43.59 cm.
   c. There are four significant digits.

Examples

Analyze each measurement.

2 10.50 inches

   a. The actual length is closer to 10.50 in. than to 10.49 in. or 10.51 in.
   b. The unit of measure is 0.01 in.
   c. There are four significant digits. The zero in the hundredths place is significant.

3 0.003 kilograms

   a. The actual mass is closer to 0.003 kg than to 0.002 kg or 0.004 kg.
   b. The unit of measure is 0.001 kg.
   c. There is one significant digit. The zeros in 0.003 are used to show only the place value of the decimal and are not counted as significant digits.

The result of a computation involving a measurement can be no more precise than the least precise measurement involved. Scientists usually round sums and differences of measures to the *same precision* as the least precise measurement. Products and quotients are rounded to the *same number of significant digits* as the least precise measurement.

Examples

In Example 4, which measurement is least precise? Why?

4 **The sides of a triangle measure 12.26 cm, 5.8 cm, and 2.125 cm. What is its perimeter?**

12.26 cm + 5.8 cm + 2.125 cm = 20.185
2 decimal places   1 decimal place   3 decimal places

Round to 20.2 cm since the least precise measurement has 1 decimal place.

The perimeter of the triangle is 20.2 cm.

In Example 5, which measurement is least precise? Why?

5 **The length of a rectangle is 12.63 cm and the width is 5.1 cm. What is the area of the rectangle?**

$12.63 \boxed{\times} 5.1 \boxed{=} 64.413$
4 significant digits   2 significant digits

Round to 64 cm since 64 has 2 significant digits.

The area of the rectangle is 64 cm^2.

Checking for Understanding

Communicating Algebra

1. "All measurements are approximate" means __?__.
2. When you say that the length of a pencil is 15.6 cm, you are stating that the actual length is between __?__ and __?__.
3. The precision of a measurement depends on __?__.

Tell how many significant digits each measurement has. Then tell the unit of measure.

4. 7.5 m

5. 21.25 ft

6. 107.50 mm

7. 0.007 kg

8. 4.0 mi

9. 0.7403 m

Exercises

For each measurement, give the measurements between which the actual length lies.

10. 146 cm

11. 63 mm

12. 3.6 cm

13. 18.2 m

14. 2.59 in.

15. 3.0 m

16. 4.90 m

17. 10 ft

Calculate. Round to the correct number of significant digits.

18. 14.38 cm + 5.7 cm + 3.9082 cm

19. 15.273 L − 8.2 L

20. 127.2 g + 42.3 g − 5.7 g

21. 29.307 m + 4.23 m + 50.93 m

22. 5.372 cm × 4.8

23. 29.78 km ÷ 3.7

24. 4.397 cm × 2.4

25. 50.3 kg ÷ 0.2937

26. Graph the equation $x + 2y = 4$.   (Lesson 8-6)

27. Using the formula $A = bh$, find the area of a parallelogram that has a base of 7 meters and a height of 5 meters.   (Lesson 13-1)

28. Name the pairs of alternate interior angles in the diagram at the right. (Lesson 12-4)

29. Using the formula $V = \frac{1}{3}Bh$, find the volume of a pyramid with a rectangular base, 7 in. by 3 in., and a height of 8 in. (Lesson 13-9)

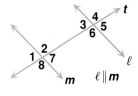

$\ell \parallel m$

30. Measurement   Sam and Della both correctly measure the same piece of metal using different measuring instruments. Sam says the length is 12.3 cm. Della says the length is 123.4 mm. Which measurement is more precise? Why?

31. Measurement   Arrange the following measurements in order of precision, from most to least precise.   0.40 mm, 40 mm, 0.4 mm, 0.0004 mm

32. Masonry   How high is a stack of 25 bricks, if each brick is 3.75 inches thick? Round to reflect the proper precision.

33. If 56.7 represents the sum of two measurements that have been rounded to the nearest tenth, what is the least possible and the greatest possible number that was rounded?

34. Write a few sentences about what precision is and why it is important in measurement.

13-11 Strategy: Make a Model or Drawing

Objective:
Solve problems by making a model or drawing.

Lee's Masonry makes concrete steps for new homes. The steps for one home are to be 5 feet long, with three steps. Each step will be 0.5 foot high and 1 foot wide. How much concrete will be needed?

Without a drawing, it is difficult to determine how to find the volume. A model or drawing can help you solve problems like this one, as well as many other problems, especially problems involving geometry.

> **Explore**

You know some of the dimensions of the steps. You need to find the volume of the steps.

> **Plan**

After making a labeled drawing of the steps using the information you have, the problem becomes clearer. Notice that the steps can be divided into three rectangular prisms. You know how to find the volume of each prism. You can find the total volume by adding.

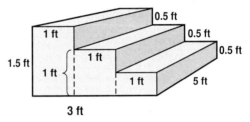

> **Solve**

Volume of lowest step $= \ell wh$
$\qquad\qquad\qquad\quad = 5 \cdot 1 \cdot 0.5$
$\qquad\qquad\qquad\quad = 2.5$
Volume of middle step $= \ell wh$
$\qquad\qquad\qquad\quad = 5 \cdot 1 \cdot 1$
$\qquad\qquad\qquad\quad = 5$
Volume of highest step $= \ell wh$
$\qquad\qquad\qquad\quad = 5 \cdot 1 \cdot 1.5$
$\qquad\qquad\qquad\quad = 7.5$
Total volume $= 2.5 + 5 + 7.5 = 15$

It will take 15 cubic feet of concrete to make the steps.

How many cubic feet would a 4th step add to the total volume?

> **Examine**

Notice that the steps can be divided into three different rectangular prisms.

Volume of bottom layer $= 5 \cdot 3 \cdot 0.5 = 7.5$
Volume of middle layer $= 5 \cdot 2 \cdot 0.5 = 5$
Volume of top layer $= 5 \cdot 1 \cdot 0.5 = 2.5$
Total volume $= 7.5 + 5 + 2.5 = 15$

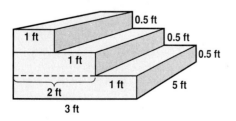

Since the volume using either method is 15 ft^3, the answer checks.

Checking for Understanding

Solve Exercises 1 and 2 by making a model or drawing.

1. There are 2 tiles. One is in the shape of a square and the other is in the shape of an equilateral triangle. The sides of the square are congruent to the sides of the triangle. The tiles are put together to form a pentagon. If one side of the triangle is 20 cm long, what is the perimeter of the pentagon?

2. A painting 15 cm by 25 cm is bordered by a matting that is 3 cm wide. The frame around the matting is 2 cm wide. What is the area of the picture including the frame and matting?

**Communicating
Algebra**

3. Describe the model or drawing you used for Exercise 1. How did it help you solve the problem?

4. Describe the model or drawing you used for Exercise 2. How did it help you solve the problem?

Exercises

**Independent
Practice**

Solve. Use any strategy.

5. A rectangular swimming pool is 4 meters wide and 10 meters long. A concrete walkway is poured around the pool. The walkway is 1 m wide and 0.1 m deep. What is the volume of the concrete?

6. A circular track has a diameter of 400 feet. Candy starts out jogging around the track at a rate of 600 feet per minute. One minute later, Andrea starts jogging at the same place. She goes in the same direction and jogs at 700 feet per minute. Will Andrea pass Candy on the first, second, third, or fourth quarter of the track?

7. A nonagon is a polygon with 9 sides. How many different diagonals can be drawn in a nonagon?

8. Half of a garden is planted in corn. Half the remaining garden is planted in strawberries. A third of the part not planted in corn or strawberries is planted in tomatoes. If 6 m^2 is planted in tomatoes, what is the area planted in corn?

9. The owner of Newland Dairy Farm plans to put a fence around the pasture behind the new dairy barn. The pasture is rectangular and measures 200 ft by 800 ft. The fence is only needed on three sides of the pasture because the barn will serve as one of the shorter sides. If a fence post is needed at every corner and at every 10 ft along the sides, how many posts are needed?

10. Alex has 36 identical cubes. Each edge of a cube is 1 in. long. How can Alex arrange the cubes to have the smallest surface area?

11. A circular fountain has a diameter of 8 feet. A flower garden is planted around the fountain. If the garden extends 12 feet beyond the fountain, what is the area of the garden?

Wrap-Up

12. **Make Up a Problem** Write a problem from your everyday life that can be solved using a model or drawing. Solve the problem.

Review

Language and Concepts

Choose the correct term to complete each sentence.

| altitude |
| cubic centimeter |
| precision |
| pyramid |
| square centimeter |
| surface area |
| volume |

1. The __?__ is a common unit of measure for area.

2. The __?__ is a common unit of measure for volume.

3. The __?__ of a solid is the sum of the areas of its faces.

4. __?__ is the amount of space that a solid contains.

5. The __?__ of a parallelogram is a segment perpendicular to the bases with endpoints on the bases.

6. A __?__ is named by the shape of its base.

7. The smaller the unit of measure, the greater the degree of __?__ of a measurement.

Skills

Find the area of each figure. (Lesson 13-1)

8.
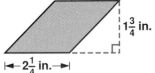
$1\frac{3}{4}$ in.
$\leftarrow 2\frac{1}{4}$ in. $\rightarrow$

9.
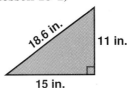
18.6 in.　11 in.
15 in.

10.
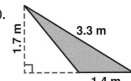
1.7 m　3.3 m
1.4 m

Find the area of each trapezoid. (Lesson 13-2)

11.

19.1 m　24 m
10 m
13 m

12.
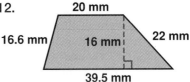
20 mm
16.6 mm　16 mm　22 mm
39.5 mm

Find the area of each circle described below. Round decimal answers to the nearest tenth. (Lesson 13-3)

13. radius, 9 mi

14. diameter, 4.2 m

Find the surface area of each prism. (Lesson 13-4)

15.
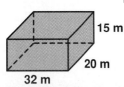
15 m
20 m
32 m

16.
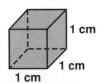
1 cm
1 cm
1 cm

17.
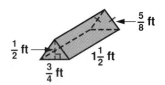
$\frac{5}{8}$ ft
$\frac{1}{2}$ ft
$1\frac{1}{2}$ ft
$\frac{3}{4}$ ft

Find the surface area of each cylinder described below. Round each answer to the nearest whole number. (Lesson 13-5)

18. height, 2.3 m; radius, 1 m

19. height, 41 cm; diameter, 40 cm

Find the surface area of each figure described below. Round decimal answers to the nearest tenth. (Lesson 13-6)

20. a pyramid with base 10 in. by 10 in. and slant height $8\frac{1}{4}$ in.

21. a cone with radius 1 m and slant height 2 m

Find the volume of each figure described below. Round decimal answers to the nearest tenth. (Lessons 13-7, 13-8, 13-9)

22. triangular prism: base of triangle, 5 cm; height, 3.4 cm; prism height, 12 cm

23. hexagonal prism: area of base, 166.25 in²; height, 20 in.

24. rectangular prism: length, 4 m; width, 4 m; height, 16 m

25. cylinder: radius, 7 ft; height, 18 ft

26. cylinder: diameter, 10 m; height, 22 m

27. rectangular pyramid: length, 8 m; width, 9 m; height, 21 m

28. cone: radius, 6 in.; height, 11 in.

Applications and Problem Solving

Determine the number of significant digits in each measurement. (Lesson 13-10)

29. 600 ft

30. 84.0 cm

31. 53.005 m

32. 64 yd

Solve by making a drawing. (Lesson 13-11)

33. The radius of a circular flower garden and its border is 2.8 meters. Without the border, the radius is 2.3 meters. Find the area of the border.

ortfolio Suggestion

Select an item from this chapter that you feel shows your best work and place it in your portfolio.

Curriculum Connection

• **History**   Write a brief report on how the Egyptian pyramids were constructed and the tools that were used.

Read More About It

Holton, Jean Laity. *Geometry: A New Way of Looking at Space.*
McCauley, David. *Pyramid.*
Razzell, Arthur G. and K.G.O. Watts. *Circles and Curves.*

Test

Find the area of each figure.

1.

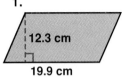

12.3 cm

19.9 cm

2.

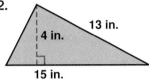

13 in.

4 in.

15 in.

3.

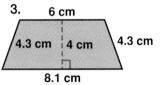

6 cm

4.3 cm   4 cm   4.3 cm

8.1 cm

Find the area of each circle described below. Round decimal answers to the nearest tenth.

4. radius, 5 km

5. radius, 1.5 cm

6. diameter, 42 in.

Find the surface area of each figure described below. Round decimal answers to the nearest tenth.

7. a rectangular prism with base 25 cm by 8 cm and height 3.4 cm

8. a cylinder with radius 2 in. and height $7\frac{1}{2}$ in.

Find the volume of each figure described below. Round decimal answers to the nearest tenth.

9. a cone with radius 8 cm and height 5 cm

10. a rectangular prism with length 7.1 m, width 4.5 m, and height 9.3 m

11. a rectangular pyramid with length 12 cm, width 15 cm, and height 20 cm

12.

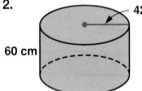

42 cm

60 cm

13.

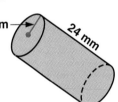

8 mm

24 mm

14.

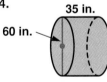

35 in.

60 in.

Determine the number of significant digits in each measurement.

15. 0.008 ft

16. 87.496 m

17. 24,000 mi

18. 8.090 cm

Solve.

19. A cylindrical can of soup has a 3-in. diameter and a 5-in. height. What is the approximate area of the label that covers the side of the can?

20. A circular part of a square field 600 m by 600 m is irrigated. The irrigated part has a radius of 300 m. If the circle is completely contained within the square, find the amount of the field that is not irrigated.

BONUS

Explain why the area of a parallelogram is not doubled when the base and height are doubled.

Academic Skills Test

Cumulative, Chapters 1-13

1. Which is equivalent to $\frac{1}{6} \times \frac{1}{6}$?

   A 2^{-6} **C** 6^2

   B 6^{-2} **D** 2^6

2. Using the formula $F = \frac{9}{5}C + 32$, which temperatures are approximately equivalent?

   A 30°C, 86°F

   B 30°C, -1°F

   C 10°C, 42°F

   D 10°C, 34°F

3. Which number line shows the solution of the equation $y = |-2|$?

   A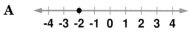
   -4 -3 -2 -1 0 1 2 3 4

   B
   -4 -3 -2 -1 0 1 2 3 4

   C
   -4 -3 -2 -1 0 1 2 3 4

   D
   -4 -3 -2 -1 0 1 2 3 4

4. On a trip, the Garcias drive 104 miles in 2 hours. If they continue at the same rate, which proportion will give t, the total time in hours, for a 550-mile trip?

   A $\frac{104}{2} = \frac{t}{550}$ **C** $\frac{2}{t} = \frac{550}{104}$

   B $\frac{104}{2} = \frac{550}{t}$ **D** $\frac{t}{2} = \frac{550}{446}$

5. Ellen's science scores were 73, 84, 78, 76, and 84. What is the mean (average) of these scores?

   A 84   **C** 78

   B 79   **D** 77

6. When a six-sided number cube is rolled, what is the probability of rolling a 5 or an even number?

   A $\frac{1}{6}$ **C** $\frac{2}{3}$

   B $\frac{1}{2}$ **D** $\frac{5}{6}$

7. Which shows a step in constructing an angle congruent to a given angle?

   A   **C**

   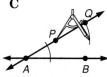

   B   **D**

   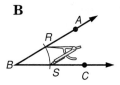
   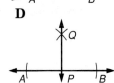

8. Which figure must contain exactly two acute angles?

   A a parallelogram

   B a rectangle

   C an acute triangle

   D an obtuse triangle

9. What is the area of the trapezoid?

   A 45 ft²

   B 64 ft²

   C 96 ft²

   D 112 ft²

   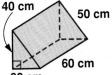

10. What is the volume of the triangular prism?

    A 3600 cm³

    B 4800 cm³

    C 30,000 cm³

    D 36,000 cm³

CHAPTER OBJECTIVES

In this chapter you will learn to:

- [ ] find square roots and squares
- [ ] solve equations by finding square roots
- [ ] use the Pythagorean Theorem
- [ ] use trigonometric ratios to solve problems involving right triangles
- [ ] use Venn diagrams to solve problems

Applying Algebra to Right Triangles

Hot-air balloons are fascinating to watch as they drift slowly by. A free-floating balloon flies in whatever direction the wind is blowing. The pilot can control the vertical movement of the balloon, but cannot steer it.

You probably don't think of a right triangle when you watch a balloon. However, you can use a right triangle to show the vertical motion controlled by the pilot and the horizontal motion of the wind. The third side of the triangle is the actual path of the balloon.

So the next time you watch a hot-air balloon, imagine the right triangle in the sky.

History Connection

Class Project
Do research to find how armies used balloons in the Civil War, World War I, and World War II.
Write a paragraph that tells how they were used.

14-1   Squares and Square Roots

Objective:
Find square roots and squares.

Key Terms:
square root
radical sign
radicand

The ability of a hang glider to glide through the air depends on its wingspan and the area of its wing. The relationship of the wingspan and the wing area is called the aspect ratio. This ratio can be expressed by the formula $R = \dfrac{s^2}{A}$, where R represents the ratio, s the wingspan, and A the wing area.

Suppose a glider has an aspect ratio of 2.7 and a wing area of 30 square feet. What is its wingspan?

$$R = \frac{s^2}{A}$$

$$2.7 = \frac{s^2}{30} \qquad \text{Replace } R \text{ with 2.7 and } A \text{ with 30.}$$

$$30(2.7) = s^2$$

$$81 = s^2$$

$$81 = s \cdot s$$

What number multiplied by itself gives 81? Since $81 = 9 \cdot 9$, $s = 9$. The wingspan is 9 feet.

Since $9 \cdot 9 = 81$, one **square root** of 81 is 9. It is also true that $-9 \cdot (-9) = 81$. This suggests that another square root of 81 is -9.

Definition of Square Root	If $x^2 = y$, then x is a square root of y.

In algebra the symbol $\sqrt{}$, called a **radical sign**, is used to indicate a nonnegative square root. Any expression inside a radical sign is called the **radicand.**

$$\text{radical sign} \rightarrow \sqrt{81} \leftarrow \text{radicand}$$

$\sqrt{81} = 9$ $\sqrt{81}$ indicates the nonnegative square root of 81.

$-\sqrt{81} = -9$ $-\sqrt{81}$ indicates the negative square root of 81.

Squaring a number and finding the square root of a number are closely related to the square.

FYI

Hang gliding has been popular since the early 1970s. The record altitude on a hang glider is 16,158 feet by Bob Calvert of Great Britain.

The square shown at the right has sides that are 5 feet long. The area of the square is found by *squaring 5.* If you know that the area of the square is 25 square feet, then the length of a side is found by *finding the square root of 25.*

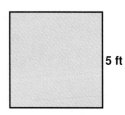

5 ft

Examples

1 Find $\sqrt{9}$.

The symbol $\sqrt{9}$ represents the nonnegative square root of 9.
Since $3 \cdot 3 = 9$, $\sqrt{9} = 3$.

2 Find $-\sqrt{4}$.

The symbol $-\sqrt{4}$ represents the negative square root of 4.
Since $2 \cdot 2 = 4$, $-\sqrt{4} = -2$.

The inverse operation of the square root operation is the squaring operation. Thus, the square of either 9 or -9 is 81. The square of a number is the product of the number and itself.

$$9^2 = 9 \cdot 9 = 81$$
$$(-9)^2 = -9 \cdot (-9) = 81$$

Numbers such as 81, that are squares of whole numbers, are called *perfect squares.*

Checking for Understanding

Communicating Algebra

1. In your own words, define *square root.*

2. Explain how squaring a number is like finding the area of a square.

3. If $n^2 = 25$, then n is a ___?___ of 25.

Guided Practice

State the square of each number.

4. 5   **5.** 8   **6.** 12   **7.** 6

Exercises

Independent Practice

Find each square root.

8. $\sqrt{4}$ **9.** $\sqrt{9}$ **10.** $\sqrt{49}$ **11.** $\sqrt{16}$

12. $\sqrt{64}$ **13.** $\sqrt{25}$ **14.** $\sqrt{100}$ **15.** $-\sqrt{9}$

Find each square root.

16. $-\sqrt{25}$　　　　**17.** $-\sqrt{64}$　　　　**18.** $-\sqrt{16}$　　　　**19.** $-\sqrt{100}$

20. $\sqrt{121}$　　　　**21.** $-\sqrt{144}$　　　　**22.** $\sqrt{169}$　　　　**23.** $-\sqrt{400}$

Find the square of each number.

24. 9　　　　　　**25.** 30　　　　　　**26.** 25　　　　　　**27.** 17

28. 23　　　　　　**29.** 14　　　　　　**30.** 40　　　　　　**31.** 34

Challenge

Find each square root.

32. $\sqrt{0.04}$　　　　**33.** $\sqrt{0.16}$　　　　**34.** $\sqrt{\frac{4}{9}}$　　　　**35.** $\sqrt{\frac{49}{81}}$

Mixed Review

36. Find the slope of the line that contains the points A(6, -2) and B(3, 7). (Lesson 8-8)

37. Statistics Find the lower and upper quartiles of the data 13, 16, 17, 14, 11, 9, 17, 21. (Lesson 10-4)

38. Personal Finance Angie allows 40% of her income for clothing, 30% for entertainment, 20% for food, and 10% for savings. Make a circle graph of these data. (Lesson 12-2)

39. Find the perimeter of the rectangle. (Lesson 3-6)

5 m

8 m

40. Find the length of the brace (x). (Lesson 12-10)

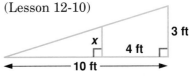

3 ft

x　4 ft

10 ft

Connection

41. Geometry Draw a square that has an area of 64 square centimeters. What is the length of a side of the square?

Applications

42. Hang Gliding Steve's hang glider has a wingspan of 15 feet and a wing area of 150 square feet. Find the aspect ratio of his hang glider. Use the formula $R = \frac{s^2}{A}$.

43. Hang Gliding Adventure Sports advertises a hang glider with an aspect ratio of 1.6. If its wing area is 160 square feet, what is its wingspan?

Critical Thinking

44. Describe a realistic situation in which $\sqrt{9}$ would be the only solution of the equation $x^2 = 9$.

Wrap-Up

Journal Entry

45. In your own words, explain how the squaring operation is related to the square root operation.

14-2 Approximate Square Roots

Objective:
Approximate square roots.

In airshows, skydivers often compete to see who can land in the smallest square region. If a target region has an area of 250 square feet, how long is the region on each side?

Let $x =$ the length of one side of the region.

Then $x^2 = 250$. Why?

$$x = \sqrt{250}$$ Why?

The number 250 is not a perfect square; that is, 250 has no square root that is a whole number. However, we know that 250 is greater than 225 or 15^2 and less than 256 or 16^2. So the square root of 250 should be greater than 15 and less than 16.

$$225 < 250 < 256$$
$$15^2 < 250 < 16^2$$
$$15 < \sqrt{250} < 16$$

Since 250 is closer to 256 than to 225, the best whole number estimate for 250 would be 16. The target region is about 16 feet long on each side.

Calculator Hint

A calculator can be used to find square roots. $\boxed{\sqrt{x}}$ is the square root key. When this key is pressed, the calculator replaces the number in the display with its nonnegative square root.

Examples

1 **What is the best whole number estimate for $\sqrt{118}$?**

100 and 121 are the closest perfect squares.

$$100 < 118 < 121$$
$$10^2 < 118 < 11^2$$
$$10 < \sqrt{118} < 11$$

Check: ⌷ ⌷8 $\boxed{\sqrt{x}}$ ⌷0.86278

Since 118 is closer to 121 than 100, the best whole number estimate is 11. This checks with the result on the calculator.

2 **What is the best integer estimate for $-\sqrt{26.79}$?**

$$-36 < -26.79 < -25$$
$$-6^2 < -26.79 < -5^2$$
$$-6 < -\sqrt{26.79} < -5$$

Check:
26.79 $\boxed{\sqrt{x}}$ $\boxed{+/-}$ -5.175906

Since -26.79 is closer to -25 than -36, the best integer estimate is -5. This checks with the result on the calculator.

FYI

The symbol $\boxed{\sqrt{x}}$ was first used in print by Rudolff in 1525.

Checking for Understanding

Communicating Algebra

1. A square root is a whole number if the radicand is a __?__ .

2. After a nonnegative square root is found on a calculator, the __?__ key will give the negative square root.

Guided Practice

Find the best whole number estimate for each of the following. Then check your estimate using a calculator.

3. $\sqrt{24}$ 4. $\sqrt{32}$ 5. $\sqrt{38}$ 6. $\sqrt{42}$

7. $\sqrt{45}$ 8. $\sqrt{52}$ 9. $\sqrt{57}$ 10. $\sqrt{65}$

Exercises

Independent Practice

Find the best integer estimate for each of the following. Then check your estimate using a calculator.

11. $\sqrt{90}$ 12. $\sqrt{15}$ 13. $-\sqrt{47}$ 14. $-\sqrt{89}$

15. $-\sqrt{29}$ 16. $-\sqrt{625}$ 17. $\sqrt{2601}$ 18. $\sqrt{97}$

19. $\sqrt{6.76}$ 20. $\sqrt{7.84}$ 21. $\sqrt{2.89}$ 22. $\sqrt{4.41}$

23. $\sqrt{13.69}$ 24. $\sqrt{20.25}$ 25. $-\sqrt{62}$ 26. $-\sqrt{71}$

Mixed Review

27. Find C(7, 3).   (Lesson 11-3)

28. Find the area of a circle with diameter 12 feet.   (Lesson 13-3)

29. Find $-\sqrt{121}$.   (Lesson 14-1)

Applications

30. **Sports**   The area of a regulation-size baseball diamond is 8100 square feet. The Cincinnati Reds cover their diamond with a tarp to protect it from the rain. If each side of the diamond is the same length, how long is the tarp on each side?

31. **Construction**   City code requires that a party house must allow 4 square feet for each person on the dance floor. Rocky's Dance Emporium wants to have a dance floor that is square and that is large enough for 100 people at a time. How long should it be on each side?

Critical Thinking

32. Use your calculator to try to find the square root of -25. What is shown on the display? Why do you think this happens?

Wrap-Up

33. In your own words, explain how to find the whole number estimate for a square root.

14-3 Strategy: Use Venn Diagrams

Objective:
Use Venn diagrams to solve problems.

Key Term:
Venn diagram

Exquisite Interiors has 40 sample floor tiles in the shapes of various polygons. If 20 tiles are regular polygons, 14 tiles are quadrilaterals, and 5 tiles are squares, how many tiles are not regular polygons, quadrilaterals, or squares?

At first this problem may seem to be fairly easy. However, you must remember that squares are a special kind of quadrilateral and are also regular polygons.

Since 20 + 14 + 5 equals 39, why is the answer not 1?

Explore
There are 40 tiles altogether. Twenty tiles are regular polygons, 14 are quadrilaterals, and 5 are squares. You must find out how many other tiles there are.

Plan
You can use a **Venn diagram** to illustrate the data. A rectangle is used to represent all the tiles. A circle is used to represent the tiles that are regular polygons and an intersecting circle is used to represent the tiles that are quadrilaterals. The intersection of the circular regions represents tiles that are both regular polygons and quadrilaterals. In other words, the intersection represents the square tiles.

Why are intersecting circles used?

Solve
There are 5 squares, so 5 is placed in the intersection of the circular regions. The number of non-square regular polygons is 20 − 5 or 15. The number of non-square quadrilaterals is 14 − 5 or 9. There are 15 + 5 + 9 or 29 polygons represented in the circular regions. There are 40 − 29 or 11 other polygons that are neither quadrilaterals nor regular polygons.

Polygons

| regular polygons 15 | both (squares) 5 | quadri-laterals 9 |

other 11

FYI

Leonhard Euler first used circles to show the relationship of sets. Later John Venn put the circles in a rectangle to form diagrams like the ones in this lesson.

Examine
Look at the Venn diagram again. Add the number of polygons in each region.

$$11 + 15 + 5 + 9 = 40$$

Since the total is 40, the answer is correct.

Checking for Understanding

Communicating Algebra Use the Venn diagrams below to complete Exercises 1-3.

Polygons

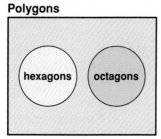

Figure 1

Quadrilaterals

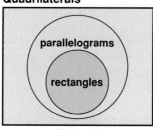

Figure 2

Polygons

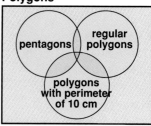

Figure 3

1. Explain why the circular regions in Figure 1 do not intersect.

2. Explain why one circular region is contained within the other in Figure 2.

3. Write a sentence describing the polygons in the region where all three circles intersect in Figure 3.

Guided Practice

The Venn diagram below represents the students involved in extra activities at Valley High School.

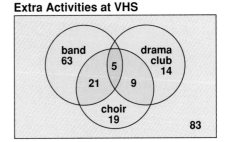

4. How many students are involved with all three of the listed activities?

5. How many students are in the band?

6. How many students are in both the drama club and the band?

7. How many students are involved in at least one extra activity?

Exercises

Independent Practice

Solve. Use any strategy.

8. Mama Sophia's Pizzeria conducted a small survey. They asked 50 customers which pizza topping they liked better, pepperoni or sausage. Twenty-eight people liked pepperoni, 25 people liked sausage, and 8 people liked both equally. How many people did not like either pepperoni or sausage?

9. Hattie must read a book for a book report. She read half of her book on Sunday. On Monday, she read another 30 pages. On Tuesday, she read 6 pages. On Wednesday, she read half of the remaining pages. If she has 20 pages yet to read, how many pages does the book have?

10. There are 26 students in a math class. The class takes a survey and finds that 14 students have pet dogs, 10 students have cats, and 5 students have birds. Four students have dogs and cats, 3 students have dogs and birds, and 1 student has a cat and a bird. If no one has all three of these animals, how many students have none of these animals?

Critical Thinking

11. A geometry teacher drew some quadrilaterals on the chalkboard. There were 5 trapezoids, 12 rectangles, 5 squares, and 8 rhombuses. What is the least number of figures the teacher could have drawn?

Wrap-Up

12. Make up your own problem that can be solved using a Venn diagram.

14-4 The Real Number System

Objectives:
Identify types of numbers.
Solve equations by finding square roots.
Simplify square roots.

Key Terms:
rational number
irrational number
real number

The ancient Greeks believed that all numbers could be expressed as ratios of whole numbers. This is true of most of the numbers found in previous chapters.

$$\frac{3}{7} \qquad -15 = \frac{-15}{1} \qquad 0.33 = \frac{33}{100} \qquad 0.111\ldots = \frac{1}{9} \qquad \sqrt{100} = 10$$

These numbers are called **rational numbers.** A rational number is any number that can be expressed as $\frac{a}{b}$, where a and b are integers and b does not equal 0. Remember that rational numbers can always be expressed by using terminating or repeating decimals.

Consider $\sqrt{2}$. $\sqrt{2} = 1.4142136\ldots$ This decimal continues forever without any pattern of repeating digits.

You cannot express a number like $\sqrt{2}$ as either a terminating or a repeating decimal. This kind of number is called an **irrational number.**

Definition of Irrational Number	An irrational number is a number that cannot be expressed as $\frac{a}{b}$, where a and b are integers and b does not equal 0.

Each of the following is an irrational number.

$$\sqrt{2} \qquad \sqrt{3} \qquad -\sqrt{11} \qquad \pi$$

Examples

Determine whether each number is a rational or irrational number.

How can you show that $0.\overline{3}$ is equivalent to $\frac{1}{3}$? Explain.

1   **0.33333 . . .**

The three dots indicate that the 3s keep repeating. This decimal can be expressed as $\frac{1}{3}$. So it is a rational number.

2   **0.75**

How can you show that 0.75 is equivalent to $\frac{3}{4}$? Explain.

This decimal is a terminating decimal. It can be expressed as $\frac{75}{100}$ or $\frac{3}{4}$. So it is a rational number.

3   **0.0101101110 . . .**

This decimal does not terminate, and it does not repeat. Notice that the number of 1s between the 0s is different each time. This decimal is an irrational number.

The set of rational numbers and the set of irrational numbers together form the set of **real numbers.** The Venn diagram at the right shows the relationships among whole numbers, integers, rational numbers, irrational numbers, and real numbers.

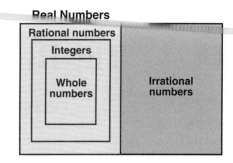

The following chart shows the sets of numbers to which several numbers belong.

Number	Whole Number	Integer	Rational	Irrational	Real
-7		✔	✔		✔
$\sqrt{15}$				✔	✔
$\sqrt{9}$	✔	✔	✔		✔
0.12121212 . . .			✔		✔
0.303303330 . . .				✔	✔
$\frac{4}{5}$			✔		✔

Throughout the text, you have solved equations that have rational number solutions. Some equations have solutions that are irrational.

You can solve some equations that involve squares by taking the square root of each side.

Examples

4 **Solve $x^2 = 144$.**

$x^2 = 144$

$x = \sqrt{144}$ or $x = -\sqrt{144}$ Take the square root of each side.

$x = 12$ or $x = -12$

5 **Solve $x^2 = 50$. Round your answer to the nearest tenth.**

$x^2 = 50$

$x = \sqrt{50}$ or $x = -\sqrt{50}$

50 [√x] 7.071068

$x \approx 7.1$ or $x \approx -7.1$

Physics Connection

When investigating traffic accidents, police officers often need to determine how fast a car was traveling before it skidded to a stop. The formula

$s = \sqrt{30fd}$ can be used to estimate the speed. In the formula, s is the speed, d is the distance the car skidded, and f is a friction factor.

Suppose you wrote a number as a product of its prime factors. Then you looked for factors that occurred twice to find the square root. Do you think the result would be the same as taking the square root of a number? Consider $\sqrt{196}$.

$$\sqrt{196} = \sqrt{2 \cdot 2 \cdot 7 \cdot 7}$$
$$= \sqrt{2^2 \cdot 7^2}$$
$$= \sqrt{2^2} \cdot \sqrt{7^2}$$
$$= 2 \cdot 7 \text{ or } 14 \qquad \text{Check: } 196 \ \boxed{\sqrt{x}} \ 14 \ \checkmark$$

The following property of square roots was used to simplify $\sqrt{196}$.

| **Product Property of Square Roots** | In words: | The square root of a product is equal to the product of the square roots of its factors. |
| | In symbols: | For any nonnegative numbers a and b, $\sqrt{a \cdot b} = \sqrt{a} \cdot \sqrt{b}$. |

You can use this property to simplify irrational numbers.

Example

6 **Simplify $\sqrt{200}$.**

$$\sqrt{200} = \sqrt{5 \cdot 5 \cdot 2 \cdot 2 \cdot 2} \qquad \text{prime factorization}$$
$$= \sqrt{5^2 \cdot 2^2 \cdot 2}$$
$$= \sqrt{5^2} \cdot \sqrt{2^2} \cdot \sqrt{2} \qquad \text{product property}$$
$$= 5 \cdot 2 \cdot \sqrt{2} \text{ or } 10\sqrt{2}$$

The simplest form of $\sqrt{200}$ is $10\sqrt{2}$. $10\sqrt{2}$ is approximately 14.142 or -14.142. Use your calculator to check this result.

Checking for Understanding

Communicating Algebra

1. A repeating decimal names a(n) __?__ number.

2. A terminating decimal names a(n) __?__ number.

3. A nonterminating, nonrepeating decimal names a(n) __?__ .

4. The set of real numbers is made up of __?__ and __?__ .

5. The property that allows $\sqrt{4 \cdot 5}$ to be simplified to $2\sqrt{5}$ is the __?__ .

Guided Practice

Name the sets of numbers to which each number belongs: the whole numbers, the integers, the rational numbers, the irrational numbers, and/or the real numbers.

6. -5   7. $-\sqrt{7}$ 8. $\frac{2}{3}$ 9. -2.7   10. $-\sqrt{16}$

Exercises

Independent Practice

Name the sets of numbers to which each number belongs: the whole numbers, the integers, the rational numbers, the irrational numbers, and/or the real numbers.

11. 6

12. $\frac{-3}{4}$

13. 1.8

14. -4.6

15. $\sqrt{4}$

16. $\sqrt{11}$

17. $-\sqrt{10}$

18. $\sqrt{25}$

19. 0.89

20. 5.454454445 . . .

21. 0.531

22. -0.666 . . .

Solve each equation. Round decimal answers to the nearest tenth.

23. $a^2 = 49$

24. $b^2 = 81$

25. $c^2 = 4$

26. $d^2 = 64$

27. $s^2 = 121$

28. $n^2 = 169$

29. $p^2 = 1$

30. $t^2 = 225$

31. $e^2 = 28$

32. $g^2 = 40$

33. $f^2 = 11$

34. $y^2 = 60$

Simplify.

35. $\sqrt{28}$

36. $\sqrt{72}$

37. $\sqrt{54}$

38. $\sqrt{80}$

39. $\sqrt{90}$

40. $\sqrt{108}$

41. $\sqrt{128}$

42. $\sqrt{242}$

Challenge

Solve each equation.

43. $m^2 + 9 = 25$

44. $t^2 - 25 = 75$

45. $x^2 + x^2 = 50$

Mixed Review

46. Write the negation of $2 + 4 < 8$. Then state whether the statement and its negation are true or false.   (Lesson 2-6)

47. Convert 8.3 kilograms into grams.   (Lesson 7-8)

48. If 5 liters of punch cost $4.90, how much do 12 liters cost?   (Lesson 9-3)

49. *True* or *false:* An equilateral triangle can never be an obtuse triangle. (Lesson 12-6)

Find the best integer estimate for each of the following.   (Lesson 14-2)

50. $\sqrt{88}$

51. $-\sqrt{19.1}$

Applications

52. **Sky Diving**   When a sky diver jumps from an airplane, the time (t) in seconds it takes to reach a given distance can be estimated by using the formula $t = \sqrt{\frac{2s}{g}}$. In this formula, s is the free-fall distance, and g is the acceleration due to gravity, 32 ft/s^2. Sally jumps from an airplane to free-fall 2500 feet. How long will it take her to reach that distance?

53. **Meteorology**   Use the formula $t^2 = \frac{d^3}{216}$ to estimate the amount of time a thunderstorm will last. In this formula, t is time in hours and d is the diameter of the storm system in miles. If a thunderstorm is 6 miles wide, find t.

Critical Thinking

True or *false.*

54. Every rational number is a real number.

55. Every rational number is an integer.

Wrap-Up

56. In your own words explain the Venn diagram on page 534.

14-5   The Pythagorean Theorem

Objective:

Use the Pythagorean Theorem to find the length of the side of a right triangle.

Key Terms:

hypotenuse
leg
Pythagorean Theorem

Softball has become a popular pastime in the United States. It is played on a diamond like the one shown at the right. The distance from one base to another is 60 feet. As you can see, the boundary of a softball diamond is square. When the grounds people lay out a softball diamond, they measure the distances between the bases carefully. Then they measure the diagonals to make certain the diamond is really a square. How long should a diagonal be if the diamond has been laid out correctly?

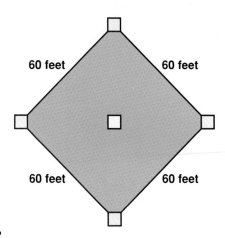

The answer to this question can be found by using a relationship involving the lengths of the sides of a right triangle that was discovered thousands of years ago. Credit for this discovery is given to a famous Greek mathematician named Pythagoras.

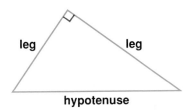

Remember that a right triangle is a triangle with one right angle. The side opposite the right angle is called the **hypotenuse**. The other two sides are called **legs**. The hypotenuse is always the longest side of a right triangle.

The relationship known as the **Pythagorean Theorem** is true for any right triangle.

Pythagorean Theorem	In a right triangle, the square of the length of the hypotenuse is equal to the sum of the squares of the lengths of the legs.

If a and b are the lengths of the legs of a right triangle and c is the length of its hypotenuse, an equation can be used to state the Pythagorean Theorem.

Pythagorean Theorem

$$c^2 = c^2 + b^2$$

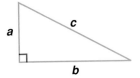

Example

1 **Use the Pythagorean Theorem to find the length of a diagonal of a softball diamond.**

A diagonal and two adjacent sides of a square form a right triangle. The legs of the right triangle are sides of the square and the hypotenuse of the right triangle is a diagonal of the square.

Let c represent the length of a diagonal (hypotenuse).

$$c^2 = a^2 + b^2 \qquad \text{Pythagorean Theorem}$$
$$c^2 = 60^2 + 60^2 \qquad \text{Replace } a \text{ with 60 and } b \text{ with 60.}$$
$$c^2 = 3600 + 3600$$
$$c^2 = 7200$$
$$c = \sqrt{7200} \text{ or } c = -\sqrt{7200}$$
$$c \approx 85 \text{ or } c \approx -85 \qquad \text{Use a calculator.}$$

The length of the diagonal should be about 85 feet.

Why can you ignore the negative value of c in this problem?

The Pythagorean Theorem can be used to find the length of any side of a right triangle if the lengths of the other two sides are known.

Example

Estimation Hint

THINK:
$60 \times 60 = 3600$
$70 \times 70 = 4900$
$\sqrt{4624}$ is between 60 and 70.

2 **Find the length of the third side of the right triangle.**

$$c^2 = a^2 + b^2$$
$$85^2 = a^2 + 51^2$$
$$7225 = a^2 + 2601$$
$$7225 - 2601 = a^2 + 2601 - 2601$$
$$4624 = a^2$$
$$\sqrt{4624} = a \text{ or } -\sqrt{4624} = a$$
$$68 = a \qquad \text{Use a calculator.}$$

The length of the side is 68 centimeters. Compare with the estimate.

You can use the Pythagorean Theorem to see if a triangle is a right triangle.

Examples

Determine whether each triangle is a right triangle.

3

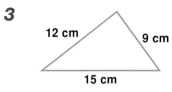

$$c^2 = a^2 + b^2$$
$$15^2 \stackrel{?}{=} 12^2 + 9^2$$
$$225 \stackrel{?}{=} 144 + 81$$
$$225 = 225$$

The triangle is a right triangle.

4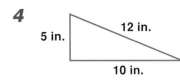

$$c^2 = a^2 + b^2$$
$$12^2 \stackrel{?}{=} 10^2 + 5^2$$
$$144 \stackrel{?}{=} 100 + 25$$
$$144 \neq 125$$

The triangle is not a right triangle.

Checking for Understanding

Communicating Algebra

1. The side opposite the right angle of a right triangle is called the __?__.

2. The Pythagorean Theorem states that __?__.

3. The Pythagorean Theorem only holds for __?__ triangles.

Guided Practice

Identify the lengths of the hypotenuse and legs in each right triangle.

4.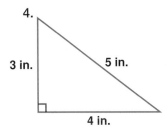

5.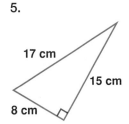

6.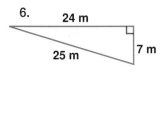

State an equation you should use to find the length of the hypotenuse of each right triangle. Then find the length of the hypotenuse.

7.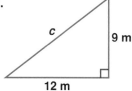

8.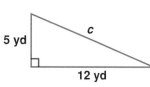

9.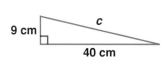

Exercises

Independent Practice

**ESTIMATION
MENTAL MATH
CALCULATOR
PAPER/PENCIL**

Use the Pythagorean Theorem to find the length of the hypotenuse of each right triangle. The lengths of the legs are given. Round decimal answers to the nearest tenth.

10. 6 mm, 8 mm

11. 9 ft, 40 ft

12. 8 m, 15 m

13. 12 in., 9 in.

14. 24 m, 7 m

15. 5 yd, 12 yd

16. 20 cm, 21 cm

17. 15 in., 36 in.

18. 16 ft, 30 ft

Solve each equation.

19. $101^2 = a^2 + 20^2$

20. $89^2 = a^2 + 80^2$

21. $109^2 = 60^2 + b^2$

22. $85^2 = 13^2 + b^2$

23. $61^2 = a^2 + 60^2$

24. $97^2 = 65^2 + b^2$

Find the missing measure for each right triangle. Round decimal answers to the nearest tenth.

25. b, 77 m; c, 85 m

26. b, 55 cm; c, 73 cm

27. a, 15 ft; c, 17 ft

28. a, 40 in.; c, 41 in.

29. a, 35 km; c, 37 km

30. b, 91 yd; c, 109 yd

31. b, 140 in.; c, 149 in.

32. a, 19 ft; c, 181 ft

The measures of the three sides of a triangle are given. Determine if each triangle is a right triangle.

33. 7 ft, 9 ft, 6 ft

34. 5 m, 12 m, 13 m

35. 9 in., 12 in., 14 in.

36. 4 m, 7 m, 5 m

37. 30 cm, 24 cm, 18 cm

38. 9 cm, 40 cm, 41 cm

Mixed Review

39. Simplify $12a + 2(3a + 4)$. (Lesson 1-4)

40. Solve $p = 82.37 - 25.5$. (Lesson 5-4)

41. Find the volume of a rectangular prism that has length 10 cm, width 7 cm, and height 3 cm. (Lesson 13-7)

42. *True* or *false*: 82 is a perfect square. (Lesson 14-1)

43. Simplify $\sqrt{56}$. (Lesson 14-4)

Solve. Round decimal answers to the nearest tenth.

Connection

44. **Geometry** The points $R(2, 4)$ and $T(6, 1)$ are graphed on the coordinate system at the right. Find the distance between R and T. (Hint: Notice the right triangle. First find the measures of $\overline{RS}$ and $\overline{ST}$.)

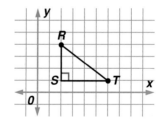

Applications

45. **Sky Diving** The radius of the canopy of a parachute is 12 feet. If the load is suspended 16 feet below, how long are the suspension lines?

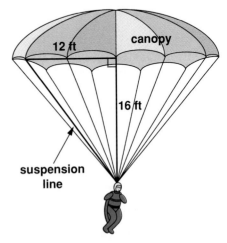

46. **Space Travel** The parachute used by the Viking Lander on Mars had a radius of 8 meters and the payload was suspended 20 meters below the canopy. How long were the suspension lines?

Critical Thinking

47. How is the length of a diagonal of a rectangle related to the length and the width of the rectangle?

Wrap-Up

48. Use an equation and the figure at the right to restate the Pythagorean Theorem.

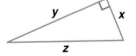

Exploration

Graphing Irrationals

Materials: compass, straightedge

In this Exploration you will graph irrational numbers on a number line.

You know how to graph integers and rational numbers on a number line. You can also graph irrational numbers on a number line. Consider the following method for graphing $\sqrt{2}$.

▶ At 1, construct a perpendicular line segment 1 unit in length. Draw the line segment shown in color. Label it c.

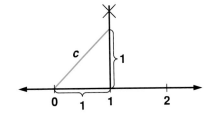

▶ The Pythagorean Theorem can be used to show that c is $\sqrt{2}$ units long.

$$c^2 = a^2 + b^2$$

$$c^2 = 1^2 + 1^2$$

$$c^2 = 2$$

$$c = \sqrt{2}$$

▶ Open the compass to the length of the segment shown in color. Construct a congruent segment on the number line.

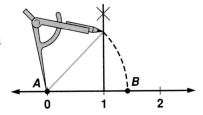

Your Turn: Graph $\sqrt{5}$ on a number line. Think of $\sqrt{5}$ as $\sqrt{2^2 + 1^2}$.

Graph $\sqrt{8}$ on a number line. Think of $\sqrt{8}$ as $\sqrt{2^2 + 2^2}$.

Analysis

1. Describe a method to graph $\sqrt{10}$.

2. Explain how the graph of $\sqrt{2}$ can be used to locate the point that represents $\sqrt{3}$.

3. Explain how to graph $-\sqrt{2}$.

4. Can all real numbers be graphed on a number line?

14-6 Using the Pythagorean Theorem

Objective:
Solve problems using the Pythagorean Theorem.

Valerie Hayes, Vice President of Express Flying Service, is planning to build a security fence around a field to use as a holding area for small planes. A diagram of the field is shown at the right. How much fencing does Ms. Hayes need to buy?

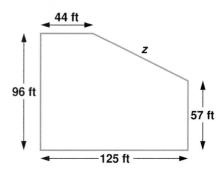

Explore

What is asked? How much fencing is needed?
What is implied? What is the perimeter of the field?

Plan

To find the perimeter, you must find the unknown length, z. You can separate the figure into more familiar shapes as shown at the left.

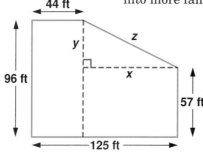

The hypotenuse of the right triangle is the side with length z. Find the measures of the legs of the right triangle and then use the Pythagorean Theorem to find the length of the hypotenuse.

Solve

First find the measures of the legs.

$$x = 125 - 44 \qquad\qquad y = 96 - 57$$
$$x = 81 \qquad\qquad\qquad y = 39$$

Now apply the Pythagorean Theorem.

$$z^2 = x^2 + y^2$$
$$z^2 = 81^2 + 39^2 \qquad \text{Replace } x \text{ with 81 and } y$$
$$z^2 = 6561 + 1521 \quad \text{with 39.}$$
$$z^2 = 8082$$

$$8082 \; \boxed{\sqrt{x}} \; 89.899944$$

$$z \approx 90$$

Finally, find the perimeter.

$$96 + 44 + 90 + 57 + 125 = 412$$

So, 412 feet of fencing is needed.

Why is the perimeter less than the perimeter of a 125 ft by 96 ft rectangle?

Examine

If the field included the corner area, its perimeter would be $2 \cdot 96 + 2 \cdot 125$ or 442 feet. The answer should be a little less than this. Thus, the answer is reasonable.

Checking for Understanding

Communicating Algebra

1. Discuss each step that was used to solve the example.

2. Find another way to separate the irregular figure on page 542 to form rectangles and a right triangle. Then explain how your plan can be used to solve this problem.

Guided Practice

State an equation that can be used to find the answer to each question. Then solve. Round decimal answers to the nearest tenth.

3.

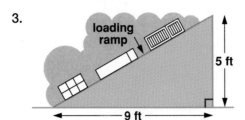

How long is the ramp?

4. How high is the television screen?

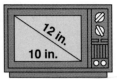

5.

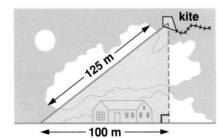

How high is the kite?

6.

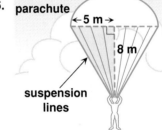

How long are the suspension lines?

Exercises

Independent Practice

Solve. Round decimal answers to the nearest tenth.

7.

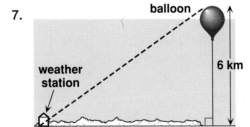

How far is the weather balloon from the weather station?

8.

At what height does the ladder touch the house?

9. The diagonal brace on a gate is 5 feet long. The height of the gate is 4 feet. How wide is the gate?

10. The members of the Campers Club hike 7 miles east, then 3 miles south to find their overnight camping spot. How far are they from their starting point?

11. Luise takes a shortcut to school by walking diagonally across an empty lot. The rectangular lot is 20 meters wide and 40 meters long. How much shorter is the shortcut than a route on the sides of the lot?

Mixed Review 12. *True* or *false*: If the corresponding parts of two triangles are congruent, then the two triangles are congruent.   (Lesson 12-8)

13. Use the Pythagorean Theorem to find the length of the hypotenuse of a right triangle that has sides 10 meters and 24 meters.   (Lesson 14-5)

Decision Making 14. For safety reasons the base of a 24-foot ladder should be at least 8 feet from the wall. Can a 24-foot ladder be used to reach a window that is 22 feet above the ground? Explain.

Critical Thinking 15. When separating the drawing of the irregularly shaped field on page 542, why is it convenient to form rectangles and right triangles?

Wrap-Up 16. Describe some real-life problems that can be solved by using the Pythagorean Theorem.

Mid-Chapter Quiz

Find each value. Round decimal answers to the nearest tenth.
(Lessons 14-1, 14-2)

1. $\sqrt{36}$
2. 13^2
3. $-\sqrt{12}$

Name the sets of numbers to which each number belongs: the whole numbers, the integers, the rational numbers, the irrational numbers, and/or the real numbers.   (Lesson 14-4)

4. $-\sqrt{49}$
5. $\sqrt{6}$

Solve. Round decimal answers to the nearest tenth.   (Lessons 14-4, 14-5)

6. $t^2 = 324$
7. $30^2 = y^2 + 24^2$

8.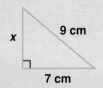

9 cm

x

7 cm

What is the value of x?

9.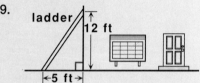

ladder 12 ft

←5 ft→

How long is the ladder?

10. There are 27 students in an algebra class. Of these students, 12 students are in the school band and 9 students are in the school choir. If 5 students are in both the band and choir, how many of these students are in neither the band nor choir? (Lesson 14–3)

14-7　Special Right Triangles

Objective:
Find missing measures in 30° - 60° and 45° - 45° right triangles.

Suppose you draw a square $ABCD$ with sides 4 cm long. Then you draw diagonal AC. You can use a protractor to find that $\angle CAB$, $\angle ACB$, $\angle CAD$, and $\angle ACD$ each measure 45°. The triangles formed are called 45° - 45° right triangles.

Now use the Pythagorean Theorem to find the length of diagonal AC.

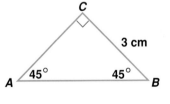

Let d represent the measure of $\overline{AC}$.

$$d^2 = (AD)^2 + (DC)^2$$
$$d^2 = 4^2 + 4^2 \qquad \text{The length of each side is 4 cm.}$$
$$d^2 = 32$$
$$d = \sqrt{32}$$
$$d = \sqrt{2 \cdot 4^2} \text{ or } 4\sqrt{2} \qquad \text{The length of the diagonal is } 4\sqrt{2} \text{ cm.}$$

Suppose the length of each side of the square were 5 cm. What would be the length of $\overline{AC}$? Using the Pythagorean Theorem, you find that the length is $5\sqrt{2}$ cm. These and other examples suggest that in a 45° - 45° right triangle, you can find the length of the hypotenuse by multiplying the length of a leg by $\sqrt{2}$.

Example

If the length of the hypotenuse of a 45° - 45° right triangle is $2\sqrt{2}$ cm, what is the length of each leg?

1　Find the length of $\overline{AB}$ in $\triangle ABC$.

$$c = a\sqrt{2}$$
$$c = 3\sqrt{2}$$

3 ⊗ 2 √x̄ ⊟ 4.2426407

The length of $\overline{AB}$ is about 4 cm.

The sides of a right triangle with a 30° angle and a 60° angle also have special relationships. In a 30° - 60° right triangle, the length of the side opposite the 30° angle is one-half the length of the hypotenuse.

$$a = \frac{1}{2}c \text{ or } c = 2a$$

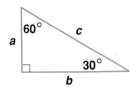

Also, in a 30° - 60° right triangle, you can find the length of the side opposite the 60° angle by multiplying the length of the other leg by $\sqrt{3}$.

$$b = a\sqrt{3}$$

Examples

2 **Find the length of $\overline{BC}$ in $\triangle ABC$.**

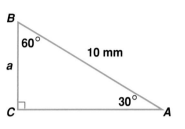

$$a = \frac{1}{2}c$$

$$a = \frac{1}{2}(10) \quad \text{Replace } c \text{ with 10.}$$

$$a = 5$$

The length of $\overline{BC}$ is 5 mm.

3 **Find the length of $\overline{QR}$ in $\triangle PQR$.**

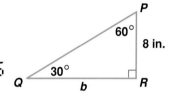

$$b = a\sqrt{3}$$

$$b = 8\sqrt{3} \quad \text{Replace } a \text{ with 8.}$$

$$8 \;\boxed{\times}\; 3 \;\boxed{\sqrt{x}}\; \boxed{=}\; 13.856406$$

The length of $\overline{QR}$ is about 14 inches.

Checking for Understanding

Communicating Algebra

1. In a 30° - 60° right triangle, the length of the side opposite the 30° angle is ___?___ .

2. In a 45° - 45° right triangle, the length of the hypotenuse can be found by ___?___ .

Guided Practice

The length of the hypotenuse of a 30° - 60° right triangle is given. Find the length of the side opposite the 30° angle.

3. 8 in.   4. $2\frac{1}{2}$ m   5. $4\frac{1}{4}$ in.   6. 8.28 m

The length of a leg of a 45° - 45° right triangle is given. Find the length of the hypotenuse. Write your answer in simplest radical form.

7. 3 ft   8. 5 in.   9. $2\frac{1}{2}$ cm   10. 14.9 mm

Exercises

Independent Practice

ESTIMATION
MENTAL MATH
CALCULATOR
PAPER/PENCIL

Find the value of x. Round decimal answers to the nearest tenth.

11.

12.
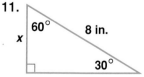

13.
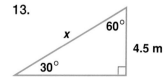

Find the value of x. Round decimal answers to the nearest tenth.

14.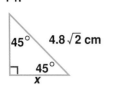

15.

16.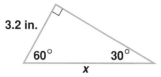

17.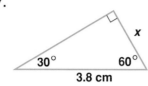

18.

19.

20.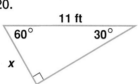

21.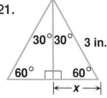

22.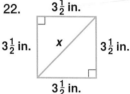

Mixed Review

23. Solve $x = \left(-4\frac{1}{8}\right)\left(\frac{1}{4}\right)$.   (Lesson 6-3)

24. Use the formula $V = \frac{1}{3}\pi r^2 h$ to find the volume of a circular cone that has radius 2 m and height 14 m.   (Lesson 13-9)

25. A big screen TV has sides of 3 feet and 4 feet. What is the length of its diagonal?   (Lesson 14-6)

Applications

26. **Sewing**   In sewing, cutting on the bias means cutting on the diagonal. Ron has a square piece of material that is 2 yards on each side. What is the length of the diagonal of this piece of material? Round your answer to the nearest tenth.

27. **Building**   Rhonda wants to get a refrigerator into her house. She cannot lift it up the steps leading to her back door, so she decides to build a ramp. She decides to have the ramp make a 30° angle with the ground. If her back door is 2 feet above the ground, how long should Rhonda cut the boards to make the ramp?

Critical Thinking

28. The length of the hypotenuse of a 30° - 60° right triangle is 7 feet. Find the length of the other two sides. Round answers to the nearest tenth.

Wrap-Up

29. Draw a 30° - 60° right triangle. Use x as the length of the hypotenuse and label the side opposite the 30° angle in terms of x.

14-8 The Tangent Ratio

Objective:
Find the tangent of an angle and find the measure of an angle using the tangent.

If you were asked to find the height of a desk or a doorway, you would probably measure it directly by using a ruler or a tape measure. However, it would be difficult to use either one of these instruments to find the height of a street lamp, a flagpole, a tree, or a lighthouse. The measures of the sides of the right triangles form special ratios that can help you find heights like these. One of these ratios is called the **tangent** ratio.

Key Term:
tangent

Definition of Tangent	If $\triangle ABC$ is a right triangle and A is an acute angle, $$\text{tangent of } A = \frac{\text{measure of the side opposite to } \angle A}{\text{measure of the side adjacent to } \angle A}.$$

What is the equation for the tangent of $\angle B$?

The abbreviation for tangent is *tan*.

In the figure at the right, $\tan A = \dfrac{a}{b}$.

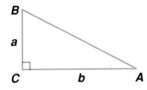

Example

1 **Find tan S.**

$$\tan S = \frac{\text{measure of the side opposite to } \angle S}{\text{measure of the side adjacent to } \angle S}$$

$$\tan S = \frac{3}{4} \text{ or } 0.75$$

So tan S is $\dfrac{3}{4}$ or 0.75.

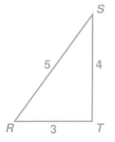

A calculator can be used to find the tangent ratio for an angle with a given degree measure or to find the degree measure of an angle if the tangent ratio is known.

Examples

2 **Find the tangent of 63°.**

$$63 \; \boxed{\text{TAN}} \quad 1.9626105$$

Therefore, tan $63° \approx 1.9626$.

3 **Find the measure of $\angle A$ given that tan $A = 4.7030$.**

$$4.7030 \; \boxed{\text{INV}} \boxed{\text{TAN}} \quad 77.995961$$

The measure of $\angle A$ is approximately 78°.

Checking for Understanding

1. Describe the procedure for using a calculator to approximate a tangent ratio for an angle with a given degree measure.

Write a fraction in simplest form for each tangent ratio.

2.

$\tan A = \underline{\ ?\ }$
$\tan B = \underline{\ ?\ }$

3.

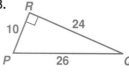

$\tan P = \underline{\ ?\ }$
$\tan Q = \underline{\ ?\ }$

4.

$\tan W = \underline{\ ?\ }$
$\tan Y = \underline{\ ?\ }$

Exercises

Use a calculator to find each tangent ratio to the nearest ten thousandth.

ESTIMATION
MENTAL MATH
CALCULATOR
PAPER/PENCIL

5. $\tan 30°$ **6.** $\tan 45°$ **7.** $\tan 60°$ **8.** $\tan 18°$

9. $\tan 52°$ **10.** $\tan 74°$ **11.** $\tan 83°$ **12.** $\tan 20°$

Use a calculator to find the angle measure that corresponds to each tangent ratio. Round answers to the nearest degree.

13. $\tan A = 0.0875$ **14.** $\tan B = 0.2680$ **15.** $\tan C = 2.7380$

16. $\tan D = 0.6495$ **17.** $\tan P = 0.8391$ **18.** $\tan Q = 8.1440$

19. Use a protractor to draw a 120° angle. Classify it as *acute*, *right*, or *obtuse*. (Lesson 12-1)

The length of the hypotenuse of a 30° - 60° right triangle is given. Find the length of the side opposite the 30° angle. (Lesson 14-7)

20. 10 cm **21.** $8\frac{1}{4}$ in.

22. Navigation Refer to the picture at the right. Write an equation that relates tan 22° with the distance to the lighthouse (*d*) and the height of the lighthouse (*h*).

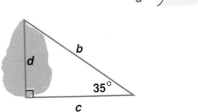

23. Surveying Refer to the picture at the right. Write an equation that relates tan 35° with the distance across Hidden Lake (*d*) and another side of the triangle.

24. Write an equation that relates tan 25° with the length of the sides of the triangle at the right. Solve the equation for *x*. Round to the nearest tenth.

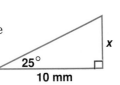

25. Draw and label a right triangle. Write an equation for the tangent of each acute angle in the right triangle.

14-9   The Sine and Cosine Ratios

Objective:
Find the sine and cosine of an angle and find the measure of an angle using the sine or cosine ratio.

The tangent ratio is a ratio that involves the measure of the legs of a right triangle. The **sine** ratio and the **cosine** ratio are two ratios that involve the length of the hypotenuse and the length of one leg. These ratios are defined below.

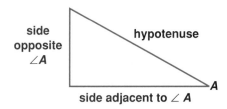

side opposite ∠A

hypotenuse

side adjacent to ∠ A

A

Definition of Sine and Cosine	If $\triangle ABC$ is a right triangle and A is an acute angle, $$\text{sine of } \angle A = \frac{\text{measure of the side opposite } \angle A}{\text{measure of the hypotenuse}}, \text{ and}$$ $$\text{cosine of } \angle A = \frac{\text{measure of the side adjacent to } \angle A}{\text{measure of the hypotenuse}}.$$

Key Terms:
sine
cosine
trigonometry
trigonometric ratios

The abbreviation for sine is *sin*, and the abbreviation for cosine is *cos*.

In the figure at the right,

$$\sin A = \frac{a}{c} \text{ and } \cos A = \frac{b}{c}.$$

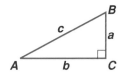

Example

1   Find sin P and cos P.

$$\sin P = \frac{\text{measure of the side opposite } \angle P}{\text{measure of the hypotenuse}}$$

$$\sin P = \frac{3}{5} \text{ or } 0.6$$

$$\cos P = \frac{\text{measure of the side adjacent to } \angle P}{\text{measure of the hypotenuse}}$$

$$\cos P = \frac{4}{5} \text{ or } 0.8$$

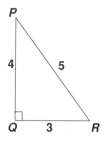

Is the sine or cosine of an angle ever greater than 1? Why?

A calculator can be used to find the sine ratio or the cosine ratio for an angle with a given degree measure. It can also be used to find the degree measure of an angle if the sine ratio or the cosine ratio is known.

Examples

How can you use a
calculator to find
cos 57°?

2 **Find the sine of 57°.**

$$57 \;\boxed{\text{SIN}} = 0.8386706$$

Therefore sin 57° ≈ 0.8387.

3 **Find the measure of ∠D given that cos D = 0.9390.**

$$0.9390 \;\boxed{\text{INV}} \;\boxed{\text{COS}} \; 20.115708$$

If sin W = 0.7980, how
would you use a
calculator to find the
measure of ∠W?

The measure of ∠D is about 20°.

The word **trigonometry** means triangle measurement. The tangent, sine, and cosine ratios are **trigonometric ratios** because they are ratios of the measures of sides of right triangles.

Checking for Understanding

Communicating Algebra

1. Explain the meaning of the sine ratio.

2. Explain the meaning of the cosine ratio.

3. Describe the procedure for using a calculator to approximate the cosine ratio for an angle with a given degree measure.

4. Describe the procedure for using a calculator to find the degree measure of the angle that corresponds to a given sine ratio.

Guided Practice

Write a fraction in simplest form for each sine ratio and each cosine ratio.

5. 6. 7.

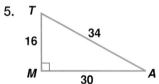

sin A = _?_   cos S = _?_   sin W = _?_
cos A = _?_   sin S = _?_   cos W = _?_
sin T = _?_   cos R = _?_   cos X = _?_
cos T = _?_   sin R = _?_   sin X = _?_

Exercises

Independent Practice

Use a calculator to find each sine ratio or cosine ratio. Round answers to the nearest ten thousandth.

8. sin 16°   9. cos 20°   10. cos 67°   11. sin 54°

12. cos 12°   13. sin 32°   14. sin 72°   15. cos 48°

Use a calculator to find the angle measure that corresponds to each sine ratio or cosine ratio. Round answers to the nearest whole number.

16. $\sin A = 0.1392$
17. $\sin B = 0.8900$
18. $\cos T = 0.9063$

19. $\cos Q = 0.7050$
20. $\sin R = 0.5740$
21. $\cos S = 0.3250$

22. $\sin B = 0.5444$
23. $\cos D = 0.2491$
24. $\sin R = 0.9165$

Mixed Review

25. Find the quotient $\dfrac{m^{11}}{m^3}$.   (Lesson 4-10)

26. Find 32.76 cm + 4.1 cm + 13.13 cm.   (Lesson 13-10)

27. Use a calculator to find tan 55° to the nearest ten thousandth.   (Lesson 14-8)

Applications

For Exercises 28-30, refer to the drawing at the right.

28. **Broadcasting**   One end of a guy wire is attached to a television tower at point A and the other end is attached to a ground anchor at point B. Write an equation that relates the 80° angle with the length of the guy wire (ℓ) and the distance of A from the ground (h).

29. **Broadcasting**   Write an equation that relates the 80° angle with the length of the guy wire (ℓ) and the distance of B from the base of the tower (d).

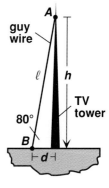

Critical Thinking

30. If the length of the guy wire (ℓ) is 120 feet, find h. Round your answer to the nearest tenth.

Wrap-Up

31. Explain the differences among the tangent, sine, and cosine ratios.

Team Problem Solving

A large cube made up of many unit cubes is painted on all six faces. If the dimensions of the large cube are 10 × 10 × 10, how many of the unit cubes are painted on three faces? two faces? one face? no faces? Suppose the cube's dimensions are $n \times n \times n$. How many of the unit cubes are painted on three faces? two faces? one face? no faces?

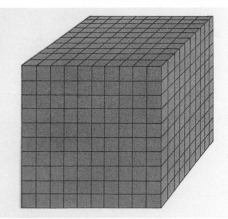

Algebra in Action-Physics

Vectors

A hiker leaves camp and walks 15 kilometers due north. The hiker then walks 6 kilometers due east. To find the direction and displacement of the hiker, we can use directed segments called vectors.

A vector is a line segment which possesses both magnitude (length) and direction.

Draw a vector to represent *15 kilometers due north.*

Next, draw a vector starting at the endpoint of the first vector to represent *6 kilometers due east.*

Finally, connect the starting point of the first vector with the endpoint of the second vector. This resultant vector will tell you the direction and displacement of the hiker.

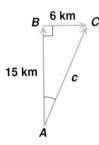

Direction:

$$\tan A = \frac{6}{15}$$
$$\tan A = 0.4$$
$$A \approx 22°$$

Displacement:

$$a^2 + b^2 = c^2$$
$$6^2 + 15^2 = c^2$$
$$36 + 225 = c^2$$
$$261 = c^2$$
$$16 \approx c$$

The direction of the hiker is about 22° northeast, and his displacement is about 16 km from the starting point.

1. A soccer player kicks the ball 12 feet to the west. Another player kicks the ball 18 feet to the north. What is the direction and displacement of the ball? Round your answer to the nearest whole number.

14-10 Using Trigonometric Ratios

Objective:
Solve problems by using the trigonometric ratios.

The Conservation District measures the distance across Hidden Valley Lake every ten years to see how much erosion has occurred. Since they cannot measure the distance directly, they use a right triangle and trigonometric ratios to find the distance.

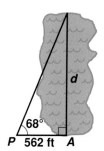

Explore

Form a right triangle with the distance across the lake as one leg. Special surveying instruments can be used to find the measure of $\angle P$. You can measure the distance from A to P directly using a tape measure or other device.

Plan

Look at the figure. You know the measure of one angle and the measure of the side adjacent to it.

Since the tangent ratio is the measure of the opposite side divided by the measure of the adjacent side, use the tangent ratio to solve for d, the distance across the lake.

Solve

$$\tan 68° = \frac{\text{measure of side opposite the } 68° \text{ angle}}{\text{measure of side adjacent to the } 68° \text{ angle}}$$

$$\tan 68° = \frac{d}{562}$$

$$562 \, (\tan 68°) = d$$

```
562 ⊠ 68 [TAN][=] 1390.9988
```

The distance across Hidden Valley Lake is about 1391 feet.

Examine

If the angle was 60°, the distance across the lake would be $562(\sqrt{3})$, or about 973 feet. Since the angle is greater than 60° the distance should be greater than 973 feet. The answer appears to be reasonable.

The trigonometric ratios and the Pythagorean Theorem can be used to find the measure of any side or angle of a right triangle if the measure of one side and any other side or acute angle are known.

Examples

1 **In $\triangle RST$ find the measure of $\angle R$.**

The measure of the side opposite $\angle R$
and the hypotenuse are known.
Therefore the sine ratio should be used.

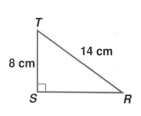

$$\sin R = \frac{\text{measure of side opposite } \angle R}{\text{measure of the hypotenuse}}$$

$$\sin R = \frac{8}{14}$$

$$\sin R \approx 0.5714$$

$$R \approx 34.8$$

The measure of $\angle R$ is about $34.8°$.

2 **In $\triangle ABC$ find the length of $\overline{AB}$.**

*Should the length of $\overline{AB}$
be greater than or less
than 14 inches?*

The measure of the side adjacent to B is
known and the measure of the hypotenuse
is needed. Use the cosine function.

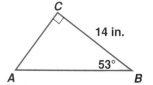

$$\cos B = \frac{\text{measure of side adjacent to } \angle B}{\text{measure of the hypotenuse}}$$

$$\cos 53° = \frac{14}{x}$$

$$x \cos 53° = 14 \quad \text{Why?}$$

$$x = \frac{14}{\cos 53°} \quad \text{Why?}$$

$$14 \;\boxed{\div}\; 53 \;\boxed{\text{cos}}\;\boxed{=}\; 23.262962$$

The length of $\overline{AB}$ is about 23.3 inches.

Checking for Understanding

**Communicating
Algebra**

1. Explain how to determine which trigonometric ratio to use when solving for an
unknown measure of a right triangle.

**Guided
Practice**

**Tell which trigonometric ratio would be the best choice for finding the
value of x. Then solve for x. Round answers to the nearest tenth.**

2.

3.

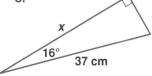

4.

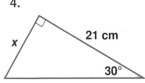

Tell which trigonometric ratio would be the best choice for finding the value of x. Then solve for x. Round answers to the nearest tenth.

5.

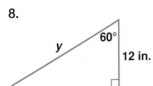

47 in. x 20 in.

6.

11 m

x
4 m

7.

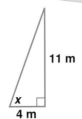

36

50 mm x

Exercises

Solve for y. Round answers to the nearest tenth.

8.

y 60° 12 in.

9.

y
35 cm 50°

10.

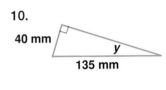

40 mm y 135 mm

11.

10 ft 35° y

12.

y 14 in. 8 in.

13.

30 cm y 70°

Mixed Review

14. Find the area of a parallelogram with base 8 inches and height 7 inches. (Lesson 13-1)

15. Use a calculator to find cos 83° to the nearest ten thousandth. (Lesson 14-9)

Applications

Solve. Round answers to the nearest tenth.

16. **Traveling** In a sightseeing boat near the base of Horseshoe Falls at Niagara Falls, a passenger estimates the angle of elevation to the top of the falls to be 30°. If Horseshoe Falls is 173 feet high, what is the distance from the boat to the base of the falls?

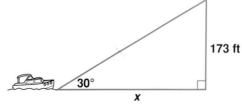

173 ft

30°
x

17. **Traveling**   While picnicking in San Jacinto Battlefield Park, a pre-algebra student 800 feet from the base of the monument estimates the angle of elevation to the top of the monument to be 35°. From this information, estimate the height of the monument.

18. **Navigation**   From a boat in the ocean, a cliff is sighted through the fog. The angle of elevation is 42° and the height of the cliff is 135 meters. How far is the boat from the cliff?

19. **Building Design**   A ramp is designed to help people in wheelchairs move more easily from one level to another. If a ramp 16 feet long forms an angle of 12° with the level ground, what is the vertical rise?

Critical Thinking

20. To *solve a triangle* means to find the measures of all the sides and all the angles of the triangle. Solve $\triangle ABC$ shown at the right.

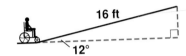

Wrap-Up

21. **Make Up a Problem**   Write three problems about right triangles. The first problem should require the tangent ratio to find the solution. The second problem should require the sine ratio and the third problem should require the cosine ratio.

History

The Pythagoreans

In the sixth century B.C., the famous mathematician Pythagoras opened a school in the southern part of the Italian peninsula. Any student who showed exceptional ability was invited into a secret society called the Pythagorean Order.

The motto of this society was "all is number." All of their doctrines were based on mathematical principles. They divided mathematics into four parts: music, arithmetic, astronomy, and geometry. These four subjects became the school curriculum from the time of Plato and Aristotle up to the Renaissance.

The best known mathematical discovery of the Pythagoreans was the Pythagorean Theorem. The Babylonians and the Egyptians knew and used this right triangle property 1500 years before Pythagoras; however, Pythagoras is credited with deriving its proof.

Review

Language and Concepts

Choose the correct term to complete each sentence.

1. The (square, square root) of 4 equals 16.

2. In the expression $\sqrt{25}$, the symbol $\sqrt{}$ is called a (radical sign, radicand).

3. The number $\sqrt{81}$ belongs to the set of (irrational, rational) numbers.

4. The longest side of a right triangle is the (hypotenuse, leg).

5. A right triangle that has a leg that is half the length of the hypotenuse is a (30° - 60°, 45° - 45°) right triangle.

Skills

Find the best integer estimate for each of the following. Then check your estimate using a calculator. (Lessons 14-1, 14-2)

6. 16^2
7. $\sqrt{49}$
8. $-\sqrt{900}$
9. $-\sqrt{39}$
10. $\sqrt{9.61}$

Simplify. (Lesson 14-4)

11. $\sqrt{48}$
12. $\sqrt{32}$
13. $\sqrt{300}$
14. $\sqrt{45}$

Name the sets of numbers to which each number belongs: the whole numbers, the integers, the rational numbers, the irrational numbers, and/or the real numbers. (Lesson 14-4)

15. 7
16. $\frac{1}{4}$
17. -2.6
18. $\sqrt{8}$

Use the Pythagorean Theorem to find the length of the hypotenuse of each right triangle. The lengths of the legs are given. (Lesson 14-5)

19. 12 m, 16 m
20. 14 ft, 48 ft
21. 30 cm, 16 cm

Find the missing measure for each right triangle. (Lesson 14-5)

22. a, 8 in.; b, 15 in.
23. b, 63 ft; c, 65 ft
24. a, 15 cm; c, 39 cm

Find the value of x. Round decimal answers to the nearest tenth. (Lesson 14-7)

25.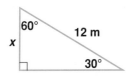

26.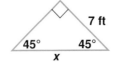

27.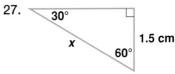

Write a fraction in simplest form for each ratio.
(Lessons 14-8, 14-9)

28. $\sin A$ **29.** $\tan B$ **30.** $\cos A$

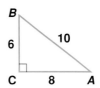

Solve for *y*. Round answers to the nearest tenth. (Lesson 14-10)

31.

32.

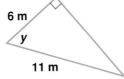

33.

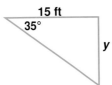

Applications and Problem Solving

Solve. Round decimal answers to the nearest tenth. (Lessons 14-3, 14-6)

34. In a recent survey of 120 students, 60 students said they play tennis and 50 students said they play softball. If 20 students play both sports, how many students do not play either tennis or softball?

35. A telephone pole is 28 feet tall. A wire is stretched from the top of the pole to a point on the ground that is 15 feet from the bottom of the pole. How long is the wire?

ⓟortfolio Suggestion

Place your favorite word problem from this chapter in your portfolio and attach a note explaining why it is your favorite.

Curriculum Connection

● **History** Learn more about Pythagoras and his followers.

● **Science** Do research to find how a hot-air balloon works.

Read More About It

Lamm, Joyce. *Let's Talk About the Metric System.*

McCauley, David. *The Way Things Work.*

Pallas, Norvin. *Calculator Puzzles, Tricks, and Games.*

Test

Find the best integer estimate for each of the following.

1. $\sqrt{81}$ **2.** 11^2 **3.** $-\sqrt{18}$ **4.** $\sqrt{3481}$

Name the sets of numbers to which each number belongs: the whole numbers, the integers, the rational numbers, the irrational numbers, and/or the real numbers.

5. 9   **6.** -1.7 **7.** $-\sqrt{11}$ **8.** $\frac{3}{8}$

Find the missing measure for each right triangle.

9. a, 13 m; c, 85 m   **10.** b, 48 cm; c, 73 cm   **11.** a, 20 ft; b, 21 ft

Write a fraction in simplest form for each ratio.

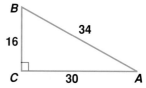

12. $\cos B$ **13.** $\tan A$ **14.** $\sin B$

Solve for x. Round decimal answers to the nearest tenth.

15.

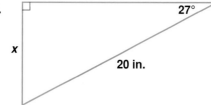

16.

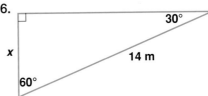

17.

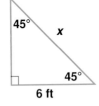

18.

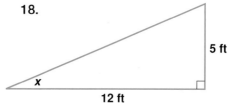

Solve.

19. There are a total of 50 triangles. Twenty triangles are right triangles and 15 are isosceles. If 7 triangles are both right and isosceles, how many are neither right nor isosceles?

20. Yana is standing 28 feet from a point directly below his kite. If the string attached to it is 53 feet long, how high is the kite flying?

BONUS

The length of each side of the cube is 2 cm. Find the length of the diagonal of the cube. Round your answer to the nearest tenth.

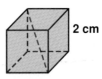

Academic Skills Test

Cumulative, Chapters 1-14

1. Which is the solution for the equation $3c - 5 = 7c + 5$?

   A $-2\frac{1}{2}$ **C** 1

   B -1 **D** $2\frac{1}{2}$

2. A basketball team had a record of 20 wins and 8 losses. What was the ratio of wins to losses?

   A 2 to 5   **C** 5 to 2

   B 4 to 5   **D** 5 to 7

3. The family size of students in a pre-algebra class is recorded in the table.

No. of Students	1	4	3	6	5	2	3
Family Size	2	3	4	5	6	7	8

   How many students have fewer than 5 members in their family?

   A 6   **C** 9

   B 8   **D** 14

4. Angie's Place sells made-to-order sandwiches. There are 3 kinds of bread, 3 kinds of cheese, and 4 kinds of meat. How many different combinations of bread, cheese, and meat (one of each) can be ordered?

   A 10   **C** 18

   B 12   **D** 36

5. $\triangle EFG$ is congruent to $\triangle KML$. What is the measure of $\angle M$?

   A $40°$

   B $60°$

   C $70°$

   D $80°$

   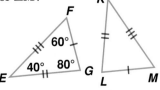

6. Which figure appears to be a trapezoid?

   A   **C**

   B   **D**

7. What is the area of the circle? Use $\pi \approx \frac{22}{7}$.

   A 44 m^2

   B 49 m^2

   C 154 m^2

   D 616 m^2

   14 m

8. How much paper is needed to cover a shoebox that is 12 inches long, 6 inches wide, and 5 inches high?

   A 360 in^2 **C** 252 in^2

   B 324 in^2 **D** 162 in^2

9. A rectangular window is 40 inches high and 30 inches wide. What is the length of a diagonal?

   A 60 in.   **C** 35 in.

   B 50 in.   **D** 25 in.

10. $\triangle ABC$ is a right triangle. What is the length of the hypotenuse $\overline{AC}$?

    A 21 in.

    B 26 in.

    C 30 in.

    D 52 in.

    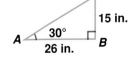

CHAPTER OBJECTIVES

In this chapter you will learn to:

- identify and classify polynomials
- add, subtract, and multiply polynomials
- solve problems by eliminating possibilities

Polynomials

There's more to mathematics in baseball than just keeping score or finding the speed at which a pitcher throws a fast ball.

Fenway Park is home to baseball's Boston Red Sox. Ninety-six meters from home plate in left field is a green wall about 18 meters high. Because it is so difficult to hit a home run over this wall, it is referred to as The Green Monster.

Mathematically, a polynomial like $x^2 - 9x + 18$ can be used to describe the speed with which the batter must hit the ball in order to clear the wall.

Science Connection

Class Project
Write a paragraph about Isaac Newton's contribution in the study of gravity. In your paragraph, mention the polynomial that describes the distance a falling object travels in t seconds.

563

15-1 Polynomials

Objectives:
Identify and classify polynomials.
Find the degree of a polynomial.

Key Terms:
monomial
degree
polynomial
like term
binomial
trinomial
constant

In biology, you study animals and plants. To distinguish common characteristics, plants and animals are classified by *genus* and *species*. A common frog, for example, is identified as *Rana pipiens,* where *Rana* is the genus and *pipiens* is the species.

In mathematics, we classify expressions that have common characteristics.

Why is $\frac{x}{2}$ considered a product rather than a quotient?

Expressions such as y, 15, $-7x^2$, $\frac{x}{2}$, and $\sqrt{3}x^3y^2$ are called **monomials.**

Notice that each expression consists of a *number,* a *variable,* or a *product* (not a quotient) of numbers and variables. Expressions like $2x + 3$, $\frac{4}{y^2}$, or $\sqrt{m}$ are *not* monomials.   Why?

Examples

State whether each expression is a monomial.

1 $\frac{4}{m}$

The expression $\frac{4}{m}$ is *not* a monomial because it contains a variable in its denominator.

2 $\frac{3}{4}x$

The expression $\frac{3}{4}x$ is a monomial because it is the product of a number and a variable.

The **degree** of a monomial is the sum of the exponents of its variables. The following chart shows you how to find the degree of a monomial.

Monomial	Variable(s)	Exponent(s)	Degree
y	y	1	1
$-6r^2$	r	2	2
$\sqrt{3}x^3y^2$	x, y	3, 2	3 + 2 or 5
$\frac{2}{3}bc^8$	b, c	1, 8	1 + 8 or 9

Remember that $y = y^1$.

A **polynomial** is a monomial or the sum or difference of two or more monomials. Expressions such as $5x + 1$, $x^2y + x$, and $5 - 3c + 6c^2$ are polynomials. Each monomial in the polynomial is called a **term** of the polynomial.

Why are $6x^3y$ and $17x^3y$ like terms? Why are $3a^2b$ and $4ab^2$ not like terms?

Two monomials that are the same or differ only by their coefficients, are **like terms.** For example, $6x^3y$ and $17x^3y$ are like terms; $3a^2b$ and $4ab^2$ are not. A polynomial with two *unlike* terms is a **binomial;** a polynomial with three unlike terms is a **trinomial.**

Examples

State whether each expression is a polynomial. If it is, identify it as a *monomial, binomial,* or *trinomial.*

3 $7y^3 + 4y^2 + y$

The expression $7y^3 + 4y^2 + y$ is a polynomial because it is the sum of three monomials. Since it has three unlike terms, it is a trinomial.

4 $5b - \dfrac{1}{b^2}$

The expression $5b - \dfrac{1}{b^2}$ is not a polynomial because $\dfrac{1}{b^2}$ is not a monomial.

You can find the degree of a polynomial by determining which term in the polynomial has the greatest degree. The degree of the polynomial is the same as that of the term with the greatest degree.

You know how to find the degree of a term, or monomial, containing variables. A monomial that does not contain a variable is called a **constant.** What do you think is the degree of a constant? Consider the constant 15.

$$15 = 15 \cdot 1$$
$$= 15 \cdot x^0 \qquad \text{Recall that } x^0 = 1.$$

So, the degree of 15 or any nonzero constant is 0.

The constant 0 has *no* degree.

Example

Find the degree of each polynomial.

5 $a^2 + 2ab + b^4$

First, find the degree of each term.

$a^2 \rightarrow 2$

$2ab \rightarrow 1 + 1$ or 2

$b^4 \rightarrow 4$ greatest degree

The degree of $a^2 + 2ab + b^4$ is 4.

Checking for Understanding

Communicating Algebra

1. In your own words, define *monomial*. Write three examples of a monomial.
2. Explain how you would find the degree of the polynomial $p^3 - p^2 m^2 + 4m - 1$.
3. Are $2x^3 y^2$ and $5x^2 y^3$ like terms? Explain why or why not.

Guided Practice

State whether each expression is a monomial.

4. $7x$

5. y^2

6. -8

7. $-5xy$

8. $3ab + c$

9. $\frac{2}{3}x$

10. $-\frac{11xy}{7}$

11. $\sqrt{mp}$

Find the degree of each monomial.

12. $11m$

13. $-6y^3$

14. 12

15. $\frac{a}{7}$

16. $3m^2 n$

17. $-2p^2 q r^4$

18. $\frac{ab^2}{3}$

19. $-b$

Exercises

Independent Practice

State whether each expression is a polynomial. If it is, identify it as a *monomial*, *binomial*, or *trinomial*.

20. $5m$

21. $2x + 1$

22. $x^3 - y^3$

23. $4 + 3a - 8a^3$

24. $\sqrt{d} - 5$

25. $\frac{3}{g}$

26. $\frac{e}{6}$

27. $\frac{6}{x + y}$

28. $\frac{4ab}{c} - \frac{2d}{x}$

29. -17

30. $5ab^2 - 2a + b^2$

31. $15n^2 + 3nt$

Find the degree of each polynomial.

32. $3y^2$

33. $-5x^3 yz^2$

34. $2x + 1$

35. $21m + 12n$

36. $-3x^2 + 5m^5$

37. $4x^3 + 7xy - 2xz^3$

38. $18a^3 b^4 - 11ab^5$

39. $19n^2 + 27n^2 t^2$

40. $32xyz - 11x^2 y + 17xz^2$

41. $a^2 + b^3 + c^{14} + d^2$

42. $2xy^2 z - 5xyz^5 + 6x^3$

43. $3x^4 y^2 z + 6xy^3 z - 12z^6$

44. $2^2 a^2$

45. $3^3 xy - 5x$

46. $-5^2 rs^3 - 2y^5$

Evaluate each polynomial if $a = -1$, $b = 2$, $c = -3$, and $d = 4$.

47. $a^3 - 2ab$

48. $b^3 - 2ac$

49. $4a^5 - 2ab^2$

50. $5abc + 2a^2 b - 6ab^2$

51. $4ab^2 - 2ab + 6bc^2$

52. $5ac^2 - 2a + b^2$

53. $\sqrt{2b}$

54. $\sqrt{d} - b^2$

55. $\sqrt{-27c} - a$

56. Solve $3.5k = 31.5$ using the inverse operation.   (Lesson 1-7)

57. *True* or *false*: $|5| = -5$.   (Lesson 2-2)

58. Name two factors of any nonzero number.   (Lesson 4-1)

59. Solve the inequality $b - 2.3 \leq 5.9$.   (Lesson 5-7)

60. Estimate 9% of 312.   (Lesson 9-6)

61. Use the formula $A = \frac{1}{2}h(a + b)$ to find the area of a trapezoid that has bases of 13 mm and 5 mm and a height of 10 mm.   (Lesson 13-2)

62. Find $-\sqrt{196}$.   (Lesson 14-1)

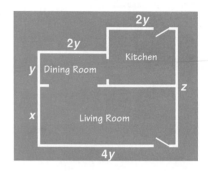

Application

63. Architecture   Mr. Snyder, an architect, draws a floor plan for the first floor of a house. Using the measurements given, write a polynomial that represents the total area of the first floor.

Critical Thinking

64. What is the degree of the polynomial $x^{n + 2} + x^{n - 3}y^3 + y^{n - 1}$?

Wrap-Up

65. From your everyday experience, think of an example in which you classify objects.

Biography: Amalie Emmy Noether

Emmy Noether was born into a distinguished German-Jewish family of mathematicians and scientists in 1882. Her interest and ability in mathematics developed when she began studying at the University of Erlangen, Germany. She went on to complete a PhD degree in 1907.

While German universities permitted women to earn degrees, job opportunities for them were limited. The best position Ms. Noether could reach in Germany was that of "unofficial associate professor." In 1933, however, when Ms. Noether came to the United States, she was appointed to an actual faculty position at Bryn Mawr College in Pennsylvania.

The focus of Emmy Noether's work in mathematics was *abstract algebra,* the study of different algebraic systems. Her work so inspired her successors that mathematicians often speak of it as the "Noether school" of mathematics.

15-2   Adding Polynomials

Objective:
Add polynomials.

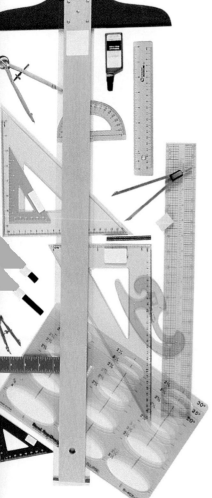

When an architect designs a building, she often makes a *model* of her plan. The model helps her to see whether her design is going to work. In mathematics, we can also make models. A model of a polynomial can be made by using tiles or drawings like the ones shown below.

Let □ be a 1 × 1 square. The square has an area of 1.

Let □ be a 1 × x rectangle. The rectangle has an area of x.

Let □ be an x × x square. The square has an area of x^2.

You can use these tiles to make a model of a polynomial.

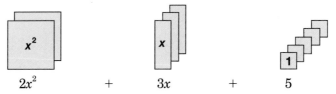

$$2x^2 \quad + \quad 3x \quad + \quad 5$$

Tiles can also help you to understand how to add polynomials.

Example

1 Find $(3x^2 + x + 2) + (x^2 + 2x + 4)$ using tiles.

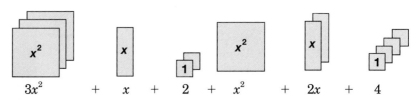

$$3x^2 \quad + \quad x \quad + \quad 2 \quad + \quad x^2 \quad + \quad 2x \quad + \quad 4$$

In this problem, both $3x^2$ and x^2 are represented by $x \times x$ tiles. This means they are like models, or in algebraic terms, they are like terms. Both x and $2x$ are represented by $1 \times x$ tiles. They are also like terms. Similarly, 2 and 4 are like terms.

To add these polynomials, combine the like terms.

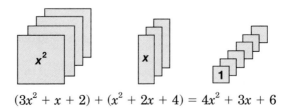

$$(3x^2 + x + 2) + (x^2 + 2x + 4) = 4x^2 + 3x + 6$$

You can also add polynomials by using some of the properties discussed in earlier chapters.

Examples

Find each sum.

2 $(3x + 1) + (2x + 5)$

$$= (3x + 2x) + (1 + 5) \qquad \text{Associative and commutative properties of addition}$$
$$= (3 + 2)x + (1 + 5) \qquad \text{Distributive property}$$
$$= 5x + 6$$

3 $(2a^2 - 4ab + b^2) + (3a^2 - 5b^2)$

$$= (2a^2 + 3a^2) + (-4ab) + [b^2 + (-5b^2)]$$
$$= (2 + 3)a^2 + (-4ab) + [1 + (-5)]b^2$$
$$= 5a^2 + (-4ab) + (-4b^2)$$
$$= 5a^2 - 4ab - 4b^2$$

What properties are used to solve this problem?

4 $(4rs^2 - r^2 - 3s^2) + (2s^2 + 5rs^2 + 7r^2)$

Sometimes it is helpful to arrange the terms in vertical columns and then add.

$$\begin{array}{l} 4rs^2 - r^2 - 3s^2 \\ + \ 5rs^2 + 7r^2 + 2s^2 \\ \hline 9rs^2 + 6r^2 - 1s^2 \end{array} \qquad \text{Notice that like terms are aligned and added.}$$

$\rightarrow \quad 9rs^2 + 6r^2 - s^2 \qquad -1s^2 = -s^2$

Checking for Understanding

Communicating Algebra

1. Identify the like terms in $x^2 + 7x + 2x^2 + x + 4$.

2. Name the polynomial represented by the model at the right.

3. Draw a model of the polynomial $3x^2 + 6x + 4$.

Guided Practice

Find each sum using tiles or drawings.

4. $(2x + 3) + (x + 4)$

5. $(5x + 1) + (3x)$

6. $(2x^2 + 3x + 1) + (3x^2 + x + 4)$

7. $(3x^2 + 5x + 2) + (4x^2 + 3)$

Exercises

Independent Practice

Find each sum.

8. $(5x - 7y) + (6x + 8y)$

9. $(7n + 2t) + (4n - 3t)$

10. $(7m - 2n) + (9m + 4n)$

11. $(17n - 5m) + (11n - 3m)$

12. $(2x^2 + 3) + (4x - 7)$

13. $(5r + 5s) + (6r - 8s)$

Find each sum.

14. $(5m + 3n) + 8m$

15. $(12x + 7y) + 8y$

16. $(5x + 8y) + (-3x + 5y)$

17. $(-5a - 8b) + (4a + 8b)$

18. $(6x^2 + 10x + 3) + (-2x^2 + 7x - 12)$

19. $(3a + 5ab - 3b^2) + (7a + 5b^2)$

20.
$$\begin{array}{r} 4a + 5b - c \\ + \ 8a - 6b + c \\ \hline \end{array}$$

21.
$$\begin{array}{r} -6b^2 + 13b - 5 \\ + \ \ 2b^2 \qquad - 10 \\ \hline \end{array}$$

22.
$$\begin{array}{r} -21y^2 + 11y - 32 \\ + \ \ 18y^2 - 8y + 10 \\ \hline \end{array}$$

23.
$$\begin{array}{r} 4a + 5b - 6c \\ 3a - 7b + 2c \\ + \ 2a - \ \ b + 7c \\ \hline \end{array}$$

24.
$$\begin{array}{r} 2x^2 - 5x + 7 \\ x^2 - \ \ x + 11 \\ + \ 5x^2 + 7x - 13 \\ \hline \end{array}$$

25.
$$\begin{array}{r} -9ax^3 - 5ax^2 + 6ax \\ -3ax^3 - 6ax^2 - 7ax \\ + \ 5ax^3 \qquad + \ ax \\ \hline \end{array}$$

Combine like terms. Then evaluate for $a = 3$ and $b = 7$.

26. $(2a + 5b) + (6a - 3b)$

27. $(-21a + 7b) + (18a - 3b)$

28. $(4a + 3b + 2) + (3a - 4b - 10)$

29. $(5a + 2b - 1) + (10a - 3b + 2)$

Mixed Review

30. Find the circumference of a circle that has a radius of 4 inches.   (Lesson 6-13)

31. Find the surface area of a rectangular prism if the length is 1.2 feet, the width is 0.3 feet, and the height is 0.9 feet.   (Lesson 13-4)

32. Solve $n^2 = 64$.   (Lesson 14-4)

33. State whether $\dfrac{a + c}{5}$ is a polynomial.   (Lesson 15-1)

Connection

34. Geometry   Write and simplify an expression for the perimeter of each figure.

a.

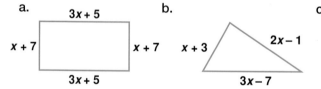

3x + 5

x + 7

3x + 5

b.

x + 7   x + 3

2x − 1

3x − 7

c.

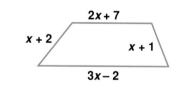

2x + 7

x + 2

x + 1

3x − 2

Application

35. Construction   A standard measurement for a window is *united inch.* You can find the united inches of a window by adding the length of the window to the width. If the length of a window is $2x - 8$ inches and the width is $x + 3$ inches, what is the size of the window in united inches?

Critical Thinking

36. Given: $(3x - 5y) + (7x + 2y) = 10x - 3y$.
Find: $(10x - 3y) - (7x + 2y)$.

Wrap-Up

37. Make Up a Problem   Write a polynomial addition problem. Add the two polynomials in three different ways (tiles, horizontal format, vertical format).

15-3 Subtracting Polynomials

Objective:

Subtract polynomials.

In the last lesson, you used tiles as a model for *adding* polynomials. Now you will use tiles as a model for *subtracting* polynomials.

Example

1 Find $(2x^2 + 5x + 4) - (2x^2 + x + 2)$ **using tiles.**

Make a model of $2x^2 + 5x + 4$. Then remove two x^2-tiles, one x-tile, and two 1-tiles.

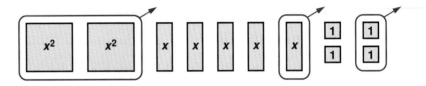

There are four x-tiles and two 1-tiles remaining.

$(2x^2 + 5x + 4) - (2x^2 + x + 2) = 4x + 2$

Remember, you can subtract two numbers by using additive inverses.

What is the additive inverse of -2?

$$\overbrace{\text{additive inverses}}$$
$$7 - 2 = 5 \qquad\qquad 7 + (\text{-}2) = 5$$
$$\underbrace{\text{same result}}$$

When 2 is multiplied by -1, the result is its additive inverse, -2. To find the additive inverse of a polynomial, the entire polynomial is multiplied by -1. Study the following table.

Polynomial	Multiply by -1	Additive Inverse
$\text{-}y$	$\text{-}1(\text{-}y)$	y
$4x - 3$	$\text{-}1(4x - 3)$	$\text{-}4x + 3$
$2xy + 5y$	$\text{-}1(2xy + 5y)$	$\text{-}2xy - 5y$
$2x^2 - 3x - 5$	$\text{-}1(2x^2 - 3x - 5)$	$\text{-}2x^2 + 3x + 5$

Example

Find each difference.

2 $(5x + 3) - (2x + 1)$ Add the additive inverse of $2x + 1$.

$= (5x + 3) + (\text{-}1)(2x + 1)$

$= 5x + 3 + (\text{-}2x) - 1$ $(\text{-}1)(2x + 1) = \text{-}2x - 1$

$= 5x + (\text{-}2x) + 3 + (\text{-}1)$ Commutative property of addition

$= 3x + 2$ Combine like terms.

Examples

What is the additive inverse of $(x^2 - 4x)$?

3 $(3x^2 - 6x + 4) - (x^2 - 4x)$

Add the additive inverse of $(x^2 - 4x)$.

$= (3x^2 - 6x + 4) + (-x^2 + 4x)$

$= 3x^2 + (-x^2) - 6x + 4x + 4$

$= 2x^2 - 2x + 4$

4 $(7a^2 - 6ab - b^2) - (2a^2 + ab - 5b^2)$

Align like terms and add the additive inverse of the second polynomial.

$$7a^2 - 6ab - b^2$$
$$\underline{+ (-2a^2) + (-ab) + 5b^2}$$
$$5a^2 - 7ab + 4b^2$$

Checking for Understanding

Communicating Algebra

1. What is the sum of a number and its additive inverse?

2. Explain how you would find the additive inverse of $3a^2 + 4ab - b^2$.

Guided Practice

Find each difference using tiles or drawings.

3. $(5x + 2) - (2x + 1)$

4. $(4x + 7) - (x + 3)$

5. $(4x^2 + 3) - (4x^2 + 2)$

6. $(3x^2 + 5x + 4) - (x^2 + 3x + 2)$

7. $(5x^2 + 4x + 2) - (4x^2 + x + 2)$

8. $(2x^2 + 6x + 2) - (x^2 + 1)$

State the additive inverse of each polynomial.

9. a

10. $-3x$

11. $12xy$

12. $x + 4$

13. $4n - 1$

14. $5x^2 - 3$

15. $4a + 3b$

16. $m^2 + 4m + 1$

17. $5g^2 - 3g - 8$

18. $-3d^2 + 2$

19. $6x^2 + 3xy - 2y^2$

20. $-a^2 - ab$

Exercises

Independent Practice

Find each difference.

21. $(7m + 3) - (3m + 1)$

22. $(3x + 5) - (2x + 9)$

23. $(x + 6) - (4x + 3)$

24. $(5x - 3) - (3x + 7)$

25. $(3x + 4y) - (2x + y)$

26. $(12a - 5b) - (3a - 9b)$

27. $(5x^2 + 11) - (5x^2 - 6)$

28. $(4d^2 - 7d + 1) - (3d^2 + 4)$

29. $(12y^2 + 6y - 1) - (3y^2 - 5y - 6)$

30. $(3z^2 + 5z + 8) - (-2z^2 + 2z + 3)$

31. $(6a^2 - 5ab + b^2) - (2a^2 + 3ab - 5b^2)$

32. $(-6x + 7y - 9) - (3x - 14y + 11)$

33.
$$5y^2 + 7y + 9$$
$$\underline{- 2y^2 + 3y + 6}$$

34.
$$7m^2 - 6m + 13$$
$$\underline{- 7m^2 + 3m - 5}$$

35.
$$a + 3b - c$$
$$\underline{- 2a + b + 7c}$$

36.
$$3a^2 + 5ab$$
$$\underline{- \qquad -2ab + b^2}$$

37.
$$11m^2n^2 + 4mn - 6$$
$$\underline{- (-5m^2n^2) - 6mn + 17}$$

38.
$$7z^2 \qquad + 4$$
$$\underline{- 3z^2 + 2z - 6}$$

Mixed Review

39. Solve $-2(3 + 4x) = 5x - 6$.   (Lesson 7-4)

40. Name the quadrant of the graph of $(-2, 9)$.   (Lesson 8-4)

41. Find the surface area of the pyramid at the right. (Lesson 13-6)

8 ft

5 ft

5 ft

42. Use the Pythagorean Theorem to find the length of the hypotenuse of the right triangle at the right.   (Lesson 14-5)

24 cm

18 cm

43. Find the sum $(4a^2 + 7a - 2) + (3a^2 - 9a - 3)$.   (Lesson 15-2)

Challenge

Savings Account	
Name	**Balance**
Bill	$6x + 7$
Jorge	$7x - 10$
Marinella	$12x + 3$
Angela	$4x + 27$

44. Which two savings accounts have a difference of $5x + 13$?

45. Which two accounts have a difference of $8x - 24$?

46. Which two accounts have a difference of $3x - 37$?

47. Suppose $x = \$10$. Which account has the greatest balance?

Application

48. Framing   Jose plans to mat and frame a picture. The area inside the frame is $4x^2 - 7$ square inches. The area of the picture is $2x^2 + 3$ square inches. How much matting will Jose need?

Critical Thinking

49. Write a polynomial subtraction problem with a difference of $3m^2 - 5m + 4$. Check your work by adding.

Wrap-Up

50. Make Up a Problem   Write a polynomial subtraction problem. Find the difference in three different ways (tiles, horizontal format, vertical format).

Mid-Chapter Quiz

State whether each expression is a polynomial. If it is, identify it as a *monomial, binomial,* or *trinomial*.   (Lesson 15-1)

1. $4a^2$

2. $-2y + 3$

3. $\frac{a}{5}$

4. $r^3 - s^2$

5. $-\dfrac{3}{(x + y)}$

6. $7mt^2 + 3m - 2t$

Find each sum or difference.   (Lessons 15-2, 15-3)

7. $(3a + 6) + (a - 1)$

8. $(7x - 2) + (-3x + 1)$

9. $(5r^2 + 4) - (2r^2 - 3)$

10. $(4x + 3y) + (-7x + 3y)$

11. $(6n^2 - 5n + 1) - (3n - 4)$

12. $(-8c^2 + 2cd - 5d^2) - (c^2 + 3cd - 4d^2)$

15-4 Powers of Monomials

Objective:
Find powers of monomials.

Amber opens a savings account with $1000. It earns 4% interest every 6 months. So, in one year, the balance is $(1000)(1.04)(1.04)$ or $(1000)(1.04)^2$. Suppose she keeps the money in the account for four years. Then the deposit is worth $1000(1.04^2)^4$. To compute this product, you need to know how to find the power of a power.

Consider the calculation for the following expressions:

$$(4^2)^3 = 4^2 \cdot 4^2 \cdot 4^2 \qquad\qquad (x^4)^2 = (x^4)(x^4)$$
$$= 4^{2+2+2} \qquad\qquad\qquad\quad = x^{4+4}$$
$$= 4^6 \qquad\qquad\qquad\qquad\quad = x^8$$

For $(4^2)^3$, how can you obtain the exponent 6 from the exponents 2 and 3? For $(x^4)^2$, how can you obtain 8 from 4 and 2? These and other similar examples suggest that you can find a *power of a power* like $(x^4)^2$ by multiplying the exponents.

Power of a Power

For any number a, and positive integers m and n,
$$(a^m)^n = a^{mn}.$$

Example

1 Simplify $(y^5)^3$.

$$(y^5)^3 = y^{5 \cdot 3} \qquad \text{Multiply the exponents.}$$
$$= y^{15}$$

Check: $(y^5)^3 = y^5 \cdot y^5 \cdot y^5$
$$= y^{5+5+5} \text{ or } y^{15} \quad \vee$$

FYI

If you put $1000 in an account that earns 4% interest every 6 months, in ten years it will be worth $1000(1.04^2)^{10}$ or $2191—more than twice what you started with!

To find the volume of a cube, you multiply its length, width, and height. The volume of the cube shown at the right is $x \cdot x \cdot x$ or x^3. Suppose you double its dimensions. Then its volume is $(2x)(2x)(2x)$ or $(2x)^3$.

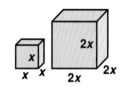

$$(2x)(2x)(2x) = (2 \cdot 2 \cdot 2)(x \cdot x \cdot x) \qquad \text{Commutative and}$$
$$= 2^3 x^3 \text{ or } 8x^3 \qquad\qquad \text{associative properties}$$

So, $(2x)^3 = 2^3 x^3$. The exponent, 3, is applied to *both* factors. $(2x)^3$ is an example of a power of a product.

These and other examples suggest that the power of a product is the product of the powers.

Power of a Product	For all numbers a and b and positive integer m, $$(ab)^m = a^m b^m.$$

Example

2 **Simplify $(3xy)^3$.**

$$(3xy)^3 = (3^1 x^1 y^1)^3$$
$$= 3^3 x^3 y^3$$
$$= 27x^3 y^3$$

The rules for the *power of a power* and the *power of a product* can be combined into one rule. This rule can be stated as follows.

Power of a Monomial	For all numbers a and b and positive integers m, n, and p, $$(a^m b^n)^p = a^{mp} b^{np}.$$

Examples

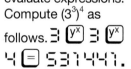

Calculator Hint

You can use $\boxed{y^x}$ to evaluate expressions. Compute $(3^3)^4$ as

follows. $3\ \boxed{y^x}\ 3\ \boxed{y^x}$

$4\ \boxed{=}\ 531441.$

3 **Simplify $(a^2 b^3)^2$.**

$$(a^2 b^3)^2 = (a^2)^2 (b^3)^2$$
$$= a^4 b^6$$

Check: $(a^2 b^3)^2 = (a^2 b^3)(a^2 b^3)$
$$= (a^2 \cdot a^2)(b^3 \cdot b^3)$$
$$= a^4 b^6 \ \checkmark$$

4 **Evaluate $(-4a^3)^2$ if $a = 2$.**

$$(-4a^3)^2 = (-4 \cdot 2^3)^2 \quad \text{Replace } a \text{ with 2.}$$
$$= (-4)^2 \cdot (2^3)^2 \quad \textbf{Check:} \quad (-4 \cdot 2^3)^2 = (-4 \cdot 8)^2$$
$$= 16 \cdot 2^6 \qquad\qquad\qquad = (-32)^2$$
$$= 16 \cdot 64 \text{ or } 1024 \qquad\qquad = 1024 \ \checkmark$$

Checking for Understanding

Communicating Algebra

1. What is the area of the square?

 3x

2. Draw and label the dimensions of a square with an area of y^8 units.

3. In your own words, explain the difference between $x^3 \cdot x^2$ and $(x^3)^2$.

4. Suppose the dimensions of a cube are tripled. What happens to its volume?

Guided Practice

Simplify.

5. $(2^3)^2$ **6.** $(a^5)^3$ **7.** $(3y)^2$ **8.** $(mn^3)^3$

9. $(x^2 y^3)^2$ **10.** $(-pq)^2$ **11.** $(n^3 r^2)^4$ **12.** $(-ab^2)^3$

Exercises

Simplify.

13. $(5^3)^3$ **14.** $[(-4)^2]^2$ **15.** $(m^2)^5$ **16.** $(-y^3)^6$

17. $(5c)^3$ **18.** $(10y)^2$ **19.** $(-3z)^3$ **20.** $(-7n)^4$

21. $(yz)^4$ **22.** $(pq)^5$ **23.** $(xy^3)^3$ **24.** $(x^3y^5)^2$

25. $(3ab^4)^3$ **26.** $(4x^2y^3)^2$ **27.** $(-2x^2)^3$ **28.** $(-3r^3)^2$

29. $3x(2x)^2$ **30.** $2b^2(-3b)^3$ **31.** $-3x(4xy)^2$ **32.** $-3b(2a)^3$

 Evaluate each expression if $a = 2$ and $b = -3$.

33. $2a^2b$ **34.** $3ab^2$ **35.** $(ab^2)^2$ **36.** $(-ab)^3$

37. $(-2b)^2$ **38.** $a(b^2)^3$ **39.** $(3b^2)^2$ **40.** $2(3a^2)^2$

Mixed Review

41. Solve $-13 + m = -5$.   (Lesson 3-1)

42. In a sample, 20 of 80 students surveyed would choose black pens. If 400 students buy pens, how many black pens should be ordered?   (Lesson 10-8)

43. Using the formula $V = \pi r^2 h$, find the volume of a cylinder that has a radius of 8 cm and a height of 15 cm.   (Lesson 13-8)

44. The length of a leg in a $45° - 45°$ right triangle is 7 inches. Find the length of the hypotenuse.   (Lesson 14-7)

45. Find the difference $(6x^2 + 2) - (3x^2 - 8)$.   (Lesson 15-3)

Applications

46. Chemistry   Some chemicals and metals such as salt and magnesium can be extracted from sea water. If a manufacturer wanted to extract salt from a cubic mile of sea water, about how many cubic feet of sea water would this be? Make an estimate by rounding 5280 feet to 5000 feet, or 5×10^3.

47. Biology   It takes 4 hours for a culture with 1 bacterium to split into two, or double to two bacteria. To grow to 1024 bacteria, the culture must double ten times, or 2^{10}, which requires 40 hours. If the culture doubles another 10 times, there will be $(2^{10})^2$ bacteria.

  a. Simplify $(2^{10})^2$.

  b. If $2^{10} = 1024$, then we could say 2^{10} is close to 10^3. Using this information, estimate $(2^{10})^2$.

Connection

48. Geometry   Find the volume of the rectangular prism at the right.

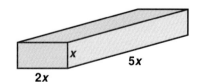

Critical Thinking

49. Are $(2^3)^4$ and $(4^3)^2$ equal? Explain why or why not. Hint: Write both expressions as a power of 2.

Wrap-Up

50. Using the information at the top of page 574, find the amount in Amber's savings account.

Multiplying Polynomials

Materials: tiles

In this Exploration, you will use tiles as a model for multiplying a polynomial by a monomial.

▶ Consider a rectangle with a width of x and a length of $x + 2$. What would this rectangle look like? Use the edges of an x-tile to mark off the dimensions of the rectangle. Then complete the rectangle by filling it in with tiles.

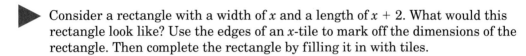
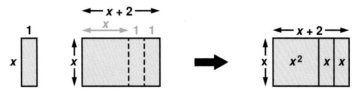

▶ Now, find the area of the large rectangle. To find the area, add the areas of all the individual tiles that make up the large rectangle. There is one x^2-tile and two x-tiles. The area is $x^2 + x + x$ or $x^2 + 2x$.

Another way to find the area of the large rectangle is to multiply the length by the width. This would be $x(x + 2)$.

▶ Since both $x^2 + 2x$ and $x(x + 2)$ represent the area of the same rectangle, you can say: $x(x + 2) = x^2 + 2x$.

Your Turn: Draw a model of $2x(x + 2) = 2x^2 + 4x$.

Tell whether each statement is *true* or *false*. Justify your answer with tiles.

1. $x(2x + 3) = 2x^2 + 3x$ **2.** $2x(3x + 4) = 6x^2 + 4x$

Multiply each expression using tiles.

3. $x(x + 5)$ **4.** $2x(x + 2)$ **5.** $3x(2x + 1)$

Analysis

6. Suppose you have a square garden plot that measures x feet on a side. If you double the length of the plot and increase the width by 3 feet, how large will the new plot be? Write two expressions for the area of the new plot.

7. If the original plot was 10 feet on a side, what is the area of the new plot?

15-5 Multiplying a Polynomial by a Monomial

Objective:
Multiply a polynomial by a monomial.

You can use tiles as a model for multiplying a polynomial and a monomial.

Would you get the same answer if you drew the rectangle with 2x at the top and x + 3 on the side?

The figure at the right is a rectangle whose width is $2x$ and whose length is $x + 3$.

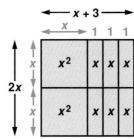

The area of the rectangle is the product of the length and the width. The expression $2x(x + 3)$ represents the area of the rectangle.

The area of the rectangle is also the sum of the areas of the tiles. The area is $2x^2 + 6x$.

Therefore, $2x(x + 3) = 2x^2 + 6x$.

You can also use the distributive property to multiply a polynomial and a monomial.

Examples

1 Find $6(3x - 2)$.

$$6(3x - 2) = 6(3x) - 6(2) \qquad \text{Use the distributive property.}$$
$$= 18x - 12$$

2 Find $-5y(2y^2 - 4)$.

$$-5y(2y^2 - 4) = -5y(2y^2) - (-5y)(4)$$
$$= -10y^3 - (-20y)$$
$$= -10y^3 + 20y$$

3 Find $2rs(-6r^2 + r - 7)$.

$$2rs(-6r^2 + r - 7) = 2rs(-6r^2) + 2rs(r) - 2rs(7)$$
$$= -12r^3s + 2r^2s - 14rs$$

Checking for Understanding

Communicating Algebra

1. Use the rectangle at the right to determine the product of x and $2x + 2$.

2. Draw a rectangle and fill it in with tiles to represent $x(2x + 1)$.

3. In your own words, explain how you would use the distributive property to multiply $4k(5k - 3)$.

Find each product using tiles or drawings.

4. $2x(2x + 1)$ **5.** $x(x + 3)$ **6.** $3x(x + 1)$ **7.** $x(3x + 2)$

Exercises

Find each product using the distributive property.

8. $5(2x + 7)$ **9.** $3m(m + 9)$ **10.** $8b(b - 7)$

11. $x(-3x + 2)$ **12.** $2x(4x^2 + 3x)$ **13.** $3a(a^3 - 2a^2)$

14. $8m(m^3 - 2m)$ **15.** $-3n(20n^4 - 6n^2)$ **16.** $-7x(-3x^2 + 6x)$

17. $7(a^2 - 2a + 3)$ **18.** $10(x^2 + 4x - 7)$ **19.** $-9(2k^3 - 3k - 6)$

20. $11(-4d^2 + 6d + 12)$ **21.** $x(7x^3 + 21x - 13)$ **22.** $a^2(-4a^4 + 3a^2 - 9)$

Mixed Review

23. Probability Using a tree diagram, find the number of possible outcomes: *a toss of a coin and a roll of a die.* (Lesson 11-1)

24. Carpentry Lauren and Miguel both measure a sheet of plywood. Lauren says the length is 3.2 m. Miguel says the length is 322 cm. Which measurement is more precise? Why? (Lesson 13-10)

25. Use a calculator to find cos 23°. Round to the nearest ten thousandth. (Lesson 14-9)

26. Simplify $(-3y^2)^2(2y^3)$. (Lesson 15-4)

Connections

27. Geometry The area of the rectangle at the right is $x(x + 7)$.

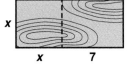

  a. What is the area of each small rectangle?

  b. If the areas of the two small rectangles are added, does this sum equal the area of the rectangle as a whole, $x(x + 7)$?

  c. What property does this illustrate?

28. Geometry For the rectangular prism at the right, compute the area of the top, bottom, front, back, left and right sides. Then add to find the total surface area.

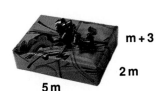

Critical Thinking

29. Find the product of $(x + 1)$ and $(x + 2)$ using tiles. Hint: You will need to use tiles having area measures of x^2, x, and 1.

Wrap-Up

30. Make Up a Problem Write a multiplication problem similar to those in Exercises 4-7. Find the product using tiles and using the distributive property.

Algebra in Action-Genetics

Punnett Squares

Punnett squares are used to show possible ways that genes can combine at fertilization. In a Punnett square, *dominant* genes are shown with capital letters. *Recessive* genes are shown with the lowercase of the same letter. Letters representing the parent's genes are placed on the outer sides of the Punnett square. Letters inside the boxes of the square show the possible gene combinations for their offspring.

	T hybrid tall **t**	
T hybrid tall	**TT**	**Tt**
t	**Tt**	**tt**

The Punnett square at the right represents a cross between two hybrid tall pea plants. A hybrid trait is the result of a combination of a dominant and a recessive gene. The plants are tall because the dominant trait masks the recessive trait.

Let T represent the dominant gene for tallness.
Let t represent the recessive gene for shortness.

The gene combinations for the offspring are as follows:
- 25% of the offspring are pure tall, TT.
- 50% of the offspring are hybrid tall, Tt.
- 25% of the offspring are pure short, tt.

Notice that the Punnett square is similar to an area model.

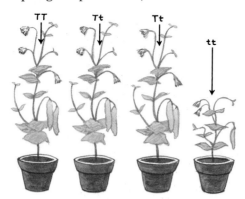

In humans, free earlobes is a dominant trait over attached earlobes. Let E represent free earlobes and e represent attached earlobes. Draw a Punnett square for each parent combination. Predict the gene combinations for the offspring.

1. EE, Ee   2. EE, EE   3. ee, ee   4. Ee, ee

15-6　Multiplying Binomials

Objective:
Multiply binomials.

You can use tiles as a model for multiplying two binomials.

Consider the binomials $x + 3$ and $2x + 1$. To multiply these binomials, mark off a rectangle that has dimensions $x + 3$ and $2x + 1$.

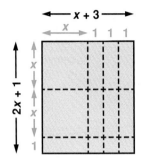

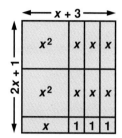

Next, fill in the rectangle with tiles.

How many terms are in $2x^2 + 7x^2 + 3$? What kind of polynomial is this?

The product $(x + 3)(2x + 1)$ represents the area of the rectangle. The area is also the sum of the areas of the tiles, $2x^2 + 7x + 3$. Therefore, $(x + 3)(2x + 1) = 2x^2 + 7x + 3$.

Checking for Understanding

Communicating Algebra

1. How is multiplying two binomials the same as multiplying a binomial by a monomial? How is it different?

2. Draw a rectangle and fill it in with tiles to represent $(x + 3)(3x + 1)$.

Guided Practice

For each model, name the two binomials being multiplied and give their product.

3.

4.

5.

6.

Exercises

Independent Practice

Find each product using tiles or drawings.

7. $(x + 2)(x + 2)$

8. $(x + 1)(x + 5)$

9. $(2x + 1)(x + 3)$

10. $(3x + 1)(x + 2)$

11. $(x + 4)(2x + 1)$

12. $(2x + 2)(2x + 1)$

13. $(2x + 1)(3x + 2)$

14. $(3x + 3)(x + 1)$

15. $(x + 3)(3x + 2)$

Mixed Review

16. If $\angle A$ and $\angle B$ are complementary and the measure of $\angle B$ is 64°, find the measure of $\angle A$.   (Lesson 12-3)

17. Find a whole number estimate for $\sqrt{97}$.   (Lesson 14-2)

18. Find the product $3r^2(r^3 - 5r^2 + 2r)$.   (Lesson 15-5)

Application

19. **Home Economics**   Mrs. Baxter makes baby quilts and full-sized quilts. The baby quilts are x feet by x feet. The full-sized quilts are 3 feet wider and 4 feet longer than the baby quilts. Draw an x^2 tile to represent the baby quilt and then draw additional tiles to increase its size to a full-sized quilt. How has the area changed? Write a binomial multiplication problem to describe the area of the full-sized quilt.

Critical Thinking

20. For each rectangle below, name the two binomials being multiplied. Then, name the product.

    a.

x^2	x^2	x
x	x	1
x	x	1

    b.

x^2	x	x
x^2	x	x
x	1	1

    What do you notice about the products for **a** and for **b**? What property does this illustrate?

Wrap-Up

21. Find the product of $(2x + 5)(x + 1)$ using tiles.

Team Problem Solving

It is true that $50\frac{1}{2} + 49\frac{38}{76} = 100$.

Can you find two other addition expressions that use each of the digits 0 to 9 exactly once and equal 100? Hint: There can be more than two addends.

Factoring

Materials: tiles

In this Exploration, you will examine how to factor polynomials from a geometric perspective.

▶ Consider the rectangle with a length of $x + 2$ and a width of $x + 1$. From the previous lesson, you know that $(x + 2)(x + 1) = x^2 + 3x + 2$.

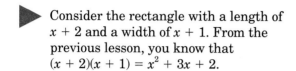

The binomials $(x + 2)$ and $(x + 1)$ are **factors** of the trinomial $x^2 + 3x + 2$.

▶ Now consider the polynomial $x^2 + 7x + 6$. Is this polynomial factorable?

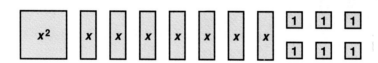

This polynomial is factorable if the tiles can be arranged into a rectangle.

The rectangle at the right shows that $x^2 + 7x + 6$ is factorable. The factors are $(x + 1)$ and $(x + 6)$.

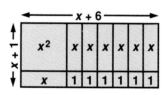

Your Turn: **Show that $x^2 + 6x + 8$ is factorable by forming a rectangle with tiles.**

Build a rectangle with the appropriate tiles or by using a drawing. Tell whether each polynomial is factorable.

1. $x^2 + 5x + 6$ **2.** $x^2 + 7x + 5$ **3.** $3x^2 + 8x + 5$

Analysis

4. Consider the multiplication $x(x + 1)(x + 2)$. Can this multiplication be represented geometrically? If so, how? Make a drawing to explain your answer.

15-7 Strategy: Eliminate Possibilities

Objective:
Solve problems by
eliminating possibilities.

Albert, Keith, Bernardo, and Ernie are friends. Each of them is on one of the following school teams: football, soccer, cross country, or golf. Use the following information to determine who is in each sport.

Albert is shorter than the boy who plays soccer.
Bernardo only likes to play games with round balls.
Ernie has a problem with his knee and cannot run.
Keith practices kicking a ball as part of his training.

Explore There are 4 boys and 4 sports. You must match each boy with the sport he plays by using the information from the 4 statements above.

Plan Make a chart to organize the information. Through the process of elimination, each boy can be matched with his sport.

Solve Put an x to show that Albert does not play soccer. Put 2 x's to show that Bernardo does not play football or run cross country. Put 3 x's to show that Ernie does not play football or soccer, or run cross country. Circle the box that shows that Ernie plays golf. Since only one student plays golf, put x's in the rest of the boxes in that row. Now, by the process of elimination, you can see that Bernardo plays soccer. Place an x to show that Keith does not run cross country, which means he plays football. Albert's sport, then, is cross country.

*How would the problem
be changed if the last
statement did not exist?*

	Albert	**Keith**	**Bernardo**	**Ernie**
football	X	◯	X	X
soccer	X	X	◯	X
cross country	◯	X	X	X
golf	X	X	X	◯

Albert runs cross country, Keith plays football, Bernardo plays soccer, and Ernie plays golf.

Examine Check the result against the statements. The first statement says that Albert does not play soccer, and the answer says that Albert runs cross country. There is no conflict here. Using the same method for each sentence, you can see that there are no conflicts.

Checking for Understanding

Guided Practice

1. Solve by eliminating possibilities:
   Darla, Amanda, Norma, and Tara are friends and each has one of the following pets: dog, cat, parrot, or gerbil. If each girl has a different pet, use the following information to match each girl with her pet.

   Norma likes to visit the girl with the gerbil.
   Tara and Amanda frequently help the girl with the dog to walk her pet.
   Darla cannot have a dog or a cat because she is allergic to them.
   Tara plans to teach her pet how to talk.

2. In problem 1, which pet did you match first?

3. How did the first match help to solve the rest of problem 1?

Exercises

Independent Practice

Solve. Use any strategy.

4. Cecilia, Greta, Lucy, Dustin, Pedro, Reuben, and Grant have started a band to make some extra money on weekends. The band has a lead guitar player, a rhythm guitar player, a bass guitar player, a keyboard player, a drummer, a lead singer, and a backup singer. If each person has one responsibility in the band, use the following information to match each person with his or her responsibility.

   Cecilia, Lucy, and the bass guitar player like pizza with anchovies.
   Reuben, Greta, and the lead singer like pizza with mushrooms.
   Lucy and the keyboard player are cousins.
   Pedro and Lucy do not have good singing voices.
   Grant is taller than the drummer and the bass guitar player.
   Greta does not play any guitar.
   Cecilia, Grant, and the lead singer live on the same street.
   Reuben wants to learn how to play the drums.
   The keyboard player and the drummer baked a cake to surprise Greta.
   Reuben's best friend plays the bass guitar.
   Cecilia and the drummer like to play tennis in their spare time.
   Grant and the lead guitar player are both freshmen.
   Grant, Cecilia, and the keyboard player are studying chemistry in school.

Connection

5. **Statistics** Jerry's Music Store took a survey of 100 people. Sixty-three people said they like rock, 49 said they like rap, and 24 said they like jazz. Twenty-five people like rock and rap, 5 like jazz and rap, and 14 like rock and jazz. If 3 people like all 3 types of music, how many people do not like rock, rap, or jazz?

Wrap-Up

6. Explain how to solve problems by eliminating possibilities.

Language and Concepts

Choose the correct term to complete each sentence.

1. The degree of a monomial is the (sum, product) of the exponents of its variables.

2. A (binomial, trinomial) is the sum or difference of three monomials.

3. The degree of a polynomial is the same as that of the term that has the (least, greatest) degree.

4. To subtract two polynomials, you add the (additive inverse, multiplicative inverse) of the second polynomial.

5. The (commutative property, distributive property) is used in multiplying a polynomial by a monomial.

Skills

State whether each expression is a polynomial. If it is, identify it as a _monomial_, _binomial_, or _trinomial_. (Lesson 15-1)

6. $3x^4 - x$

7. $\dfrac{4}{ax}$

8. ax^2

9. $9b^2 + b - 1$

10. $8x - \dfrac{5}{3}$

11. $k^3 - \dfrac{2}{k}$

Find the degree of each polynomial. (Lesson 15-1)

12. $4x$

13. $5a^2b$

14. $3x + y^2$

15. $19m^2n^3 - 14mn^4$

16. $x^2 - 6xy + xy^2$

17. $12rs^2 + 3r^2s + 5r^4s$

Find each sum. (Lesson 15-2)

18. $(2x^2 - 5x) + (3x^2 + x)$

19. $(a^2 - 6ab) + (3a^2 + ab)$

20. $(x^2 - 5x + 3) + (4x - 3)$

21. $(-3y^2 + 2) + (4y^2 - 5y - 2)$

Find each difference. (Lesson 15-3)

22. $(7a - 11b) - (3a + 4b)$

23. $(6y - 8z) - (6y + 4z)$

24. $(3a^2 - b^2 + c^2) - (a^2 + 2b^2)$

25. $(14a^2 - 3a) - (6a^2 + 5a + 17)$

Simplify. (Lesson 15-4)

26. $(a^2)^3$

27. $(-2x)^3$

28. $(p^2q)^3$

29. $-5c(2cd)^3$

30. $4y(y^2z)^3$

31. $6a(-ab)^7$

Find each product. (Lesson 15-5)

32. $4d(2d - 5)$

33. $x(-5x + 3)$

34. $a^2(2a^3 + a - 5)$

35. $3y(-y^2 - 8y + 4)$

36. $-2g(g^3 + 6g + 3)$

37. $-3az(2z^2 + 4az + a^2)$

Find each product using tiles. (Lesson 15-6)

38. $(x + 3)(x + 1)$

39. $(2x + 1)(x + 1)$

40. $(3x + 2)(2x + 2)$

Applications and Problem Solving

41. Solve by eliminating possibilities:

Lindsay, Lee, Anna, and Marcos formed a study group. Each one has a favorite subject that is different from the others. The subjects are art, math, music, and physics. Use the following information to match each person with his or her favorite subject.

Lindsay likes subjects where she can use her calculator.
Lee does not like music or physics.
Anna and Marcos prefer classes in the cultural arts.
Marcos plans to be a professional cartoonist.

ortfolio Suggestion

Review the items in your portfolio. Make a list of the items, noting why each item was chosen. Replace any items that are no longer appropriate.

Curriculum Connection

- **Science** Research the scientific names used to classify a lion and a tiger. How are these names similar? How are these names different?

- **Science** Find an example of a polynomial in your science book.

Read More About It

Livingston, M.C., ed. *Poems of Lewis Carroll.*

Selby, Peter H. *Practical Algebra: A Self-Teaching Guide.*

Wyler, Rose and Gerald Ames. *It's All Done with Numbers.*

Test

State whether each expression is a polynomial. If it is, identify it as a *monomial*, *binomial*, or *trinomial*.

1. $-\dfrac{3}{y}$

2. $4x^2 + 5x + 1$

3. x^2y

4. $rs^3 - \dfrac{3}{r}$

5. $\dfrac{3a^5b^4}{7}$

6. $1 - 4mt + 2m^2t^3$

Find the degree of each polynomial.

7. $4x^2 + 3$

8. $-9ab^3$

9. $12ct^2 + 9c^3t^2$

10. $\dfrac{r^2s^4}{3}$

Find each sum or difference.

11. $(-4a^2 + 3) - (3a^2 - 7)$

12. $(7x - 6y) + (-4x + y)$

13. $(10b^2 - 7bc) - (5b^2 - 4bc)$

14. $(-5n^2 - 6n) - (-5n^2 + 4n + 2)$

15. $\begin{array}{r} 6x^2 - 5x + 4 \\ + (-2x^2) - 4x - 8 \\ \hline \end{array}$

16. $\begin{array}{r} 9w^2 + 5w - 8 \\ - 6w^2 + 8w - 9 \\ \hline \end{array}$

Simplify.

17. $(ab^3)^3$

18. $2(-3x^2)^3$

19. $(3x^2y^3)^2$

Find each product.

20. $2x(3x^2 + 1)$

21. $5a^3(-3a^2 - 8a + 2)$

22. $-4rs(1 - 3r + r^3s^2)$

Find each product using tiles or drawings.

23. $(2x + 1)(x + 2)$

24. $(x + 1)(3x + 4)$

Solve.

25. If the length of a window is $7x + 4$ inches and the width is $3x - 11$ inches, what is the size of the window in united inches? (Remember: United inches equals the length of the window plus the width.)

BONUS

Factor the polynomial $2x^2 + 9x + 4$.

1. A clock has a diameter of 14 inches. Which expression shows the length of trim needed to circle the edge of the clock? (Use $\pi \approx \frac{22}{7}$.)

   A $\frac{22}{7} \times 7^2$ **C** $\frac{22}{7} \times 28$

   B $\frac{22}{7} \times 14$ **D** Not Here

2. Which is the solution of the system of equations $x + y = 4$ and $y = -3x$?

   A $(-6, 2)$
   B $(1, 3)$
   C $(-2, -6)$
   D $(3, 1)$

3. The price of a CD player Lois wants to buy is listed at $139.99. If the sales tax rate is 6%, a reasonable estimate for the total cost is—

   A $0.80   **C** $140
   B $8.00   **D** $150

4. Julio is 6 feet tall. At 10:00 he casts a shadow 4 feet long. At the same, the flagpole casts a shadow 42 feet long. How tall is the flagpole?

   A 63 feet   **C** 28 feet
   B 44 feet   **D** Not Here

5. What is the area of the triangle?

   A 480 cm^2
   B 360 cm^2
   C 240 cm^2
   D 74 cm^2

   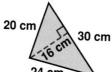

   20 cm   16 cm   30 cm   24 cm

6. A cylindrical can is 12 cm tall and the ends each have a diameter of 10 cm. Which expression can be used to find the area the can label covers?

   A $3.14 \times 10 \times 12$
   B $3.14 \times 5 \times 5 \times 12$
   C $2(3.14 \times 5^2) + 3.14 \times 10 \times 12$
   D $3.14 \times 5^2 + 2(3.14 \times 10 \times 12)$

7. Jane's kite is flying directly over a spot 120 feet from where she is standing. The kite string is 130 feet long. How high is the kite?

   A 500 feet
   B 177 feet
   C 125 feet
   D 50 feet

   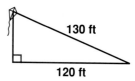

   130 ft   120 ft

8. In a right triangle, the tangent ratio is the measure of the leg opposite an angle to the measure of the leg adjacent to the angle. What is the tangent ratio of angle D?

   A $\frac{3}{5}$

   B $\frac{4}{5}$

   C $\frac{4}{3}$

   D $\frac{5}{3}$

   D 15 in. E 9 in.   12 in. F

9. Which is equivalent to $(7x^2 + 3y) - (3x^2 + 5y)$?

   A $4x^2 - 2y$ **C** $4x^4 - 2y^2$
   B $4x^2 + 8y$ **D** $4x^4 + 8y^2$

10. Which is equivalent to $(-2x)^3$?

    A $-8x^3$ **C** $-2x^3$
    B $8x^3$ **D** $2x^3$

Mathematical Symbols

÷   divide
·   times
=   is equal to
−   negative
+   positive
±   positive or negative
≠   is not equal to
>   is greater than
<   is less than
≯   is not greater than
≮   is not less than
≥   is greater than or equal to
≤   is less than or equal to
°   degrees

⊥   is perpendicular to
∥   is parallel to
≈   is approximately equal to
≅   is congruent to
~   is similar to
√   nonnegative square root
%   percent
π   pi
$\overleftrightarrow{AB}$ line AB
$\overline{AB}$ line segment AB
$\overrightarrow{AB}$ ray AB
$\triangle ABC$ triangle ABC
$\angle ABC$ angle ABC
$m \overline{AB}$ measure of line segment AB
$m\angle ABC$ measure of angle ABC

Metric System

Prefixes

kilo-	hecto-	deka-	no prefix	deci-	centi-	milli-
thousands	hundreds	tens	ones	tenths	hundredths	thousandths

Length
1 centimeter (cm) = 10 millimeters (mm)
1 meter (m) = 100 centimeters
1 meter = 1000 millimeters
1 kilometer (km) = 1000 meters

Area
1 square centimeter (cm^2) = 100 square millimeters (mm^2)
1 square meter (m^2) = 10,000 square centimeters

Volume
1 cubic centimeter (cm^3) = 1000 cubic millimeters (mm^3)

Capacity
1 liter (L) = 1000 milliliters (mL)

Mass
1 gram (g) = 1000 milligrams (mg)
1 kilogram (kg) = 1000 grams

Time
1 minute (min) = 60 second (s)
1 hour (h) = 60 minutes
1 day (d) = 24 hours
1 year = 365 days

Formulas

$C = \pi d$ or $2\pi r$ circumference of a circle

$A = \ell w$ area of a rectangle

$A = bh$ area of a parallelogram

$A = \frac{1}{2}bh$ area of a triangle

$A = \pi \cdot r \cdot r$ or πr^2 area of a circle

$I = p \cdot r \cdot t$ interest

$d = r \cdot t$ distance

$A = \frac{1}{2}h(a + b)$ area of a trapezoid

$V = Bh$ volume of a prism or cylinder

$V = \frac{1}{3}Bh$ volume of a pyramid or cone

$c^2 = a^2 + b^2$ Pythagorean Theorem

$C = \frac{5}{9}(F - 32)$ Fahrenheit to Celsius

$F = \frac{9}{5}C + 32$ Celsius to Fahrenheit

Trigonometric Ratios

Angle	sin	cos	tan	Angle	sin	cos	tan
0°	0.0000	1.0000	0.0000	45°	0.7071	0.7071	1.0000
1°	0.0175	0.9998	0.0175	46°	0.7193	0.6947	1.0355
2°	0.0349	0.9994	0.0349	47°	0.7314	0.6820	1.0724
3°	0.0523	0.9986	0.0524	48°	0.7431	0.6691	1.1106
4°	0.0698	0.9976	0.0699	49°	0.7547	0.6561	1.1504
5°	0.0872	0.9962	0.0875	50°	0.7660	0.6428	1.1918
6°	0.1045	0.9945	0.1051	51°	0.7771	0.6293	1.2349
7°	0.1219	0.9925	0.1228	52°	0.7880	0.6157	1.2799
8°	0.1392	0.9903	0.1405	53°	0.7986	0.6018	1.3270
9°	0.1564	0.9877	0.1584	54°	0.8090	0.5878	1.3764
10°	0.1736	0.9848	0.1763	55°	0.8192	0.5736	1.4281
11°	0.1908	0.9816	0.1944	56°	0.8290	0.5592	1.4826
12°	0.2079	0.9781	0.2126	57°	0.8387	0.5446	1.5399
13°	0.2250	0.9744	0.2309	58°	0.8480	0.5299	1.6003
14°	0.2419	0.9703	0.2493	59°	0.8572	0.5150	1.6643
15°	0.2588	0.9659	0.2679	60°	0.8660	0.5000	1.7321
16°	0.2756	0.9613	0.2867	61°	0.8746	0.4848	1.8040
17°	0.2924	0.9563	0.3057	62°	0.8829	0.4695	1.8807
18°	0.3090	0.9511	0.3249	63°	0.8910	0.4540	1.9626
19°	0.3256	0.9455	0.3443	64°	0.8988	0.4384	2.0503
20°	0.3420	0.9397	0.3640	65°	0.9063	0.4226	2.1445
21°	0.3584	0.9336	0.3839	66°	0.9135	0.4067	2.2460
22°	0.3746	0.9272	0.4040	67°	0.9205	0.3907	2.3559
23°	0.3907	0.9205	0.4245	68°	0.9272	0.3746	2.4751
24°	0.4067	0.9135	0.4452	69°	0.9336	0.3584	2.6051
25°	0.4226	0.9063	0.4663	70°	0.9397	0.3420	2.7475
26°	0.4384	0.8988	0.4877	71°	0.9455	0.3256	2.9042
27°	0.4540	0.8910	0.5095	72°	0.9511	0.3090	3.0777
28°	0.4695	0.8829	0.5317	73°	0.9563	0.2924	3.2709
29°	0.4848	0.8746	0.5543	74°	0.9613	0.2756	3.4874
30°	0.5000	0.8660	0.5774	75°	0.9659	0.2588	3.7321
31°	0.5150	0.8572	0.6009	76°	0.9703	0.2419	4.0108
32°	0.5299	0.8480	0.6249	77°	0.9744	0.2250	4.3315
33°	0.5446	0.8387	0.6494	78°	0.9781	0.2079	4.7046
34°	0.5592	0.8290	0.6745	79°	0.9816	0.1908	5.1446
35°	0.5736	0.8192	0.7002	80°	0.9848	0.1736	5.6713
36°	0.5878	0.8090	0.7265	81°	0.9877	0.1564	6.3138
37°	0.6018	0.7986	0.7536	82°	0.9903	0.1392	7.1154
38°	0.6157	0.7880	0.7813	83°	0.9925	0.1219	8.1443
39°	0.6293	0.7771	0.8098	84°	0.9945	0.1045	9.5144
40°	0.6428	0.7660	0.8391	85°	0.9962	0.0872	11.4301
41°	0.6561	0.7547	0.8693	86°	0.9976	0.0698	14.3007
42°	0.6691	0.7431	0.9004	87°	0.9986	0.0523	19.0811
43°	0.6820	0.7314	0.9325	88°	0.9994	0.0349	28.6363
44°	0.6947	0.7193	0.9657	89°	0.9998	0.0175	57.2900
45°	0.7071	0.7071	1.0000	90°	1.0000	0.0000	∞

Extra Practice

Example

Evaluate $3x - (5y + c)$ if $x = 8$, $y = 2$, and $c = 6$.

$$
\begin{aligned}
3x - (5y + c) &= 3 \cdot 8 - (5 \cdot 2 + 6) \\
&= 3 \cdot 8 - (10 + 6) && \text{Multiply inside parentheses first.} \\
&= 24 - 16 && \text{Substitute 24 for } 3 \cdot 8 \text{ and 16 for } 10 + 6. \\
&= 8 && \text{Substitute 8 for } 24 - 16.
\end{aligned}
$$

Evaluate each expression.

1. $7 + 8 + 12 \div 4$
2. $12 + 20 \div 4 - 5$
3. $36 \div 9 + 7 - 6$
4. $(25 \cdot 3) + (10 \cdot 3)$
5. $(40 \cdot 2) - (6 \cdot 11)$
6. $40 \cdot (6 - 2)$
7. $\dfrac{96 - 11}{11 + 6}$
8. $\dfrac{84 + 12}{13 + 11}$
9. $\dfrac{3 \cdot 3 + 3}{3 \cdot 3 - 3}$

Evaluate each expression if $a = 4$, $b = 2$, and $c = 3$.

10. $ab - bc$
11. $4a + b \cdot b$
12. $9 \cdot c - ab$
13. $4a - (b + c)$
14. $4(a + b) - c$
15. $6a + 6b$
16. $5a + 6 - 8c$
17. $36 - 12c$
18. $7a - (2b + c)$
19. $\dfrac{6(a + b)}{3c}$
20. $\dfrac{4ac}{b}$
21. $\dfrac{3(a + b)}{c - 1}$

Example

Simplify $p + 2(r + 8p)$.

$$
\begin{aligned}
p + 2(r + 8p) &= p + 2r + 2 \cdot 8p && \text{Distributive property} \\
&= p + 2r + 16p && \text{Substitution property of equality} \\
&= p + 16p + 2r && \text{Commutative property of addition} \\
&= (1 + 16)p + 2r && \text{Distributive property} \\
&= 17p + 2r && \text{Substitution property of equality}
\end{aligned}
$$

Simplify each expression.

1. $3a + 7a$
2. $9c + 7c$
3. $13a + 5a$
4. $21c + 10c$
5. $30x + 20x$
6. $24y + 16y$
7. $10a + 30a - 7d$
8. $16b + 17b + b$
9. $5a + 10b + 7a$
10. $5(x + y) + 3y$
11. $9(a + 2) + 14a$
12. $3a + 4y + 8a + 5f$
13. $3(b + 8) + 9b$
14. $2(3 + x) + 5(3 + 2y)$
15. $5(r + s) + 4(2r + 3s)$

SET 1C *(Lessons 1-5, 1-7, 1-10)* Solve equations and inequalities.

Example

Using inverse operations, solve $48 = 6x$.

$$48 = 6x$$
$$48 \div 6 = x \qquad \text{Write the related division sentence.}$$
$$8 = x \qquad \text{Substitute 8 for } 48 \div 6.$$

Solve each equation from the given set of numbers.

1. $y - 4 = 7$; {10, 11, 12}

2. $x + 34 = 56$; {12, 22, 32}

3. $20 \div a = 5$; {4, 6, 8}

4. $39 = 3b$; {11, 12, 13}

Solve each equation by using inverse operations.

5. $4 + x = 16$

6. $37 = z + 22$

7. $8 = 12 - a$

8. $r - 19 = 3$

9. $4x = 24$

10. $48b = 192$

11. $99 = 3x$

12. $p \div 7 = 4$

13. $12 = \frac{x}{3}$

State whether each inequality is true for the given value.

14. $2r - 12 \le 5$; $r = 11$

15. $26 > 3n + 2$; $n = 3$

16. $6s - 18 > 0$; $s = 3$

17. $7 + 4b < 25$; $b = 4$

SET 1D *(Lessons 1-8, 1-9)* Translate verbal phrases into algebraic expressions.

Example

Write an algebraic expression for *four years older than Bart*.

Let y represent the number of years.

The words *years older* suggest addition.

The expression is $y + 4$.

Translate each phrase into an algebraic expression.

1. the sum of x and 5

2. y multiplied by 12

3. 58 less than s

4. twice a certain number

5. quotient of 63 and y

6. the difference of x and 6

7. 3 less pieces of pizza than Sandra

8. twice the distance that Pat ran

9. four more points than Ed

10. 5 times the cost of a pen

Translate each algebraic expression into a verbal phrase.

11. $x - 12$

12. $9s$

13. $5y - 2$

14. $4(m + 3)$

15. $\frac{a - 5}{3}$

16. $7 + \frac{c}{2}$

17. $9 + 2k$

18. $\frac{6}{5g}$

SET 2A *(Lesson 2-1)*

Example

Graph {-3, -1, 0, 4} on a number line.

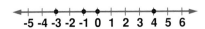

Name the coordinates of the points that are graphed.

1.

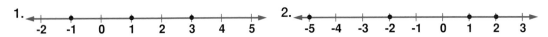

Graph each set of numbers on a number line.

3. {0, 4, 5}   **4.** {1, 7, 9}   **5.** {-1, 2, 4}

6. {-4, -3, -1}   **7.** {-3, -1, 1, 3}   **8.** {-2, 0, 4, 5}

9. {-1, -2, -6}   **10.** {5, -1, 3}   **11.** {-4, 0, -5}

SET 2B *(Lesson 2-2)* Compare and order integers.

Example

Use the integers graphed on the number line to write two inequalities.

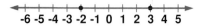

Since -2 is to the left of 3, write -2 < 3.

Since 3 is to the right of -2, write 3 > -2.

Use the integers graphed on each number line to write two inequalities.

1.

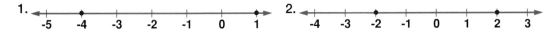

Replace each ■ with <, >, or = .

3. 12 ■ 4   **4.** -5 ■ 3   **5.** -9 ■ -7   **6.** |-5| ■ 2

Order the numbers in each set from least to greatest.

7. {8, -2, 0}   **8.** {3, -4, -1}   **9.** {-2, -9, 5, 2}

SET 2C *(Lessons 2-3, 2-4, 2-5, 2-7, 2-8)* Compute with integers.

Example

Simplify the expression $32ab - (-11ab)$.

$$32ab - (-11ab) = 32ab + 11ab \qquad \text{Subtract -11}ab \text{ by adding its additive inverse.}$$
$$= (32 + 11)ab \qquad \text{Use the distributive property.}$$
$$= 43ab \qquad \text{Substitute 43 for 32 + 11.}$$

Solve each equation.

1. $x = 13 + (-12)$
2. $y = -2 + (-4) + 18$
3. $-3 - 12 = p$
4. $r = -8 \cdot 4$
5. $11 - 3 = t$
6. $w = (10)(-3)(6)$
7. $r = 1 + (-43) + 2$
8. $h = (-4)(-2)(0)$
9. $v = \frac{-54}{9}$
10. $(-4)(7)(3) = x$
11. $z = \frac{78}{-2}$
12. $\frac{-344}{-86} = b$

Simplify each expression.

13. $-3a + 7a$
14. $9m + (-32m)$
15. $-2x - 8x$
16. $-14 \cdot (6x)$
17. $14ab + 5ab$
18. $(5y)(13z)$
19. $(-8a)(-b)$
20. $4(-5)(-2xy)$

Evaluate each expression if $a = 2$, $b = -5$, and $c = 14$.

21. $-18 + b$
22. $a - b - c$
23. $3ab$
24. $-15b$
25. $\frac{c}{-7}$

SET 3A *(Lessons 3-1, 3-2, 3-3, 3-4)* Solve equations using the properties of equality.

Example

Solve $17 = \frac{d}{-9}$.

$$17 = \frac{d}{-9}$$
$$17 \cdot (-9) = \frac{d}{-9} \cdot (-9) \qquad \text{Multiply each side by -9.}$$
$$-153 = d$$

Solve each equation.

1. $f + 14 = 19$
2. $p + (-3) = 11$
3. $y - 28 = -23$
4. $-16 = -7 - r$
5. $-13a = -91$
6. $9b = -144$
7. $5 + x = -14$
8. $18 = -2y$
9. $15 - (-4) = d$
10. $\frac{a}{-4} = 7$
11. $-19 = \frac{f}{5}$
12. $4 = \frac{c}{-116}$
13. $-12b = 144$
14. $18 = t - (-4)$
15. $14c = -154$
16. $t - 25 = -13$
17. $\frac{x}{-6} = 15$
18. $\frac{b}{-120} = -5$

SET 3B   (Lessons 3-5, 3-6)   Apply equation-solving techniques to solve problems involving formulas.

Example

Solve $I = \frac{V}{R}$, if R = 13 ohms and I = 50 amperes.

$$I = \frac{V}{R}$$

$$50 = \frac{V}{13} \qquad \text{Replace } I \text{ with 50 and } R \text{ with 13.}$$

$$50 \times 13 = \frac{V}{13} \times 13 \qquad \text{Multiply each side by 13.}$$

$$650 = V \qquad \text{The voltage is 650 volts.}$$

Solve. Use the given formula.

1. $s = g + c$, if s = 199 and g = 174   **2.** $f = t - h$, if t = 108 and h = 12

3. $d = rt$, if d = 130 and t = 5   **4.** $s = g + c$, if s = 135 and c = 13

5. $I = \frac{V}{R}$, if V = 210 and R = 7   **6.** $I = \frac{V}{R}$, if I = 33 and R = 12

7. $d = rt$, if d = 1200 and t = 24   **8.** $f = t - h$, if f = 89 and t = 126

9. $s = g + c$, if g = 167 and c = 23   **10.** $d = rt$, if d = 275 and r = 25

SET 3C   (Lessons 3-7, 3-8)   Solve inequalities.

Example

Solve $-6d \le 42$.

$$-6d \le 42$$

$$\frac{-6d}{-6} \ge \frac{42}{-6} \qquad \text{Divide each side by } -6 \text{ and reverse the order symbol.}$$

$$d \ge -7$$

Solve each inequality.

1. $w + 8 < 16$ **2.** $r - 4 > -7$ **3.** $a + (-3) \ge 9$

4. $x - (-5) > 14$ **5.** $10 \le -5 + b$ **6.** $-6z < 18$

7. $-16 \ge -8b$ **8.** $-46 \le 23a$ **9.** $y - 11 < -3$

10. $\frac{s}{-4} < 12$ **11.** $-9 \ge \frac{k}{3}$ **12.** $\frac{r}{-13} \le -30$

13. $a + (-27) > 3$ **14.** $24 < 6d$ **15.** $47a \le 423$

16. $\frac{s}{3} < 19$ **17.** $r - (-3) > -16$ **18.** $\frac{n}{-5} \le 116$

19. $3 - c \ge 18$ **20.** $-4m < -120$ **21.** $4 + d < -8$

SET 4A *(Lessons 4-1, 4-4)* Find factors of numbers and monomials.

Example

Factor $30a^2b$ completely.

$$30a^2b = 2 \cdot 15 \cdot a^2 \cdot b$$

$$= 2 \cdot 3 \cdot 5 \cdot a \cdot a \cdot b$$

Factor each number or monomial completely.

1. 5
2. 16
3. 22
4. 35
5. 92

6. 27
7. 34
8. 42
9. 50
10. 56

11. $6cd^2$
12. $21x$
13. $19abc$
14. $14x^3y$
15. $9a^3$

16. $18r^2s^3$
17. $74m^2k$
18. 112
19. $44p^3q^3$
20. 273

SET 4B *(Lessons 4-5, 4-6, 4-7)* Use prime factorization to find the GCF and LCM.

Example

Use prime factorization to find the LCM for 14 and 20.

$$14 = 2 \cdot 7$$
$$20 = 2 \cdot 2 \cdot 5$$

The greatest power of 2 is 2^2.

The greatest power of 5 is 5.

The greatest power of 7 is 7.   The LCM is $2^2 \cdot 5 \cdot 7$ or 140.

Find the GCF for each set of numbers or expressions.

1. 12, 18
2. 14, 63
3. 112, 144
4. $15k, 35k^2$

5. $33r, 121$
6. $6ab, 12ab, 14ab$
7. $4xy, 16x$
8. $-7z, 49z^2$

9. $18ab, 9a, 45ab^2$
10. $8, -28k$
11. $-13z, 39yz, 52y$
12. $-16w, -28w^3$

Find the LCM for each set of numbers.

13. 3, 12
14. 4, 14
15. 6, 21
16. 7, 9

17. $20k, 30k^2$
18. $8cd, 36cd$
19. $6, 8, 12x$
20. $4r, 14, 35r^2$

21. $4pq, 6p^2$
22. $5c, 35$
23. $7ab, 11a$
24. $12x^3, 16xy$

SET 4C (Lesson 4-8) Use the LCM to write equivalent fractions.

Example

Which is larger, $\frac{4}{15}$ or $\frac{3}{10}$?

Write $\frac{4}{15}$ and $\frac{3}{10}$ as equivalent fractions with the same denominator.

The LCM of 15 and 10 is $2 \cdot 3 \cdot 5$ or 30.

Find equivalent fractions with the LCM as the denominator.

$\frac{4}{15} = \frac{\blacksquare}{30} \rightarrow \frac{4}{15} = \frac{8}{30}$ $\qquad$ $\frac{3}{10} = \frac{\blacksquare}{30} \rightarrow \frac{3}{10} = \frac{9}{30}$

Since $\frac{9}{30} > \frac{8}{30}$, then $\frac{3}{10} > \frac{4}{15}$.

Replace each ● with <, >, or = to make a true statement.

1. $\frac{1}{3}$ ● $\frac{3}{8}$
2. $\frac{4}{6}$ ● $\frac{9}{12}$
3. $\frac{5}{12}$ ● $\frac{2}{5}$
4. $\frac{1}{4}$ ● $\frac{2}{9}$

5. $\frac{7}{8}$ ● $\frac{10}{12}$
6. $\frac{7}{9}$ ● $\frac{3}{4}$
7. $\frac{5}{6}$ ● $\frac{15}{18}$
8. $\frac{3}{11}$ ● $\frac{2}{5}$

9. $\frac{5}{6}$ ● $\frac{20}{24}$
10. $\frac{5}{16}$ ● $\frac{1}{3}$
11. $\frac{4}{5}$ ● $\frac{9}{12}$
12. $\frac{13}{15}$ ● $\frac{15}{18}$

SET 4D (Lessons 4-2, 4-9, 4-10) Multiply and divide powers.

Example

Simplify $\frac{(-a)^5}{(-a)^2}$.

$\frac{(-a)^5}{(-a)^2} = (-a)^{5-2}$

$\qquad = (-a)^3$

$\qquad = -a^3$

Simplify. Assume no denominator equals zero.

1. $7^3 \cdot 7^2$
2. $n^5 \cdot n^2$
3. $(3x^3)(6x)$
4. $(-4x^5y)(8x^2)$

5. $(ab^3)(8a^2b^2)$
6. $x^3(xy^5)$
7. $\frac{a^7}{a^6}$
8. $\frac{c^7}{c^3}$

9. $\frac{(-4)^4}{(-4)^3}$
10. $\frac{c^{50}}{c^{47}}$
11. $\frac{m^9}{m^9}$
12. $(ab^2)(4a^2b)$

13. $\frac{(-2)^6}{(-2)^4}$
14. $\frac{n^{19}}{n^3}$
15. $(-3r^2)(-4r)$
16. $k^3(m^2k^3)$

SET 5A   *(Lessons 5-1, 5-2, 5-3, 5-4, 5-5, 5-6)*   Add and subtract rational numbers.

Example

Solve $t = 4\frac{7}{8} - 7\frac{1}{3}$.

$$t = 4\frac{7}{8} - 7\frac{1}{3}$$

$\quad = 4\frac{21}{24} - 7\frac{8}{24}$ The LCD of 8 and 3 is 24.

$\quad = 4\frac{21}{24} + \left(-7\frac{8}{24}\right)$ Subtract $7\frac{8}{24}$ by adding its inverse, $-7\frac{8}{24}$.

$\quad = 4\frac{21}{24} + \left(-6\frac{32}{24}\right)$ Rename $-7\frac{8}{24}$ as $-6\frac{32}{24}$.

$\quad = -2\frac{11}{24}$

Estimate each sum or difference to the nearest whole number.

1. $7.4 + 5.86$ **2.** $3.55 - 0.82$ **3.** $2.7 + 3.1 + 4.6$ **4.** $11.5 - 2.7$

Solve each equation. Write each solution in simplest form.

5. $\frac{7}{3} + \frac{2}{3} = a$ **6.** $\frac{5}{12} - \frac{2}{12} = x$ **7.** $h = \frac{15}{8} + \frac{-4}{8}$

8. $48.6 - 3.04 = p$ **9.** $-13.8 - 22.3 = s$ **10.** $x = -5.8 + 16.12$

11. $3\frac{3}{5} + \frac{7}{10} = s$ **12.** $b = -4\frac{1}{6} + 2\frac{1}{8}$ **13.** $t = -33.3 + 9.51$

SET 5B   *(Lesson 5-7)*   Solve equations and inequalities with rational numbers.

Example

Solve the inequality $n - 13.53 > 5.48$.

$$n - 13.53 > 5.48$$

$$n - 13.53 + 13.53 > 5.48 + 13.53 \qquad \text{Add 13.53 to each side.}$$

$$n > 19.01$$

Any number greater than 19.01 is a solution.

Solve each equation or inequality. Write the solutions in simplest form.

1. $a - 5.6 = 4.9$ **2.** $8.2 + b = 2.4$ **3.** $p + 7 = 1.1$

4. $c + 1\frac{3}{5} = 2$ **5.** $\frac{3}{8} + z = 4\frac{1}{4}$ **6.** $g - \frac{4}{9} = -\frac{1}{3}$

7. $x - 1\frac{3}{5} = -5\frac{1}{10}$ **8.** $s + \frac{1}{3} = 6\frac{1}{5}$ **9.** $x + 3.1 = 1.6$

10. $a + 4.2 > 6.9$ **11.** $1.3 + d \leq 0.5$ **12.** $4.8 - y < 9$

13. $p + \frac{3}{4} > 5\frac{1}{4}$ **14.** $s - \frac{1}{9} > -4\frac{2}{3}$ **15.** $n + 4\frac{7}{8} \geq 6\frac{1}{6}$

Example

Use the expression $a + (n - 1)d$ to find the eighth term in the arithmetic sequence 5, 9, 13, 17, ...

$$a + (n - 1)d = 5 + (8 - 1)4 \qquad \text{The first term: } a = 5$$
$$= 5 + (7)4 \qquad \text{The number of terms: } n = 8$$
$$= 5 + 28 \qquad \text{The common difference: } d = 4$$
$$= 33 \qquad \text{The eighth term is 33.}$$

Write the next three terms of each sequence.

1. 3, 6, 9, . . .

2. 1, 3, 7, 15, 31, . . .

3. 10, 5, 0, -5, . . .

4. 1.1, 2.3, 3.5, 4.7, . . .

5. 1, 4, 9, 16, . . .

6. -6, -5, -3, 0, 4, . . .

7. 2.4, 4.6, 6.8, 9.0, . . .

8. 7, 4, 1, -2, . . .

9. 109, 120, 131, 142, . . .

Find the twelfth term in each arithmetic sequence.

10. 5, 7, 9, 11, . . .

11. 0, 8, 16, 24, . . .

12. 4, 1, -2, -5, . . .

13. 199, 187, 175, 163, . . .

14. -12, -10, -8, -6, . . .

15. 48, 39, 30, 21, . . .

SET 5D   *(Lesson 5-10)*   Convert within the customary system.

Example

$$\begin{array}{r} 9 \text{ lb} \quad 3 \text{ oz} \\ -2 \text{ lb } 15 \text{ oz} \end{array}$$

$$\begin{array}{r} \overset{8}{\cancel{9}} \text{ lb} \quad \overset{19}{\cancel{3}} \text{ oz} \\ -2 \text{ lb } 15 \text{ oz} \\ \hline 6 \text{ lb} \quad 4 \text{ oz} \end{array}$$

Since 15 oz is greater than 3 oz, rename 9 lb 3 oz as 8 lb 19 oz. Then subtract.

Add or subtract.

1. $\begin{array}{r} 8 \text{ ft } 9 \text{ in.} \\ + 3 \text{ ft } 5 \text{ in.} \\ \hline \end{array}$

2. $\begin{array}{r} 11 \text{ gal } 3 \text{ qt} \\ - 10 \text{ gal } 1 \text{ qt} \\ \hline \end{array}$

3. $\begin{array}{r} 4 \text{ lb } 9 \text{ oz} \\ + 5 \text{ lb } 8 \text{ oz} \\ \hline \end{array}$

4. $\begin{array}{r} 8 \text{ ft } 2 \text{ in.} \\ - 4 \text{ ft } 7 \text{ in.} \\ \hline \end{array}$

5. $\begin{array}{r} 2 \text{ gal } 2 \text{ qt} \\ + 9 \text{ gal } 2 \text{ qt} \\ \hline \end{array}$

6. $\begin{array}{r} 12 \text{ lb} \\ - 1 \text{ lb } 7 \text{ oz} \\ \hline \end{array}$

7. $\begin{array}{r} 1 \text{ yd } 2 \text{ ft} \\ + 8 \text{ yd } 1 \text{ ft} \\ \hline \end{array}$

8. $\begin{array}{r} 10 \text{ ft} \quad 1 \text{ in.} \\ - 5 \text{ ft } 10 \text{ in.} \\ \hline \end{array}$

SET 6A   *(Lessons 6-1, 6-11, 6-12)*   Compare and order rational numbers expressed as fractions and decimals.

Example

Replace the ● with <, >, or = to make a true sentence.

$$-\frac{5}{8} \; ● \; -0.7$$

$-0.625 \; ● \; -0.7$ Express $-\frac{5}{8}$ as a decimal.

Since $-0.625 > -0.7$, $-\frac{5}{8} > -0.7$.

Replace each ● with <, >, or = to make a true sentence.

1. $9.3 \; ● \; 9.33$
2. $-4.2 \; ● \; -5.1$
3. $1.05 \; ● \; -1.1$
4. $-18.2 \; ● \; -18.9$

5. $\frac{8}{9} \; ● \; \frac{13}{15}$
6. $\frac{2}{3} \; ● \; \frac{5}{8}$
7. $-\frac{1}{4} \; ● \; -\frac{2}{7}$
8. $-\frac{3}{4} \; ● \; -\frac{9}{12}$

9. $8\frac{3}{5} \; ● \; 8\frac{2}{3}$
10. $2\frac{3}{8} \; ● \; 2.4$
11. $-7.5 \; ● \; -7\frac{1}{3}$
12. $-13\frac{5}{8} \; ● \; -13.7$

13. $645 \; ● \; 6.45 \times 10^{2}$
14. $-3.05 \; ● \; -3\frac{1}{20}$
15. $-\frac{5}{12} \; ● \; -\frac{8}{18}$
16. $9\frac{2}{3} \; ● \; 9.6 \times 10^{-1}$

SET 6B   *(Lessons 6-2, 6-3, 6-4, 6-5, 6-6, 6-7)*   Multiply and divide rational numbers.

Example

Solve x = $6\frac{2}{3} \div \left(-1\frac{2}{3}\right)$.

$x = 6\frac{2}{3} \div \left(-1\frac{2}{3}\right)$

$x = \frac{20}{3} \div \left(-\frac{5}{3}\right)$ Rewrite the mixed numbers as fractions.

$x = \frac{20}{3} \cdot \left(-\frac{3}{5}\right)$ Dividing by $-\frac{5}{3}$ is the same as multiplying by $-\frac{3}{5}$.

$x = \frac{\overset{4}{\cancel{20}}}{\cancel{3}} \cdot \left(-\frac{\cancel{3}}{\cancel{5}}\right)$
     $\underset{1}{} \quad \underset{1}{}$

$x = -4$

Solve each equation. Write each solution in simplest form.

1. $y = \frac{1}{4} \cdot \frac{8}{9}$
2. $t = -\frac{3}{4}\left(\frac{2}{5}\right)$
3. $b = -3\left(-5\frac{1}{3}\right)$
4. $a = \left(-\frac{3}{7}\right)^{2}$

5. $z = (1.4)(3)$
6. $c = (-1.9)(5.3)$
7. $p = (0.7)(-2.6)$
8. $g = (-0.05)(-8.02)$

9. $x = \frac{8}{9} \div \frac{2}{3}$
10. $d = -2\frac{1}{5} \div \frac{4}{5}$
11. $r = 6 \div \left(-\frac{1}{3}\right)$
12. $w = -7\frac{1}{2} \div \left(-1\frac{2}{3}\right)$

13. $h = 1.44 \div 8$
14. $y = 22.5 \div 2.5$
15. $m = -9.54 \div (-5.3)$
16. $k = -0.027 \div 0.45$

SET 6C *(Lessons 6-8, 6-13)* Solve equations and inequalities with rational numbers.

Example

Solve $\frac{a}{-3.2} < -2.6$.

$$\frac{a}{-3.2} < -2.6$$

$$(-3.2)\frac{a}{-3.2} > (-3.2)(-2.6)$$ Multiply each side by –3.2.
Remember to reverse the order symbol.

$$a > 8.32$$ Any number greater than 8.32 is a solution.

Solve each equation or inequality. Check your solution.

1. $3a = 2.1$

2. $-\frac{1}{2}n = 3$

3. $\frac{x}{1.5} = -6.2$

4. $\frac{3}{5}y = -1\frac{1}{10}$

5. $\frac{n}{3} > -2.4$

6. $-\frac{2}{3}a \leq \frac{1}{6}$

7. $\frac{1}{3}b \leq 0.45$

8. $-3.5k = 0.28$

9. $\frac{n}{1.6} \geq -4.3$

10. $\frac{5}{8}x = -\frac{3}{4}$

11. $-\frac{2}{3}m \geq -7$

12. $-0.33d < -8.25$

13. $-10\frac{1}{2}r \leq 1\frac{3}{4}$

14. $-5.5t > -0.11$

15. $\frac{5}{6}c = -2\frac{1}{9}$

16. $9y > 3\frac{3}{5}$

SET 6D *(Lesson 6-10)* Find terms of a geometric sequence.

Example

Use the expression $ar^{(n-1)}$ to find the eighth term of the sequence 64, 32, 16, 8, . . .

The first term: $a = 64$

The common ratio: $r = \frac{1}{2}$

The number of terms: $n = 8$

$$ar^{(n-1)} = 64 \cdot \left(\frac{1}{2}\right)^{(8-1)}$$

$$= 64 \cdot \left(\frac{1}{2}\right)^{7}$$

$$= 64 \cdot \frac{1^7}{2^7}$$

$$= 64 \cdot \frac{1}{128}$$

$$= \frac{1}{2}$$ The eighth term is $\frac{1}{2}$.

Write the next three terms of each sequence.

1. 3, 6, 12, 24, . . .

2. -2, 6, -18, 54, . . .

3. 9, 3, 1, $\frac{1}{3}$, . . .

4. 2500, 500, 100, 20. . . .

5. $\frac{1}{2}$, -2, 8, -32, . . .

6. 144, 72, 36, 18, . . .

Use the expression $ar^{(n-1)}$ to find the seventh term of each sequence.

7. 1, 4, 16, 64, . . .

8. 3, -9, 27, -81, . . .

9. 729, 243, 81, 27, . . .

10. 32, 16, 8, 4, . . .

11. 96, -48, 24, -12, . . .

12. $\frac{1}{8}$, $\frac{1}{4}$, $\frac{1}{2}$, 1, . . .

SET 7A *(Lessons 7-2, 7-3, 7-4, 7-7)* Write and solve two-step equations and problems.

Example

Solve $3x + 7 = -2 + 8x$.

$$3x + 7 = -2 + 8x.$$
$$3x - 8x + 7 = -2 + 8x - 8x \qquad \text{Subtract } 8x \text{ from each side.}$$
$$-5x + 7 = -2$$
$$-5x + 7 - 7 = -2 - 7 \qquad \text{Subtract 7 from each side.}$$
$$-5x = -9$$
$$\frac{-5x}{-5} = \frac{-9}{-5} \qquad \text{Divide each side by -5.}$$
$$x = \frac{9}{5}$$

Solve each equation.

1. $4y + 9 = 13$
2. $5s - 7 = -23$
3. $-3 = 18 - x$
4. $\frac{z}{4} + 17 = 9$
5. $-4 = \frac{d + 5}{9}$
6. $\frac{x - 7}{-6} = 4$
7. $9x - 3 = 4x + 12$
8. $17d = 8d - 3$
9. $a + 11 = 21 - a$
10. $13y + 11 = -5y + 35$
11. $4b = -b - 85$
12. $30c + 14 = 19c - 8$
13. $3x + 1 = 4x - 11$
14. $11x - 18 = 3x + 6$
15. $-5 - x = 3x + 5$

SET 7B *(Lessons 7-5, 7-6)* Solve multi-step equations and inequalities.

Example

Solve $y - 23 < -8 + 4y$.

$$y - 23 < -8 + 4y$$
$$y - 4y - 23 < -8 + 4y - 4y \qquad \text{Subtract } 4y \text{ from each side.}$$
$$-3y - 23 < -8$$
$$-3y - 23 + 23 < -8 + 23 \qquad \text{Add 23 to each side.}$$
$$-3y < 15$$
$$\frac{-3y}{-3} > \frac{15}{-3} \qquad \text{Divide each side by -3 and reverse the order symbol.}$$
$$y > -5$$

Solve each equation or inequality.

1. $3(b + 1) = 4b - 1$
2. $-5(1 - x) = 3x + 2$
3. $11(c - 2) = 3(c + 6)$
4. $\frac{2}{3}n + 8 = \frac{1}{3}n - 1$
5. $3(s + 1) - 8 = 5(s - 3)$
6. $2(2r + 5) + 1 = 5 - 2(3 - r)$
7. $3k - 8 > 16$
8. $-4a + 7 \le 13$
9. $6a + 9 < -4a + 29$
10. $9q + 3 \le 7q - 25$
11. $-2(f + 8) > -4$
12. $-13 \le \frac{d}{11} - 9$

SET 7C   *(Lesson 7-8)*   Convert within the metric system.

Example

1200 cm = ■ km

Smaller units to larger units means fewer units.

First, divide by 100.   Next, divide by 1000.

$1200 \div 100 = 12$ $12 \div 1000 = 0.012$

$1200 \text{ cm} = 12 \text{ m}$ $12 \text{ m} = 0.012 \text{ km}$ So, 1200 cm = 0.012 km.

Complete.

1. 8 km = ■ m

2. 16 g = ■ kg

3. 56 L = ■ mL

4. 4300 mg = ■ g

5. 2.5 L = ■ mL

6. 55 cm = ■ m

7. 0.2 km = ■ cm

8. 6700 mL = ■ L

9. 4800 cm = ■ m

10. 31 g = ■ kg

11. 0.5 L = ■ mL

12. 36,000 mg = ■ kg

13. 9.4 kg = ■ g

14. 43,000 cm = ■ km

15. 970 mL = ■ L

16. 0.8 km = ■ mm

17. 7.9 m = ■ km

18. 98 cm = ■ mm

SET 8A *(Lessons 8-1, 8-2)* Graph the solution of equations and inequalities on a number line.

Example

Solve -4r + 5 < 13 and graph the solution.

$$-4r + 5 < 13$$

$$-4r + 5 - 5 < 13 - 5 \qquad \text{Subtract 5 from each side.}$$

$$-4r < 8$$

$$\frac{-4r}{-4} > \frac{8}{-4} \qquad \text{Divide each side by -4.}$$

$$r > -2$$

Solve each equation and graph the solution.

1. $3x + 5 = 2$
2. $4a - 1 = 11$
3. $-3 + b = 5$
4. $2x - 4 = 3x + 1$
5. $\frac{c}{-4} = -1$
6. $-1.5g = 6$

Solve each inequality and graph the solution.

7. $a - 4 < -2$
8. $3d + 1 > 7$
9. $3.5r \geq 14$
10. $2z + \frac{1}{4} \geq \frac{3}{4}$
11. $3c - 1 \leq 4c + 2$
12. $-\frac{1}{2}t < -2$

SET 8B *(Lessons 8-3, 8-4, 8-5, 8-6, 8-11)* Graph equations and inequalities in the coordinate plane.

Example

Graph y < 3x + 4.

Find four solutions for the equation $y = 3x + 4$. Then graph the line. Remember: Since this line is not part of the solution, use a dashed line.

The graph of $y < 3x + 4$ is the region *below* the boundary line. Shade this region.

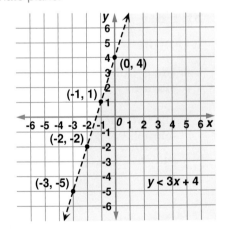

Graph each equation.

1. $y = 3x + 2$
2. $y = -\frac{1}{3}x$
3. $y = -2x - 5$
4. $y = -x$
5. $x + y = 2$
6. $x - y = 1$

Graph each inequality.

7. $y < x - 1$
8. $y > 2x + 2$
9. $y \geq -3$
10. $y < -3x + 5$
11. $y \leq \frac{1}{2}x - 7$
12. $y \geq -x - 6$

Example

Find the coordinates of the points of the *x*-intercept and the *y*-intercept for $y = 2x + 4$.

Find the *y*-intercept.
Let $x = 0$ and solve for *y*.
$$y = 2(0) + 4$$
$$y = 4$$
The *y*-intercept is 4.
The ordered pair is (0,4).

Find the *x*-intercept.
Let $y = 0$ and solve for *x*.
$$0 = 2x + 4$$
$$0 - 4 = 2x + 4 - 4$$
$$-4 = 2x$$
$$-2 = x$$
The *x*-intercept is -2.
The ordered pair is (-2,0).

Find the slope of the line that contains each pair of points.

1. A(3,2), B(5,8)
2. R(-1,1), S(2,7)
3. X(2,9), Y(-7,6)
4. L(-1,-3), M(-3,1)
5. C(7,-3), D(-1,1)
6. $G(\frac{1}{2},-1)$, $H(-\frac{1}{2},5)$

Graph each equation using the *x*-intercept and the *y*-intercept.

7. $y = x + 3$
8. $y = 2x - 8$
9. $y = -5x - 10$
10. $y = -7x - 2$
11. $y = \frac{1}{3}x - 4$
12. $y = 6 - 3x$

Example

Use the graph at the right to solve $y = -2x$ and $y = x + 3$.

The two lines intersect at A(-1,2). The solution, then, is the ordered pair (-1,2).

Check: $y = -2x$ $y = x + 3$
$$2 = -2(-1)$$ $$2 = -1 + 3$$
$$2 = 2 \quad \text{✔}$$ $$2 = 2 \quad \text{✔}$$

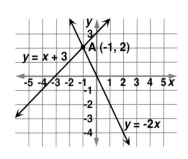

Use a graph to solve each system of equations.

1. $y = 2x - 1$
   $y = x + 1$
2. $y = x - 3$
   $y = 2x$
3. $y = -x + 2$
   $y = 5x - 10$
4. $y = \frac{1}{2}x$
   $y = 3x + 5$
5. $y = -4x + 15$
   $y = x$
6. $x + y = 5$
   $2x - y = -8$
7. $y = x + 7$
   $y = -3x - 9$
8. $y = -5x$
   $y = 2x + 14$
9. $-2x - y = 5$
   $3x - y = 0$

SET 9A *(Lessons 9-1, 9-5)* Write equivalent expressions for ratios, decimals, and percents.

Example

Express $\frac{7}{9}$ as a percent. Round to the nearest tenth.

$$7 \;\boxed{\div}\; 9 \;\boxed{=}\; 0.777777$$

$$\frac{7}{9} \approx 0.777 \to 77.7\%$$

Express each ratio as a fraction in simplest form.

1. 5 out of 80   **2.** 18 to 3   **3.** 36 to 144   **4.** 108:9

Express each decimal or fraction as a percent. Round to the nearest tenth.

5. 0.07   **6.** 0.305   **7.** 1.98   **8.** 0.003

9. $\frac{2}{5}$ **10.** $\frac{3}{8}$ **11.** $\frac{7}{4}$ **12.** $\frac{11}{12}$

Express each percent as a fraction in simplest form.

13. 42%   **14.** 5%   **15.** $87\frac{1}{2}\%$ **16.** 112%

SET 9B *(Lessons 9-2, 9-3, 9-4)* Use ratios and proportions to solve problems.

Example

What number is 75% of 185?

$$\frac{P}{B} = \frac{r}{100} \;\to\; \frac{P}{185} = \frac{75}{100}$$ Replace B with 185 and r with 75.

$$P \cdot 100 = 75 \cdot 185$$ Find the cross products.

$$75 \;\boxed{\times}\; 185 \;\boxed{\div}\; 100 \;\boxed{=}\; 138.75$$

$$P = 138.75$$ 138.75 is 75% of 185.

Solve each proportion.

1. $\frac{7}{8} = \frac{a}{64}$ **2.** $\frac{3}{5} = \frac{m}{65}$ **3.** $\frac{9}{4} = \frac{r}{36}$ **4.** $\frac{1.8}{5.4} = \frac{14.4}{p}$

Write a proportion that could be used to solve for each variable. Then solve.

5. 12 ounces at $1.40
18 ounces at x dollars

6. 16 by 20 inches reduced
to x by 8 inches

7. $\frac{1}{4}$-inch represents 1 mile
6 inches represents y miles

Use a proportion to solve each problem.

8. What is 30% of 270?   **9.** Find 80% of 72.

10. 18 is what percent of 90?   **11.** 60 is what percent of 240?

SET 9C *(Lessons 9-6, 9-7)* Solve equations involving percents.

Example

17 is what percent of 68?

$$17 = R \cdot 68 \qquad \text{Write in } P = R \cdot B \text{ form.}$$

$$17 = 68R \qquad \text{Solve for } R.$$

$$\frac{17}{68} = \frac{68R}{68} \qquad \text{Divide each side by 68.}$$

$$17 \;\boxed{\div}\; 68 \;\boxed{=}\; 0.25$$

$$0.25 = R$$

$$25\% = R$$

Solve.

1. 192 is what percent of 240?
2. Find 5% of $95.00.
3. 50 is 40% of what number?
4. 57 is 60% of what number?
5. Find 18% of 230.
6. 120 is what percent of 160?
7. What is 20% of $52.00?
8. $630 is 35% of how many dollars?
9. 9 is 1% of what number?
10. 96 is what percent of 600?

SET 9D *(Lessons 9-8, 9-10)* Solve problems involving percents.

Example

Find the percent of change from $150 to $162.

$$162 - 150 = 12 \qquad \text{Find the amount of increase.}$$

$$\frac{12}{150} = \frac{r}{100} \qquad \text{Write the percent proportion.}$$

$$12 \cdot 100 = 150r \qquad \text{Find the cross products.}$$

$$1200 = 150r \qquad \text{Divide each side by 150.}$$

$$1200 \;\boxed{\div}\; 150 \;\boxed{=}\; 8$$

$$r = 8 \qquad \text{The percent of increase is 8\%.}$$

Find the discount or interest to the nearest cent.

1. $10 hat, 35% off
2. $38 jeans, 20% off
3. $185 bike, 25% off
4. $5000 at 6% for 6 months
5. $450 at $5\frac{3}{4}\%$ for 8 months
6. $8.99 CD, 15% off

Find the percent of change in the prices below. Round to the nearest whole percent.

7. old: $25.00
   new: $23.00
8. old: $62.00
   new: $78.50
9. old: $118.00
   new: $99.98
10. old: $34.80
    new: $56.60
11. old: $1900.00
    new: $1590.00
12. old: $95.99
    new: $105.25

SET 10A   *(Lesson 10-1)*   Gather and record data.

Example

What percent of the sample have fewer than 2 pets?

Number of Pets	Tally	Frequency
0	IIII I	6
1	IIII IIII	9
2	IIII IIII I	11
3	IIII	5
4 or more	II	2

6 + 9 or 15 people have fewer than 2 pets. The number of people surveyed is 33.

$\frac{15}{33} = 0.4545\ldots$

About 45% of the sample have fewer than 2 pets.

The number of magazine subscriptions sold by each student is given below.

1. Make a frequency table for the set of data.
2. What is the lowest number of subscriptions sold?
3. What is the highest number of subscriptions sold?
4. What is the frequency of selling exactly 28 subscriptions?
5. Each student who sells 50 or more subscriptions receives a prize. How many students receive a prize?
6. What percent of the students sold over 40 subscriptions?
7. Make a histogram of the set of data.

21	50	29	42
36	23	38	51
28	42	22	19
55	34	31	39
22	28	30	35
37	37	29	44
46	52	24	28

SET 10B   *(Lessons 10-2, 10-4)*   Describe data using measures of central tendency and measures of variation.

Example

Given the set of data 4, 9, 6, 1, 4, 3, 8, 3, 4, 5, find the interquartile range.

First, order the scores.   1, 3, 3, 4, 4, 4, 5, 6, 8, 9

The median is (4 + 4) ÷ 2 = 4.
The upper quartile is 6. The lower quartile is 3.
The interquartile range is 6 − 3 or 3.

Use the set of temperature data below to answer each of the following.

1. What is the mode?
2. What is the mean?
3. What is the median?
4. What is the range of temperatures?
5. What are the upper and lower quartiles?
6. What is the interquartile range?

65	53	68	45
59	59	67	62
52	64	62	68
42	64	71	59

Example

Make a stem-and-leaf plot of these test score data: 88, 79, 93, 75, 59, 72, 71, 99, 78, 83, 81, 83, 79, 94, 79, 52. Then, determine the range.

```
5 | 2 9
6 |
7 | 1 2 5 8 9 9 9      The range is 99 − 52 or 47.
8 | 1 3 3 8
9 | 3 4 9
```

The ages of people visiting the zoo are listed at the right.

1. Construct a stem-and-leaf plot.
2. What is the age range?
3. What age range seemed most represented?
4. Construct a box-and-whisker plot.
5. Are there any outliers? If so, what are they?

10	32	33	16	17	55
52	9	8	11	22	67
28	30	37	12	5	3
3	7	47	28	41	6
26	14	2	30	82	52

Zoo admission prices are $4.00 for adults (over 18), $2.00 for students (ages 5-18), and free for children under age 5.

6. Draw a scatter plot of the ages listed above and their admission prices.
7. What type of relationship, positive, negative, or none, is shown by the scatter plot?

SET 10D (Lessons 10-8, 10-9) Use statistics to predict and recognize misleading statistics.

Example

Use the sample data on bus fares to predict how many people in a city of 100,000 would be willing to pay up to $1.50 per ride.

Amount Willing to Pay for Bus Ride	
up to $0.75	50
up to $1.00	160
up to $1.50	220
up to $2.00	70

The sample size is 500.
220 + 70 or 290 people are willing to pay up to $1.50.
$\frac{290}{500} \cdot 100,000 = 58,000$ 58,000 people would be willing to pay up to $1.50.

Use the bus fare data above to answer each of the following.

1. How many people in a city of 600,000 would be willing to pay up to $1.00 per bus ride?
2. How many people in a city of 1,500,000 would be willing to pay up to $1.50 per bus ride?

SET 11A *(Lessons 11-1, 11-2)* Count outcomes.

Example

A quiz has three true-false questions and two multiple-choice questions. If each multiple-choice question has four choices, how many outcomes are possible?

Number of outcomes for each true-false question: 2
Number of outcomes for each multiple-choice question: 4

$$\overbrace{2 \times 2 \times 2}^{\text{3 true-false questions}} \times \overbrace{4 \times 4}^{\text{2 multiple-choice questions}} = 128 \qquad \text{There are 128 possible outcomes.}$$

Find the number of possible outcomes.

1. Each spinner is spun once.

2. A die is rolled and the spinner is spun twice.

3. Marilyn has a choice of toast, muffin, or bagel with a choice of tea or coffee.

4. A quiz has six multiple-choice questions. Each question is answered with a, b, c, or d.

5. Glen has 5 shirts, 3 pairs of pants, and 2 belts. Any combination of outfits matches.

6. A quiz has two true-false questions and three multiple-choice questions. Each multiple-choice question has five choices.

SET 11B *(Lesson 11-3)* Find permutations and combinations.

Example

How many three letter "words" can be made from the letters W, S, O, and R?

Order is important, so you want to find the number of *permutations* of 4 things taken 3 at a time.

$P(4,3) = 4 \cdot 3 \cdot 2 = 24$ There are 24 three-letter words that can be made from W, S, O, and R.

Find each value.

1. 7!   **2.** $\frac{6!3!}{5!0!}$ **3.** P(5,2)   **4.** C(7,3)   **5.** C(9,6)

Answer each question.

6. How many ways can you select 5 magazines off a shelf of 8 magazines?

7. How many 4-place numbers can be formed with the digits 2, 3, 4, and 5 if no digit is used more than once in a number?

8. How many ways can 3 students be selected from a committee of 9 students to attend a school board meeting?

9. How many ways can first, second, and third place prizes be awarded to twelve different entries in a baking contest?

Example

A card is drawn from the cards at the right. Find P(5 or odd).

These events are not mutually exclusive.
Therefore, P(5 or odd) = P(5) + P(odd) − P(5 and odd). $\boxed{1}$ $\boxed{2}$ $\boxed{3}$ $\boxed{4}$ $\boxed{5}$ $\boxed{6}$

$$= \frac{1}{6} + \frac{3}{6} - \frac{1}{6} \cdot \frac{3}{6}$$

$$= \frac{4}{6} - \frac{3}{36}$$

$$= \frac{24}{36} - \frac{3}{36} \qquad \text{Rename } \frac{4}{6} \text{ as } \frac{24}{36}.$$

$$= \frac{21}{36} \text{ or } \frac{7}{12}$$

In one bag, there are 4 blue marbles, 5 red marbles, and 3 green marbles. In another bag, there are 2 white marbles and 6 yellow marbles.

1. You select one marble at random from the second bag. Find P(white).

2. You select one marble at random from the first bag. Find P(blue or green).

3. You select one marble at random from each bag. Find P(red and white).

4. You select one marble at random from each bag. Find P(red and black).

5. You select two marbles from the first bag without replacing the first one. Find the probability of selecting a red marble and then a blue marble.

SET 12A   *(Lesson 12-1)*   Use some basic terms of geometry.

Example

**Use the figure at the
right to find m∠PQR.**

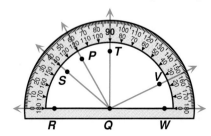

$m \angle PQR = 60°$ Use the scale that begins with 0 at $\overrightarrow{QR}$.

Use the figure above to find each measure. Classify each angle as *acute*, *right*, or *obtuse*.

1. $m\angle VQW$ 2. $m\angle TQW$ 3. $m\angle SQW$ 4. $m\angle SQR$ 5. $m\angle VQR$

**Use the figure at the right to
name an example of each term.**

6. point   7. line segment

8. ray   9. acute angle

10. line   11. obtuse angle

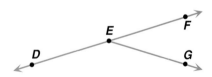

SET 12B   *(Lesson 12-2)*   Construct a circle graph.

Example

An angle of a circle graph represents 20%. Find the measure of the angle. Draw the angle on a circle graph.

$\frac{20}{100} \times 360 = 72$ Rename 20% as $\frac{20}{100}$.

There are 360° in a circle.

The measure of the angle is 72°.

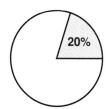

Make a circle graph to display each set of data.

1.

Davis Family Monthly Expenditures	
Housing	$800
Taxes	$600
Food	$350
Clothing	$200
Insurance	$150
Savings	$100

2.

Items Donated for Food Drive	
Canned Goods	169
Pasta	86
Cereal	70
Peanut Butter	42
Condiments	18

SET 12C   *(Lessons 12-3, 12-4)*   Identify the relationships of intersecting and parallel lines.

Example

In the diagram at the right, ℓ is parallel to m. The measure of $\angle 2$ is 38°. Find $m\angle 7$.

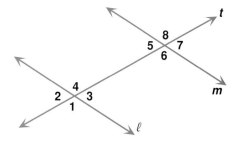

$\angle 2$ and $\angle 7$ are alternate exterior angles.

$m\angle 2 = m\angle 7$ Alternate exterior angles

$m\angle 7 = 38°$ are congruent.

The measure of $\angle 7$ is 38°.

Angles C and D are complementary. Find $m\angle C$ when $m\angle D$ has each measure.

1. 60°   2. 75°   3. 28°   4. 5°

Angles X and Y are supplementary. Find $m\angle X$ when $m\angle Y$ has each measure.

5. 140°   6. 45°   7. 95°   8. 9°

Refer to the diagram in the Example above. If the measure of $\angle 2$ is 38°, find the measure of each angle below.

9. $\angle 1$ 10. $\angle 3$ 11. $\angle 4$ 12. $\angle 5$ 13. $\angle 6$ 14. $\angle 8$

Example

Find the value of *x*.

$\angle A \cong \angle A$, $\angle D \cong \angle E$, $\angle B \cong \angle C$

So, $\triangle AEB \cong \triangle ADC$ because their corresponding angles are congruent.

$$\frac{m\overline{AE}}{m\overline{AD}} = \frac{m\overline{EB}}{m\overline{DC}}$$ Corresponding sides of similar triangles are congruent.

$$\frac{6}{9} = \frac{4}{x}$$

$6 \cdot x = 9 \cdot 4$ Cross multiply.

$6x = 36$

$x = 6$ Divide each side by 6.

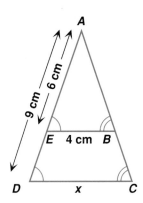

If $\triangle JKL \cong \triangle DGW$, name the part congruent to each angle or segment given.

1. $\angle K$
2. $\overline{WG}$
3. $\angle D$
4. $\overline{KL}$
5. $\overline{DG}$

Solve.

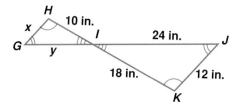

6. Find the value of *x*.

7. Find the value of *y*.

SET 12E *(Lessons 12-11, 12-12)* Classify and draw polygons.

Example

Find the sum of the measures of the angles of a heptagon.

$n = 7$ The number of sides is 7.

$(n - 2)180 = (7 - 2)180$ Replace *n* with 7.

$\qquad = 5 \cdot 180$

$\qquad = 900$ The sum of the measures of the angles of a heptagon is $900°$.

Classify each quadrilateral.

1.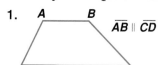
$\overline{AB} \parallel \overline{CD}$

2.
$\overline{JK} \parallel \overline{LM}$
$\overline{JL} \parallel \overline{KM}$

3.

Classify each polygon. Then find the sum of the measures of the angles of each polygon.

4.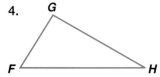

5.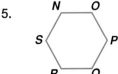

6.

SET 13A *(Lessons 13-1, 13-2, 13-3)* Find the area of polygons and circles.

Example

Find the area of a circle that has a diameter of 10 inches. Round your answer to the nearest whole number.

$A = \pi r^2$

$A = \pi \cdot 5^2$ The radius, *r*, is one-half the diameter, or 5 inches.

$A \approx 78.5$ The area is about 79 square inches.

Find the area of each region shown or described below. Round to the nearest whole number.

1.

12 m   14 m   9 m

2.
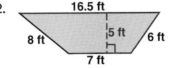
16.5 ft   8 ft   5 ft   6 ft   7 ft

3.

8 mm

4. triangle: base, 6 cm; height, 11 cm

5. parallelogram: base, 4.5 in.; height, 3 in.

6. circle: radius, 7 ft

7. circle: diameter, 15.4 m

SET 13B *(Lessons 13-4, 13-5, 13-6)* Find the surface area of prisms, cylinders, pyramids, and cones.

Example

Find the surface area of the circular cone shown at the right. Round to the nearest hundredth.

Area of base:

$A = \pi r^2$

$A = \pi \cdot 3^2$

$A \approx 28.27$

Area of curved surface:

$A = \pi r s$

$A = \pi \cdot 3 \cdot 5$

$A \approx 47.12$

3 ft   5 ft

total surface area = (area of base) + (area of curved surface)

=   28.27   +   47.12

=   75.39   The surface area is about 75.39 ft².

Find the surface area for each figure below. Round to the nearest hundredth.

1.
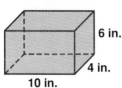
6 in.   4 in.   10 in.

2.
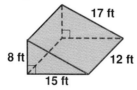
17 ft   8 ft   12 ft   15 ft

3.

5 cm   3.5 cm

4.
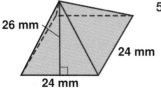
26 mm   24 mm   24 mm

5.
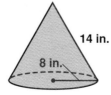
14 in.   8 in.

6.
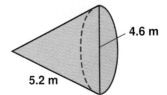
4.6 m   5.2 m

Example

Find the volume of the pyramid.

$V = \frac{1}{3}Bh$ Formula for volume of a pyramid

$V = \frac{1}{3}\ell wh$ Replace B with ℓw.

$V = \frac{1}{3} \cdot 4 \cdot 3 \cdot 7$ Replace ℓ with 4, w with 3, and h with 7.

$V = 28$

The volume is 28 cm^3.

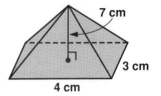

Find the volume of each figure below. Round to the nearest hundredth.

1.

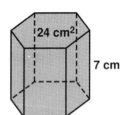

2.

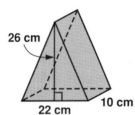

3.

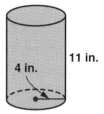

4.

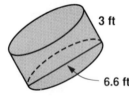

5.

6.

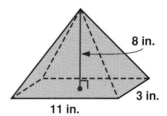

SET 14A   *(Lessons 14-1, 14-2)*   Find square roots and squares.

Example

What is the best integer estimate for $-\sqrt{38.4}$?

$$-49 < -38.4 < -36$$
$$-7^2 < -38.4 < -6^2$$
$$-7 < -\sqrt{38.4} < -6$$

Check:

$$38.4 \; \boxed{\sqrt{x}} \; \boxed{+/-} \; -6.196773$$

Since -38.4 is closer to -36 than -49, the best integer estimate is -6.
This checks with the result on the calculator.

Find the square of each number.

1. 4   2. 11   3. 15   4. 24   5. 30   6. 35

Find each square root. If the number is not a perfect square, find the best integer estimate for each square root. Then check your estimate using a calculator.

7. $\sqrt{36}$ 8. $\sqrt{81}$ 9. $\sqrt{196}$ 10. $-\sqrt{289}$ 11. $\sqrt{21}$

12. $-\sqrt{85}$ 13. $\sqrt{7.3}$ 14. $-\sqrt{484}$ 15. $-\sqrt{1600}$ 16. $\sqrt{1.99}$

17. $-\sqrt{62}$ 18. $\sqrt{74.1}$ 19. $\sqrt{625}$ 20. $-\sqrt{88.8}$ 21. $\sqrt{8100}$

SET 14B   *(Lesson 14-4)*   Solve equations by finding square roots.

Example

Solve $a^2 = 62$. Round your answer to the nearest tenth.

$$a^2 = 62$$
$$a = \sqrt{62} \text{ or } a = -\sqrt{62} \qquad \text{Find the square root of each side.}$$
$$a \approx 7.9 \text{ or } a \approx -7.9$$

Solve each equation. Round decimal answers to the nearest tenth.

1. $x^2 = 16$ 2. $y^2 = 36$ 3. $c^2 = 144$ 4. $k^2 = 225$

5. $m^2 = 324$ 6. $d^2 = 441$ 7. $r^2 = 2500$ 8. $p^2 = 34$

9. $b^2 = 89$ 10. $h^2 = 55$ 11. $j^2 = 119$ 12. $n^2 = 120$

13. $t^2 = 37$ 14. $q^2 = 361$ 15. $s^2 = 167$ 16. $a^2 = 111$

SET 14C (Lessons 14-5, 14-6) Use the Pythagorean Theorem.

Example

Find the length of the third side of the right triangle.

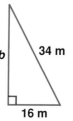

$$c^2 = a^2 + b^2 \qquad \text{Pythagorean Theorem}$$
$$34^2 = 16^2 + b^2 \qquad \text{Replace } c \text{ with 34 and } a \text{ with 16.}$$
$$1156 = 256 + b^2$$
$$1156 - 256 = 256 - 256 + b^2 \qquad \text{Subtract 256 from each side.}$$
$$\sqrt{900} = b^2$$
$$\sqrt{900} = b \text{ or } -\sqrt{900} = b$$
$$30 = b \text{ or } -30 = b$$

The length of the side is 30 meters.

Find the missing measure for each right triangle. Round decimal answers to the nearest tenth.

1.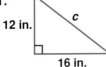

2.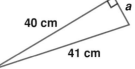

3.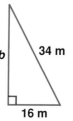

4. a, 7 m; b, 24 m

5. a, 18 in.; c, 30 in.

6. b, 10 ft; c, 20 ft

7. a, 3 cm; c, 9 cm

8. b, 8 m; c, 32 m

9. a, 36 ft; b, 42 ft

10. a, 30 km; b, 45 km

11. a, 32 yd; c, 65 yd

12. b, 11 cm; c, 19 cm

SET 14D (Lessons 14-7, 14-8, 14-9, 14-10) Use trigonometric ratios to solve problems involving right triangles.

Example

Solve for x. Round your answer to the nearest tenth.

The measure of the side adjacent to P is known and the measure of the hypotenuse is needed. Use the cosine function.

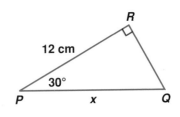

$$\cos P = \frac{\text{measure of the side adjacent to } \angle P}{\text{measure of the hypotenuse}}$$
$$\cos 30° = \frac{12}{x}$$
$$x \cos 30° = 12 \qquad \text{Multiply each side by } x.$$
$$x = \frac{12}{\cos 30°}$$
$$x \approx 13.9 \qquad \text{The length of } \overline{PQ} \text{ is about 13.9 cm.}$$

Solve for y. Round answers to the nearest tenth.

1.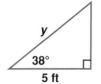

2.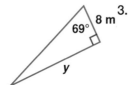

3.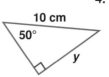

4.

SET 15A *(Lesson 15-1)*   Identify and classify polynomials.

Example

State whether $5a^2 - \frac{a}{4}$ is a polynomial. If it is, identify it as a *monomial*, *binomial*, or *trinomial*.

The expression $5a^2 - \frac{a}{4}$ is a polynomial because it is the sum of two monomials. Since it has two unlike terms, it is a binomial.

State whether each expression is a polynomial. If it is, identify it as a *monomial*, *binomial*, or *trinomial*.

1. x^2
2. $5a - 3$
3. $5 - 4d + d^2$
4. $\frac{9}{r^2}$
5. $\frac{t^5}{7}$
6. $\sqrt{11ab}$
7. $4pq^2 - pq + 9p$
8. $\frac{2}{a} + 9$
9. $\frac{r^3 - s}{3}$
10. $\frac{2}{3} - 4k^2$
11. $\frac{5n}{m} + \frac{2a}{5}$
12. -41

Find the degree of each polynomial.

13. $4x$
14. $a^2 - 6$
15. $11r + 5s$
16. $3y^2 + 4y - 2$
17. $9cd^3 - 5$
18. $-5p^3 + 8q^2$
19. $w^2 + 2x - 3y^3 - 7z$
20. $\frac{x^3}{6} - x$
21. $-17n^2p - 11np^3$

SET 15B   *(Lessons 15-2, 15-3, 15-4, 15-5, 15-6)*   Add, subtract, and multiply polynomials.

Example

Find $-5x(2x^2 - 3x + 1)$.

$-5x(2x^2 - 3x + 1) = (-5x)(2x^2) - (-5x)(3x) + (-5x)(1)$ Use the distributive property.

$\qquad\qquad\qquad\quad = -10x^3 + 15x^2 - 5x$

Find each sum, difference, or product.

1. $(3a + 4) + (a + 2)$
2. $(8m - 3) + (4m + 1)$
3. $(5x - 3y) + (2x - y)$
4. $(3n + 2) - (n + 1)$
5. $(-3c + 2d) - (7c - 6d)$
6. $(8p^2 - 2p + 3) + (-3p^2 - 2)$
7. $(4x^2 + 1) - (3x^2 - 4)$
8. $(5a - 4b) - (-a + b)$
9. $(6c^3 + 3) - (3c^3 - 1)$
10. $(-11r^2 + 3s) + (5r^2 - s)$
11. $4n(5n - 3)$
12. $-3x(4 - x)$
13. $6m(-m^2 + 3)$
14. $7r(r^2 - 3r + 7)$
15. $(2x + 2)(3x + 1)$
16. $(x + 4)(3x + 1)$

17. $\begin{array}{r} 3a^2 + 5a + 1 \\ + 2a^2 - 3a - 6 \\ \hline \end{array}$

18. $\begin{array}{r} 6x^2 - 4x + 11 \\ - 5x^2 + 5x - 4 \\ \hline \end{array}$

19. $\begin{array}{r} 8n^2 + 3mn \\ - 4n^2 + 2mn - 9 \\ \hline \end{array}$

Extra Practice   619

Glossary

absolute value of a number (55) The number of units the number is from zero on the number line.

acute angle (440) Any angle that measures between 0° and 90°.

acute triangle (457) A triangle having three acute angles.

addition property of inequalities (110) For any numbers a, b, and c: 1. If $a > b$, then $a + c > b + c$. 2. If $a < b$, then $a + c < b + c$.

addition property of equality (92) For any numbers a, b, and c, if $a = b$, then $a + c = b + c$.

additive inverse (66) The sum of a number and its additive inverse is zero. For example, the additive inverse of 65 is -65.

additive inverse property (66) For any numbers a, $a + (-a) = 0$.

adjacent angles (446) Two angles are adjacent if they have a common side and the same vertex, but they do *not* overlap.

algebraic expression (19) A combination of variables, numbers, and at least one operation.

algebraic fraction (144) Fractions with variables in the numerator and/or denominator.

alternate interior angles (449) In the figure, transversal t intersects lines l and m. $\angle 3$ and $\angle 5$, and $\angle 4$ and $\angle 6$ are alternate interior angles.

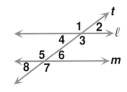

angle (439) Two rays with a common endpoint form an angle.

angle bisector (453) A ray through the interior of an angle that divides the angle into two congruent angles.

area (106, 484) The number of square units needed to cover a surface.

arithmetic sequence (190) A sequence in which the difference between any two consecutive terms is the same.

associative property of addition (22, 174) For any numbers a, b, and c, $(a + b) + c = a + (b + c)$.

associative property of multiplication (22, 220) For any numbers a, b, and c, $(a \cdot b) \cdot c = a \cdot (b \cdot c)$.

bar notation (169) In repeating decimals the line or bar placed over the digits that repeat. For example, $2.\overline{63}$ indicates that the digits 63 repeat.

base (131) In 10^3, the base is 10. The base is used as a factor as many times as given by the exponent (3). That is, $10^3 = 10 \times 10 \times 10$.

base (334) In a percent proportion, the number to which the percentage is compared.

base (484–485, 488) The base of a rectangle, a parallelogram, or a triangle is any side of the figure. The bases of a trapezoid are the parallel sides.

base (495) The bases of a prism are the two parallel congruent sides.

BASIC (21) One type of computer language. BASIC stands for **B**eginner's **A**ll-purpose **S**ymbolic **I**nstruction **C**ode.

binomial (565) A polynomial with two unlike terms.

bisect (452) To divide something into two congruent parts.

boundary (315) A line that separates a plane into two regions.

box-and-whisker plot (378) A diagram that summarizes data using the median, the upper and lower quartiles, and the extreme values. A box is drawn around the quartile value and whiskers extend from each quartile to the extreme data points.

brackets (16) A grouping symbol [ ].

capacity (195) The amount of liquid or dry substance a container can hold.

Celsius (260) The temperature scale of the metric system.

center of a circle (243) The given point from which all points on the circle are the same distance.

change in *x* (305) The horizontal change of a line in a coordinate plane.

change in *y* (305) The vertical change of a line in a coordinate plane.

circle (243) The set of all points in a plane that are the same distance from a given point called the center.

circle graph (442) A type of statistical graph used to compare parts of a whole.

circular cone (502) A shape in space that has a circular base and one vertex.

circular cylinder (498) A shape in space that has two parallel and congruent circular bases.

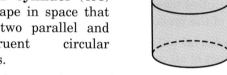

circumference (243) The distance around a circle.

closure property of addition (174) For all rational numbers *x* and *y*, *x* + *y* is a rational number.

closure property of multiplication (220) For all rational numbers *x* and *y*, *x* · *y* is a rational number.

combination (408) An arrangement or listing in which order is *not* important.

common difference (190) The difference between any two consecutive terms in an arithmetic sequence.

common factor (140) A number that is a factor of two or more numbers.

common multiple (146) A number that is a multiple of two or more numbers.

common ratio (233) The ratio between any two successive terms in a geometric sequence.

commutative property of addition (22, 174) For any numbers *a* and *b*, *a* + *b* = *b* + *a*.

commutative property of multiplication (22, 220) For any numbers *a* and *b*, *a* · *b* = *b* · *a*.

compass (452) An instrument used for drawing circles or parts of circles.

complementary angles (446) Two angles are complementary if the sum of their measures is 90°.

composite number (136) Any whole number greater than one that has more than two factors.

compound statement (116) One statement formed by connecting two simple statements with words such as *and*, *or*, *if*, *then*, and *if and only if*.

computer program (139) A series of statements that gives directions to a computer.

congruent (446, 461) Congruent figures match or fit exactly and have the same size and shape.

conjunction (116) Two statements connected by the word *and*.

constant (565) A monomial that does not contain a variable.

coordinate of a point (54, 284) A number associated with the point on a number line.

coordinate system (293) Two perpendicular number lines form a coordinate system.

corresponding angles (449) Corresponding angles are angles that hold the same position on two different parallel lines cut by a transversal.

cosine (550) If △ *ABC* is a right triangle and *A* is an acute angle,

$$\text{cosine of } \angle A = \frac{\text{measure of the side adjacent to } \angle A}{\text{measure of the hypotenuse}}.$$

cross products (327) In the proportion $\frac{3}{6} = \frac{4}{8}$, the cross products are 8 × 3 and 4 × 6.

customary units (195) Units of measure frequently used in the United States. The customary units of length are inches, feet, yards, and miles. The customary units of weight are ounces, pounds, and tons. The customary units of capacity are fluid ounces, cups, pints, quarts, and gallons.

data (362) Numerical information gathered for statistical purposes.

decagon (474) A polygon having ten sides.

liter (218, 276)   The basic unit of capacity in the metric system.

logic (70)   The study of formal reasoning.

lower quartile (375)   The median of the lower half of a set of numbers.

mass (276)   The amount of matter an object contains.

mean (367)   The sum of the numbers in a set of data divided by the number of pieces of data.

measures of central tendency (367)   Numbers or pieces of data that can represent the whole set of data.

median (368)   In a set of data, the median is the number in the middle when the data are organized from least to greatest.

meter (218, 275)   The basic unit of length in the metric system.

metric system (218)   A system of measurement using the basic units: meter for length, gram for mass, and liter for capacity.

midpoint (452)   A point on a line segment that divides the line segment into two congruent line segments.

mode (367)   The number or item that appears most often in a set of data.

monomial (129, 558)   An expression that is either a constant, a variable, or a product of numbers and variables.

multiple of a number (146)   The product of the number and any whole number.

multiplication property of equality (98)   For any numbers a, b, c, if $a = b$, then $ac = bc$.

multiplicative inverse (219)   A number times its multiplicative inverse is equal to 1. The multiplicative inverse of $\frac{3}{4}$ is $\frac{4}{3}$.

multiplicative property of zero (22)   For any number a, $a \cdot 0 = 0 \cdot a = 0$.

mutually exclusive events (424)   Two or more events such that no two events can happen at the same time.

negation (70)   If a statement is represented by p, then not p is the negation of that statement.

negative integers (54)   Whole numbers to the left of zero on the number line or numbers less than 0.
$$-1, -2, -3, \ldots$$

null set (265)   A set with no elements.

obtuse angle (440)   Any angle that measures between 90° and 180°.

obtuse triangle (457)   A triangle having one obtuse angle.

octagon (474)   A polygon having eight sides.

open sentence (30)   A sentence containing at least one variable. The sentence is neither true nor false.

opposites (66)   Two integers are opposites if their sum is zero.

ordered pair (290)   A pair of numbers in which the order is specified.

origin (293)   The point of intersection of the x-axis and y-axis in a coordinate system.

outcome (400)   One possible result of a probability event. For example, 4 is an outcome when a die is rolled.

outliers (378)   Data that are more than 1.5 times the interquartile range from the quartiles.

parallel lines (449)   Lines in the same plane that do not intersect.

parallelogram (472)   A quadrilateral with two pairs of parallel sides.

parentheses (16)   A grouping symbol ( ).

pentagon (474)   A polygon having five sides.

percent (334)   A ratio with a denominator of 100. For example, 7% and $\frac{7}{100}$ name the same number.

percent proportion (334)
$$\frac{\text{Percentage}}{\text{Base}} = \text{Rate or } \frac{P}{B} = \frac{r}{100}$$

percentage (334)   In a percent proportion, a number that is compared to another number called the base.

perfect square (527) A rational number whose square root is a rational number.

perimeter (105) The distance around a geometric figure.

permutation (407) An arrangement or listing in which order is important.

perpendicular lines (446) Intersecting lines that form right angles.

pi (π) (243) The ratio of the circumference of a circle to the diameter of a circle. Approximations for π are 3.14 and $\frac{22}{7}$.

plane (438) A flat surface that has no boundaries. A plane can be named by three non-collinear points of the plane.

point (438) A specific location in space.

polygon (474) A simple closed figure in a plane formed by three or more line segments.

polynomial (565) A monomial or the sum or difference of two or more monomials.

positive integers (54) Whole numbers to the right of zero on the number line or numbers greater than 0.

power (131) A number that can be written using an exponent.

precision (515) The precision of a measurement depends on the unit of measure. To improve the precision of a measurement, a measuring instrument that provides smaller units of measure should be chosen.

prime factorization (136) Every composite number can be expressed as the product of prime numbers. For example, the prime factorization of 63 is $3 \times 3 \times 7$.

prime number (136) A prime number is a whole number greater than 1 that has exactly two factors, 1 and itself.

principal (350) The amount of an investment or a debt.

prism (495) A figure in space that has two parallel and congruent bases in the shape of polygons.

probability (411) The probability of a certain outcome is the ratio of the number of ways the certain outcome can occur to the number of possible outcomes.

product property of square roots (535) For any nonnegative numbers a and b,
$$\sqrt{a \cdot b} = \sqrt{a} \cdot \sqrt{b}$$

property of proportions (327) If $\frac{a}{b} = \frac{c}{d}$ then $ad = bc$. If $ad = bc$, then $\frac{a}{b} = \frac{c}{d}$.

proportion (327) A sentence that states that two ratios are equivalent.

protractor (440) An instrument used to measure angles.

pyramid (501) A figure in space with three or more triangular faces and a base in the shape of a polygon.

Pythagorean Theorem (537) In a right triangle, the square of the hypotenuse is equal to the sum of the squares of the length of the legs.

quadrant (293) One of the four regions into which two perpendicular number lines separate the plane.

quadrilateral (471) A polygon having four sides.

radical sign (526) The symbol used to indicate a nonnegative square root. $\sqrt{}$

radicand (526) Any expression inside a radical sign.

radius (243) The distance from the center of a circle to any point on the circle.

range (374) The difference between the least and greatest numbers in a set of numbers.

rate (320) A ratio of two measurements having different units.

rate (334) In a percent proportion, the ratio of a number to 100.

rate of interest (350) The percent charged or paid for the use of money.

ratio (143, 324) A comparison of two numbers. The ratio of 2 to 3 can be stated as 2 out of 3, 2 to 3, 2:3, or $\frac{2}{3}$.

rational numbers (166, 533) Numbers of the form $\frac{a}{b}$ where a and b are integers and $b \neq 0$.

ray (439) A part of a line that extends indefinitely in one direction.

real numbers (534)   Irrational numbers together with rational numbers form the set of real numbers.

reciprocal (219)   Another name for a multiplicative inverse.

rectangle (472)   A parallelogram with all angles congruent.

rectangular prism (495)   A prism with rectangles as bases.

regular polygon (474)   A polygon having all sides congruent and all angles congruent.

relative frequency (417)   The fraction of the events that have positive outcomes.

repeating decimal (169)   A decimal whose digits repeat in groups of one or more. Examples are 0.181818 . . . and 0.83333. . . .

rhombus (472)   A parallelogram with all sides congruent.

right angle (440)   An angle that measures 90°.

right triangle (457)   A triangle having a right angle.

sample (363)   A randomly selected group chosen for the purpose of collecting data.

sample space (411)   The set of all possible outcomes.

scatter plot (383)   A graph that shows the general relationship between two sets of data.

scientific notation (237)   A way of expressing a number as the product of a number that is at least 1 but less than 10 and a power of 10.

sequence (190)   A list of numbers in a certain order.

side (474)   A side of a polygon is any of the line segments that form the polygon.

significant digits (515)   The digits that are recorded when a measurement is made. These digits indicate the precision of the measurement.

similar figures (465)   Figures having the same shape.

simulation (429)   The process of acting out a problem.

sine (550)   If $\triangle ABC$ is a right triangle and A is an acute angle,

$$\text{sine of } \angle A = \frac{\text{measure of the side adjacent to } \angle A}{\text{measure of the hypotenuse}}.$$

slant height (501)   The length of the altitude of a lateral face of a regular pyramid.

slope (305)   The slope of a line is the ratio of the change in y to the corresponding change in x.

$$\text{slope} = \frac{\text{change in } y}{\text{change in } x}$$

solution (30)   A replacement for a variable for which a true sentence results. The solution for $y = 8 + 9$ is 17.

square (472)   A parallelogram with all sides congruent and all angles congruent.

square (527)   The product of a number and itself.

square pyramid (501)   A pyramid with a square base.

square root (526)   One of the two equal factors of a number. A square root of 144 is 12 since $12^2 = 144$.

standard form (131)   The standard form for seven hundred thirty nine is 739.

statement (70)   Any sentence that is either true or false, but not both.

statistics (363)   The study of collecting, analyzing, and presenting data.

stem-and-leaf plot (371)   A system used to condense a set of data where the greatest place value of the data forms the stem and the next greatest place value forms the leaves.

straightedge (452)   Any object that can be used to draw a straight line.

substitution property of equality (19)   For all numbers a and b, if $a = b$, then a may be replaced by b.

subtraction (35)   For all numbers a, b, and c, $a - b = c$ and $a - c = b$ if $b + c = a$.

subtraction property of inequalities (110)   For any numbers a, b, and c:
    1. If $a > b$, then $a - c > b - c$.
    2. If $a < b$, then $a - c < b - c$.

subtraction property of equality (89)   For any numbers a, b, and c, if $a = b$, then $a - c = b - c$.

supplementary angles (446)   Two angles are supplementary if the sum of their measures is 180°.

surface area (495)   The sum of the areas of all the surfaces (faces) of a 3-dimensional figure.

system of equations (311)   A set of equations with the same variables.

tangent (548)   If $\triangle ABC$ is a right triangle and A is an acute angle,

tangent of $\angle A = \dfrac{\text{measure of the side opposite to } \angle A}{\text{measure of the side adjacent to } \angle A}$.

term (26, 565)   A number, a variable, or a product of numbers and variables.

term of a sequence (190)   A number in a sequence.

terminating decimal (169)   A decimal whose digits end. Examples are 0.25 and 0.125.

tessellation (477)   A repetitive pattern of polygons that fit together with no holes or gaps.

transversal (449)   A line that intersects two lines to form eight angles.

trapezoid (472)   A quadrilateral with exactly one pair of parallel sides.

tree diagram (400)   A diagram used to show the total number of possible outcomes in a probability experiment.

triangle (456)   A polygon having three sides.

triangular prism (496)   A prism with triangles as bases.

trigonometric ratios (551)   Ratios that involve the measures of the sides of right triangles. The tangent, sine, and cosine ratios are three trigonometric ratios.

trigonometry (551)   The study of triangle measurement.

trinomial (565)   A polynomial with three unlike terms.

unit rate (320)   A rate with a denominator of 1.

upper quartile (375)   The median of the upper half of a set of numbers.

variables (19)   Placeholders in mathematical expressions or sentences.

variation (374)   The divergence in the values of a set of data.

Venn diagram (531)   A diagram consisting of circles inside a rectangle which is used to show the relationships of sets.

vertex (439)   A vertex of an angle is the common endpoint of the rays forming the angle.

vertex (456, 474)   A vertex of a polygon is a point where two sides of the polygon intersect.

vertex (501)   The vertex of a pyramid is the point where all the faces except the base intersect.

vertical angles (446)   Congruent angles formed by the intersection of two lines. In the figure, the vertical angles are $\angle 1$ and $\angle 3$; $\angle 2$ and $\angle 4$.

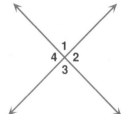

volume (506)   The number of cubic units needed to fill a space.

x-axis (293)   The horizontal line of the two perpendicular number lines in a coordinate plane.

x-coordinate (293)   The first number of an ordered pair.

x-intercept (305)   In a linear equation, the value of x when y is 0.

y-axis (293)   The vertical line of the two perpendicular number lines in a coordinate plane.

y-coordinate (293)   The second number of an ordered pair.

y-intercept (308)   In a linear equation, the value of y when x is 0.

Selected Answers

Chapter 1   The Language of Algebra

Page 17   Lesson 1-1
5. division; 48   **7.** division; 1   **9.** addition; 20
11. addition; 17   **13.** subtraction; 7   **15.** addition; 3
17. 47   **19.** 0   **21.** 16   **23.** 14   **25.** 16   **27.** 27   **29.** 2
31. 94   **33.** 72   **35.** 189   **37.** false   **39.** true   **41.** false
43. $(18 + 4) \times 3 = 66$ **45.** $24 \div 2 - (4 + 8) = 0$
47. $50 \times 6 + 8 \times 12$; $396

Pages 20–21   Lesson 1-2
5. 3   **7.** 25   **9.** 16   **11.** 15   **13.** 10   **15.** 73   **17.** 18
19. 20   **21.** 36   **23.** 8   **25.** 20   **27.** 30   **29.** 4   **31.** 19
33. 64   **35.** 12   **37.** division   **38.** subtraction   **39.** 26
40. true   **41.** $8 \cdot 7$; $(8)(7)$

Page 24   Lesson 1-3
5. commutative property of addition   **7.** multiplicative
property of zero   **9.** associative property of multiplication
11. identity property of multiplication   **13.** commutative
property of multiplication   **15.** multiplicative property of
zero   **17.** associative property of multiplication
19. multiplicative property of zero   **21.** commutative
property of addition   **23.** commutative property of
addition   **25.** 40   **27.** 130   **29.** 200   **31.** $12 + 9$
33. $2z + 7$ or $7 + z \cdot 2$ **35.** $(12 + 15) + x$; $27 + x$ **37.** $n(6 \cdot 5)$;
$30n$ **39.** $(3 \cdot 2)z$; $6z$ **41.** $(5 \cdot 4)x$; $20x$ **43.** 149
44. subtraction   **45.** 16   **46.** $ab \div 2$ **47.** 17   **48.** 9
49. Sample answer: wash dishes, dry dishes

Pages 27–28   Lesson 1-4
5. $8 \cdot 5 + 8 \cdot 9$ **7.** $5 \cdot 11 + 5 \cdot 12$ **9.** $5x + 5y$ **11.** $8(3 + 6)$
13. $9(x + y)$ **15.** $3(x + 2y)$ **17.** $8x$ **19.** $13k + 23$
21. $38x + 45$ **23.** $22a + 8$ **25.** $3y + 23$ **27.** $15f + 8g + 7q$
29. $33y + 35$ **31.** $22b + 42$ **33.** $14x + 18y$ **35.** $3y + 5y$
37. $2 \times 7 + 1 \times 7 + 1.25$ **39.** 184   **41.** 156   **43.** 430
45. 240   **46.** 24   **47.** $2 \times a \times b$; $2 \cdot a \cdot b$ **48.** 6
49. $x + (5 + 1)$ **50.** multiplicative property of zero   **51.** $12z$

Page 31   Lesson 1-5
5. 3   **7.** 29   **9.** 2   **11.** 3   **13.** 6   **15.** 14   **17.** 5   **19.** 24
21. 9   **22.** commutative property of addition   **23.** $10m + 1$
24. $11y + 21$ **25.** 180 ft^2

Page 34   Lesson 1-6
3. 30; 31   **5.** 33   **7.** 6 quarters   **9.** 11 2-point; 8 1-point

Page 34   Mid-Chapter Quiz
1. 33   **2.** 50   **3.** 2   **4.** 3   **5.** 144   **6.** $x + 12$ **7.** $15x$
8. $12x$ **9.** 29   **10.** 10

Pages 37–38   Lesson 1-7
5. $21 = 36 - 15$ or $15 = 36 - 21$ **7.** $14 = 11 + 3$
9. $c = 6 + 18$ **11.** $b \div 6 = 7$ or $b \div 7 = 6$ **13.** $28 = 4 \cdot 7$
15. $a = 7 \cdot 19$ **17.** 12   **19.** 24   **21.** 20   **23.** 3   **25.** 135
27. 12   **29.** 7   **31.** 0.83   **33.** 19.98   **35.** 3.77   **37.** 1
38. $5(x + y)$ **39.** $30y$ **40.** $53c + 6$ **41.** 9   **42.** $7 + x = 7$
49. 96   **51.** 2.5 m/s

Pages 41–42   Lesson 1-8
5. $w + 4$ **7.** $t - 9$ **9.** $11 + x$ **11.** $6 + z$ **13.** $20 - r$
15. $80 + x$ **17.** $n \div 7$ **19.** $z \cdot 6$ **21.** $j + 10$ **23.** $a + 300$

25. $2f$ **27.** $3n - 8$ **29.** the sum of a number and three
31. the product of 9 and a number   **33.** two times the sum
of a number and three   **35.** 0   **36.** $48a + 30$ **37.** 4
38. 26   **39.** true   **41.** Sandy Koufax   **43.** $k + 2$

Pages 44–45   Lesson 1-9
3. $2x = 6$ **5.** $x + 4 = 12$ **7.** $x - 5 = 7$ **9.** $x - 42 = 60$
11. $x - 9 = 15$; $24   **13.** $x - 18 = 72$; 90 games
15. $35x = 175$; $5   **17.** $x + 61 = 117$; 56 home runs   **18.** 45
19. $2n$ **20.** three times the sum of a number and two
21. no

Page 47   Lesson 1-10
5. false   **7.** true   **9.** open   **11.** false   **13.** true
15. false   **17.** true   **19.** true   **21.** false   **23.** less than
25. equal to   **27.** greater than   **28.** 9   **29.** $18 \div y$
30. $x - 6 = 15$ **31.** $4t = 56$; 14 seconds   **33.** $s < 200{,}000$

Pages 48–49   Chapter 1 Review
1. solved   **3.** multiply   **5.** $x \leq 2$ **7.** 18   **9.** 11   **11.** 133
13. 15   **15.** 15   **17.** 15   **19.** identity of addition
21. identity property of multiplication   **23.** distributive
property   **25.** $7 + x$ **27.** $5a + 10$ **29.** $14a + 12$ **31.** 5
33. 36   **35.** 5   **37.** 129   **39.** 36   **41.** 32   **43.** $5b$
45. $n \div 6$ **47.** true   **49.** false

Chapter 2   Integers

Pages 55–56   Lesson 2-1
7. -4, -3, -2, -1, 0   **9.** 3   **11.** 16
13.

15.
17. 15   **19.** 0   **21.** 5   **23.** 39   **24.** 19   **25.** 7   **26.** 3
27. $x + 23 = 41$; $18   **29.** 1   **31.** -1   **33.** fraction,
decimal   **35.** -6, 120

Pages 57–58   Lesson 2-2
7. $-5 < -2$ **9.** less than   **11.** greater than   **13.** greater
than   **15.** less than   **17.** {-3, 0, 10}   **19.** {-8, -4, -3, 0}
21. $144 > 135$ or $135 < 144$ **23.** 16   **25.** 8
24.

Pages 61–62   Lesson 2-3
5. negative   **7.** positive   **9.** 4   **11.** 15   **13.** 16   **15.** -18
17. 15   **19.** 26   **21.** -29   **23.** 16   **25.** -15   **27.** 24
29. 5   **31.** 1   **33.** -26   **35.** -27   **37.** $-40 + (-23.95)$;
-63.95   **39.** $16 + (-25)$; -9   **41.** $150 + (-275)$; -125

43. never **45.** $24y + 30$ **46.** $|-3|$ means that -3 is 3 units from zero on the number line. **47.** $3 < 8$ **48.** $30 < 31$
49. $n \div 3$ **51.** -123°C

Page 64 Lesson 2-4
3. positive **5.** negative **7.** positive **9.** -26 **11.** 6
13. -31 **15.** -26 **17.** 63 **19.** $-16x$ **21.** $-7y$
23. $-20d$ **25.** $-29f$ **27.** 1 **29.** -5 **31.** -71 **33.** false
34. Sample answers: 0, 2 **35.** 15 **36.** negative
37. $100 + (-15.95) + 18 + (-52.87)$; $49.18

Pages 67–68 Lesson 2-5
5. -6 **7.** 13 **9.** $-b$ **11.** $-cd$ **13.** $5x$ **15.** $7 + (-13) = x$
17. $-17 + (-9) = b$ **19.** $-18 + 16 = y$ **21.** -8 **23.** -6
25. 5 **27.** -18 **29.** -38 **31.** -15 **33.** -80 **35.** 0
37. 26 **39.** -5 **41.** -55 **43.** 6 **45.** $-10a$ **47.** $-29ab$
49. $-45d$ **51.** $41y$ **53.** 32 **54.** 21 **55.** 28 **56.** positive
57. 6 **58.** $-11x$ **59.** -8 **61.** -3 **63.** -$1475

Page 71 Lesson 2-6
5. true **7.** $3(11 - 5) = 8$; false; true **9.** false **11.** false
13. true **15.** Mozart was a poet; true; false
17. $3y - 2y \neq 1$; false; true **19.** $45 \leq 2(16 - 9)$; true; false
21. -25 **22.** -4

Pages 74–75 Lesson 2-7
3. negative **5.** positive **7.** negative **9.** negative
11. negative **13.** negative **15.** 66 **17.** -162 **19.** 68
21. -850 **23.** -270 **25.** -126 **27.** -48 **29.** 12
31. $-4b$ **33.** $-15x$ **35.** $30y$ **37.** $35yz$ **39.** $-3ab$
41. $40x$ **43.** 6 **44.** {-6, -2, 3} **45.** $22y$ **46.** -40
47. Some horses are not brown.

Page 75 Mid-Chapter Quiz
1. 2 **2.** 7 **3.** 5 **4.** greater than **5.** greater than
6. greater than **7.** -11 **8.** -14 **9.** 6 **10.** -8 **11.** -45
12. 15

Pages 77–78 Lesson 2-8
3. negative **5.** positive **7.** negative **9.** positive
11. negative **13.** negative **15.** 7 **17.** -12 **19.** 6
21. 3 **23.** -7 **25.** 49 **27.** -4 **29.** 10 **31.** 4 **33.** -12
35. -4 **37.** -3 **39.** 2 **40.** 166 pages **41.** 0 **42.** true
43. negative **44.** $-36x$ **45.** -50 students

Pages 80–81 Lesson 2-9
3. 111°F **5.** -4 **7.** 6194 m **9.** 630 ft **11.** 8th floor
13. positive **14.** -38

Pages 82–83 Chapter 2 Review
1. g **3.** a **5.** b **7.** 2 **9.** 3 **11.** 59 **13.** greater than
15. less than **17.** -8 **19.** -8 **21.** -5 **23.** 7 **25.** -5
27. -7 **29.** $-4a$ **31.** $-16c$ **33.** 136 **35.** -5 **37.** -6
39. 2 **41.** 1 **43.** 48 **45.** Violins are not members of the string family. **47.** 17°F

Chapter 3 Solving One-Step Equations

Pages 90–91 Lesson 3-1
5. -8 **7.** -38 **9.** -3 **11.** -75 **13.** -93 **15.** -19
17. -35 **19.** 43 **21.** 22 **23.** -25 **25.** -13 **27.** -38
29. 20 **31.** -23 **33.** $x + 850 = 10,935$; $10,085 **34.** 15

35. $24k + 3$ **36.** -7, -2, 9 **37.** -20 **38.** positive
39. 18 degrees

Pages 93–94 Lesson 3-2
5. $x - 9 = 15$ **7.** $p + 7 = -18$ **9.** $18 = t - 4$ **11.** $p + 17 = 2$
13. $m - 8 = -15$ **15.** -1 **17.** 79 **19.** -23 **21.** 34
23. -4 **25.** -74 **27.** -36 **29.** 76 **31.** -36
33. $x - 150 = 300$; $450 **34.** 12 more than 3 times x
35. 131 **36.** -22 **37.** no **38.** $g = 19$
39. $x + 34 = 21$; -13° **45.** -700

Pages 96–97 Lesson 3-3
5. -4 **7.** 8 **9.** -9 **11.** 4 **13.** -8 **15.** -13 **17.** -13
19. 19 **21.** -4 **23.** -4 **25.** -3 **27.** -5 **29.** -15
31. 48 pounds **33.** false **34.** 9 **35.** -52 **36.** 13
37. $y - 19 = 14$ **38.** -9 **39.** 14 in. **41.** 3 hours

Pages 99–100 Lesson 3-4
7. 12; 36 **9.** -16; -64 **11.** -7; -161 **13.** 13; -208
15. -98 **17.** -156 **19.** 126 **21.** 147 **23.** -210
25. -704 **27.** -408 **29.** -2889 **31.** -2535 **33.** -1704
35. 21,204 **37.** -148, 973 **39.** Add 3 and 4.
40. $45 < 55$ **41.** $x + 2900 = 5140$; 2240 feet **42.** 3 **43.** -7
44. -5 **45.** about 36 million gallons; about 900,000 barrels
47. 3 weeks

Pages 102–103 Lesson 3-5
5. 240 miles **7.** 400 mph **9.** 221 **11.** $49 **13.** 160
15. 30 **16.** 6 **17.** -6 **18.** 70 **19.** 8.5 gallons
21. 53 mph **23.** 1376 meters or about 1.4 kilometers

Page 103 Mid-Chapter Quiz
1. -182 **2.** -37 **3.** 643 **4.** -15 **5.** -8 **6.** 57 **7.** 4
8. -13 **9.** -35 **10.** 143 **11.** -405 **12.** 245 **13.** c

Pages 107–108 Lesson 3-6
5. 8 mi; 4 m² **7.** 16 cm; 15 cm² **9.** 28 ft; 33 ft²
11. 26 m; 36m² **13.** 322 yd; 4128 yd² **15.** 15 cm
17. 10 rs **18.** -532 **19.** $r = d \div 2$ **20.** 6 hours
23.   **25.** 84 ft

21 ft

21 ft

Pages 111–112 Lesson 3-7
5. $m > -1$ **7.** $16 < t$ **9.** $k \geq -15$ **11.** $y < 17$ **13.** $m > -5$
15. $z > 22$ **17.** $b < 2$ **19.** $w < -4$ **21.** $y > 27$ **23.** $z < 39$
25. $t > -11$ **27.** $r < -49$ **29.** $k < -16$ **31.** $x \geq -25$
33. $p \geq 19$ **35.** $x > 15$ **37.** $x \geq 13$ **39.** identity for
multiplication **40.** Some integers are not even. **41.** -7
42. 105 hours **43.** 16 miles **44.** 72 square feet
45. $6909 + x < 7687$; 778 teams

Pages 114–115 Lesson 3-8
5. 6; no **7.** -4; yes **9.** 9; no **11.** -6; yes **13.** $y > 13$
15. $z < 13$ **17.** $r \geq -18$ **19.** $x \geq 4$ **21.** $w \geq -146$
23. $m > 168$ **25.** $p \geq -168$ **27.** $h < 360$ **29.** $r \geq -18$
31. $n \geq -17$ **33.** $400 \geq a$ **35.** $x \leq -102$
37. $7x > -56$; $x > -8$ **39.** 3 **40.** 10 **41.** 135 square yards
42. $-6 < p$ **43.** $s \leq 19$ **45.** $x \leq 300 + 30$

Page 118 Lesson 3-9
3. false **5.** false **7.** $0 \leq n \leq 10$ **9.** false **11.** true
13. true **15.** true **17.** $3 < p \leq 11$ **19.** $-13 < m < (-6)$
20. false **21.** $-12ab$ **22.** no **23.** $30 \leq t \leq 85$

Pages 120–121   Lesson 3-10
3. $x + 4 = 16$ **5.** $\frac{x}{7} = 49$ **7.** $21x \le -84$ **9.** $4x > 76; x > 19$
11. $\frac{x}{8} > 72; x > 576$ **13.** $3x = 75; \$25$ **15.** $3x < 36; x < 12$
17. $x + 17 = 75; 58$ boxes   **18.** $f \ge (-27)$ **19.** $-4 < s < (-2)$

Pages 122–123   Chapter 3 Review
1. h **3.** d **5.** b **7.** j **9.** 5   **11.** -4   **13.** -26
15. -12   **17.** 110   **19.** 288   **21.** 3 mph   **23.** 72 mm
25. 80 ft   **27.** 7500 yd^2 **29.** $a > 2$ **31.** $m < 40$
33. $c < (-6)$ **35.** $d < 84$ **37.** $k \ge 84$ **39.** false   **41.** true
43. $5x = 60; 12$ times   **45.** $\frac{x}{5} = 4; \$20$

Chapter 4   Factors and Fractions

Pages 129–130   Lesson 4-1
5. yes   **7.** yes   **9.** 2   **11.** 2, 3, 6   **13.** yes   **15.** no
17. yes   **19.** no   **21.** 3891   **23.** 183   **25.** 120   **27.** 1
29. 1 and the number   **30.** 22   **31.** 5   **32.** 72   **33.** true
34. -7   **35.** 7 mph   **36.** $t > 60$ **37.** 1600, 2000

Pages 132–133   Lesson 4-2
7. $m \cdot m$ **9.** $a \cdot a \cdot a \cdot a \cdot a$ **11.** 2^3 **13.** 1^4 **15.** 6^3 **17.** t^5
19. a^3 **21.** false   **23.** true   **25.** 81   **27.** 48   **29.** 250
31. 64   **33.** 121   **35.** 78, 125   **37.** 7   **38.** 19
39. $0.3 \le b < 1.6$ **40.** true   **41.** yes
42. $2 + 3 + 4 + 5 = 14$, so 3 is not a factor   **47.** s^3

Page 135   Lesson 4-3
5. 30 people   **7.** 15 handshakes   **9.** $(4 + 3) \times 6 + 3 = 45$

Pages 137–138   Lesson 4-4
3. composite   **5.** prime   **7.** 2   **9.** 7   **11.** $3 \cdot 3 \cdot 7$
13. $2 \cdot 7 \cdot 7$ **15.** $2 \cdot 3 \cdot 5$ **17.** 7^2 **19.** $2^2 \cdot 7$ **21.** $-1 \cdot 2 \cdot 13$
23. $2^4 \cdot 5$ **25.** $-1 \cdot 2^4 \cdot 17$ **27.** $2^2 \cdot 7 \cdot x^2 \cdot y$
29. $-1 \cdot 2^3 \cdot 3^2 \cdot a \cdot b^3$ **31.** $2 \cdot 3 \cdot 7 \cdot x \cdot y^2$ **33.** $2 \cdot 3 \cdot 5 \cdot 7 \cdot a \cdot b^3$
35. $\{-6, -2, 0, 3\}$ **36.** 66   **37.** 8^3 **38.** 4

Pages 141–142   Lesson 4-5
5. 4   **7.** 10   **9.** $12ab$ **11.** $9a$ **13.** 1   **15.** 12   **17.** 14
19. 12   **21.** 5   **23.** 72   **25.** 3   **27.** $14k$ **29.** 26 m^2
31. 9   **33.** $8y$ **35.** 6   **37.** $6m$ **38.** $-19xy$ **39.** 30 cm
40. $78 > m$ **41.** no   **42.** 108   **43.** 14 miles
45. 7 years old

Pages 144–145   Lesson 4-6
3. $\frac{7}{8}$ **5.** $\frac{9}{10}$ **7.** yes   **9.** no, $\frac{5}{7}$ **11.** no, $\frac{17}{20}$
13. no, $\frac{1}{2y}$ **15.** no, $\frac{z}{2}$ **17.** $\frac{1}{5}$ **19.** $\frac{3}{4}$ **21.** $\frac{3}{5}$
23. $\frac{1}{3}$ **25.** $\frac{2}{3}$ **27.** $\frac{5}{7}$ **29.** $\frac{1}{2}$ **31.** $\frac{6}{7}$ **33.** $\frac{5}{8}$
35. $\frac{9}{22}$ **37.** $\frac{4m}{5}$ **39.** $\frac{3s}{10}$ **41.** $\frac{7}{13}$ **43.** $\frac{k^2}{3}$
45. The GCF of 10 and 15 is not 5.   **46.** -22   **47.** 17
48. composite   **49.** 2   **51.** $\frac{4}{7}$

Page 145   Mid-Chapter Review
1. no   **2.** yes   **3.** yes   **4.** no   **5.** 15   **6.** 1   **7.** $\frac{41}{56}$

8. -4   **9.** 28   **10.** 0^3 **11.** $3 \cdot 5^2$
12. $-1 \cdot 2^2 \cdot 3^2 \cdot 5$ **13.** $2 \cdot 7 \cdot 11 \cdot a^2 \cdot b \cdot c^3$ **14.** 18
15. $5x$ **16.** 12   **17.** $12xy$

Pages 147–148   Lesson 4-7
3. 0, 12, 24, 36   **5.** 0, 30, 60, 90   **7.** 10   **9.** 72   **11.** 60
13. yes   **15.** yes   **17.** yes   **19.** no   **21.** yes   **23.** yes
25. 30   **27.** 24   **29.** 90   **31.** 35   **33.** 84   **35.** 288
37. $84y$ **39.** $48xy$ **41.** $24k$ **43.** 84   **45.** -45   **46.** false
47. 11 amps   **48.** $3 \cdot 3 \cdot 5 \cdot a \cdot a \cdot b \cdot b \cdot b$ **49.** 13   **51.** 60 in.

Pages 150–151   Lesson 4-8
3. 10   **5.** 8   **7.** a^2 **9.** $25a^2$ **11.** less than   **13.** less
than   **15.** 4   **17.** 35   **19.** $120a$ **21.** $24k^2m$
23. greater than   **25.** greater than   **27.** greater than
29. greater than

31.

←—+——+——●——●——+——+——●——+——→
$\;-2\quad -1\quad 0\quad 1\quad 2\quad 3\quad 4$

32. $-33 < x$ **33.** false
34. 9   **35.** yes   **36.** $\frac{7r}{s^2}$ **37.** $\frac{1}{2}$

Pages 153–154   Lesson 4-9
5. $8 \cdot 4 = 32$ **7.** $32 \cdot 4 = 128$ **9.** $9 \cdot 81 = 729$ **11.** y^9
13. 3^{10} **15.** 8^6 **17.** m^4 **19.** $12a^5$ **21.** $-15x^6$
23. $-20x^5y$ **25.** a^3b^4 **27.** m^7b^2 **29.** $7^4 \cdot 4^5$ **31.** 13
32. 11   **33.** $8a$ **34.** false   **35.** 63

Pages 157–158   Lesson 4-10
5. $64 \div 4 = 16$ **7.** $625 \div 625 = 1$ **9.** y^2 **11.** x^{-1} or $\frac{1}{x}$
13. $(-2)^1$ or -2   **15.** $(-7)^2$ **17.** a^4 **19.** c^0 or 1
21. m^4 **23.** y^0 or 1   **25.** 1   **27.** $\frac{1}{10^2}$ **29.** $\frac{1}{x}$ **31.** $\frac{t^3}{s^2}$
32. false   **33.** 48 ft^2 **34.** $\frac{x}{4} \ge 9$ **35.** $\frac{1}{7}$ **36.** 36
37. $\frac{3}{11}$ **38.** $36r^2$ **39.** $\frac{21}{82} \ne \frac{1}{4}$ **41.** 10^{-2} has a negative
exponent.   **43.** 2   **45.** 10^6 or 1 million

Pages 160–161   Chapter 4 Review
1. true   **3.** false, $\frac{2}{3}$ **5.** true   **7.** false, prime   **9.** true
11. yes   **13.** yes   **15.** 2, 3, 6   **17.** 2, 3, 5, 6, 10   **19.** 4^2
21. 3^3 **23.** 48   **25.** 250   **27.** $-1 \cdot 2 \cdot 5^2$
29. $2^3 \cdot 5^2 \cdot a \cdot a \cdot b$ **31.** 16   **33.** 12   **35.** $6a^2b$ **37.** $\frac{a}{4}$
39. $\frac{9x}{5y}$ **41.** 42   **43.** $48ab$ **45.** less than   **47.** less than
49. $6x^5$ **51.** x^6y^4 **53.** y^0 or 1   **55.** $\frac{1}{4^2}$ **57.** 3 hours

Chapter 5   Rational Numbers

Page 168   Lesson 5-1
5. Rational   **7.** Rational   **9.** equal to   **11.** Integer,
Rational   **13.** Whole, Integer, Rational
15. $\frac{2}{1}, \frac{4}{2}$

←——+——+——+——●——+——→
$\;-1\quad 0\quad 1\quad 2\quad 3$

17. $-3, \frac{-6}{2}$

19. greater than **21.** greater than **23.** less than
25. -23 **26.** 6 or more hours **27.** prime **28.** $-21m^5$
29. $\frac{11}{16}$-inch

Page 171   Lesson 5-2
11. 7 **13.** 53 **15.** 408; 51 **17.** 100 **19.** $\frac{1}{2}$ **21.** $\frac{8}{25}$
23. $\frac{21}{25}$ **25.** $\frac{49}{50}$ **27.** $-\frac{33}{50}$ **29.** $-2\frac{13}{50}$ **31.** $-9\frac{16}{25}$
33. $5\frac{31}{100}$ **35.** $-\frac{93}{125}$ **37.** $1\frac{51}{100}$ **39.** $-\frac{281}{500}$ **41.** $\frac{243}{500}$
43. $9\frac{313}{500}$ **45.** $\frac{1}{3}$ **47.** $1\frac{4}{9}$ **49.** $\frac{4}{33}$ **51.** $\frac{5}{33}$
53. $2\frac{5}{11}$ **56.** $8+x$ **57.** $u=\frac{t}{w}$ **58.** 28 **59.** $\frac{4}{-5}, -\frac{4}{5}$
60. false **61.** $\frac{23}{1000}$-inch

Page 173   Lesson 5-3
3. 7 **5.** 12 **7.** 40 **9.** $45 **11.** 33, no **13.** 40, yes
15. 6.0 **17.** 24.7 **19.** 95.5 **21.** 49.6 **23.** $3 **25.** $6
27. no **30.** 9^3+2
31.

32. $-\frac{17}{50}$ **33.** $2+4+1+2=9$; yes

Pages 175–176   Lesson 5-4
5. 2.31 **7.** 4.22 **9.** 88.55 **11.** Commutative Property of
Addition **13.** Closure Property of Addition **15.** 104.77
17. 8.61 **19.** 0.2051 **21.** 2.7 **23.** -8.81 **25.** -6.48
27. $7.3s-4$ **29.** $14.29w$ **31.** $3.08y$ **33.** 5.18 **35.** 9.08
37. 13.72 **38.** $k\le 11$ **39.** 128 **40.** $3\frac{1}{5}$ **41.** 24.6
42. 23 miles **43.** $4c$ **45.** $-\$21.49$

Pages 178–179   Lesson 5-5
5. $\frac{7}{8}$ **7.** $\frac{7}{17}$ **9.** $\frac{12}{25}$ **11.** $\frac{8}{9}$ **13.** $\frac{2}{13}$ **15.** $1\frac{1}{3}$
17. $1\frac{7}{18}$ **19.** $1\frac{1}{3}$ **21.** $\frac{7}{10}$ **23.** $3\frac{1}{2}n$ **25.** $9x$ **27.** $-t$
29. false **30.** 8 **31.** 100 **32.** -23.4 **33.** $18.41d$
35. $21\frac{1}{4}$ inches

Page 179   Mid-Chapter Review
1. equal to **2.** less than **3.** greater than **4.** greater
than **5.** $\frac{3}{5}$ **6.** $\frac{7}{20}$ **7.** $-2\frac{2}{5}$ **8.** $\frac{2}{3}$ **9.** $1\frac{5}{9}$
10. $20 **11.** 30 **12.** 32 **13.** 13.4 **14.** 18.4 **15.** -0.7
16. 2 **17.** $\frac{1}{2}$ **18.** 4

Page 182   Lesson 5-6
3. 8, $\frac{5}{8}$ **5.** 14, $\frac{3}{7}$ **7.** 8, $\frac{1}{8}$ **9.** 12, $\frac{1}{3}$ **11.** 7
13. 38 **15.** $1\frac{9}{14}$ **17.** $\frac{1}{2}$ **19.** $-8\frac{1}{2}$ **21.** $-4\frac{1}{4}$
23. $4\frac{6}{7}$ **25.** $3\frac{1}{4}$ **27.** $-1\frac{1}{5}$ **29.** $5\frac{19}{20}$ **30.** 21.34
31. -12.74 **32.** $1\frac{2}{7}$ **33.** $3\frac{1}{2}$ **34.** $\frac{1}{10}$ **35.** $8\frac{3}{4}$ inches

Pages 186–187   Lesson 5-7
5. -7.9 **7.** -9.2 **9.** -15 **11.** $a<5$ **13.** 8.4 **15.** $-4\frac{1}{2}$
17. 14 **19.** -21 **21.** $4\frac{1}{8}$ **23.** $5\frac{1}{2}$ **25.** $a>3\frac{1}{2}$

27. $n<-4.3$ **29.** $f\ge -10.1$ **31.** $p>-3.5$ **33.** $a\ge -\frac{1}{2}$
35. $f\ge 8\frac{1}{4}$ **37.** $x\le -4.2$ **39.** $b=3.78$ **40.** -5
41. 8 ft **42.** -16 **43.** $-2\frac{3}{5}$ **44.** $-2\frac{3}{4}$ **45.** 2.93
46. $\frac{11}{15}$ **47.** $-\frac{5}{12}$ **48.** $1\frac{5}{56}$ **49.** $2\frac{1}{2}+x\ge 5\frac{1}{3}$; $2\frac{5}{6}$ lb
51. $269.5+x=298.2$; 28.7 miles **53.** 44.2°

Page 189   Lesson 5-8
3. 123,454,321 and 12,345,654,321 **5.** red **7.** 15, 20, 26;
9th **9.** no **11.** 4 games won so far

Pages 192–193   Lesson 5-9
5. 11, 13, 15 **7.** 89, 84, 79 **9.** 78, 74, 70 **11.** 7.45, 7.6,
7.75 **13.** no **15.** yes **17.** 19, 23, 27 **19.** 16, 20, 24
21. 9.01, 10.13, 11.25 **23.** 26, 37, 50 **25.** 125, 216, 343
27. 0; 3 **29.** 78 **31.** -8 **32.** 72
33. $z\ge -1\frac{3}{5}$ **34.** at least $5\frac{3}{4}$ yards **35.** $810

Pages 196–197   Lesson 5-10
5. 2 **7.** 8800 **9.** 12,000 **11.** 8 **13.** 6600 **15.** 12 yd
17. 11 lb 4 oz **19.** 7 gal **21.** 3 lb 11 oz **23.** 40 ft
25. 2400 pounds **27.** 2 ft 5 in. **28.** $p<-35$ **29.** $\frac{1}{3^2}$
30. $3\frac{7}{8}$ **31.** $-10, -16, -23$

Pages 198–199   Chapter 5 Review
1. e **3.** h **5.** a, b, c, d, e **7.** less than **9.** equal to
11. $\frac{17}{20}$ **13.** $\frac{2}{25}$ **15.** $-2\frac{9}{25}$ **17.** $\frac{3}{11}$ **19.** $7 **21.** $3
23. $16 **25.** 140 **27.** Commutative Property of Addition
29. Closure Property of Addition **31.** 1.11 **33.** 5.8
35. -5.5 **37.** 24 **39.** 14 **41.** 1 **43.** $2\frac{1}{4}$ **45.** $\frac{1}{6}$
47. $8\frac{19}{40}$ **49.** Add 9 to each side. **51.** 32, 35, 38; yes
53. 8 gal 1 qt **55.** 7 yd 2 ft **57.** 212 yd

Chapter 6   Multiplying and Dividing Patterns

Pages 206–207   Lesson 6-1
11. $-0.\overline{3}$ **13.** 0.125 **15.** $-0.1\overline{6}$ **17.** $-2.8\overline{3}$ **19.** 1.7
21. 0.7 **23.** -0.375 **25.** $0.\overline{1}$ **27.** $0.3\overline{8}$ **29.** 0.4375
31. $2.\overline{5}$ **33.** $0.\overline{5}$ **35.** $0.5\overline{1}$ **37.** 2.5625 **39.** 0.775
41. 0.8888888 **43.** .89 **44.** no **45.** $m^2 n^5 p^3$ **46.** true
47. 26.5 **48.** $\frac{6}{7}$ **49.** $b<4\frac{1}{8}$ **51.** $1.625 or $1.63

Page 210   Lesson 6-2
3. $35\div 7$; 5 **5.** $\frac{1}{3}\times 6$; 2 **7.** $\frac{1}{2}\times 10$; 5 **9.** 4 **11.** $6
13. 4 **15.** 3 **17.** 8 **19.** 10 **21.** 20 **23.** 7 **24.** -0
25. 3.375 **27.** less than

Pages 213–214   Lesson 6-3
5. positive **7.** positive **9.** $\frac{3}{5}$ **11.** $-\frac{5}{3}$ **13.** $-\frac{11}{5}$ **15.** 15
17. $-\frac{3}{2}$ **19.** $-\frac{35}{2}$ **21.** -3 **23.** -4 **25.** 10 **27.** $-\frac{133}{18}$
29. $\frac{1}{4}$ **31.** $\frac{25}{64}$ **33.** $\frac{16}{25}$ **35.** $\frac{9}{8}$ **37.** $\frac{4}{9}$ **39.** -4

41. $-1\frac{2}{3}$ **43.** -12 **44.** $\frac{a^2}{3b^2}$ **45.** $-2\frac{11}{25}$ **46.** $15.5d$

47. -5.46 **48.** 4 **49.** $1\frac{1}{4}$ cups

Pages 217–218   Lesson 6-4
9. 108.60 **11.** 357.24 **13.** 14.08 **15.** 59 **17.** 20.88
19. 36.54 **21.** -0.588 **23.** 1160 **25.** 127.9 **27.** -30.5
29. 3.015 **31.** 172.97 **33.** 5.13 **35.** 2.7824
37. 81.095406 **39.** 0.01564 **41.** 20 **43.** -15 **45.** 004
47. $18b^2$ **48.** $3\frac{5}{6}$ **49.** 53.5 inches **50.** 3 **51.** $-6\frac{3}{7}$
52. $\frac{3}{4}$ **53.** 67, 567 pesos **55.** \$3.2 billion

Page 221   Lesson 6-5
5. no **7.** yes **9.** $-\frac{9}{8}$ **11.** $\frac{5}{4}$ **13.** -1 **15.** $\frac{y}{x}$
17. Closure Property of Multiplication **19.** Inverse Property of Multiplication **21.** Distributive Property
23. 14 **25.** 68 **27.** $3\frac{1}{8}$ **29.** $12\frac{1}{2}$ **30.** true **31.** -4
32. -0.312 **33.** 41.41

Pages 223–224   Lesson 6-6
5. $\frac{5}{6} \cdot \frac{11}{10}; \frac{11}{12}$ **7.** $\frac{13}{15} \cdot \frac{1}{6}; \frac{13}{90}$ **9.** $\frac{13}{5} \cdot \frac{7}{27}; \frac{91}{135}$ **11.** $\frac{68}{9} \cdot \frac{-1}{8}; \frac{-17}{18}$
13. -4 **15.** $\frac{10}{3}$ **17.** $\frac{21}{16}$ **19.** $-\frac{9}{2}$ **21.** -10 **23.** 6 **25.** $\frac{5}{2}$
27. 6 **29.** $-\frac{25}{4}$ **31.** $-\frac{3}{2}$ **33.** $\frac{1}{2}$ **35.** $\frac{81}{256}$ **37.** 54
38. 44.0 **39.** \$54.70 **40.** $-\frac{4}{13}$ **41.** $\frac{3}{5}$ **43.** $2\frac{1}{4}$ inches

Pages 226–227   Lesson 6-7
7. 2.4 **9.** 90 **11.** .009 **13.** 80 **15.** -31 **17.** 3.21
19. 20 **21.** -15 **23.** 5 **25.** 2.1 **27.** 1.6 **29.** 39.0
31. 20 **32.** $9y$ **33.** Distributive Property **35.** 24 table tops

Page 227   Mid-Chapter Review
1. 0.6 **2.** 14 **3.** $8 \times 3 + 8 \times \frac{1}{4}$ **4.** 20 **5.** 0.0072
6. $-\frac{1}{2}$ **7.** $\frac{16}{81}$ **8.** -4.5 **9.** $\frac{9}{16}$

Page 230   Lesson 6-8
3. 20 **5.** $4 < c$ **7.** $-\frac{18}{25}$ **9.** 1.65 **11.** $t < -\frac{25}{22}$ **13.** -0.5
15. -13.6 **17.** -0.3 **19.** -1.08 **21.** $r \geq -1\frac{1}{9}$
23. $h < -2.3$ **25.** $x \geq -8$ **27.** 32 **28.** $f < 1.5$ **29.** $-\frac{3}{8}$
30. 4.38 **31.** about $\frac{2}{5}$ **33.** 50; 60

Page 232   Lesson 6-9
5. deductive **7.** inductive **9.** deductive **11.** 20, 25
13. 1,000, 10,000

Page 235   Lesson 6-10
5. no **7.** yes; $\frac{1}{2}$ **9.** yes; $\frac{1}{2}$ **11.** 17, 23, 30 **13.** 19, 23, 27
15. $-3\frac{1}{2}$, $-5\frac{1}{2}$, $-7\frac{1}{2}$ **17.** $\frac{5}{16}$, $-\frac{5}{32}$, $\frac{5}{64}$ **19.** 40 **21.** 2
22. 6.11 **23.** $f < -\frac{8}{21}$ **25.** \$2881.20

Page 239   Lesson 6-11
3. $1.59; 10^2$ **5.** $1.800; 10^3$ **7.** $8.5; 10^{-1}$ **9.** $5.93; 10^{-5}$
11. 520,000 **13.** $-61,000$ **15.** -0.005765 **17.** 239,000
19. 0.00005 **21.** 5.04×10^{-1} **23.** 9.7×10^4
25. 6.23×10^{-3} **27.** 8.92×10^{-6} **29.** 3.0×10^{-4} **31.** 4
32. no **33.** $t < -0.92$ **34.** inductive **35.** 2.997928 EE 10

Pages 241–242   Lesson 6-12
5. -5.049 **7.** $1\frac{3}{5}$ **9.** 1.2×10^4 **11.** less than
13. greater than **15.** greater than **17.** less than

19. less than **21.** less than **23.** equal to **25.** greater than **27.** greater than **28.** $21x^2$ **29.** $1\frac{1}{8}$ **30.** no
31. 2.4×10^{-2} **33.** Mets

Pages 244–245   Lesson 6-13
5. $24 \cdot 3.14$ **7.** $2 \cdot 2\frac{1}{2} \cdot \frac{22}{7}$ **9.** $18 \cdot 3.14$ **11.** $2 \cdot 1.3 \cdot 3.14$
13. 34.54 mm **15.** 198 ft **17.** 51.496 km **19.** 88 yd
21. -7.0×10^3 **22.** $-\frac{2}{3}$ **23.** 45.4 **25.** about 7962 mi

Pages 246–247   Chapter 6 Review
1. repeating **3.** reciprocals **5.** inverse **7.** 4.75 **9.** $0.\overline{2}$
11. 10 **13.** 130 **15.** $-2\frac{1}{4}$ **17.** 3 **19.** $-3\frac{1}{2}$ **21.** 0.2
23. 175 **25.** $2\frac{1}{4}$ **27.** -0.8 **29.** 6.8 **31.** $a \leq -18$
33. -25 **35.** no **37.** yes; 12, 3, $\frac{3}{4}$ **39.** 6.5×10^4
41. 2.1×10^{-3} **43.** equal to **45.** greater than
47. 37.68 cm **49.** 88 ft **51.** deductive

Chapter 7   Solving Equations and Inequalities

Pages 252–253   Lesson 7-1
1. 12 **3.** 12 **5.** 36 exercises **7.** 216 pages
9. 25 bacteria

Pages 257–258   Lesson 7-2
9. Add 7; -2 **11.** Subtract 4; 98 **13.** Add 19; 144
15. Multiply by -5; -9 **17.** -4 **19.** -12 **21.** 17
23. -15 **25.** -12 **27.** 132 **29.** 84 **31.** 168 **33.** -19
35. -231 **37.** 252 **39.** 33 **41.** 10 **43.** 16
45. $20 + 2c = -30$ **47.** $\frac{e}{4} - 5 = 7$ **48.** $15 \div 3$
49. true **50.** 12.6 **51.** 12 ft 6 in. **52.** 0.72 **53.** $\frac{4}{9}$
54. $\frac{7}{12}$ **55.** 50 chips

Pages 260–261   Lesson 7-3
3. Let x = number of years; $4800 - 35x = 4520$; 8 years
5. $50 + 15s = 170$; 8 scoops **7.** $1.10 + 1.40m = 12.30$;
8 miles **9.** $95 = 2t - 15$; 55 **11.** 7.5
12. $3(n + 5) = 27$

Pages 265–266   Lesson 7-4
5. $5d + 15$ **7.** subtract $2k$ from each side; subtract 10 from each side; -31 **9.** subtract $5y$ from each side; add 6 to each side; 6 **11.** subtract $2p$ from each side; subtract 8 from each side; 2 **13.** 53 **15.** -21 **17.** 3 **19.** 3.5
21. $\varnothing$ **23.** -2.5 **25.** all numbers **27.** 15 **29.** 10
31. 1.5 **33.** -2 **35.** 2 **37.** false **38.** $\frac{5}{11}$ **39.** inductive reasoning **40.** -11 **41.** 12 years

Page 266   Mid-Chapter Review
1. 10 **2.** 13 **3.** 2 **4.** -20 **5.** 4 **6.** 5 **7.** 24 caps
8. Let n = number; $3 + 2n = -9$; -6

Page 268   Lesson 7-5
3. -6 **5.** 1.5 **7.** -16 **9.** $\frac{35}{2}$ or $17\frac{1}{2}$ **11.** -4 **13.** 2

15. -1 17. 2 19. -3 21. 63 23. 10 25. 7 hours
26. 3.36×10^{-4} 27. 5 29. $2y - 682{,}742$

Page 271   Lesson 7-6
3. $x > 2$ 5. $y > -6$ 7. $-1 > y$ 9. $a > 11$ 11. $g \le 7$
13. $z < -93$ 15. $t > -4$ 17. $y < 6$ 19. $1.17 < t$
21. $-11 \ge k$ 23. $m < -2$ 25. $x > -50$ 27. $x < 0.25$
29. $y > -5$ 31. $3n - 5 > 16$ 33. $\frac{1}{2}m - 5 > 16$ 34. 36
35. $w \le 13.2$ 36. about 4 37. true 38. -8 39. 6
40. 3 41. $2(3.49) + s \le 10;\ s \le 3.02$

Page 274   Lesson 7-7
3. $8x + 10 \ge 60$ 5. $55c + 35 \le 400$ 7. $12n - \frac{1}{20}n <$
3250 9. $3n + n \le 120;\ \$90$ 11. $250(24) + 1200 \ge x;$
$\$7200$ 13. -1.2 14. $-5 > b$ 15. $x \ge 4$

Pages 276–277   Lesson 7-8
5. 2 7. 9400 9. 3 11. 5 13. 60,000 15. 3000
17. 5000 19. 2 21. 0.018 23. 7.3 25. 0.053
27. 2 29. 0.25 31. 10,600 33. 11,000 35. -7, -10,
-13 36. $5\frac{1}{6}$ 37. $2 > y$ 38. $n + (n + 2) \ge 12$
39. $p \le 238$ 41. 65 cm, 55 cm

Pages 278–279   Chapter 7 Review
1. multiply 3. $\varnothing$ 5. greater than or equal to 7. -18
9. -8 11. -0.6 13. -9 15. 56 17. -1 19. 2
21. -3 23. all numbers 25. $m < 4$ 27. $6 \le x$
29. $-3.2 < z$ 31. 500 33. 2.9 35. 700 37. 100 boxes
39. $\frac{c}{2} + 30 = 150;\ \$240$ 41. $8x - 2 < 15;\ x < \frac{17}{8}$

Chapter 8   Graphing Equations and Inequalities

Page 285   Lesson 8-1
7. -2 9. 3 11. $\frac{5}{3}$ 13. -8 15. 5 17. -6 19. 4
21. -2 23. a 25. $(12 + 7) + m$ 26. 81
27. $6x + 11 \le 35$

Pages 287–288   Lesson 8-2
7. $x < 1$ 9. $x \le 0$ 11. $x > -2$ 15. $x < 4.0$
13. $a \le 0$;

17. $x \le 2$

19. $x > 0.5$ 23. $m < -3$
21. $c > -4$

25. $r < 1$

27. $x > -6$ 30. 5 31. false 32. $110
29.

33. -6 35. $x < 9.3$

Pages 291–292   Lesson 8-3
5. no 7. yes 9. (1, 0) 11. Z 13. F 15. L 17. C
19. Y 21. (4, 1) 23. (8, 3) 25. (5, 6) 27. (9, 8)
29. (6, 0) 31. (0, 9) 32. $-22ab$ 33. -8 34. 1
35. $x > -2$ 37. triangle

Page 295   Lesson 8-4
5. $\frac{1}{2}$; -5; quadrant 4 7. -1; -4.7; quadrant 3 9. -2.25; 0;
none 11. 0; 0; none 25. three
13.–23.
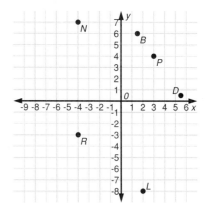

27. one 28. -12 29. $2 \cdot 2 \cdot 7 \cdot a \cdot a \cdot b \cdot b \cdot b$
30. 5.875 31. $m \ge -14$ 33. triangle

Pages 297–298   Lesson 8-5
5. (-3, 2), (0, 5), (4, 9), (9, 14) 7. (2, 2), (-1, -13), (0, -8),
(4, 12) 9. (0, 3.2), (1, 4.2), (-1, 2.2), (-2, -1.2) 11. (0, 0),
(1, 2), (-1, -2), (2, 4) 13. (0, 0), (1, 6), (-1, -6), (2, 12)
15. (0, -3), (1, -1), (-1, -7), (2, 1) 17. (0, 3), (1, -2), (-1, 8),
(2, -7) 19. $(0, 5), \left(1, 5\frac{1}{2}\right), (2, 6), (-2, 4)$ 21. $(0, 1), \left(1, 1\frac{2}{3}\right),$
(3, 3), (-3, -1) 23. (0, 6), (1, 5), (-1, 7), (6, 0) 25. (0, 5),
(1, 3), (-1, 7), (2, 1) 27. (4, 0), (1, -3), (5, 1), (-2, -6)
29. (1, -1), (-1, 1), (4, -4), (0, 0) 31. (1, 10.75), (0, 8.5),
(2, 13), (-2, 4) 33. $r < -55$ 34. $k < -10$ 35. 3
36.–37.
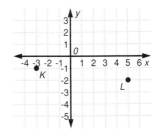

39. (-2, -5), (0, -1), (2, 3)

41. $(45, 45), (30, 60), (10, 80), (70, 20), (55, 35)$

Pages 300–301 Lesson 8-6

5.
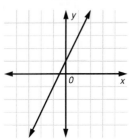

9.
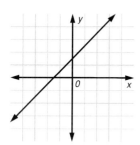

13.
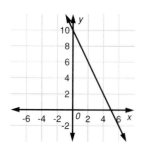

17.
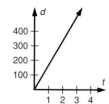

23. $x = y + 3$ **25.** $a + b = 0$ **26.** $\frac{3}{11}$ **27.** $\frac{3 + x}{5} = 11$

28. $(0, 1), (1, -2), (-1, 4), (2, -5)$ **29.** 6

31.
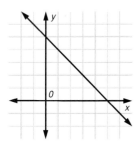

Pages 301 Mid-Chapter Quiz

1.

3.
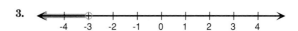

5.–7.
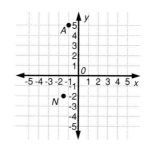

9.
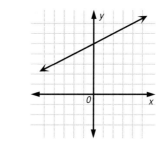

Pages 303–304 Lesson 8-7

7. $\left(\frac{1}{2}, 25\right), (2, 100)$ **9.** 156 lb **11.** $4 per hour; $24

Pages 306–307 Lesson 8-8

5. $-\frac{3}{4}$ **7.** $\frac{1}{3}$ **9.** 0 **11.** 1 **13.** $2\frac{2}{3}$ **15.** $\frac{3}{8}$ **17.** $-\frac{1}{2}$

19. $9x^3$ **20.** $2m + c$ **21.** 10 **29.** 0.1

22.
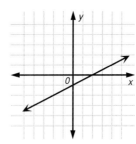

31. a. Oct. 25 b. 5 weeks c. Aug. 30

Pages 309–310 Lesson 8-9

5. 3, -3 **7.** -2, -2 **9.** $\frac{4}{3}$, -4 **11.** $\frac{1}{3}$, $-\frac{2}{3}$ **13.** 0.6, -4.2

15. 1, -1   **17.** $-\frac{3}{2}$, 3   **19.** 2, 10   **21.** 6, 3   **23.** 9.3
25. $-\frac{7}{8}$

24.

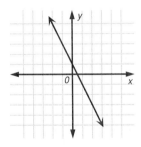

Pages 312–313   Lesson 8-10
5. (-2, 2)   **7.** (-6, -3)   **9.** (3.5, 1)   **11.** (4.5, 0)
13. (-1, -3)   **15.** (3, -2)   **17.** (-1, -1)   **19.** (0, 0)
21. (-1, 2)   **23.** $\left(\frac{1}{2}, 2\right)$ **24.** 2   **25.** $\frac{1}{2}$ **27.** (300, 1500)

26.

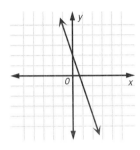

Pages 316–317   Lesson 8-11
7.

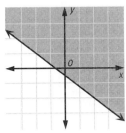

9.

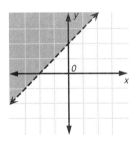

13.

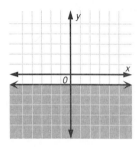

17.

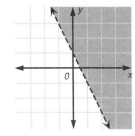

21. $\frac{2}{3}$ **22.** -3; -3
23. (2, -6)

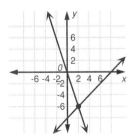

Pages 318–319   Chapter 8 Review
1. c   **3.** e   **5.** a   **7.** i   **9.** j   **11.** -4   **13.** -9
15. $x \geq -5$ **17.** $x \leq 1$ **19.** $a \geq -2$ **21.** $x \geq -6$
23.–25

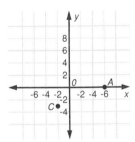

27. (0, 0), $\left(1, \frac{3}{2}\right)$, (2, 3), (-2, -3)

29. $(0, -3), (1, 2), (2, 7), (-1, -8)$ **35.** -6

31.

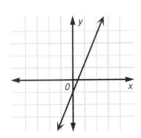

37. $\frac{2}{5}$

39.

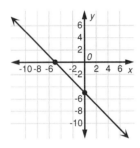

45.

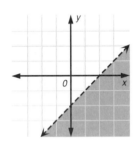

49. $83°$ F

Chapter 9   Proportion and Percent

Page 325   Lesson 9-1

3. $\frac{5}{7}$ **5.** $\frac{28}{1}$ **7.** $\frac{3}{5}$ **9.** $\frac{100 \text{ miles}}{4 \text{ hours}}$ **11.** $\frac{1}{11}$ **13.** $\frac{3}{1}$ **15.** $\frac{7}{13}$
17. $\frac{33}{40}$ **19.** $\frac{4}{1}$ **21.** $\frac{9}{5}$ **23.** 35.2 mph   **25.** $7/ticket
27. 2.4 inches/hour   **29.** $250/employee   **31.** sample
anwer: $n + 1 = 8$ **32.** 80¢   **35.** $0.18/ounce

Pages 328–329   Lesson 9-2

7. $\neq$ **9.** $=$ **11.** 18   **13.** 2   **15.** $\neq$ **17.** $\neq$ **19.** $\neq$
21. $=$ **23.** $=$ **25.** $\neq$ **27.** 36   **29.** 1   **31.** 3.5   **33.** 16
35. 2.1   **37.** 5.1   **38.** 21   **39.** 7   **40.** x represents the x
coordinate and y represents the y coordinate.   **41.** $\frac{4}{3}$

49. 9.5 gallons

Pages 332–333   Lesson 9-3

5. $\frac{40}{3} = \frac{25}{x}$; $1.875   **7.** $\frac{20}{30} = \frac{25}{x}$; 37.5 cm   **9.** $\frac{\frac{1}{4}}{1} = \frac{3}{x}$; 12 ft
11. $\frac{250}{2} = \frac{x}{5}$; 625 bushels   **13.** $\frac{15}{21} = \frac{20}{x}$; 28 cm
15. false   **16.** sample answer: $(0, -3), (1, -1), (2, 1), (3, 3)$
17. $\frac{1.5 \text{ inches}}{\text{hour}}$ **18.** 1.5   **19.** 22.5 pounds   **21.** 135 cm
23. 1.5 pounds

Pages 336–337   Lesson 9-4

5. 4%   **7.** 35%   **9.** c   **11.** b   **13.** 25.92   **15.** 40
17. 12.5%   **19.** 87.5%   **21.** 375%   **23.** 24   **25.** 60
27. 65%   **29.** $24   **30.** 4, -4   **31.** true   **32.** 15
33. $66\frac{2}{3}$%   **35.** $151.68   **37.** $3.25

Pages 340–341   Lesson 9-5

5. 36%   **7.** 47.5%   **9.** 31%   **11.** 90%   **13.** $\frac{3}{5}$ **15.** $\frac{8}{25}$
17. 0.28   **19.** 0.8   **21.** 81%   **23.** 113%   **25.** 40.7%
27. 309.3%   **29.** 20%   **31.** 225%   **33.** 62.5%
35. 58.3%   **37.** $\frac{14}{25}$ **39.** $\frac{73}{100}$ **41.** $\frac{2}{3}$ **43.** $\frac{1}{6}$ **45.** 0.82
47. 0.41   **49.** 0.485   **51.** 0.334   **53.** 12.5%   **55.** 53%
57. 18.75%   **59.** 98   **60.** 38   **61.** -23   **63.** $\frac{1}{5}$ **64.** 3
65. 20%   **66.** 120   **67.** $\frac{2}{3}$ **69.** yes; $0.7 < 0.75$

Pages 343–344   Lesson 9-6

5. 75%   **7.** 25%   **9.** 67%   **11.** c   **13.** c   **15.** $\frac{1}{4}$ **17.** $\frac{2}{3}$
19. $\frac{1}{3}$ **21.** $\frac{1}{20}$ **23.** $1\frac{1}{2}$ **25.** $\frac{1}{3}$ **27.** $\frac{1}{2}$ **29.** $\frac{1}{100}$ **31.** 20
33. 60   **35.** $5.50   **37.** 100   **39.** 25%   **41.** $33\frac{1}{3}$%
43. 20%   **45.** inductive   **46.** $\frac{5}{8}$ **47.** $c > 2$ **48.** P, 14; B,
$x; r$, 70   **49.** 82.7%   **50.** $\frac{19}{50}$ **51.** Texas: 33%; Montana: 20%;
Kansas: 17%; Nebraska: 17%; New Mexico: 15%   **53.** $3

Page 345   Mid-Chapter Quiz

1. $\frac{5}{3}$ **2.** $\frac{3}{5}$ **3.** $\frac{9}{11}$ **4.** $1.28/gallon   **5.** 0.75 pound/week
6. 6   **7.** 9   **8.** 2   **9.** 2.8 pounds   **10.** $1\frac{2}{3}$ gallons   **11.** 9
12. 37.5%   **13.** 88%   **14.** $33\frac{1}{3}$%   **15.** 87.5%   **16.** 0.2; $\frac{1}{5}$
17. 0.008, $\frac{1}{125}$ **18.** 1.1, $\frac{11}{10}$ **19.** 9   **20.** $2

Pages 347–348   Lesson 9-7

5. $15 = 0.03 \cdot B$; 50   **7.** $18 = R \cdot 60$; 30   **9.** $16 = R \cdot 64$;
25   **11.** $P = 0.31 \cdot 14$; 4.34   **13.** 25%   **15.** $242.80
17. 62.5%   **19.** $6540   **21.** 125   **23.** 50%   **25.** $c \geq 18$
26. 2   **27.** 125%   **28.** $15   **29.** $48,700   **31.** 6%

Pages 350–351   Lesson 9-8

5. $10.50   **7.** $174.38   **9.** $7.53   **11.** $650   **13.** $945
15. $21   **17.** 25%   **19.** 15%   **20.** 24   **22.** 20
23. $68.64   **25.** $1068.52   **27.** $34.09

Pages 352–353   Lesson 9-9

3. 12 ways   **5.** 35 cards   **7.** $153.90   **9.** $255

Page 355   Lesson 9-10

5. 25%   **7.** 6%   **9.** 14%   **11.** 7%   **13.** 24%   **15.** 8%
17. 4%   **19.** $\frac{1}{2}$ **20.** 20%   **21.** 20%   **22.** about 36%
23. 1%   **25.** 20%

Pages 356–357   Chapter Review

1. ratio   **3.** proportion   **5.** 882   **7.** $\frac{2}{3}$ **9.** $\frac{1}{2}$

Extensions

1. Research and write a report on Galileo's experiments on the speed of falling objects.
2. Choose an Olympic sport in which the winning speed has changed greatly during the past century. Analyze the trend and use it to predict the speeds of future Olympic winners.
3. Research and explain the term *light year*.

Culminating Activities

Show what you have learned in this project by completing one of the following activities.

1. Write a report summarizing your findings. Explain how you found the answer to each of the three questions you chose for your project. Include the data you gathered and describe your efforts to predict the speed of the object whose speed you had not previously measured.
2. Design a checklist entitled "Finding the Speed of an Object." Your checklist should be detailed enough that another student could follow it to approximate the speed of any object of his or her choosing.
3. In the project, you identified factors that affected the speed of the objects you studied. Draw graphs showing the motions of the objects in relation to each of the factors you identified.

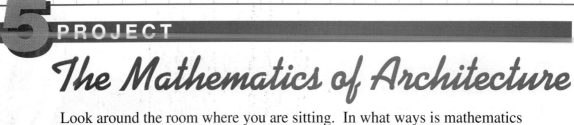
The Mathematics of Architecture

Look around the room where you are sitting. In what ways is mathematics related to what you see? If you did not know any mathematics, would you be able to construct the building that you are sitting in now?

Few things require the use of more mathematics in their construction than buildings. One architect has gone so far as to say that architecture is geometry. Even the most complex buildings are made up of arrangements of rectangular prisms, and sometimes cylinders, spheres, and pyramids. From the first sketch to the last coat of paint, each step in the construction of a building requires the use of mathematics.

- Most architects study mathematics at least to the level of calculus.
- Surveyors rely heavily on trigonometry.
- Building contractors use computers extensively in their work.
- Carpenters must make precise measurements.
- Electricians study physics and use complex formulas in their work.

Think of some of the famous buildings with which you are familiar. What geometrical shapes were combined to create the buildings?

Dismantle a Building

In this project, you will use mathematics to "take apart" a building of your choice. Through the use of trigonometry, indirect measurement, the Pythagorean Theorem, estimation, and other mathematical tools, you will determine the dimensions of the building. You will calculate the volume and surface area of the building and then, in the culminating activity, put the building back together again.

Getting Started

Follow these steps to carry out your project.

- Choose a building that you would like to study. The building can be your school or it can be an office building, sports stadium, or some other building in your town. Be sure that the building is in a safe location to which you will have free access so that you can make the measurements you will need.

- Work out a plan for finding the measurements of the building with your group. Many of the ground measurements you will be able to pace off by foot. You will need to find the height, roof measurements, and measurements of other inaccessible spots using indirect measurement. Your goal is to find as many of the building's inside and outside measurements as you will need to "put it back together."

- Take your measurements. Take measurements several times to ensure accuracy and allow each member of the group to take measurements to avoid errors. Keep accurate records.

- Find the volume and surface area of the building. Estimate the precision of your answers.

Extensions

1. Discuss the importance of mathematics to architecture with a local architect. Arrange for the architect to discuss the subject with your class.
2. Research surveying. Find out and report on how surveyors use trigonometry in their work.
3. Research the work of Frank Lloyd Wright, perhaps the twentieth century's most famous architect, or the work of I.M. Pei. Pay particular attention to the architect's views on geometry and the use of space.

Culminating Activities

"Put your building back together" by completing one of the following activities.

1. Make a scale model of the building.
2. Make a scale drawing of the building.
3. Write a report summarizing the work of your group.